Science Networks · Historical Studies
Founded by Erwin Hiebert and Hans Wußing
Volume 28

Edited by Eberhard Knobloch and Erhard Scholz

Editorial Board:

K. Andersen, Aarhus
H.J.M. Bos, Utrecht
U. Bottazzini, Roma
J.Z. Buchwald, Cambridge, Mass.
K. Chemla, Paris
S.S. Demidov, Moskva
E.A. Fellmann, Basel
M. Folkerts, München
P. Galison, Cambridge, Mass.
I. Grattan-Guinness, London
J. Gray, Milton Keynes
R. Halleux, Liège

S. Hildebrandt, Bonn
D. Kormos Buchwald, Pasadena
Ch. Meinel, Regensburg
J. Peiffer, Paris
W. Purkert, Leipzig
D. Rowe, Mainz
A.I. Sabra, Cambridge, Mass.
Ch. Sasaki, Tokyo
R.H. Stuewer, Minneapolis
H. Wußing, Leipzig
V.P. Vizgin, Moskva

Dennis E. Hesseling

# Gnomes in the Fog

The Reception of Brouwer's Intuitionism
in the 1920s

Birkhäuser Verlag
Basel · Boston · Berlin

Author's address:

Dennis E. Hesseling
Kon. Emmakade 155b
NL-2518 JL Den Haag
The Netherlands

email: dehessel@xs4all.nl


2000 Mathematics Subject Classification: 01A60, 03-03, 03F55, 03B20

QA
9
.H454
2003

A CIP catalogue record for this book is available from the Library of Congress, Washington D.C., USA.

Bibliographic information published by Die Deutsche Bibliothek
Die Deutsche Bibliothek lists this publication in the Deutsche Nationalbiografie; detailed bibliographic data is available in the internet at http://dnb.ddb.de.

## ISBN 3-7643-6536-6 Birkhäuser Verlag, Basel – Boston – Berlin

This work is subject to copyright. All rights are reserved, whether the whole or part of the material is concerned, specifically the rights of translation, reprinting, re-use of illustrations, broadcasting, reproduction on microfilms or in other ways, and storage in data banks. For any kind of use whatsoever, permission of the copyright owner must be obtained.

© 2003 Birkhäuser Verlag, P.O.Box 133, CH-4010 Basel, Switzerland
Member of the BertelsmannSpringer Publishing Group
Cover design: Micha Lotrovsky, Therwil, Switzerland
Cover illustration: L. E. J. Brouwer in the hall of the central building of the Amsterdam University, at the occasion of his inaugural lecture, 1912 (Courtesy: Brouwer Archive)
Printed on acid-free paper produced from chlorine-free pulp. TCF ∞
Printed in Germany
ISBN 3-7643-6536-6

9 8 7 6 5 4 3 2 1                                              www.birkhauser.ch

*Voor mijn ouders*

L'homme est celui qui avance dans le brouillard. Mais quand il regarde en arrière pour juger les gens du passé il ne voit aucun brouillard sur leur chemin. De son présent, qui fut leur avenir lointain, leur chemin lui paraît entièrement clair, visible dans toute son étendu. Regardant en arrière, l'homme voit le chemin, il voit les gens qui s'avancent, il voit leurs erreurs, mais le brouillard n'est pas là. Et pourtant, tous, Heidegger, Maïakovski, Aragon, Ezra Pound, Gorki, Gottfried Benn, Saint-John Perse, Giono, tous ils marchaient dans le brouillard, et on peut se demander: qui est le plus aveugle? Maïakovski qui en écrivant son poème sur Lénine ne savait pas où mènerait le léninisme? Ou nous qui le jugeons avec le recul des décennies et ne voyons pas le brouillard qui l'enveloppait?[1]

Milan Kundera[2]

---

[1] 'Man proceeds in the fog. But when he looks back to judge people of the past, he sees no fog on their path. From his present, which was their faraway future, their path looks perfectly clear to him, good visibility all the way. Looking back, he sees the path, he sees the people proceeding, he sees their mistakes, but not the fog. And yet all of them - Heidegger, Majakovski, Aragon, Ezra Pound, Gorki, Gottfried Benn, St-John Perse, Giono - all were walking in fog, and one might wonder: who is more blind? Majakovski, who as he wrote his poem on Lenin did not know where Leninism would lead? Or we, who judge him decades later and do not see the fog that enveloped him?', English translation cited from [Kundera 1995, p. 240].
[2] [Kundera 1993, p. 287]

# Contents

**Foreword**     xiii

**Introduction**     xv
- 0.1 Introduction . . . . . . . . . . . . . . . . . . . . . . . . xv
- 0.2 Audience, academic discipline and personal background . . . . . . xiii
- 0.3 Methodological remarks . . . . . . . . . . . . . . . . . . . xix
- 0.4 Sources, structure and presentation . . . . . . . . . . . . . . xx
- 0.5 Suggestions for further research . . . . . . . . . . . . . . . xxii

**1 Kronecker, the semi-intuitionists, Poincaré**     1
- 1.1 Introduction . . . . . . . . . . . . . . . . . . . . . . . . 1
  - 1.1.1 Mathematical prerequisites . . . . . . . . . . . . . . . 2
- 1.2 Kronecker . . . . . . . . . . . . . . . . . . . . . . . . . 4
  - 1.2.1 Kronecker's conflicts . . . . . . . . . . . . . . . . . 5
  - 1.2.2 Kronecker's views . . . . . . . . . . . . . . . . . . . 6
- 1.3 The French semi-intuitionists . . . . . . . . . . . . . . . . . 8
  - 1.3.1 The French semi-intuitionists' main conflict . . . . . . . 9
  - 1.3.2 The French semi-intuitionists' views . . . . . . . . . . 12
- 1.4 Poincaré . . . . . . . . . . . . . . . . . . . . . . . . . . 18
  - 1.4.1 Poincaré's conflicts . . . . . . . . . . . . . . . . . 18
  - 1.4.2 Poincaré's views . . . . . . . . . . . . . . . . . . . 20
- 1.5 Conclusion . . . . . . . . . . . . . . . . . . . . . . . . . 23

**2 The genesis of Brouwer's intuitionism**     25
- 2.1 Introduction . . . . . . . . . . . . . . . . . . . . . . . . 25
- 2.2 The early years . . . . . . . . . . . . . . . . . . . . . . . 26
  - 2.2.1 Brouwer's youth . . . . . . . . . . . . . . . . . . . . 26
  - 2.2.2 Brouwer's profession of faith . . . . . . . . . . . . . 26
  - 2.2.3 Mannoury . . . . . . . . . . . . . . . . . . . . . . . 28
  - 2.2.4 Brouwer's mysticism . . . . . . . . . . . . . . . . . . 30
- 2.3 The first act of intuitionism . . . . . . . . . . . . . . . . . 34
  - 2.3.1 Brouwer's dissertation . . . . . . . . . . . . . . . . . 35

|   |   | 2.3.2 | The unreliability of the logical principles | 46 |
|---|---|---|---|---|
|   | 2.4 | Topology | | 48 |
|   | 2.5 | Intuitionism and formalism | | 52 |
|   | 2.6 | The second act of intuitionism | | 58 |
|   |   | 2.6.1 | Intuitionistic set theory | 60 |
|   |   | 2.6.2 | Further development of intuitionistic mathematics | 67 |
|   | 2.7 | The Brouwer lectures | | 75 |
|   |   | 2.7.1 | Berlin | 75 |
|   |   | 2.7.2 | Amsterdam | 77 |
|   |   | 2.7.3 | Vienna | 79 |
|   | 2.8 | The *Mathematische Annalen* and afterwards | | 81 |
|   | 2.9 | Brouwer's personality | | 86 |
|   | 2.10 | Conclusion | | 89 |
| **3** | **Overview of the foundational debate** | | | **91** |
|   | 3.1 | Introduction | | 91 |
|   | 3.2 | Quantitative inquiry | | 93 |
|   |   | 3.2.1 | The *Fortschritte* | 93 |
|   |   | 3.2.2 | 'All' public reactions to intuitionism | 95 |
|   | 3.3 | Qualitative inquiry | | 97 |
|   |   | 3.3.1 | Themes | 97 |
|   |   | 3.3.2 | Tone | 99 |
|   |   | 3.3.3 | Currents and schools | 104 |
|   |   | 3.3.4 | People | 111 |
|   |   | 3.3.5 | Languages and media | 113 |
|   | 3.4 | Conclusion | | 115 |
| **4** | **Reactions: existence and constructivity** | | | **117** |
|   | 4.1 | Introduction | | 117 |
|   |   | 4.1.1 | Mathematical existence | 117 |
|   |   | 4.1.2 | A short history of constructivism | 120 |
|   | 4.2 | The beginning of the debate | | 124 |
|   |   | 4.2.1 | Weyl's *Grundlagenkrise* | 125 |
|   |   | 4.2.2 | Hilbert's first reactions | 133 |
|   |   | 4.2.3 | Becker's phenomenology | 145 |
|   |   | 4.2.4 | Fraenkel's early commentaries | 149 |
|   |   | 4.2.5 | Baldus' rector's address | 152 |
|   | 4.3 | The debate widened | | 154 |
|   |   | 4.3.1 | Existence in a central position | 158 |
|   |   | 4.3.2 | Existence as a minor subject | 175 |
|   | 4.4 | Later reactions | | 183 |
|   |   | 4.4.1 | The Königsberg conference | 183 |
|   |   | 4.4.2 | Wittgenstein | 190 |
|   |   | 4.4.3 | Others | 198 |

|       | 4.5   | Conclusion | 212 |
| ----- | ----- | ---------- | --- |

# 5 Reactions: logic and the excluded middle — 217
- 5.1 Introduction — 217
  - 5.1.1 A short history of classical logic — 218
- 5.2 The beginning of the debate — 220
  - 5.2.1 Weyl's *Grundlagenkrise* — 222
  - 5.2.2 Hilbert's first reactions — 225
  - 5.2.3 Addresses: Wolff, Finsler and Baldus — 229
  - 5.2.4 Fraenkel's early commentaries — 231
- 5.3 The debate widened — 231
  - 5.3.1 The excluded middle in a central position — 236
  - 5.3.2 The excluded middle as a minor subject — 268
- 5.4 Later reactions — 271
  - 5.4.1 Glivenko, Heyting and Kolmogorov — 271
  - 5.4.2 Gödel — 281
  - 5.4.3 Barzin and Errera — 289
- 5.5 Conclusion — 296

# 6 The foundational crisis in its context — 301
- 6.1 Introduction — 301
- 6.2 Metaphors — 302
  - 6.2.1 Crisis and revolution — 302
- 6.3 Philosophy — 311
  - 6.3.1 *Lebensphilosophie* — 312
  - 6.3.2 Mathematical and philosophical intuitionism: a comparison — 316
  - 6.3.3 Contemporaries' remarks — 320
  - 6.3.4 Göttingen and Hilbert — 322
  - 6.3.5 Spengler — 323
  - 6.3.6 Summary — 325
- 6.4 Physics — 326
  - 6.4.1 Theory of relativity — 327
  - 6.4.2 Quantum mechanics — 329
- 6.5 Art — 333
  - 6.5.1 Constructivism — 334
- 6.6 Politics — 335
  - 6.6.1 Mathematics and the rise of the Third *Reich* — 335
  - 6.6.2 Bieberbach's racial interpretation of the foundational debate — 338
- 6.7 *Moderne* and *Gegenmoderne* — 340
- 6.8 Conclusion — 342

# Conclusion — 345

# A Chronology of the debate — 355

| | |
|---|---:|
| B  Public reactions to Brouwer's intuitionism | 363 |
| C  Logical notations | 371 |
| Glossary | 373 |
| Bibliography | 377 |
| Index | 431 |
| Dankwoord/ Acknowledgements | 445 |

# Foreword

Writing history is a way of travelling. Travelling in a search for different cultures, different people, different ideas. Travelling with the aim, also, to obtain a better understanding of one's own situation. The only difference being that a historical travel is not undertaken through space, but through time.

The joy of carrying out historical research above reading a history book compares to the pleasures of travelling by oneself as opposed to going on an organised tour. In doing historical research, one has both the freedom and the obligation to find one's own way. To decide for oneself which way to choose, which things to see, which story to tell afterwards.

Following this metaphor, the dissertation which lies before you is the scientific report of the travel which I undertook to an area some seventy years in the past. This journey, too, had its highlights. I have photos in the form of citations. They illustrate the narrative. They were chosen for their role in the story, for their beauty, or for both reasons. They reveal something of what Henk Bos called the wonder of history.[3]

Ever since I heard about the foundational crisis in mathematics, in the beginning of my mathematics studies here in Utrecht, I have been fascinated by the subject. I wanted to know more about that period in the history of mathematics when mathematics and philosophy, usually so far away from each other, seemed to meet. That period which showed the rare characteristic of open controversy and debate inside mathematics. I wanted to read a book which would tell me all about it. As it turned out, such a book did not exist. Now, I have written it myself.

Utrecht, December 1998

Dennis Hesseling

---

[3][Bos 1987]

## At the commercial edition

This book is an improved version of my dissertation.[4] I have refined or clarified parts throughout the book in reaction to suggestions, critical remarks and new or previously unused publications, notably Van Dalen's Brouwer biographies and the translations in Ewald's and Mancosu's source books.[5]

Two reactions to Brouwer's intuitionism which I did not know of when writing the dissertation have been included, namely Rivier's *L'empirisme dans les sciences exactes*[6] and Study's *Prolegomena*. Also, I have included materials from the Church and Errera archives, which had previously not been researched.

I have specified the use of the term 'formalism', using it to indicate Hilbert's position only if the publication under discussion does so.

Finally, I have added a short appendix in which logical notations are explained, and I have significantly extended the index to allow for easy reference.

Brussels, October 2002

Dennis Hesseling

---

[4] [Hesseling 1999]
[5] [Van Dalen 1999A], [Van Dalen 2001], [Ewald 1996] and [Mancosu 1998].
[6] [Rivier 1930]

# Introduction

> It is difficult to overestimate the significance of these events. In the third decade of the twentieth century two mathematicians [Brouwer and Weyl, DH] – both of them of the first magnitude, and as deeply and fully conscious of what mathematics is, or is for, or is about, as anybody could be – actually proposed that the concept of mathematical rigor, of what constitutes an exact proof, should be changed!
>
> John von Neumann[7]

## 0.1 Introduction

The significance of the foundational debate in mathematics that took place in the 1920s seems to have been recognized only in circles of mathematicians and philosophers. In their classic 'A history of the modern world', Palmer and Colton, for all their proclaimed attention paid to the history of ideas, mention no mathematician more modern than sir Isaac Newton.[8] Kline, in his standard work 'Mathematical Thought from Ancient to Modern Times', presents the standard interpretation of the foundational debate in the final chapter.[9] And Mehrtens, in his recent and influential *Moderne – Sprache – Mathematik*, devoted specifically to the foundations of mathematics, pays substantial attention to Brouwer and the foundational debate.[10]

This study is about the so-called foundational crisis[11] in mathematics, more specifically the foundational crisis in the 1920s. According to the *Historisches Wörterbuch der Philosophie* ('Historical dictionary of philosophy'), a foundational crisis[12] arises in a field of science[13]

---

[7][Von Neumann 1947, p. 188]
[8][Palmer & Colton 1995]; of Einstein, only his work on physics is mentioned. This indicates that there is a serious gap between the historiography of mathematics and general historiography.
[9][Kline 1972, pp. 1192–1207]
[10]See 6.7.
[11]Throughout the book, 'foundational crisis' and 'foundational debate' are used as synonyms.
[12]*Grundlagenkrise*
[13][Ritter & Grunder 1971–1995, vol. 3, pp. 910–911]

> wenn gewisse über Einfluß auf die Wissenschaftsorganisation verfügende Gruppen (...) auf den Wissenschaftsbetrieb des betreffenden Bereiches reflektieren, an der Gültigkeit gewisser dort erarbeiteter Ergebnisse (...) oder der zu ihrer Gewinnung angewandten Verfahren begründete Zweifel anmelden und Änderungen im Wissenschaftsbetrieb dieses Bereiches verlangen. Ein *Grundlagenstreit* ist im Gange, wo einflußreiche Gruppen von Wissenschaftlern miteinander unverträgliche Vorschläge zur Behebung einer Grundlagenkrise ihrer Wissenschaft durchzusetzen versuchen.[14]

This definition applies very well to the foundational crisis in mathematics at the beginning of the 20th century, which is presented as a paradigmatic case.[15] However, the description is rather wide and can be applied to more situations which we normally would not consider to represent a crisis. Here, I narrow the definition just given by adding the demand that there has to be a sense of crisis among the participants to the debate, expressing itself for instance by emotional or polemical contributions.

The standard interpretation of the foundational crisis in the 1920s has it that the set theoretical paradoxes led to the development of three schools in the foundations of mathematics: intuitionism, formalism and logicism.[16] Mehrtens has already attacked the role ascribed to the paradoxes in the standard interpretation and called it a founding myth of modernism.[17] In this analysis, I go further and claim that, first, logicism played only a marginal role in the discussion that developed in reaction to intuitionism. Second, whereas formalism was seen as a dominant current in the discourse on mathematics in the 1920s, I will argue that there were actually very few formalists among those who contributed to the foundational debate. The intuitionistic critique of classical mathematics and the debate it evoked were pivotal in the development and spreading of formalism. In this way, it was counter-modernism which gave rise to modernism, rather than the other way round.[18]

The purpose of my research was to study the reactions, mainly by mathematicians and philosophers, to Brouwer's intuitionism. The central questions were how they reacted to it and why they reacted in such a way. What caused the controversy about intuitionism? Why could a debate develop which continued for several years?

---

[14]'if certain groups which possess influence on the organisation of science (...) reflect on the scientific process of the field in question, express motivated doubts concerning the validity of certain results which were achieved (...) or about the methods employed to achieve them, and request changes in the scientific process of that field. A *foundational fight* is going on, if influential groups of scientists attempt to drive through mutually irreconcilable proposals to eliminate a foundational crisis in their science.'

[15][Ritter & Grunder 1971–1995, vol. 3, p. 911]. For a discussion of the definition, see [Thiel 1972, pp. 6–28].

[16]Cf. [Kline 1972, pp. 1183–1207]

[17][Mehrtens 1990, pp. 150–151; 298]

[18]The terms counter-modernism and modernism are explained in 6.7.

## 0.1. INTRODUCTION

Why could the debate go so far that mathematical journals published papers proclaiming a revolution in mathematics, or later re-naming it a *Putsch*?[19] Why were so many people interested in foundational questions, and why were some of the responses so agitated and emotional? Why did mathematicians not regard questions such as what kind of mathematical objects exist as mere issues of personal preference, as they had done some fifteen years earlier and as we tend to do now?[20] What did the foundational crisis mean for the self-understanding of mathematics?

In order to answer such questions, it does not suffice to confine oneself to the mathematical level; one also has to take non-mathematical circumstances into account. Thus, I also looked at culture, philosophy, physics, and politics, as far as they might have interacted with the foundational debate in mathematics.

The fact that it took mathematicians of name several years to develop, understand and accept a plurality of views on the foundations of mathematics which can now be explained in a few hours to first-year mathematics students should not be misunderstood. It marks the profound change that the self-understanding of mathematics underwent.

The foundational debate is presented with all its brilliant contributions and its shortcomings, its new ideas and its misunderstandings, its main characters and its dead ends. Some of the contributions to the debate stand almost completely separated from the rest; others do not contain ideas that we consider worthwhile nowadays, or only contain a brief remark about Brouwer's intuitionism. Many of such contributions neither support nor refute any specific contention defended here. They are simply part of the foundational debate and were included for the sake of completeness.

**Re-writing and more** The value of this research extends, I hope, beyond the writing and re-writing of a certain chapter in the history of mathematics. In the first place, it reveals a feature of mathematics that is often forgotten or neglected, namely its human character. In the foundational debate in the 1920s, such very human peculiarities as emotional arguments, misunderstanding and unwillingness, chance and power politics played an important and sometimes even decisive role. Secondly, this study does not only reveal aspects of the self-understanding and the self-presentation of past mathematicians. Since many mathematicians today are

---

[19] [Weyl 1921, p. 226] and [Hilbert 1922, p. 160], resp.

[20] The present-day view on the need for foundations in mathematics is probably best voiced by Mehrtens, who uses a bridge metaphor: '*Brücken werden auf Fundamente gesetzt, die den klimatischen und geologischen Bedingungen so gut angepaßt werden, daß sie allen erwartbaren Belastungen standhalten. Brücken brauchen keine 'absoluten' Fundamente. Die Mathematik braucht sie ebensowenig; auch ihre Fundamente sind historisch lokale Konstruktionen, die den Belastungen zumeist gut standhalten.*' ('Bridges are put on foundations which are so well adapted to the climatological and geological demands that they resist all loads that can be expected. Bridges do not need 'absolute' foundations. Just as little does mathematics; its foundations are historically local constructions, too, which mostly resist the loads well.'), [Mehrtens 1998, p. 468].

formalists, at least on Sundays,[21] it at the same time shows some historical roots of one of the important current philosophies of mathematics, with its successes and shortcomings. It is in the foundational debate in the 1920s that one finds most roots of today's conception of mathematics.

## 0.2 Audience, academic discipline and personal background

The history of mathematics, as an academic discipline, is part of intellectual history rather than of mathematics. Its methods are historical rather than mathematical. Its aim is to improve our understanding of the past rather than to help the current working mathematician.

However, most historians of mathematics are mathematicians by training, many of them work in mathematics departments, and a number of European historians of mathematics will typically do their teaching to a large degree in mathematics proper. In this respect, I am no exception.

This situation puts considerable strain on a work in the history of mathematics. The pressure is even stronger for the present one, since it is on the junction not only of history and mathematics, but also of philosophy. Even if the boundaries between the different disciplines are somewhat arbitrary at times, the fact that most academics were trained within one of them makes the distinction relevant in any case. In this way, academic borders, like state borders, tend to reinforce themselves. This book is an attempt to cross some of these borders.

Its seems hardly possible to write a work in the history of philosophy of mathematics which satisfies mathematicians, philosophers and historians alike. By using too much mathematics, or by assuming too much mathematics known, historians and philosophers are driven away, and vice versa. I tried to find a balance by explaining (sometimes basic) historical, philosophical or mathematical concepts in footnotes and in the glossary. In this way, the information is available for those who need it, but it can also easily be skipped by those who already know.

My first introduction to the field of history of mathematics was in Copenhagen, when I took Jesper Lützen's course *Matematikkens historie* at Københavns universitet. When I returned to Utrecht to finish my mathematics study, I had the chance to obtain more knowledge about both history of mathematics and intuitionism, since both of these are relatively well represented in Utrecht. Besides a Ph.D. in mathematics, I have an additional background in philosophy and history.

As to my personal preference in the foundational debate, I feel a natural sympathy for those who stood up and pleaded for something they believed in, even

---

[21] Reuben Hersch criticized the working mathematician for being 'a Platonist on weekdays [when it matters] and a formalist on Sundays [when it does not]', [Hersch 1979]; cited from: [Rowe 1996, p. 11].

if this meant arguing against established traditions and against the vast majority of the working mathematicians – thus, for the intuitionists.

## 0.3 Methodological remarks

**Anachronism and historicism** It is clear that, from the point of view of present-day mathematics, all past attempts aimed at achieving another form of mathematics than the one now dominant will be condemned as experiments that did not lead to the 'right' result, i.e., to mathematics as we know it now. In such a view, intuitionism would be considered a 'failed experiment', which at best contributed to the margins of modern mathematics. The danger of present-mindedness especially arises when writing history of mathematics, since mathematics still has an aura of objective truth. In such an anachronistic and un-historical way, however, no proper evaluation of intuitionism or the foundational debate can be accomplished.

In this analysis, I expressly took into account the intuitionistic view on the foundational debate. Seen in this light, the reactions of the mathematical community to a new conception of mathematics turn out to be one-sided. They focused on the critical side of intuitionism, while paying little attention to innovative intuitionistic contributions. Moreover, many mathematicians who took part in the debate had difficulties in understanding intuitionism. Without taking the full intuitionistic point of view into account, however, it is difficult to obtain a proper understanding of the historical situation.

**Internalism and externalism** In doing historical research, one looks for explanations for what we see as past events. Whether these explanations are internal or external to mathematics is, in my view, of secondary importance at most. I find it dogmatic to assume both that all explanations for mathematical events lie inside the field of mathematics, and that the explanations, or parts of them, necessarily lie outside it. I prefer to join the trend in the historiography of mathematics of the last decades which incorporates both internal and external factors.[22] Thus, I place mathematics in a broader context in which factors *may* be found that influenced the development and interpretation of mathematics.

In general, it seems to me that Droysen's hermeneutic scheme of interpretations is still of much value.[23] More specifically regarding the history of science, it seems natural to start looking for explanations within science, moving slowly further away from the 'hard core' of science until the explanations satisfy us. The choice which explanation is given to a certain past event becomes in this way part of a pragmatic process, where different historians may make different decisions.

---

[22] Cf. [Kitcher & Aspray 1988, p. 23]
[23] As explained in [Lorenz 1994, p. 94]. Droysen (1808–1886) distinguished between four interpretations in historical research: the pragmatic interpretation of the sources, the interpretation of the circumstances, the psychological interpretation, and the interpretation of ideas (*Zeitgeist*).

The diversity of interpretations to which this, ideally, gives rise, presents the best opportunity for understanding past events. This study provides one such colour in the historical palette.

**Construction, reconstruction and interpretations**  Like an intuitionistic proof, this historical study is more a construction than a reconstruction, in this case of history rather than of mathematics. Now, we can get an extensive overview of the foundational debate in a way the participants to the debate themselves never had.[24] This should be kept in mind when reading the exposition.

Following Derrida, who claims that there is no such thing as *the* meaning of a story and that the author is not the only one who determines what meaning a text has,[25] I wrote the book with the idea to present the foundational debate in a way which was open to various interpretations. Therefore, the expositions in the various chapters are mostly descriptive; conclusions are generally only drawn at the end of a chapter. In this way, the book can also be of value to people who do not agree with the main conclusions.[26] To put it in an intuitionistic metaphor: the book can be read as a choice sequence, where the reader can choose which way to go through the historical material. However, this can never be a lawless sequence, since a selection of the historical material is presented.

Concerning modern trends in historiography, Hayden White put forward a post-modern philosophy of history according to which there are only differences in degree between stories of historians and those of fiction writers.[27] I myself, however, continue to adhere to the idea that there are such things as facts, and that one can reasonably argue about these in a historical discourse – even though facts are open to various interpretations.

## 0.4  Sources, structure and presentation

**Sources**  To carry out this study, I analysed written sources from, mainly, the 1920s and early 1930s. These included published papers in both mathematical and non-mathematical journals and newspaper articles, as well as archive materials such as correspondence and manuscripts. Altogether, I used more than 1,000 of these primary sources, of which over 250 were public reactions to intuitionism. In order to collect archival materials, I visited archives held in Amsterdam, Berlin, Brussels, Göttingen, Jerusalem, Konstanz, Lausanne, Münster, Paris, Utrecht, and Zürich. These included the archives of Bernays, Brouwer, Carnap, Einstein, Errera,

---

[24]The person who came closest to it was Fraenkel, who was very well informed on foundational literature.

[25][Lorenz 1994, p. 138]

[26]A good example of a book presenting most interesting material, but full of Freudian interpretations that would satisfy few people nowadays, is Theweleit's *Männerphantasien* ('Men's phantasies'), [Theweleit 1977–1978].

[27][Lorenz 1994, pp. 139–140]

## 0.4. SOURCES, STRUCTURE AND PRESENTATION

Fraenkel, Gonseth, Heyting, Hilbert, Klein, Ramsey, Reidemeister, Scholz, and Weyl. In this way, I covered most of the important actors in the debate.

There are some archives which I did not use in carrying out this study. Bieberbach's archive is still closed at the time of writing, due to legal constraints. Unfortunately, Becker's *Nachlaß* seems to contain only few materials. The Skolem materials are supposedly somewhere in the mathematical institute in Oslo (although nobody seems to know exactly where), but, since Skolem was not one of the main persons involved, I did not take the time to go and look for them.

Biographical information on mathematicians was generally taken from the *Lexikon bedeutender Mathematiker*,[28] unless stated otherwise.

**Presentation** In the presentation of the results, I decided to make much use of citations. In the first place, I think using a well-chosen quote provides the best opportunity for staying as closely as possible to the original source. Furthermore, some of the original texts that were used for this analysis are written in a style that would get lost if a mere paraphrase of the contents would be given. Often, the style is more polemic than other texts in mathematical journals, the main source of this study. Therefore, they are worth exposing in full.

I essentially followed the line of the debate, including the confusion that regularly arose. In my view, this confusion is an essential part of the foundational debate.

Occasionally, sentences or small paragraphs were used identically in the chapters 4 and 5. This was done when they served as an introduction to a person and were so short that I thought it more reader-friendly to duplicate the passage than to give a cross-reference.

**Languages** All full quotes are given in the original language, with an English translation appearing in the footnote right at the end of the quote whenever the citation is not in English. These translations are my own, unless stated otherwise. There are several reasons why I chose to put the original language in the text and an English translation in the footnotes, and not vice versa.

In the first place, the use of translated quotes in the text would have given a wrong impression of the debate. Even if it is mentioned that the quote was translated, the reader might still easily end up with the idea of a certain linguistic uniformity which did not appear in the debate. By using the original quotes in the text, such an impression is countered. Now, readers can experience the diversity for themselves, even though it may cost some of them more effort. Secondly, some characteristics of the original texts always get lost in a translation, even if the translation was done by a professional translator – which I am not. By highlighting the original citations, any negative effects translations may have should be diminished. Finally, I think some readers may enjoy reading the original texts, as I did.

---

[28] [Gottwald, Ilgauds & Schlote 1990]

**Structure and suggestions for reading** The kernel of the book lies in the chapters 4 and 5, in which individual reactions to Brouwer's intuitionism are discussed. Chapter 3 gives characteristics of the debate as a whole. In chapter 6, the context in which the debate took place is sketched, and more externalist factors are treated. Chapter 2 serves as an introduction to intuitionism. Chapter 1, finally, gives a historical introduction to Brouwer's most important predecessors. Generally spoken, the contents of the chapters 1 and 2 is already known in the literature, whereas the contents of the other chapters has not been analysed in such an extensive way as is done here.

Thus, readers looking for a shortcut through the book could proceed as follows. The most important part of chapter 1 is the section on Kronecker (1.2), since he was the person that people referred to most when pointing out some of Brouwer's predecessors. Those who have no knowledge of intuitionism should read chapter 2, consulting at least section 2.3 and the rest of the chapter from 2.5 on. It is not necessary to understand all the mathematical details in order to understand the debate that is described in later chapters. Chapter 3 should be read in full, with the exception of section 3.2.1. The chapters 4 and 5 are so extensive that one can make a selection oneself, in order to get an impression of how people reacted to intuitionism. Cross-references should suffice in finding one's way through the explanations. Such a selection should in any case include the sections on Weyl's *Grundlagenkrise* (4.2.1, 5.2.1), on Hilbert's first reactions (4.2.2, 5.2.2), and the parts on Hilbert's 1925 lecture (in 4.3.1 and 5.3.1), since these are crucial for an understanding of the debate. The reader is also advised to include Fraenkel's reactions in his reading, since he was the main commentator of the foundational debate. Chapter 6 is indispensable for anyone interested in the cultural context of the foundational debate.

**Lectures** Parts of the book were presented in lectures given in Amsterdam (NL), Luminy (F), Mainz (D), Oberwolfach (D), Palermo (I), Roskilde (DK), Utrecht (NL), and Washington, D.C. (USA), and I profited from comments which I received after the lectures.

## 0.5 Suggestions for further research

The analysis presented here could be expanded in two ways. Firstly, the appearance and influence of social and intellectual networks could be investigated. Whereas I focused on the history of ideas as presented in (mostly) mathematical papers, with some additional biographical data on the participants, such research could embed the foundational debate more into its social environment.

Secondly, one could focus more on popular presentations of the foundational debate. In the 1920s, it was not uncommon to find reports about mathematical events in ordinary newspapers. I included some of these in the book, but only those that I happened to come across in some of the archives, most notably in

## 0.5. SUGGESTIONS FOR FURTHER RESEARCH

the Fraenkel archive. I did not do any systematic research in this respect, and I think that looking into for instance local newspapers in Germany at times when important lectures were delivered (Hamburg in July 1921, Berlin from January to March 1927) could provide more information on how the debate was presented before a non-scientific audience.

# Chapter 1

# Kronecker, the semi-intuitionists, Poincaré

Und da möchte ich Sie nun dringend bitten, dass Sie nicht die Expropriation kontinuieren, die die deutsche referierende mathematische Literatur an mir verübt hat, indem sie mich dasjenige, was mein ausschliessliches persönliches geistiges Eigentum ist, mit Poincaré, Kronecker und Weyl teilen lässt.[1]

L.E.J. Brouwer[2]

## 1.1 Introduction

Modern foundational research got its full start with Cantor's publications on set theory from 1874 onwards. The tradition which developed in this style was to include as its contributors some of the most famous mathematicians of that time. With the publication of *Was sind und was sollen die Zahlen?* ('What are numbers and what should they be?') in 1887, Dedekind was considered to have given a secure foundation for the theory of natural numbers, based on the laws of logic. Hilbert gave a strictly axiomatic foundation to geometry in his *Grundlagen der Geometrie* ('Foundations of geometry') in 1899, and later extended this to his general proof theory. Zermelo axiomatised set theory. It was a tradition of rigorous proofs with an appeal to logic rather than to intuition. And it was against this tradition that several mathematicians protested, who, by so doing, held views more or less similar to the ones Brouwer was to take later on. This chapter is about these

---

[1]'I would strongly like to request you not to continue the expropriation that the German mathematical review literature has perpetrated on me by making me share with Poincaré, Kronecker and Weyl what is my exclusive personal spiritual property.'
[2]Letter from Brouwer to Fraenkel, 28/01/1927; cited from: [Van Dalen 2000, p. 304]

mathematicians. In chronological order, they are: Kronecker, the French semi-intuitionists Borel, Baire and Lebesgue, and Poincaré.

**Practical information** This chapter is divided into three parts. Firstly, the necessary mathematical theory of sets is explained. Secondly, in the three following sections the individuals or groups of individuals involved are treated: Kronecker, the French semi-intuitionists and Poincaré. These descriptions are all divided into two divisions. To start with, the main conflicts regarding the foundations of mathematics in which these people were involved are summarized. For sure, none of these disputes was as serious as the one that centered around Brouwer and Hilbert in the 1920s. Nevertheless, I think it worthwhile to compare these arguments to the one between Brouwer and Hilbert, in order to put the latter into perspective. After that, the views that brought these mathematicians into conflict with their colleagues are treated. In doing so, I restricted myself to describing what their views concerning the foundations of mathematics were. Lastly, I summarize what we have found and compare Brouwer's views to the ones presented here.

### 1.1.1 Mathematical prerequisites

To understand some of the controversies described below, a general knowledge of Cantorian set theory is required. In Cantorian set theory, two kinds of transfinite numbers are distinguished: ordinal numbers and cardinal numbers. Cantor saw set theory as embodying the laws of the infinite, which he regarded as having deep ontological significance.[3]

**Ordinal numbers** Ordinal numbers are used to indicate the number of elements in a well-ordered set.[4] This number, however, depends on which ordering is used. As an example, imagine the natural numbers N with the first natural number 1 put behind all the others in the ordering: $2, 3, 4, \ldots, 1$. This gives rise to an equally well-ordered set, whose ordering differs by having one natural number after the denumerably many elements. The denumerably many before the 1 are just as many as the natural numbers with the standard ordering. Therefore, N with the new ordering has a bigger ordinal number than the standard N.

Cantor used two so-called principles of generation to build up the collection of ordinal numbers. The first of these states that one is always allowed to define new ordinal numbers by successive addition of units to an already defined ordinal number. The other principle says that, given any limitless sequence of defined ordinal numbers, a new transfinite ordinal number can be generated. This number can be thought of as the limit of the sequence, i.e., it is the smallest ordinal number larger than all the numbers in the sequence considered.[5]

---

[3][Rowe 1997, p. 541]
[4]A set is called well-ordered if it is an ordered set in which every non-empty subset has a first element.
[5][Dauben 1979, pp. 97–98], [Cantor 1883, p. 577]

## 1.1. INTRODUCTION

Cantor then defined the ordinal number (or order type) of a well-ordered set in the following way. The first ordinal numbers are the natural numbers. They are created by the first principle of generation, starting from 1. Since this is a limitless sequence of defined ordinal numbers, one can apply the second principle of generation, which gives rise to a new ordinal number larger than all the natural numbers. Cantor called this the first transfinite ordinal number $\omega$, to which we shall refer as $\omega_0$. Carrying on in the same way as the natural numbers were created, that is, by adding a 1 to the last number created and appending this new number as the last one in the ordering, one subsequently obtains $\omega_0 + 1$, $\omega_0 + 2$, etcetera. This second limitless sequence can then be taken as a whole, which gives rise to the new ordinal number $2\omega_0$, etcetera. In general, a whole system of transfinite ordinal numbers can thus be created, indicated as $\Omega$ (nowadays called ORD). To illustrate the structure of the system of ordinal numbers, some examples of ordinal numbers in the so-called second number class are given below:

| | | | |
|---|---|---|---|
| $\omega_0$ | $\omega_0 + 1$ | ... | $\omega_0 + n$ ... |
| $2\omega_0$ | $2\omega_0 + 1$ | | ... |
| ... | | | ... |
| $m\omega_0$ | ... | | ... |
| ... | | | ... |
| $\omega_0^2$ | ... | | ... |
| ... | | | ... |
| ... | $k_l\omega_0^l + k_{l-1}\omega_0^{l-1} + ... + k_1\omega_0 + k_0$ | ... | ... |
| ... | | | ... |
| $\omega_0^{\omega_0}$ | ... | | ... |
| ... | | | ... |

In modern terms, two well-ordered sets $A$ and $B$ are said to have the same ordinal number if there is a bijection between $A$ and $B$ that respects the ordering. If there is a bijection between $A$ and a proper part of $B$ which respects the ordering, then the ordinal number of $A$ is less than the ordinal number of $B$.

**Cardinal numbers** The other important concept of Cantorian set theory is that of a cardinal number. A cardinal number (or power) indicates the number of elements of a set regardless of its order. For finite sets ordering the elements of the set in a different way does not change its ordinal number; for infinite sets, as we have seen, it does. Therefore, ordinal numbers only differ from cardinal ones for infinite numbers.

Note that, if we add 0 to $\Omega$, for all $\gamma$ in $\Omega$ the set of the elements in $\Omega$ from 0 to $\gamma$ has ordinal number $\gamma$. The cardinal number $||\gamma||$ of a set of ordinal number $\gamma$ is defined as the smallest ordinal number of all sets which can be brought into one-to-one correspondence with the set of ordinal number $\gamma$, regardless of ordering. Thus, following the above observation, $||\gamma||$ is the cardinality of the set from 0 to $\gamma$. The cardinal number belonging to $\omega_0$ is called $\aleph_0$.

Both the ordinal and the cardinal numbers can be dealt with in a similar way as the natural numbers. Sums, products, exponentiations and order relations can be defined on them. In this way, a (double) arithmetisation of the infinite is achieved.[6]

In the above system $\Omega$ Cantor built in distinctions as well. He called the natural numbers the first number class, then went on with the second number class which included all ordinal numbers of which the set of numbers preceding it had power $\aleph_0$. The first number of the third number class is the first transfinite ordinal number which does not belong to the second number class, and it is designated by $\omega_1$. The corresponding cardinality of this class is $\aleph_1$, etcetera. In this way, a sequence of number classes and corresponding alephs is created which has no end. Cantor called the system of alephs TAW.

Now the question arises: does the system TAW include all the cardinal numbers? Or, to put it differently, is it possible that there is a set of which the cardinal number is not an aleph? Cantor did not think so. In particular, he claimed that the power of the continuum was the one following on the power of the natural numbers; in formula: $2^{\aleph_0} = \aleph_1$. This is Cantor's famous Continuum Hypothesis. But he could prove neither of the claims. These were to become controversial issues.

## 1.2 Kronecker

Leopold Kronecker (1823–1891) studied philosophy and mathematics in Berlin. His doctoral examination was on the history of legal philosophy. In 1845, he received his doctorate for a dissertation on number theory written under the supervision of Dirichlet. In 1855, having spent numerous years managing a family estate, Kronecker returned to Berlin as an amateur mathematician and a private scholar. He was elected to the Berlin Academy in 1861, by which he was entitled to teach at university.[7] In those days, the Berlin university, where also Kummer and Weierstraß worked, was the leading mathematical center in Germany. Kronecker was one of the main mathematicians in the field of number theory.[8] He was especially known for his work in arithmetic and algebra, mostly in the field of elliptic functions.[9] In later years, Kronecker more and more stressed his views on philosophical questions in mathematics. Already in the early 1870s Kronecker started opposing such mathematical concepts as upper and lower limits, the Bolzano-Weierstraß theorem, and the use of irrational numbers in general.[10]

---

[6][Serfati 1995, pp. 208–209]
[7][Ewald 1996, pp. 941–942]
[8][Schoenflies 1922, p. 103]
[9][Biermann 1973, p. 508]
[10][Dauben 1979, p. 67]

## 1.2. KRONECKER

### 1.2.1 Kronecker's conflicts

Two major controversies with regard to the foundations of mathematics may be distinguished in which Kronecker was engaged.[11]

**Set theory** The first conflict was about Cantor's theory of transfinite numbers. In 1874, after a ten-year hesitation to put forth his results, Georg Cantor (1845–1918) started publishing on set theory, including the theory of transfinite numbers. Kronecker, who had been one of Cantor's professors in Berlin, objected because the numbers Cantor had introduced could not be constructed from the natural numbers. To him, these numbers had no real meaning whatsoever. He even went so far as to ensure that Cantor's papers would only be published in the prestigious Crelle's Journal, of which he was the editor, after a considerable delay.[12] The dispute surpassed purely mathematical matters. In order to annoy Kronecker, Cantor, who lectured in Halle, applied for a position in Berlin. Kronecker answered by announcing that he would publish a paper explaining his views on the foundations of mathematics in the Acta Mathematica (but this never happened). Acta Mathematica was the only journal of which the editor, Mittag-Leffler, had reacted positively to Cantor's work.[13] Cantor felt very much attacked by Kronecker's criticism; the bitterness can be read in the letters he wrote to Mittag-Leffler on this subject in 1884.[14] These feelings even seem to have contributed to Cantor's nervous breakdown.[15]

It should be noted, however, that there never was an open confrontation in print between Kronecker and Cantor.[16] This means that the whole conflict described above was constructed from second-hand information by people like Schoenflies who are said to have known the informal circuit. Recently, objections to this image of Kronecker have been raised by Edwards, who studied Kronecker's work intensively. In Edward's view, Cantorian set theory did not form part of what Kronecker saw as the foundations of mathematics. This could also very well, maybe even better, explain why Kronecker never reacted to it in print.[17] However, the important thing for us is the way in which mathematicians saw Kronecker in the beginning of the 20th century. At that time, the picture described above was not disputed.

It was also the item of set theory which brought Kronecker into conflict with his most famous colleague at the university of Berlin, Karl Weierstraß (1815–1897). Their opinions on the value of Cantor's creation were diametrically opposed. Kro-

---

[11]The first one was taken from [Aspray 1981, pp. 57–61], the second from [Dauben 1979, p. 68].
[12][Schoenflies 1922, pp. 98–99]
[13][Schoenflies 1927]
[14]Parts of the letters were published in [Schoenflies 1927].
[15][Schoenflies 1927, p. 8, 16]
[16][Dauben 1979, p. 162]
[17][Edwards 1989, p. 67]

necker expressed his feelings in the following prophetic way in a letter to Schwarz:[18]

> Wenn mir noch Jahre und Kräfte genug bleiben, werde ich selber noch der mathematischen Welt zeigen, daß nicht bloß die Geometrie, sondern auch die Arithmetik der Analysis die Wege zeigen kann — und sicher die strengeren. Wenn ich's nicht mehr thue, so werden's die thun, die nach mir kommen, und sie werden auch die Unrichtigkeit aller jener Schlüsse erkennen, mit denen jetzt die sogenannten Analysis arbeitet.[19]

The 'wrongness' of the new analysis can be interpreted as what we would now call its non-constructive, and therefore ill-founded, character.[20] To see his analysis described as 'so-called' was a bitter and hurting experience for Weierstraß. For this reason, he even seriously considered leaving Berlin in 1885. The only ground why he did not do this, so it seems, was his fear of Kronecker getting too much influence.[21]

**Irrational numbers** Secondly, Kronecker also protested against Cantor and Dedekind's efforts to establish a completely general theory of irrational numbers. Again, his main objection was the absence of constructions on the basis of arithmetical laws. Edwards argued that here the stress should be laid on the words 'completely general'. Kronecker thought that one should only work with specific irrational numbers; starting to talk about the totality of irrational numbers would mean leaving the foundations on which the irrational number was constructed.[22]

However, most mathematicians do not seem to have put too much weight on Kronecker's radical opinions on these subjects. They were mostly seen as a peculiarity of a temperamental but great man.[23]

### 1.2.2 Kronecker's views

Kronecker's position[24] is often summarized by the dictum '*Die ganzen Zahlen hat der liebe Gott gemacht, alles andere ist Menschenwerk*'.[25]

In Kronecker's view, there is a fundamental difference between arithmetic and algebra on the one hand, and analysis on the other. The objects of arithmetic

---

[18][Biermann 1988, p. 138]
[19]'If enough years and power will be left to me, I will personally show the mathematical world that not only geometry, but also arithmetic can show analysis the way — and for sure the stricter one. If I will not do it, then those who come after me will, and they will recognize the inaccuracy of all those conclusions with which the so-called analysis nowadays works.'
[20][Edwards 1989, p. 70]
[21][Biermann 1988, p. 138]
[22][Edwards 1989, p. 75]
[23][Kneser 1925, p. 221]
[24]The information in this section was drawn from [Molk 1885], unless stated otherwise.
[25]'God created the integers, everything else is the work of man'. Kronecker is said to have made the remark at a lecture at the *Berliner Naturforscher-Versammlung* in 1886; cf. [Weber 1891-92, p. 19].

## 1.2. KRONECKER

and algebra are positive integers and polynomial functions with positive integers as coefficients on them, which are entities that *exist*; whereas analysis is about rational, negative, imaginary, irrational and transcendental numbers, which are mere *symbols*. The only symbols that can be transferred from analysis to arithmetic and algebra are the rational numbers, since these can always be re-written into integers in a finite number of operations. Others can only serve as a means of help. To give an example: $\sqrt{2}$ is nothing but a symbol; the only thing that is real about it is the equation $x^2 = 2$ that defines it. In modern terms, this means that we transfer computations from the polynomial ring $\mathbb{Q}[x]$ to $\mathbb{Q}[x]/(x^2 - 2)$. In the latter $\sqrt{2}$ is represented by an $x_1$ satisfying $x^2 - 2 = 0$. Thus, e.g., $\sqrt{2} \cdot \sqrt{2}$ is represented by $x_1 \cdot x_1 = x_1^2 = 2$, therefore $\sqrt{2} \cdot \sqrt{2} = 2$.

Connected to this view is Kronecker's opinion on mathematical definitions. When defining mathematical objects, Kronecker demands the definition to be not just logical, but algebraic. The difference, in his view, lies in the fact that an algebraic definition requires a method that will result in the defined object in a finite number of steps, whereas logic restrains itself to saying that an object is, or is not. In modern terms, one could say that the first way of defining an object is more combinatorial. More specifically, Kronecker requires a definition to show us how to decide, in every possible situation, whether the definition is fulfilled or not. If this requirement is not met, he considers the object to be not well-defined.[26]

Many authors believe that Kronecker wanted to recast analysis without the use of the irrational numbers.[27] However, he never explicitly stated this intention. Indeed, seeing the frequent use Kronecker himself made of irrational numbers, it seems more consistent to assume, as Edwards does, that Kronecker's main objection was against the use of an *arbitrary* irrational number, and against the totality of irrational numbers. As long as the irrationals were given explicitly and constructively, Kronecker did not object. He uses the same line of thought regarding infinite mathematical objects: to accept objects that are given constructively, but to reject any 'arbitrary' one.[28]

Kronecker never described what analysis according to his views would look like. It seems clear that, following his opinions, Cantorian set theory and the infinite cardinal numbers would disappear, just as infinite sets as defined by Dedekind, among other things. It is not clear if Kronecker envisaged anything filling up these holes.[29]

A final remark about Kronecker's stance is that he explicitly wants to have certain mathematical concepts explained, before starting to work on mathematics. In *Über den Zahlbegriff*, one of the papers in which Kronecker most clearly formulated his views on the foundations of mathematics, he states it as follows:[30]

Auf dem freien Plane philosophischer Vorarbeit, aus welchem man in

---
[26][Edwards 1989, p. 72]
[27][Molk 1885], [Kneser 1925, p. 221], [Biermann 1973, p. 507]
[28][Edwards 1989, pp. 71–75]
[29][Kneser 1925, p. 222]
[30][Kronecker 1887, p. 337]

8     CHAPTER 1. KRONECKER, THE SEMI-INTUITIONISTS, POINCARÉ

die eingehegten Gebiete der verschiedenen Wissenschaften gelangt, sind auch die Begriffe der Zahl, des Raumes und der Zeit zu entwickeln, von welchen in der Mathematik Gebrauch gemacht wird. Und es erscheint zweckmässig, die Entwicklung dort so weit zu führen, dass die Begriffe schon mit ihren Grundeigenschaften ausgestattet sind, wenn die specialwissenschaftliche Behandlung beginnt.[31]

We will see this view re-appear on the constructivist side.

## 1.3  The French semi-intuitionists

The French semi-intuitionists, as they were later called,[32] consisted basically of three persons: Borel, Baire and Lebesgue. Of these, Borel was the one who expressed himself most on foundational affairs; therefore, it is mainly his ideas which are presented below.

**Borel**  Émile Borel (1871–1956) became famous during the last years of the nineteenth century. This was mostly due to his work on the foundations of measure theory, his creation of the theory of divergent series and the so-called quasi-analytic functions. His name is still remembered in the so-called Heine-Borel theorem. His proof for Picard's theorem in 1896 was a real sensation, since it had been pursued ever since its formulation, some seventeen years before.[33] Borel started editing the famous *Collection de monographies sur la théorie des fonctions*, known in English as the 'Borel Tracts'. Later, Lebesgue and Baire published some of their work in this series, too. After the First World War Borel turned to probability theory and politics.[34]

**Baire**  René Baire (1874–1932) did not publish much, but what he wrote was of great value. In his doctoral thesis in 1899 he developed, among other things, the concept of semi-continuity. The most successful part of his work consisted in the application of transfinite set theory to resolve problems in function theory. He created the class of Baire functions, which is explained below, and provided a

---

[31]'In the open field of preliminary philosophical labour, from which one reaches the fenced-in domains of the various sciences, the concepts of number, space, and time, which are used in mathematics, are also to be developed. And it seems expedient to carry the development so far that, when the treatment in the special sciences begins, the concepts will already have been equipped with their basic properties.' English translation cited from [Ewald 1996, vol. II, p. 948].

[32]Heyting gave them the name in [Heyting 1934]; the term *'halbintuitionistisch'* also appears in [Becker, O. 1927, p. 461], but there it is used to refer to Weyl's *Das Kontinuum*.

[33]Picard's theorem states that any single-valued analytic function of a complex variable assumes any finite complex value, with the possible exception of one value, in an arbitrary neighbourhood around an isolated essential singular point; cf. [Solomentsev 1991, pp. 157–158].

[34][Collingwood 1959, pp. 488–492]

## 1.3. THE FRENCH SEMI-INTUITIONISTS

framework for independent research into the theory of functions of real variables. In this field, he influenced French mathematics profoundly.[35]

**Lebesgue** Henri Lebesgue (1875–1941) acquired a reputation in the field of integration theory, in which he used works of both Borel and Baire. In his doctoral thesis in 1902, he presented an integration theory which took into account the discontinuous functions given by Baire. Later, he completed Borel's definitions of measure and used them in order to obtain a generalisation of the Riemann integral. He also applied these new definitions to mathematical problems. In this way, he was able to prove, e.g., that the fundamental theorem of the calculus, $\int_a^b f'(x)dx = f(b) - f(a)$, to which exceptions were known, holds in general if the integral is taken in Lebesgue's sense. Also, he was highly successful in applying the new integral definition in the theory of trigonometric series. He worked at the Sorbonne from 1910 on.[36]

### 1.3.1 The French semi-intuitionists' main conflict

As was the case with Kronecker, the French semi-intuitionists, too, were critical of Cantor's set theory. At the beginning of the 20th century, Cantor's name was well-known in mathematical circles, and the importance of his set theory was widely acknowledged. In 1897, at the First International Congress for Mathematicians, the important contributions made by applying Cantorian set theory to function theory were clearly stated in Hurwitz's opening address.[37] More importantly still, Hilbert, in his famous address at the Second International Mathematicians' Congress in Paris in 1900, highlighted the continuum problem by presenting Cantor's Continuum Hypothesis together with the well-ordering of the continuum as the first in his list of unsolved problems.[38]

At the same time, however, doubts concerning the status of Cantorian set theory also became more frequent. Following on the Burali-Forti[39] and Russell paradox,[40] König launched an attack at the Third International Congress of Math-

---

[35][Costabel 1970]
[36][Hawkins 1973]
[37][Hurwitz 1898, p. 97]
[38][Hilbert 1901, pp. 298–299]
[39]The Burali-Forti paradox comes into being if one allows all ordinals to be joined in a set. Since the set of all ordinals $\Omega$ is a well-ordered set, $\Omega$ has an ordinal number $o$, which by definition is bigger than all the ordinals in $\Omega$. However, since $\Omega$ is the set of all ordinals, $o$ has to be in $\Omega$, whereby $o$ is bigger than itself.
The paradox was named after Burali-Forti and for decades ascribed to him. However, Burali-Forti did not conclude that there is a paradox, but that the set of all ordinal numbers is not well-ordered, cf. [Burali-Forti 1897].
[40]Russell originally formulated his paradox with regard to Frege's *Grundgesetze der Arithmetik*. Consider the predicate $p$ 'to be a predicate which does not apply to itself'. Then $p$ applies to itself if and only if it does apply to itself, hence we have a contradiction. Since $p$ was allowed in Frege's system, the paradox destroyed Frege's attempt at a foundation of mathematics. The Russell paradox can easily be transferred to set theory if one allows sets to have sets as an ele-

ematicians in 1904. He claimed that the power of Cantor's continuum was not an aleph at all. Even though Hausdorff shortly afterwards succeeded in pointing out a mistake in the proof,[41] the general feeling of inconvenience with Cantor's creation did not disappear. For it might always turn out possible to find a reparation of König's proof. There were only two possible ways of countering this definitively: either by proving that the continuum could be well-ordered, or by proving that every transfinite cardinal number was an aleph. In the same year, Zermelo produced a proof of the former.[42]

**Axiom of choice**  Zermelo's 1904 paper, which was published in the *Mathematische Annalen*, contained a proof for the theorem that every set can be well-ordered, based on the axiom of choice. The axiom of choice states, in Zermelo's original formulation, that 'for every subset $M'$ [of an arbitrary set $M$, DH], one imagines an arbitrary element $m'_1$ grouped to it, which appears in $M'$ and may be designated the 'distinguished' element of $M'$'.[43] It had been used before quite naturally by several mathematicians in analytical proofs, but nobody had stated it explicitly as a principle. Zermelo used the axiom of choice in order to find an ordering of the entire set by starting from orderings of its subsets. He stressed that the importance of his result lay in the fact that it was now proved that the power of every set had to be an aleph. The axiom of choice, however, is purely existential, and no constructive way of actually finding the well-ordering of a given set was presented. More specifically, no explicit well-ordering of the continuum was given. It is questionable whether this satisfied Hilbert's demand for a *direct* proof of the well-ordering.[44]

Zermelo's paper came under considerable criticism. The next issue of the *Mathematische Annalen*, in 1905, contained reactions to his usage of the axiom of choice by Bernstein, Jourdain, Schoenflies and Borel.[45] Borel's paper led to a correspondence between several French mathematicians, which was published in the *Bulletin de la Société mathématique de France*.[46] It should be noted that the title given to these letters, *Cinq lettres sur la théorie des ensembles* ('Five letters on set theory'), was rather unfortunate because, as both Hadamard and Lebesgue remarked, the key issue is mathematical existence rather than set theory.[47] Of these

---

ment. Let $U$ be the universe of all sets. By considering the collection of all sets in $U$ which are not an element of itself, one obtains a similar paradox.

[41] [Hausdorff 1904]

[42] [Dauben 1979, pp. 241–250]

[43] '*Jeder Teilmenge $M'$ denke man sich ein beliebiges Element $m'_1$ zugeordnet, das in $M'$ selbst vorkommt und das 'ausgezeichnete' Element von $M'$ genannt werden möge.*', [Zermelo 1904, p. 514], cited from [Mehrtens 1990, p. 445].

[44] [Dauben 1979, pp. 250–253], [Aspray 1981, pp. 62–65], [Hilbert 1901, p. 299]

[45] The debate raged further until 1908, and included mathematicians from Britain, France, Germany, Hungary, Italy, the Netherlands and the United States; cf. [Moore 1982, p. 85].

[46] [Baire et al. 1905]

[47] Lebesgue: '*La question revient à celle-ci, peu nouvelle: peut-on démontrer l'existence d'un être mathématique sans le définir?*' ('The question amounts to the following, not very new one: can one prove the existence of a mathematical entity without defining it?'), [Baire et al. 1905,

## 1.3. THE FRENCH SEMI-INTUITIONISTS

mathematicians, Hadamard was the only one who defended Zermelo's position; the other ones, Borel, Baire and Lebesgue, were sharply opposed to the axiom of choice.

Borel protested against the absence of even a theoretical means of carrying out the choice required by the axiom of choice. He claimed that whenever an arbitrary choice is to be made a non-denumerably infinite number of times, as is the case for example when ordering the continuum, we have left the domain of mathematics. The only thing Zermelo, in Borel's view, had proven was the equivalence of the following two problems:

1. to well-order an arbitrary set $S$, and

2. to choose a distinguished element from each non-empty subset of $S$ in a determined way.

He compared the working of the axiom of choice to the ordering of a set by first choosing an arbitrary element, assigning it with rank 1, then taking a second one which gets rank 2, and so on transfinitely many times. Nobody, Borel claimed, would accept such a reasoning for ordering a set.[48]

Baire went further than Borel in his criticism. In his view, if a set is given, one cannot consider its subsets as given as well. *A fortiori*, a choice made from all these subsets is devoid of sense. All mathematics, Baire maintained, should be constrained to definable domains. He admitted that Zermelo's result was consistent, but he also found it meaningless.[49]

Lebesgue objected to the choice function that had not been defined uniquely. He interpreted the word 'to choose' as 'to nominate', and required the correspondence to be determined once and for all. Zermelo's proof did not establish this. In Zermelo's way, one could never know if two mathematicians, in using the axiom of choice, were talking about the same choice set. Furthermore, Lebesgue asked what it means to make an infinite number of choices. It could not be choosing one element after the other; thus, it would have to be a law. But then the law should be given.[50] To sum up: Lebesgue found Zermelo's proof not 'Kroneckerian enough'.[51]

It should be noted that the discussion was carried out in a very friendly, reasonable and non-polemical atmosphere. Lebesgue explicitly mentioned that the debate was in fact about conventions in mathematics.[52] Mutual respect can also

---

p. 265]; Hadamard: 'La question me paraît tout à fait claire maintenant (...). De plus en plus nettement, elle tient tout entière dans la distinction (...) entre ce qui est *déterminé* et ce qui peut être *décrit*.' ('The question now seems to me to be very clear (...). It is more and more clearly completely contained in the distinction (...) between what is *determined* and what can be *described*.'), [Baire et al. 1905, p. 269]

[48] [Borel 1905, pp. 194–195]
[49] [Baire et al. 1905, p. 264]
[50] [Baire et al. 1905, p. 268]
[51] [Dauben 1979, pp. 253–259], [Bockstaele 1949]
[52] [Baire et al. 1905, p. 265]

be found in the following words by Hadamard:[53]

> Ce sont deux conceptions des Mathématiques, deux mentalités qui sont en présence. Je ne vois, dans tout ce qui a été dit jusqu'ici, aucun motif de changer la mienne. Je ne prétends pas l'imposer.[54]

This pragmatic tone was in sharp contrast to the way in which the foundational debate in the 1920s was conducted.[55]

One last aspect of the French semi-intuitionists' attitude should be pointed out. Moore showed that the pervasiveness of the axiom of choice was far greater than generally perceived: all of the semi-intuitionists had implicitly used it in their earlier research, often in crucial parts of proofs. Baire once even explicitly stated that he thought it beside the point to ask oneself whether a correspondence that he had proved in a non-constructive way could be established effectively.[56] Whether it was Zermelo's explicit formulation of the axiom of choice that made them change their point of view remains unclear. Even after the debate some of them continued using it.[57]

In 1908, after the whole discussion, Zermelo published a new proof of the well-ordering theorem based on his axiomatisation of set theory. This time, he refrained from mentioning the human activities of thinking and ordering, and replaced them by abstract notions. His new formulation of the axiom of choice read: 'A set $S$, which consists of a set of disjoint parts $A$, $B$, $C$, ..., which all contain at least one element, has at least one subset $S_1$ which has exactly one element in common with each of the considered parts $A$, $B$, $C$, ...'.[58] Thus, he hoped to avoid the psychologistic interpretations that had been put forward in the debate.[59]

## 1.3.2 The French semi-intuitionists' views

Borel, Baire and Lebesgue can be grouped together as the French semi-intuitionists because they share the same, what we would now call, constructivistic ideas about

---

[53][Baire et al. 1905, p. 270]

[54]'These are two opposing conceptions of mathematics, two opposing mentalities. In all that has been said up to this point, I do not see any reason for changing mine. I do not mean to impose it.' English translation based on the translation in [Ewald 1996, vol. II, p. 1084].

[55]See 3.3.2.

[56][Moore 1982, pp. 64–70]

[57][Moore 1982, p. 103]

[58]'*Eine Menge S, welche in eine Menge getrennter Teile A, B, C, ... zerfällt, deren jeder mindestens ein Element enthält, besitzt mindestens eine Untermenge $S_1$, welche mit jedem der betrachteten Teile A, B, C, ... genau ein Element gemein hat.*', [Zermelo 1908, p. 110], cited from [Mehrtens 1990, p. 453].

[59][Moore 1982, p. 143]. For a more detailed discussion of Zermelo's two formulations of the Axiom of Choice, the reader is referred to [Mehrtens 1990, pp. 445–453].

During the foundational debate in the 1920s, Zermelo mostly remained silent. Actually, he published not a single paper between 1914 and 1927. The story goes that this was because he could not expect to offend anybody any more, [Fraenkel 1967, p. 149]. However, his bad health must have played a role, too, [Pinl 1969, p. 221]. Zermelo did publish some papers related to foundational questions from the end of the 1920s on, but he did not react to intuitionism publicly.

## 1.3. THE FRENCH SEMI-INTUITIONISTS

the foundations of mathematics. However, it should be remarked that none of them formulated a coherent philosophy of mathematics. Their comments were rather piecemeal observations.[60] The more important pieces are described below.[61]

**Mathematics**  In Borel's view, the only real science is subjective science; that is to say: knowledge that we can accumulate and express. All the rest is metaphysical abstraction. Applied to mathematics, this means that the only important part for us is that part which we can think of, can execute and know well enough to be able to treat it without fear for error. We should also be sure that other mathematicians think the same, when talking about the same object. This last demand can make the situation change over time: mathematical concepts which were considered too vague, can become accepted after clarification.

Thus, in Borel's view mathematics is not a formal matter, but a human mental activity, expressed in some language. Therefore, both contents and form are relevant. Borel attaches more value to the role of language than Brouwer was to do later.[62]

Effective realisability and calculability are key notions in Borel's view on mathematics. To him, this shows that mathematics is linked to concrete things. He describes the general tendency in his mathematical work as follows:[63]

> Je tache d'y montrer que les Mathématiques ne sont pas un jeu purement abstrait de l'esprit, mais sont, au contraire, en étroite connexion avec la réalité concrète.[64]

**Logic**  Borel remarks that logic supplies us with an unlimited number of possibilities, but that those form nothing but the material from which to select. When discussing mathematical proofs, he writes:[65]

> (...) on est obligé d'avouer que le seul rôle de la logique (...) a été d'en fournir les matériaux, et l'on ne confond pas le tailleur avec l'architecte.[66]

One can compare the role of logic to that of nature in physics: it supplies the physicist with material, but that material itself is not physics.

Lebesgue seems to precede Brouwer in his criticism of the principle of the excluded middle. In his letter to Borel he writes:[67]

---

[60][Moore 1982, p. 92]
[61]Unless stated otherwise, information in this section was drawn from [Bockstaele 1949].
[62]See 2.2.4 and 2.3.1.
[63][Fréchet 1965, p. 25]
[64]'I attempt to show from it that mathematics is not a purely abstract mind game, but, on the contrary, that it is in close connection with concrete reality.'
[65][Borel 1907, p. 279]
[66]'(...) one has to admit that the role of logic (...) has been limited to supplying the material for it, and one does not confuse the mason with the architect.'
[67][Baire et al. 1905, p. 269]

> Bien que je doute fort qu'on nomme jamais un ensemble qui ne soit
> ni fini, ni infini, l'impossibilité d'un tel ensemble ne me paraît pas
> démontré.[68]

**Mathematical existence** All mathematicians taking part in the discussion agree that consistency is a necessary demand for mathematical existence; whether it is sufficient is a different question.

An argumentation with the negative property of being free from contradictions does not necessarily have mathematical contents. What is needed, in Borel's view, is that it be supported by the deeper mathematical reality. What exactly that is, Borel most often leaves to philosophers. Sometimes he describes the extra demand as intuition, meaning experiencing a mathematical object as a concrete mental reality. That is to say: the criterion Borel uses is that we should know some essential properties of it, so as to be able to distinguish it from all other mathematical objects.

More specifically, Borel regards the works of mathematicians such as Cantor and Zermelo as logically possible. However, the question that remains to be answered is: do the objects which they defined bear any relation to traditional parts of mathematics?[69] Commenting on Hilbert's axiomatic way of doing mathematics, Borel writes, in a way that resembles Brouwer's words from about the same period:[70]

> Cette manière de procéder est évidemment légitime, en ce sens que
> l'on a toujours le droit de créer un vocabulaire et de construire avec ce
> vocabulaire un édifice logique; mais l'absence de contradiction logique
> ne suffit pas à caractériser la construction scientifique.[71]

In Borel's view, the criterion for existence is the question whether one knows the object one talks about completely and whether one can calculate with it or not; only then, the definition can be accepted. Also, a mathematical object should be what he calls effectively definable; i.e., the definition should be possible with a finite number of words. If not, it is to be considered non-existent. As an example of a non-existent number, Borel treats the case in which a denumerable number of persons one after the other choose a digit. The number that is thus made cannot be defined in a finite number of words; for therefore one should have to describe the denumerable number of choices. Thus, the number cannot be considered realised,

---

[68]'Although I strongly doubt that one could ever name a set which is neither finite nor infinite, I do not consider the impossibility of such a set to have been proved.'

[69][Borel 1914, pp. 176–177]

[70][Borel 1914, p. 175]; on Brouwer, see 2.3.1.

[71]'This way of proceeding is evidently legitimate, in the sense that one always has the right to create a vocabulary and to construct a logical building with that vocabulary; but the absence of a logical contradiction does not suffice to characterise the scientific construction.'

## 1.3. THE FRENCH SEMI-INTUITIONISTS

and therefore it does not exist.[72] Lebesgue is less demanding than Borel and just wants the mathematical object to be defined.[73]

The demand of using only a finite number of words in a definition does not mean that words such as 'always', 'infinite', etcetera, are forbidden in defining mathematical objects. The important thing, Borel claims, is that, even if a definition includes an infinite number of operations, these should be clearly described by a law in a finite number of words, together with the order in which to execute them. Only in this way can mathematicians be sure to be talking about the same object. The desired description can only be reached, Borel maintains, by using infinite sets or infinite numbers of operations that are effectively enumerable. That is: there should be a one to one correspondence between the given set and the set of natural numbers, and this correspondence should be effectively known to us.

It is also in these terms that Borel explains Richard's paradox. Richard's paradox is the following. Take all the decimal numbers that can be defined by a finite number of words, and provide them with an ordering. The ordering can be made by first taking all the definitions that take exactly $n$ words; within this group, one uses the alphabetical order. Then, all the groups of definitions of length $n$ are ordered in order of rising $n$. This automatically gives rise to an ordering on the collection of the thus defined decimal numbers, a denumerable sequence $a_1, a_2, a_3, \ldots, a_n, \ldots$ referred to as $(a)$. Now define the decimal number $\alpha$ as follows: the $n$th decimal of $\alpha$ is defined as $n' + 1$ mod 10, where $n'$ is the $n$th decimal of the $n$th number in the sequence $(a)$. It is immediately clear that $\alpha$ is different from all numbers in the sequence $(a)$; nevertheless, we have just defined it in a finite number of words, therefore, by definition, it should belong to $(a)$. Borel explains this in the following way: the set $(a)$ is denumerable, but not effectively denumerable, and therefore it can never be seen as completely defined. Since $(a)$ is needed in order to define $\alpha$, this is were the problem arises.

**Natural numbers** Borel accepts both the concept of a natural number and the set of natural numbers as *notions claires* ('clear notions'). When asked why these notions can be accepted, he answers:[74]

> Nous n'avons pas à résoudre ces questions difficiles; il nous suffit de constater l'accord *pratique* des mathématiciens dans l'usage qu'ils font de ces notions.[75]

Borel does not want to go too deeply into the question why this notion is so clear to mathematicians; in his view, this is the work of philosophers and psychologists. The certainty to know that it is an intuitively clear notion suffices.

---

[72][Borel 1928, p. 154]. The example resembles the construction of a choice sequence, the concept of which was crucial in de development of Brouwer's intuitionism, with the important difference that a choice sequence is constructed by one individual; see 2.6.
[73][Baire et al. 1905, p. 265]
[74][Borel 1914, p. 179]
[75]'We do not have to resolve these difficult questions; for us it suffices to note the *practical* agreement among mathematicians in the use of these notions.'

Baire disagrees with Borel on this point. He does not consider the term 'denumerable infinite' to be well-founded; in his view, mankind can never know more than finite systems. Everything that goes beyond that stage is virtual and consists of nothing but conventions.[76]

**Set theory** Borel was one of the first mathematicians to show the success with which set theory can be applied in function theory and measure theory. Also, the definition of what we nowadays call a Lebesgue integral relied heavily on Cantor's work.[77]

Borel admits that he had at first been seduced by reading Cantor's work on set theory. But, he adds, he had always thought of it not as a goal in itself, but only as a means.[78] For Borel, a set can be seen as existing only if a correspondence can be defined between this set and an already known set. Then the question arises which are the sets one starts from. For Borel these are not only the set of natural numbers, but also the continuum. As a justification for this position, he again uses the argument that mathematicians seem to agree on what they are talking about when they mention the continuum. This, he claims, can be explained by using the geometrical intuition.[79] Note that the justification here is a different one from the one used for the natural numbers. There, each element was accepted separately, and the set of natural numbers as its collection; the continuum is seen more as a whole.

Borel does see the weakness of his position. Therefore, he defines what he calls the practical continuum, which consists of all real numbers that can be defined finitely. Contrary to the geometrical continuum, the practical continuum is denumerable. This shows that there are elements in the geometrical continuum that cannot be defined. The practical continuum, Borel claims, is what mathematicians really use; the theoretical continuum is nothing but a metaphysical concept.

As to the usefulness of reasonings on non-definable entities, for instance transfinite sets, Borel makes the following remark. The value of these reasonings can be compared to that of theories in mathematical physics. By those, we do not claim to express reality, but they rather serve as a guide by which we can find new phenomena. The question why these reasonings fulfil this task would require an amount of work that would, Borel maintains, be out of proportion to the answer found.[80]

So, in the end, Borel does not accept any other sets than the denumerable ones, since they are the only ones that can be really defined. These, he claims, are the only reality that we can reach.[81]

---

[76] [Baire et al. 1905, p. 263]
[77] [Schoenflies 1922, pp. 102–103]
[78] [Fréchet 1965, p. 25]
[79] At the same time, however, he warns against relying too much on this intuition; cf. [Borel 1914, p. 177].
[80] [Baire et al. 1905, p. 273]
[81] [Fréchet 1965, pp. 34–35]

## 1.3. THE FRENCH SEMI-INTUITIONISTS

Baire, Borel and Lebesgue give a construction of well-ordered sets, from which they derive the transfinite numbers as mere symbols. Of all the transfinite numbers in the second number class, only a small number can be defined, so that the phrase 'the set of all transfinite numbers of the second number class' is without meaning.

**Higher analysis** The semi-intuitionists try to build up higher analysis by what can be called a constructive method. That is to say: only certain methods, that are considered to be well-known, are used in the creation of mathematics. We will now briefly look at how they applied this view to analysis.

For this purpose, Borel studies a certain type of sets. He starts with sets $\{x_1, x_2, \ldots, x_n\}$ in $[0,1]^n$ of the form $a_1 \leq x_1 \leq b_1, a_2 \leq x_2 \leq b_2, \ldots, a_n \leq x_n \leq b_n$.[82] To these sets, he applies the two following operations a finite or denumerably infinite number of times:

1. take the difference between two sets $S_1$ and $S_2$, where $S_2 \subset S_1$;

2. join a finite or denumerably infinite number of already defined, disjoint sets.

He calls the sets obtained in this way measurable; since Lebesgue we know them as B(orel)-measurable sets. Note that the definition is given constructively.

Lebesgue thought that with these B-measurable sets the goal of a closed area within mathematics, from which one cannot depart by means of the normal mathematical construction methods, had been reached. However, Lusin and Souslin proved this to be wrong: applying a projection to a B-measurable set does not, in general, lead to another B-measurable set.

Baire presents a constructive approach to the role of functions in mathematics which is as follows. Let $C$ be a closed set in the $n$-dimensional space, and let $f_1, f_2, \ldots, f_m, \ldots$ be functions on $C$. Then we call a function $F$ the limit of the sequence $f_1, f_2, \ldots, f_m, \ldots$, if for any point $p$ in $C$ $\lim_{m \to \infty} f_m(p) = F(p)$. With this limit concept, Baire's class of functions is defined as follows. We define a correspondence between classes of functions and ordinal numbers of the first or second number class (starting with zero) as follows:

1. a continuous function belongs to class zero;

2. a function belongs to class $\alpha$ ($\alpha > 0$) if it is the limit of a sequence of functions of lower classes, and if it does not belong to any previous class.

An interesting connection between these two constructive areas of analysis, point sets and function theory, was given by Lebesgue. In order to understand it, we need one more definition. Let $B$ be the collection of B-measurable sets, and let $f(x)$ be a function $\mathbb{R} \to \mathbb{R}$. Then $f(x)$ is called B-measurable if the inverse image under

---

[82][Borel 1898, p. 46]. Borel does not specify whether the interval $[0,1]$ should be seen as part of the geometrical or of the practical continuum; it is not necessary to do so, since the definition can be applied in both cases.

$f$ of a B-measurable set is again a B-measurable set.[83] What Lebesgue proved is that the functions of Baire's class are identical to the B-measurable functions. Therefore, the classifications in both sections run parallel.[84]

## 1.4 Poincaré

Jules Henri Poincaré (1854–1912) was a mathematician of universal abilities. His mathematical activities cover a wide range of fields, including the theory of functions, Lie groups, algebraic topology, algebraic geometry, non-Euclidean geometry, differential equations, number theory, mathematical physics and celestial mechanics. From 1881 on, he worked as a professor of mathematics at the university of Paris. As his international recognition increased, he began to discuss mathematics for a wider audience.[85]

### 1.4.1 Poincaré's conflicts

Poincaré was involved in several conflicts, the main ones being about set theory, about the nature of mathematics, and about impredicative definitions.

**Set theory** Poincaré sharply criticised Cantorian set theory for its use of impredicative definitions[86] and of the actual infinite.[87] One of Poincaré's famous remarks which echoed through the history of mathematics is that set theory is a disease, from which mathematicians would later consider themselves recovered. However, as Jeremy Gray showed, the ascription of the remark to Poincaré is the result of several small mistakes, as well as of historians not checking their primary sources. Most probably, Poincaré never made the remark, and his attitude is better represented by the statement that set theory was a 'beautiful pathological case'[88] which would bring joy to the doctor who followed it.[89]

**Nature of mathematics** The best-known fight in which Poincaré got involved was the discussion with Bertrand Russell on the nature of mathematics. In a series of papers published between 1905 and 1912 in the *Revue de Métaphysique et de Morale* and other journals, they debated the nature of mathematical reasoning.[90]

---

[83][Kolmogorov & Forin 1975, pp. 279–280]

[84]For more details, cf., e.g., [Halmos 1950].

[85][Dieudonné 1975, p. 60], [Ewald 1996, p. 972]. The information in this section was drawn from [Bockstaele 1949], unless stated otherwise.

[86]An impredicative definition is a definition in which the object to be defined is defined in terms of the collection of all the elements of a certain kind, of which the object to be defined itself is one.

[87][Dauben 1979, p. 266]

[88]'un beau cas pathologique', [Poincaré 1909, p. 182]

[89][Gray 1991]

[90]The papers were later brought together in [Heinzmann 1986].

## 1.4. POINCARÉ

Russell believed in the complete reduction of mathematics to logic, something to which Poincaré was strongly opposed. One of the focal differences of opinion was the question whether mathematical reasoning was analytic or synthetic in character.[91] Poincaré, a neo-Kantian, stressed the synthetic character of mathematics, and he specifically put forward the principle of mathematical induction as showing the non-analytic character of mathematical reasoning.[92] Russell, though in word also advocating the synthetic character of mathematics, broadened the definition of synthetic so much that it included all kinds of reasoning that would normally be considered analytic. Moreover, the way in which Russell tried to discredit Poincaré's views would never work. What Russell showed was that logical inferences existed for intuitive reasoning. This was a valid argument for someone taking the logicist view that mathematical reasoning is completely deductive. But Poincaré saw mathematical reasoning as something with deeper epistemological contents than its logical counterpart. Therefore, Russell's argumentation could not possibly make Poincaré change his point of view.[93]

**Impredicative definitions** The third conflict that should be mentioned is the one between Poincaré on the one hand and especially Zermelo on the other on the issue of impredicative definitions. This one took place between 1906 and 1909. It started with Poincaré's rejection of the logicists' definition of a finite number. In this definition, a number is classified as finite if it belongs to every inductive set, that is to say, to every set that contains the element 0 and, for every $n$ in the set, contains the element $n + 1$. In predicate logic this would read: a number $f$ is finite if $\forall V : (0 \in V, n \in V \Rightarrow n + 1 \in V) \Rightarrow f \in V$. Poincaré now claims that, in order to avoid a vicious circle, the collection of inductive sets that is used in defining a finite number may not contain sets that are defined by means of the set of finite numbers, since the latter is the set to be defined. In other words, Poincaré wants to restrict the scope of the 'for all' quantifier in the definition so as to exclude sets that make use of the set that is to be defined by it. Only by this restriction, he argues, can logical proofs be seen as one big tautology in which the conclusion follows necessarily from the premises. However, once this restriction is adopted, mathematical induction cannot be proved for the finite numbers thus defined, and therefore the logicist definition fails.[94]

The main objection of Zermelo against this point of view is his claim that the definition of the vicious circle principle itself is impredicative, so it fails to meet its own requirements. For take as a formal definition of the vicious circle principle the requirement that the class of $P$ is the class of all objects that satisfy $P$, but that do not presuppose the class of $P$ for its definition or for checking

---

[91] An analytic judgement is a judgement that is necessarily true on purely logical grounds, because the meaning is already implicit in the subject; a synthetical judgement gets its meaning from non-logical sources as, e.g., experience.
[92] See 1.4.2.
[93] [Detlefsen 1993], [Dauben 1979, pp. 266–267]
[94] [Goldfarb 1988, pp. 72–73]

whether they satisfy $P$. As one can see, in order to define the class of $P$ without a vicious circle, one has to use exactly the class of $P$ in its definition. Therefore, this clearly is an impredicative definition. Furthermore, Zermelo argues, if one would accept this restriction, important parts of mathematics would have to be given up, e.g. the fundamental theorem of algebra. This argument is extremely useful against Poincaré, since he had claimed that foundational arguments would not make mathematicians abandon their results. Therefore, Poincaré replies by sketching a proof of the fundamental theorem which does not make use of impredicative definitions. In the end, Zermelo completely rejected the vicious circle principle.[95]

### 1.4.2 Poincaré's views

**Mathematical existence** Poincaré claims that existence is freedom from contradiction and nothing else, especially in the case of indirect definitions:[96]

> Il ne faut pas oublier que le mot existence n'a pas le même sense quand il s'agit d'un être mathématique et quand il est question d'un objet matériel. Un être mathématique existe, pourvu que sa définition n'implique pas contradiction, soit en elle-même, soit avec les propositions antérieurement admises.[97]

Even within the 'existence is freedom from contradiction' group, however, one should distinguish between two points of view. One holds that, once a definition concerning a class of objects has been proved free from contradiction, every object belonging to that class has its existence proved. The other one, and that is the one Poincaré belongs to, demands that every individual mathematical object has to have its definition proved non-contradictory in order to be considered existing. For what is needed is not only a definition of the whole class, but also a means of distinguishing between the different objects in that class. Moreover, the definition should be possible in a finite number of words.[98]

Poincaré also treats more philosophical questions concerning mathematical existence. In his book *Science et méthode* ('Science and method'), he complains about the formal character of Hilbert's definitions, where it does not become clear what the 'things' one reasons about are and where one is not even allowed to try to find this out.[99]

---

[95] [Goldfarb 1988, pp. 73–75]

[96] [Poincaré 1902, p. 70 (1968 ed.)]

[97] 'One should not forget that the word 'existence' does not mean the same thing when dealing with a mathematical object and a material object. A mathematical object exists provided that its definition does not imply a contradiction, be it in itself, be it with propositions that have been accepted before.'

[98] [Goldfarb 1988, p. 78]

[99] [Mehrtens 1990, pp. 245–246]; on Hilbert's early view on the existence of mathematical objects, see 4.2.2.

## 1.4. POINCARÉ

**Logic and intuition**  In *La valeur de la science* ('The value of science'), Poincaré compares the work of a logician to that of a physiologist. What logic does is decomposing mathematical proofs into elementary operations, but the sequence of elementary operations one thus obtains is not the whole proof. The same happens to the physiologist who studies an elephant by means of a microscope, but by doing so does not know the whole animal sufficiently.[100]

In Poincaré's view, one cannot construct mathematics without mathematical intuition. That is to say: in mathematics principles are needed which cannot be based on logic. In *Sur la nature du raisonnement mathématique* ('On the nature of mathematical reasoning'), the first major exposition of his philosophical views, he writes:[101]

> Le syllogisme ne peut rien nous apprendre d'essentiellement nouveau (...). Il faut bien concéder que le raisonnement mathématique a par lui-même une sorte de vertu créatrice et par conséquent qu'il se distingue du syllogisme.[102]

He clarifies this by pointing out that logic can never conclude a general statement from a particular one; whereas in mathematics one can prove that $a + b = b + a$, by starting from $a + 1 = 1 + a$. Therefore, he concludes, mathematics can never be reduced to logic. What is needed, apart from logic, is above all the principle of complete induction, or, in other words, the intuition of the sequence of natural numbers. Only by using this principle can mathematics make the step from the particular to the general case, which is the basis of all science.

Poincaré describes this as the main synthetical judgement a priori in the Kantian sense.[103] Therefore, he calls it 'le raisonnement mathématique par excellence' ('the pre-eminent mathematical argument').[104]

Furthermore, Poincaré claims a special status for the principle of complete induction on the basis of his theory of mathematical existence. In Poincaré's view, mathematical existence requires a proof of the freedom from contradiction of the system in question. Such a proof will generally involve an infinite number of statements, of which one has to prove that none of these contradict each other. The only way in which such a general proof can be given is by means of the principle of complete induction.[105] Therefore, the principle of complete induction can never be postulated as a mere axiom, because also then a proof for the freedom from contradiction should be given, which would again involve the same principle of complete induction.

---

[100][Poincaré 1905, pp. 35–36]
[101][Poincaré 1894, pp. 371–372]
[102]'The syllogism cannot teach us anything essentially new (...). It must be conceded that mathematical reasoning has of itself a sort of creative virtue, and consequently differs from the syllogism.' English translation partially cited from [Ewald 1996, vol. II, p. 974].
[103]A synthetical judgement a priori is a fundamental and true judgement of our mind, that cannot be proved analytically nor be derived from experience.
[104][Poincaré 1902, p. 19]
[105]Note that, at the time, model theory was non-existent.

Poincaré distinguishes between three different meanings given to the term 'intuition'. Sometimes it is seen as a sensory intuition, sometimes more as a generalisation of sensory intuition by a kind of induction. Neither of these offer any help in the search for mathematical certainty and exactness. What does work, however, is intuition seen as a purely mental understanding of some fundamental principles or relations. Without these, Poincaré claims, mathematics is impossible.

Poincaré uses two arguments to support his claim. In the first place, even though all theorems in mathematics are proved by logical deductions starting from the axioms, one can still ask why exactly *these* axioms were chosen. In Poincaré's view, this cannot be explained otherwise than by an appeal to intuition. The second argument is that, with logic alone, one could make heaps of new mathematical proofs, most of which, however, would be worthless to mathematics. Logic points out a million possible ways that can be followed; but in order to know how to obtain our goal in the far distance, we need our intuition to show us the way. Even the understanding of an already given proof is impossible without the intuition, because only in that way we can see the different steps in the proof as a whole. By reducing mathematical thought to its logical form, as the logicists do, one mutilates it, so Poincaré claims.[106]

Thus, logic and intuition have their own role, and both are needed in mathematics. As Poincaré himself says:[107]

> La logique qui peut seule donner la certitude est l'instrument de la démonstration: l'intuition est l'instrument de l'invention.[108]

**Natural Numbers** Poincaré regards the natural numbers as part of the basis of mathematics. He points out that expressions like 'in no case' or 'a class with one member' are used in the Peano-style definitions of the natural numbers. What then, he asks, is the progress? He calls this kind of reasoning a *petitio principii*: the objects that are to be defined are used in the definition itself. The logicists look at it differently, since they see the numbers to be defined as formal objects, which are defined by means of words. For Poincaré, however, there is no difference between the numbers used in defining the numbers and the numbers defined. This is part of his psychologicistic attitude towards mathematics. He describes the Peano method as defining the clear by the obscure.[109]

**Impredicativity and set theory** When the paradoxes had become known in, among other things, set theory, mathematicians' attention was drawn to the so-called impredicative definitions. Especially Poincaré thinks these definitions are to blame for the antinomies in set theory. He also points out that this kind of definition

---

[106][Goldfarb 1988, pp. 63–64]
[107][Poincaré 1905, p. 37]
[108]'Logic, which can by itself give certainty, is the instrument of proving; intuition is the instrument of inventing.'
[109][Goldfarb 1988, pp. 65–67]

## 1.5. CONCLUSION

is only possible if one accepts the actual infinite; for only then can one consider the whole collection as already existing before one has defined all its elements. If one rejects the actual infinite and sees infinite collections as never completely realised, as Poincaré does, it is quite clear that one cannot accept impredicative definitions.

Poincaré spells out his opinion on the impredicativity of a definition by looking at Richard's paradox.[110] The error, Poincaré claims, lies in the fact that the sequence $(a)$ is used in order to define $\alpha$, while at the same time $\alpha$ belongs to $(a)$. In his view, a definition of a sequence is only correct if it does not have recourse to the sequence itself. Poincaré extends his argumentation on this so-called vicious circle principle to explain the set theoretical paradoxes. In this way, he uses it as a universal restriction on definitions in set theory.[111]

For Poincaré, an infinite set only means a collection to which always new elements can be added; the actual infinite does not exist. In particular, the continuum cannot be seen as a closed whole. The only things that exist are the general definition of a real number, and the real numbers that have been finitely defined.

As to the axiom of choice, Poincaré admits that mathematicians are inclined to follow their own intuition. He judges it to be a synthetical judgement a priori and thus seems to accept it.[112]

## 1.5 Conclusion

The general picture arising from the above description is the following. During the period from the 1870s until the First World War, the foundations of mathematics were never beyond doubt. The use of set theory, logic and impredicative definitions was criticised, and more stress was laid on intuition and procedures which can be effectively realised. Moreover, the criticism did not come from some remote corner of philosophy, but it was spread by mathematicians of fame like Kronecker, Borel and Poincaré. Their fame had been achieved outside foundational matters, in the area of mathematics proper. In these respects, what Brouwer was to do did not constitute a new contribution, and can be seen as a continuation, even if a more radical one, of this critical constructivist attitude. A difference with respect to the three mathematicians mentioned here is that Brouwer explicitly acknowledged Borel and Poincaré as earlier intuitionists,[113] whereas Kronecker was presented as a prominent forerunner of intuitionism primarily by Hilbert.[114]

It should also be noted that the above-mentioned criticism was directed primarily against *new* concepts in mathematics. Apart from the Kroneckerian criticism of the irrational numbers, which was not taken too seriously, what Kronecker,

---

[110] Richard's paradox is explained in 1.3.2.
[111] [Goldfarb 1988, pp. 71-72]
[112] [Bockstaele 1949, p. 99]
[113] See 2.5.
[114] See 6.2.1; Weyl made the same comparison, but he did not stress the point, [Weyl 1921, p. 223].

Borel and Poincaré denounced were concepts such as Cantorian set theory and Zermelo's axiom of choice — subjects which were new to mathematics at the end of the 19th, beginning of the 20th century. Moreover, the alternative they offered (and which they never fully worked out) was nothing more than a restriction of mathematics as it then existed.[115] In these respects, it must be stressed that Brouwer's attitude was completely different. Brouwer attacked principles that had been accepted in mathematics since Antiquity, and he put forward an alternative mathematics based on new principles. How he developed these views is shown in the next chapter.

---

[115]It was this aspect that caused Mehrtens to use the term *Gegenmoderne* for these mathematicians. On Mehrtens, see 6.7.

## Chapter 2

# The genesis of Brouwer's intuitionism

> Ik heb dat vak lief, en waarom het dan niet dienen ook in de samenleving; wat is een God zonder altaren op aarde? En als ik meer philosoof [sic] dan mathematicus mocht zijn, dan zal het ook door die dwangbuis nog wel heenbreken.[1]
>
> L.E.J. Brouwer[2]

## 2.1 Introduction

In this chapter, I treat Brouwer's intuitionism as it developed over the years. The starting point is formed by Brouwer's earlier, more philosophical writings around the turn of the century. I stop roughly in 1933, since I also treat the foundational debate up until that year.[3] As a rule, I follow Brouwer's original descriptions; only occasionally I felt it necessary to provide modern explanations.

Readers looking for a more modern explanation of intuitionism could consult Dummett's *Elements of intuitionism*,[4] which includes the philosophical ideas of intuitionism and should be comprehensible to a wide audience. Other useful sources include Troelstra and Van Dalen's *Constructivism in Mathematics*,[5] which

---

[1] 'I cherish that subject, so why not serve it also in society; what is a God without altars on Earth? And if I would be more of a philosopher than a mathematician, then it will probably break through that straitjacket as well.'

[2] Letter to Adama van Scheltema, 24/06/1908; cited from [Van Dalen 1984A, p. 85] Brouwer wrote this to his friend Adama van Scheltema when his teachers Korteweg and De Vries were trying to get him accepted as *privaat-docent* at the university of Amsterdam.

[3] See 3.1.

[4] [Dummett 1977]

[5] [Troelstra & Van Dalen 1988]

is broader and more technical, and Heyting's somewhat older *Intuitionism, an introduction*.[6] Many details of Brouwer's life and work covered in this chapter were taken from Van Dalen's Brouwer biography *Mystic, Geometer, and Intuitionist*, and the Dutch, more popular version *L.E.J. Brouwer. Een biografie*,[7] which any interested reader should consult.

## 2.2 The early years

### 2.2.1 Brouwer's youth

Luitzen Egbertus Jan Brouwer was born in Overschie, close to Rotterdam, on February 27, 1881. His family was of Frisian origin, and he had two brothers. Brouwer's talent became clear at an early age. When Brouwer was nine, he had already completed primary school in Medemblik, North-Holland, which children normally leave at the age of twelve. He then attended the HBS,[8] first in Hoorn, later in Haarlem, where he scored top grades despite his age. Before finishing secondary school, his parents and teachers agreed to transfer him to the gymnasium, since this was the standard way to enter university afterwards. In the meantime, a special programme enabled him to finish the HBS as well, which he did at the age of fourteen. A scholarship he obtained from the St. Job's foundation, which was directed at students of Frisian descent, provided financial support and continued to do so until the end of his university studies. When Brouwer was sixteen, he obtained both the diplomas Gymnasium $\alpha$ and $\beta$ (the literary and the science part), two years ahead of other youngsters, who normally leave school with only one of these. In 1897, he entered the *Gemeente Universiteit* ('Municipal University') of Amsterdam to study mathematics and science.[9]

### 2.2.2 Brouwer's profession of faith

Intuitionism as we now know it was not developed until the end of the first World War. However, there is a constant line in Brouwer's thought, which dates back to his earliest, more philosophical writings. In this section, the idealistic roots of Brouwer's philosophy of mathematics which were present in his early work are pointed out.

One of the first completed writings of Brouwer we have is his profession of faith. The declaration, written in 1898 at the age of seventeen, was made upon demand of the Remonstrant Church.[10] It marks Brouwer as an exceptional per-

---

[6][Heyting 1956]
[7][Van Dalen 1999A] and [Van Dalen 2001], resp.
[8]The *Hogere Burgerschool* ('Higher Citizen's School'), a secondary school type created by the Dutch 1863 educational reform, focused on middle class youngsters.
[9][Van Stigt 1990, pp. 21–23]
[10]The Remonstrant Church was a liberal break-off from mainstream Calvinism which came into being in the northern Netherlands in the beginning of the 17th century. The Grand Pensionary Johan van Oldenbarnevelt had adopted the case of the Remonstrants, [Van Dalen 1999A, p. 17].

## 2.2. THE EARLY YEARS

sonality already at that age, since he refused to follow the order in which he was asked to write the profession, and even dismissed many of the questions put to him as irrelevant.[11]

In the profession, we can clearly recognize the metaphysical idealism Brouwer would stick to all his life.[12] At the beginning of the declaration, Brouwer writes:[13]

> (...) het eenige ware voor mij is mijn eigen ikheid van het oogenblik, omgeven door een schat van voorstellingen waaraan de ikheid gelooft, en die haar doen leven. Een vraag of die voorstellingen 'waar' zijn heeft geen zin, voor mijn ikheid bestaan alleen de voorstellingen en zijn als zodanig reëel; van een tweede, onafhankelijk van mijn ikheid, daaraan beantwoordende realiteit is geen sprake.[14]

Thus, Brouwer professes a form of idealism generally described as ontological solipsism: the self is the only existing entity, all the rest are images.[15] In this way, the world is seen as essentially spiritual. This idea is not far removed from the mystic conception of the basic unity of the world.

Already in this writing, Brouwer criticises what he sees as the inadequacy of language, claiming that language is 'too clumsy an instrument' to describe the feeling of God. This feeling, in Brouwer's view, cannot even be thought, let alone be written down.[16] Later, Brouwer was to expand upon this remark several times.[17]

Brouwer did not study full time. In July 1898, he joined the Dutch army as a volunteer. Van Dalen conjectures that he did so in order to get rid of his army

---

[11] [Van Stigt 1990, pp. 387–389]

[12] Kreisel argued, not without reason, that such a firm position was an advantage: 'It is (...) highly likely that, at an early stage, his [Brouwer's, DH] own work benefited greatly from two very usual consequences of any doctrinaire position. He was able to develop his ideas vigorously, first because he had put out of his mind all but the matter in hand; and second because weaknesses of a position are less 'disturbing' if (one thinks) there is no alternative.' [Kreisel & Newman 1969, p. 41]

[13] [Van Stigt 1990, pp. 387–388]

[14] '(...) the only thing that is real to me is my own self at this moment, surrounded by a wealth of images in which the self believes and which make the self live. The question whether these images are 'factual' is devoid of meaning: for my self only the images exist and are, as such, real. A second reality, independent of my self and corresponding to these images, is out of the question.' This translation, which is my own, differs from the one given by Van Stigt, who translates 'het eenige ware' by 'the only truth', [Van Stigt 1990, p. 391]. In Dutch, however, the word 'waar' can mean both 'true' and 'real'. It is important to know in which sense Brouwer used the word, since the first meaning would point at an epistemological stance, whereas the second is about metaphysics. Since Brouwer speaks about reality in the second Dutch sentence quoted here, I think the latter is the more appropriate translation.

[15] [Mittelstraß 1980–96, Band 3, p. 389]. The most positive judgement Mittelstraß is willing to give to this form of solipsism, 'absurd, but unrefutable', marks the resistance that this stance still meets. Later, Heyting somewhat relaxed the metaphysical character of intuitionism and put more stress on epistemological questions; see 4.4.1.

[16] [Van Stigt 1990, pp. 388–389]

[17] See 2.2.4, 2.3.1 and 2.3.2.

obligations before his real career took off.[18] He first spent more than half a year in the service, then alternated military life with his studies. In December 1900, he received his *candidaats* degree, concluding the first part of his academic study, *cum laude*. Until January 1903, Brouwer would spend many a university holiday in the army. The most lasting effect of his military training was that it ruined his health and his nerves.[19]

### 2.2.3 Mannoury

Gerrit Mannoury (1867–1956) was probably one of the persons who influenced Brouwer most in developing ideas towards intuitionistic mathematics. Brouwer met Mannoury, a self-taught Dutch mathematician, at the Amsterdam Mathematical Society. Mannoury had obtained his teacher's diploma, which normally took four years, in only three months. The mathematics professor Korteweg recognised Mannoury's gifts and for some time gave him private tutorials at home.[20] In 1903 Mannoury became *privaat-docent*[21] in the logical foundations of mathematics at the university of Amsterdam and thus was allowed to teach. Brouwer was one of the first to attend his lectures.[22] Mannoury's lectures were probably on the borderline between mathematics and philosophy, as can be seen from his published lectures in *Methodologisches und Philosophisches zur Elementar-Mathematik* ('Methodological and philosophical remarks on elementary mathematics').[23] Brouwer, in retrospect, declared that what had strongly attracted him in Mannoury's lectures was that mathematics had acquired a new character:[24]

> De ondertoon van Mannoury's verhandelingen had namelijk niet gefluisterd: 'Ziehier eenige nieuwe aanwinsten voor ons museum van onwrikbare waarheden', maar ongeveer het volgende: 'Ziehier wat ik voor u gebouwd heb uit de structuurelementen van ons denken. (...)'[25]

Even if we keep in mind that Brouwer gave this description in the *laudatio* at the occasion of Mannoury's honorary degree and that Brouwer might, in retrospect, describe his reaction more in line with his own later work than it may have been at the time, the citation makes it clear that Brouwer's intuitionistic view on mathematics was strongly stimulated by Mannoury's lectures.

---

[18] At the time, the Netherlands had a system of conscription in which a lottery was used to determine which persons had to fulfil their military service.
[19] [Van Dalen 1999A, pp. 22–27]
[20] [Van Dalen 1999A, pp. 43–45]
[21] A *privaat-docent* (as a *Privatdozent* in Germany) did not receive a salary, but was allowed to lecture at the university and obtain fees from the students who attended.
[22] [Brouwer 1947A, pp. 474–475]
[23] [Mannoury 1909]
[24] [Brouwer 1947A, p. 193]
[25] 'For the undertone of Mannoury's argument had not whispered: 'Behold, some new acquisitions for our museum of immovable truths', but approximately the following: 'Look what I have built for you out of the structural elements of our thinking. (...)''

## 2.2. THE EARLY YEARS

In the *laudatio*, Brouwer refers to three French papers of Mannoury he had read at the time in the *Nieuw Archief voor Wiskunde*.[26] Of these, I find *Lois cyclomatiques* ('Cyclomatic laws'), the paper that introduced topology in the Netherlands, the most telling. This is the paper that differs most from classical mathematical papers, mainly, one might say, in style. In the paper, one frequently encounters such expressions as 'I have called these numbers', 'one can choose', 'one can establish', 'we only have to add', etc.[27] At the time, mathematicians tended to prefer more impersonal expressions, such as 'these numbers are called' or 'it is possible to establish'. One might be inclined to think that Mannoury diverged from the stylistic norm because he was a self-taught man. However, if we look at other writings of Mannoury, it is probable that the difference was more than mere style.[28] In *Methodologisches und Philosophisches zur Elementar-Mathematik*, Mannoury speaks about a mathematical object 'which can only be a product of our mind'.[29] Thus, it seems that the way in which Mannoury wrote mathematics reflects his view on the ontological status of mathematical objects.

To Brouwer, too, Mannoury's 'style' meant more. As Brouwer later called it, he owed 'the awakening of [his] mathematical consciousness'[30] in a large degree to Mannoury's lectures. Mannoury's descriptions revealed to him what mathematics really was: not a realm which already existed independently of human beings, but a human creation. This was in accordance with Brouwer's general idealistic philosophy. Also Mannoury's habit of being critical of established principles must have appealed to Brouwer's non-conformist character. It should be noted, however, that, despite this general agreement, Brouwer and Mannoury held different opinions on various — if not most — questions concerning the philosophy of mathematics. Brouwer therefore used to refer to Mannoury as his 'dialectic partner'.[31]

In the case of Brouwer and Mannoury, it is very hard — if not impossible — to tell who influenced whom on any specific point. What may seem a reaction to Brouwer in one of Mannoury's works may well be an old point which Mannoury

---

[26][Mannoury 1898–99], [Mannoury 1900A] and [Mannoury 1900B].
[27][Mannoury 1898–99, pp. 126–128]
[28]Later, Mannoury explicitly attacked the classical style in mathematics. In his inaugural lecture *Over de sociale betekenis van de wiskundige denkvorm* ('On the social significance of the mathematical way of thinking'), delivered on October 8, 1917, which still makes good reading today, Mannoury said: '*En wiskunst is gevoelloos, is onwezenlijk, is dood. Wil het althans zijn, en verbant daartoe zoveel doenlijk uit haar woordeboek al wat aan de waarneming ontleend is, wat aan waarneming herinnert, en gevoelt zich het veiligst als zij zich in letters en in cijfers uitdrukt.*' ('And the art of mathematics is insensitive, is inessential, is dead. At least it wants to be so, and therefore it banishes from its dictionary as much as possible all that is derived from observation or that reminds of observation, and it feels safest when it expresses itself in letters and numbers.') [Mannoury 1917, p. 8]

A modern echo of the criticism of the use of objectivistic language in scientific reports can be found with Rupert Sheldrake, [Sheldrake 1994, pp. 154–155], whose critical remarks are in general more valuable than the alternative he offers.

[29]'welche doch nur ein Produkt unseres Geistes sein kann', [Mannoury 1909, p. 8]
[30]'mijn mathematische bewustwording', letter from Brouwer to Mannoury, 30/3/1917, [MI Brouwer, CB.GMA.7]
[31][Brouwer & Mannoury 1946, p. 3]

restated, and which had been taken up by Brouwer. Nevertheless, I chose to take Mannoury's comments to Brouwer's work at face value, which means that most of them are classified as reactions to Brouwer and therefore appear in later chapters.

In 1904, Brouwer finished his studies, again with a *cum laude* degree. He decided to stay at the university of Amsterdam to write a dissertation on the foundations of mathematics. His supervisor was Diederick Korteweg (1848–1941), an applied mathematician specialised in fields suchs as algebra, geometry, the theory of oscillation, voting theory, and electricity. Korteweg was a student of Van der Waals, and held the chair of mathematics, mechanics and astronomy at Amsterdam university.[32] Brouwer's choice of supervisor was not a difficult one: the alternative was Van Pesch, who was not exactly at the cutting edge of contemporary research.[33]

In the same year, Brouwer married Elizabeth (Lize) de Holl. She was eleven years his senior and was divorced from Hendrik Peijpers, a former army doctor whom she had married when she was a young girl. This first marriage was far from happy; Peijpers forced Lize to have an abortion whenever she became pregnant. Lize had managed to stay out of her husband's control only once, and a child, Anna Louise, was born in 1893. Lize and Brouwer had known each other for two years when they got married. Lize was to take over her father's pharmacy, first managing it by contracting a provisor (a licensed pharmacist), which only worsened their financial situation, but later, after she had finished her special pharmacists exam in the end of 1907, as a fully licensed pharmacist herself. The whole situation – a divorced woman with a child, a marriage in which the woman was substantially older than the man, and without all the normal financial precautions – was quite exceptional at the time, but that presumably did not worry Brouwer too much.[34]

### 2.2.4 Brouwer's mysticism

During the first years of the 20th century, Brouwer sought solitude in walks. He 'pilgrimaged' to Italy three times, walking all the way to Florence and Rome. In this way, he could 'turn into himself'.[35]

In 1905, at the age of 24, Brouwer published a monograph under the title *Leven, kunst en mystiek* ('Life, Art and Mysticism'). It contained the lectures he had given in Delft the same year in reply to lectures the Dutch Hegelian Bolland had delivered there.[36] Compared to Brouwer's profession of faith, the tone of the monograph is distinctively more pessimistic, with chapter titles such as 'The sad

---

[32] A short biographical sketch can be found in [Willink 1998, pp. 81–88].
[33] [Van Dalen 1999A, p. 86]
[34] [Van Dalen 1999A, pp. 52–55; 199]
[35] [Van Stigt 1990, p. 26]
[36] Brouwer was invited to lecture by the *Society of Free Study*, after having established himself as a main challenger of Bolland through a series of articles in the well-known student magazine of Amsterdam University *Propria Cures*, [Van Stigt 1996, p. 386].

## 2.2. THE EARLY YEARS

world' and 'Man's downfall caused by the Intellect'.[37] However, one should not exaggerate this negative, romantic tendency. The negative view applies primarily to the outer world, where Brouwer fulminates against the artificiality and meaninglessness of many human activities and the servility with which many people continue carrying them out. Opposed to this, Brouwer places the inner strength of the self:[38]

> Ge voelt u almachtig, want ge wilt alleen, wat in de Richting ligt, en daarbij zullen de bergen voor u wijken (...).[39]

Looking at the history of the foundational debate that was to follow, it is not exaggerated to assume that Brouwer here described a strong feeling within himself.

Characteristic of *Leven, kunst en mystiek*, save its disdain for women,[40] is Brouwer's rejection of the intellect and his adherence to mysticism.[41] In those days, mysticism was a popular current, even though many took it in a wide sense as a general love for the mysterious.[42] Brouwer was a mystic in a stricter sense, where an awareness of unity takes a central place.

The main reason for Brouwer's dislike of the intellect seems to be that it hinders introspection.[43] The intellect renders people the 'devil's service'[44] of linking the means and the end. In this way, they are enabled to strive for fulfilling their lusts. But many people know something originally intended as a means only as an end in itself, and since most people only imitate others, the whole human enterprise soon looses sense.

Brouwer's mysticism followed in the footsteps of a number of Christian, Buddhist and Hindu mystics whom Brouwer admired for their reliance on their personal experience and their acceptance of an 'inner vision' as their supreme authority.[45]

---

[37] *'De droeve Wereld'*, *'De val door het Intellect'*, [Brouwer 1905, p. 5; p. 17]. The former is a play of words on Van Eeden's 'Joyous World'.

[38] [Brouwer 1905, p. 16]

[39] 'You feel all-powerful, for you only desire that which follows the Direction, and in that the mountains shall make way for you (...).'; translation based on [Brouwer 1996, p. 394].

[40] *'de vrouw, die zonder den man niet kan, wier voldongen Karma in niets is dan haar sekse, zóó, dat tusschen de intiemste natuur van een vrouw en een leeuwin minder verschil is, dan tusschen twee tweelingbroeders, die mannen zijn.'*, ('the woman, who cannot be without the man, whose accomplished Karma is nothing but her sex in such a way that there is less difference between the most intimate nature of a woman and a lioness than between two twin brothers, who are male.') [Brouwer 1905, p. 52]. In practice, however, Brouwer valued women highly; for example, he was one of the first Dutch mathematicians to employ a female assistant in the 1920s, cf. [Van Dalen 1999A, p. 74].

[41] I treat Brouwer's mysticism only superficially here. The best introduction to Brouwer's mysticism for non-mystics I know is [Heyerman 1981].

[42] [Van Nes 1901, pp. 1–12]

[43] For example, Brouwer writes: *'Het hooggeschatte Intellect dan is tegelijk het vermogen en de dwang, om dóór te leven in Begeerte en Vrees, en niet uit heilzame verlegenheid tot Zelfinkeering terug te vluchten (...).'* ('The highly esteemed Intellect, then, has both enabled and forced man to continue living in Desire and Fear, rather than – from a salutary sense of timidity – take refuge in self-reflection (...).'), [Brouwer 1905, p. 18]. Cf. [Heyerman 1981, p. 30]

[44] 'duivelsdienst', [Brouwer 1905, p. 17]

[45] [Van Stigt 1996, p. 385]

He frequently cites and refers to mystics such as Meister Eckehart, Jakob Böhme and the Bhagavad Gîta. To those who do not feel an affinity with mysticism, it is hard to give meaning to these parts of *Leven, kunst en mystiek*. Brouwer recognizes the limited scope of his writing and acknowledges that the mystical parts will not make any converts.[46]

*Leven, kunst en mystiek* clearly shows Brouwer's concern with living a meaningful life:[47]

> Van al de boomen zien de menschen der cultuur het bosch niet meer, nee, weten niet eens meer dat er een bosch is; wie vraagt, waarvoor hij eigenlijk leeft, wordt in het practische leven, waar die vraag eigenlijk juist alleen zin heeft, voor gek versleten (...).[48]

It was the same concern that made him appreciate intuitionistic over formalistic mathematics.

**Language** Certain elements of Brouwer's thought identified earlier, recur in *Leven, kunst en mystiek*. Thus, Brouwer confirms his metaphysical idealism.[49] In the meantime, his attack on the value and usefulness of language has become stronger and more elaborated. The chapter devoted to language opens with the following words:[50]

> Het intellect gaat direct vergezeld van de taal. Met het leven in het intellect komt de onmogelijkheid, om zich op directe wijze — door gebaar en blik van oogen instinctief, of nog materieloozer, door alle afstandsscheidingen heen — met elkaar in betrekking te stellen, en gaan ze[51] zich en hun nakroost dresseeren op een teekenverstandhouding door grove klanken, moeitevol en — vrij machteloos, want nooit nog heeft door de taal iemand zijn ziel aan een ander meegedeeld; alleen een verstandhouding, die toch reeds is, kan door de taal worden begeleid (...).[52]

---

[46][Brouwer 1905, pp. 75–77]
[47][Brouwer 1905, p. 21]
[48]'These cultured people cannot see the wood for the trees anymore, worse, they have forgotten that there is such wood; anyone who raises the question of the real meaning of life is declared insane in our practical life, the only place where the question actually makes sense.'; translation based on [Brouwer 1996, p. 397].
[49][Brouwer 1905, p. 16]
[50][Brouwer 1905, p. 37]
[51]Brouwer does not mention who 'they' should be.
[52]'Language is the direct companion of the intellect. By living in the intellect it becomes impossible to communicate directly with each other — by gesture and look of the eyes, instinctively, or even more spiritually, through all spatial separation. People then drill themselves and their offspring in sign understanding by means of crude sounds, strenuously and — quite powerlessly, for nobody has ever communicated his soul to someone else by means of language. Only an understanding, that already is, can be accompanied by language (...).'

## 2.2. THE EARLY YEARS 33

Only in very restricted areas, Brouwer continues, can people communicate reasonably well. But even in mathematics and logic, 'which in fact cannot be separated sharply',[53] no two persons will think the same when using basic concepts. The role of logic is as restricted as that of language in general:[54]

> (...) ridicuul wordt het gebruik van de taal, waar wordt gehandeld over fijne wilsschakeringen, zonder dat er in die wilsschakeringen wordt geleefd; zoo, als zoogenaamde wijsgeeren of metaphysici handelen onder elkaar over moraal, over God (...); menschen, die elkaar niet eens liefhebben, (...) ja die soms elkaar zelfs niet persoonlijk kennen; dán praten ze òf langs elkander heen, òf ze bouwen een logisch systeempje, dat alle verband met de werkelijkheid mist; want logica is leven in de hersenen, begeleiden kan ze het leven daarbuiten, richten uit eigen kracht nooit.[55]

Brouwer was later to specify his criticism of logic and point to parts of classical logic which, in his view, were not universally valid.[56]

Finally, Brouwer points out that language should always be linked to (expressions of) life:[57]

> Een taal, die geen vastheid aan den wil ontleent, die op zichzelf wil voortleven in het reine 'begrip', is een onding; een tijdlang door te kunnen spreken, en te worden betrapt nòch op tegenstrijdigheden, nòch op stilzwijgenden, in den wil wortelende, vooronderstellingen, is een groote kunst, (...) maar te schatten op de waarde van een acrobaat.[58]

Here, Brouwer speaks about language. Later, he was to develop a similar idea on the use of axiomatic systems in mathematics, when he claimed that consistency does not suffice to establish a mathematical system.[59]

Mathematics only figures marginally in *Leven, kunst en mystiek*. The most important thing Brouwer has to say about it is that he considers it unimportant.[60]

---

[53]'die eigenlijk niet scherp te scheiden zijn', [Brouwer 1905, p. 37]

[54][Brouwer 1905, p. 38]

[55]'(...) the use of language becomes ridiculous when subtle nuances of the will are being treated, without living in these nuances; for example, when so-called philosophers or metaphysicians deal among themselves with morality, with God (...); people, who do not even love each other, (...) who sometimes do not even know each other personally. Then they are either talking at cross-purposes, or they are building a little logical system that lacks any relation with reality. For logic is life in the brains, it may accompany life outside it, but never direct it by itself.'

[56]See 2.3.2.

[57][Brouwer 1905, p. 40]

[58]'A language which does not derive its solidity from the will, which wants to live on by itself in pure 'understanding' (the word 'begrip' could also mean 'concept', DH), is an absurdity; to be able to continue speaking for a while, without being caught in contradictions or in tacit presuppositions rooted in the will, is definitely an art, (...) but to be appreciated only as one appreciates an acrobat.'

[59]See 2.3.1.

[60][Brouwer 1905, p. 18]

*Leven, kunst en mystiek* certainly appealed to some people. The author Frederik van Eeden was impressed, and the Dutchman Giltay wrote to Brouwer that the work was always on his desk. The latter considered it 'the most formidable accusation against our 'civilization' written in the Netherlands.'[61] However, in general the work remained unnoticed, apart from a rather negative review in one of the daily newspapers.[62]

Summarizing, one could say that in his early writings Brouwer put forward two main ideas on the value of language. Firstly, its usefulness as a means of communication is played down. In Brouwer's view, language can only function in very restricted areas, such as mathematics and logic, and even there it will never function perfectly. Secondly, Brouwer maintains that language by itself is of no significance. Language, and logic as a special case, cannot by itself create anything new, even if its use does not lead to contradictions, but can only accompany what already is there. Thus, we see a great stress on the individual person and his or her pre-linguistic thoughts.

It is interesting to note that Brouwer later did not do, the precise thing he in this work claimed should be done. The last words of the book seem to be inspired by Schopenhauer and read:[63]

> Maar alleen, die weet, niets te bezitten, niet te kunnen bezitten en geen vastheid bereikbaar, en in berusting zich overgeeft, die alles opoffert, die alles geeft, die niets meer weet en niets meer wil en niets meer weten wil, die alles laat gaan en verwaarloost, hem wordt alles gegeven en opent zich de wereld van vrijheid, van pijnlooze contemplatie, van — niets.[64]

What Brouwer proclaims is that one should not claim any knowledge, nor strive to achieve things, but accept what contemplation would bring. What Brouwer was to do, however, was to claim that he knew what real mathematics was, and to strive for the world to accept it.

## 2.3 The first act of intuitionism

In 1905, Brouwer moved to '*de Hut*', a cottage that Rudolf Mauve, a friend of Brouwer and son of the famous painter Anton Mauve, built for him on a strip of

---

[61]'niemand in ons land heeft zùlk een geweldige aanklacht tegen onze 'beschaving' doen hooren', letter from Giltay to Brouwer, 19/12/1925; [MI Brouwer]
[62][Van Dalen 1999A, p. 76]
[63][Brouwer 1905, p. 99]
[64]'But only he who knows that he does not own anything, that he cannot own anything and that no solidity is attainable, who surrenders in acceptance, who sacrifices everything, who gives everything, who no longer wants and no longer knows and no longer wants to know, who lets everything go and neglects everything, to him everything shall be given and another world shall open itself, the world of freedom, of painless contemplation, of — nothing.'

## 2.3. THE FIRST ACT OF INTUITIONISM

land Brouwer owned in Blaricum, in the region 't Gooi (situated between Amsterdam and Utrecht).[65] At the time, the region was a popular place for painters, poets and idealistic cummunes called 'colonies', such as Frederik van Eeden's 'Walden'.[66] Brouwer, however, preferred being alone. Now, he could work on his thesis without being disturbed.

In his 1927 Berlin lectures,[67] Brouwer distinguished between two so-called acts (*'Handlungen'*) in the development of intuitionism: one, the separation between mathematics and mathematical language, and two, the construction of sets.[68] I follow Brouwer's distinction, since it is an appropriate way of presenting the historical development of intuitionism.

### 2.3.1 Brouwer's dissertation

On February 19, 1907, Brouwer defended his doctoral dissertation. This was a small miracle, considering the fact that by September 1906 he had not written any part of the dissertation worth mentioning.[69] Although he had planned to write it on 'The value of mathematics', his thesis bore the more neutral title *Over de grondslagen der wiskunde* ('On the foundations of mathematics').[70]

Brouwer's objective was not to describe various views on the foundations of mathematics or to provide foundations for mathematics as it was then practised, but to work out his own ideas in the philosophy of mathematics. This becomes clear from a letter Brouwer sent to Korteweg in 1906. In the letter, Brouwer writes that he still adheres to his convictions from two years ago, but that he is glad to find that he can now support them better with mathematical arguments.[71] Thus, Brouwer started from his own ideas and looked for mathematics that fitted in, instead of working the other way round. At the same time, the period the letter refers back to is the time when Brouwer was working on both his dissertation and on *Leven, Kunst en Mystiek*, suggesting that the former was developed out of the same ideas as the latter.[72] Already at that time, Brouwer did not tolerate different forms of mathematics, claiming that 'no mathematics can exist, which has not been constructed intuitively'.[73]

As happens more often with dissertations, Brouwer's original set-up was more extended than the one that was published. In the end, Brouwer reduced the structure of his work to three chapters: 'The construction of mathematics', 'Mathematics and experience', and 'Mathematics and logic'. The second chapter was the one about which Korteweg was most critical. He and Brouwer differed on the is-

---

[65] [Van Dalen 2001, p. 30]
[66] Named after Thoreau's experiment.
[67] See 2.7.1.
[68] [Brouwer 1992, pp. 21–23]
[69] [Van Dalen 1999A, p. 90]
[70] [Van Dalen 1981, p. 2]
[71] Letter from Brouwer to Korteweg, 7/9/1906; cited from [Van Dalen 1981, p. 5]
[72] Cf. [Heyerman 1981, p. 36]
[73] 'geen wiskunde, die niet (...) intuitief is opgebouwd, kan bestaan', [Brouwer 1907, p. 118]

sue to what extent a mathematical dissertation could contain more philosophical material.[74] As a consequence, Brouwer had to leave out some parts of this section. These were mainly about items such as causality, mysticism, language, space, time, and science,[75] and would also nowadays generally be considered outside the domain of mathematics.

**Chapter 1: The building of mathematics**

When Brouwer wrote his dissertation, he did not yet use the word 'intuitionism'. In a letter to Korteweg, he characterised his view on mathematics as *opbouwende wiskunde* ('constructive mathematics').[76] Beginning the dissertation with a chapter on the construction of mathematics is therefore only natural.

Brouwer constructs mathematics in the following way. He starts with the sequence of natural numbers, which in his view is intuitively clear. It should be noted that Brouwer presents the natural numbers as a sequence characterised by a law, not as an infinite set. From the natural numbers, he constructs the negative, the rational and the irrational numbers. By speaking about the stock of known numbers being denumerable 'at any point of the development',[77] Brouwer indicates that the construction of these mathematical entities takes place in time. The continuum is introduced by a special continuum intuition. On this, a measurable continuum is defined, the individual points of which can only be approximated by an approximating sequence, which is ever unfinished. The approximation takes place by means of a so-called dual scale, i.e., a system of finite sequences of rational numbers written in binary form. Since the scale thus constructed need not be everywhere dense, Brouwer contracted 'by brute force' segments in which there was no point of the scale: two points are seen as distant if their dual approximation differs after a finite number of digits.[78]

Brouwer frequently uses the term *bouwen* ('to build', 'to construct'), and also variations, such as *opbouwen* ('to build up') and *bijbouwen* ('to add on').[79] Contrary to what the English translation might suggest, these expressions would not remind the (Dutch) reader of geometry. In that field, the word '*construeeren*' ('to construct') would have been used in Dutch, too.[80] Thus, Brouwer used terms

---

[74]This becomes clear from the remaining part of the Brouwer-Korteweg correspondence, published in [Van Dalen 1981]. Probably, Brouwer and Korteweg also differed on what to consider philosophy; Brouwer claimed that in the dissertation he remained on grounds that 'in my own view are constantly mathematical' (*'dat in mijn eigen oogen voortdurend wiskundig is'*), letter from Brouwer to Korteweg, 7/11/1906, [Van Dalen 1981, p. 11].

[75]The original unpublished fragments were published by Van Dalen in a re-edition of Brouwer's dissertation, [Van Dalen 1981, pp. 25–35]; an English translation without the Dutch original can be found in [Van Stigt 1990, pp. 405–415].

[76]Letter from Brouwer to Korteweg, 16/10/1906; [Van Dalen 1981, p. 6]

[77]'op elk punt van ontwikkeling', [Brouwer 1907, p. 48]

[78][Brouwer 1907, pp. 44–52], [Van Dalen 1999A, p. 102]

[79]Cf., e.g., [Brouwer 1907, p. 105]

[80]Sometimes, Brouwer did use '*construeeren*' as an alternative for '*bouwen*', but he clearly preferred the latter, which implies more a buidling upwards from the ground; cf. [Van Stigt 1998, p. 7].

## 2.3. THE FIRST ACT OF INTUITIONISM

which seem to serve mostly as metaphors for mental constructions.

At one place in the chapter, Brouwer seems to anticipate the idea of a choice sequence, which was to play a fundamental role in his second act of intuitionism.[81] Choice sequences had been considered before by Du Bois-Reymond and Borel.[82] Discussing possible point sets, Brouwer arrives at the question whether the point set in an arbitrary segment of the continuum is dense or not. The technicalities are not of interest to us here. In answering the question, Brouwer uses a dual (i.e., binary) scale in which each time either the next dual number is determined by the preceding one, or there is a choice between two dual numbers. In the case of a choice, Brouwer speaks about a 'multiplying bifurcation'.[83] The main difference with a choice sequence is that with the latter, the choice is exclusively put with the mathematician constructing the choice sequence. The drawing Brouwer makes to illustrate the dual scale is exactly the one used later for choice sequences.[84]

Brouwer concludes the chapter by stating that mathematics has to be constructed from basic 'mental elements'[85] by juxtaposition or by making sequences of certain types. This is the only possible foundation of mathematics, Brouwer maintains. Still, in the construction process one should each time strictly check what intuition allows one to do and what not.[86]

### Chapter 2: Mathematics and experience

Brouwer opens the second chapter of his thesis with a sketch of mathematics. In Brouwer's view, seeing things in a mathematical way means seeing things as repetitions of causal systems in time. Thus, Brouwer's idea of mathematics is much broader than the traditional account and seems to include all sorts of theoretical reasoning based on causal systems. Since seeing things in a mathematical way turned out to be useful and gave man power, Brouwer maintains, man developed pure mathematics, thus having mathematical systems available to be projected upon reality whenever this seemed functional.[87]

Brouwer returns to some of the themes he had treated in *Leven, kunst en mystiek*.[88] However, this time the tone is much more neutral, due to Korteweg's intervention. For example, Brouwer simply notes that many people prefer to concentrate on what he calls (without further explanation) 'observed' mathematical sequences instead of 'experienced' ones.[89] There is only an undertone of the author having more esteem for the latter, which he associates with instinct, than

---

[81] See 2.6.
[82] [Du Bois-Reymond, P. 1882, pp. 89–92] and [Borel 1912, pp. 309–310]; on Borel, see 1.3.2.
[83] 'zich vermenigvuldigende tweevertakking', [Brouwer 1907, p. 106]
[84] [Brouwer 1907, pp. 105–106]
[85] *voorstellingseenheden*, [Brouwer 1907, p. 118]
[86] [Brouwer 1907, p. 118]
[87] [Brouwer 1907, pp. 122–124]
[88] See 2.2.4.
[89] 'waargenomen', 'gevoelde', [Brouwer 1907, p. 122]

# 38   CHAPTER 2. THE GENESIS OF BROUWER'S INTUITIONISM

for the former, made by the intellect. In this context, Brouwer speaks about 'intuitive mathematics'. In a footnote, he adds that in fact the building of intuitive mathematics as such is 'an *act*, and not *science*'.[90]

The goal of this chapter, Brouwer writes, is to rectify Kant's view of the a priori. This had become necessary because of the introduction of non-Euclidian geometries, which clearly did not fit into the Kantian a priori. In Brouwer's opinion, the creation of the image of space is a free act of the intellect, and thus cannot be part of the a priori.[91] The only a priori element in science is the intuition of time, by which it becomes possible to see, in Brouwer's words, 'repetition as 'thing in time and thing again' (...) and by which life moments fall apart as sequences of qualitatively different things'.[92] Since the primordial intuition coincides with becoming aware of time as change 'by itself',[93] Brouwer concludes that the only a priori element in science is time.[94]

Finally, Brouwer maintains that the only synthetical judgements a priori are the ones that can be seen as construction possibilities following from the primordial intuition. Among these, he reckons the principle of complete induction, just as Poincaré had done.[95]

**Chapter 3: Mathematics and logic**

The third and last chapter of Brouwer's dissertation is the one that most points the way towards the further development of intuitionism. Here, Brouwer takes a consistent stand on the relationship between mathematics, mathematical language and logic, and uses this to criticise the ideas of such famous colleagues as Cantor, Russell and Hilbert.

**Mathematics, mathematical language and logic**  Brouwer starts by stating that his goal in this chapter is to show that mathematics is independent of 'the so-called *laws of logic*'.[96] In order to support this claim, he presents a logician who maintains that in doing mathematics, one has to use logical principles. Whereupon Brouwer replies:[97]

> De woorden van uw wiskundig betoog zijn slechts de begeleiding van een woordloos wiskundig bouwen (...).[98]

---

[90]'een *daad*, en geen *wetenschap*', [Brouwer 1907, p. 139]
[91][Brouwer 1907, pp. 154–156]
[92]'herhaling als 'ding in den tijd en nog eens ding' (...) en op grond waarvan levensmomenten uiteenvallen als volgreeksen van qualitatief verschillende dingen', [Brouwer 1907, p. 122]
[93]'zonder meer', [Brouwer 1907, p. 140]
[94][Brouwer 1907, pp. 139–140]
[95]See 1.4.2.
[96]'de zoogenaamde *logische wetten*', [Brouwer 1907, p. 164]
[97][Brouwer 1907, p. 166]
[98]'The words of your mathematical discourse are merely the accompaniment of wordless mathematical building (...).'

## 2.3. THE FIRST ACT OF INTUITIONISM

Thus, Brouwer makes it clear that to him, the activity of mathematical construction should be separated from the language in which the activity is described afterwards. Logicians only speak about the latter, whereas it is the former that constitutes real mathematics. The mental constructing of mathematics is a languageless activity.[99] This is what Brouwer later was to identify as the 'first act of intuitionism'.[100] Already in *Leven, kunst en mystiek* Brouwer had argued against an overestimation of the role of language in human communication, claiming that 'only an understanding, that already is, can be accompanied by language'.[101] Now, he argues that only mathematics that has already been constructed, can be accompanied by mathematical language.

Therefore, it is not surprising that Brouwer inverts the relationship: it is not mathematics that is dependent on logic, but logic is dependant on mathematics. Mathematical language follows upon mathematical activity, and logic consists of looking at that language in a mathematical way. Brouwer uses a comparison to clarify what he means: the language of logical argumentation is just as little an application of theoretical logic as the human body is an application of anatomy. It is a description of regularities in the language, and hence follows afterwards.[102]

In the dissertation, Brouwer's description of logic is still quite neutral, but behind it was a deep resentment against logic. In a letter to Korteweg, written shortly before the defence of his thesis, Brouwer expresses his rejection of logic in a more pronounced way:[103]

> De theoretische logica leert in de tegenwoordige wereld niets, en men weet dit, tenminste de verstandige menschen; zij dient nog alleen voor advocaten en volksleiders, om andere menschen niet te beleeren, maar te bedriegen, en dat dat kan, komt, doordat het vulgus onbewust redeneert: die taal met logische figuren is er, ze zal dus ook wel bruikbaar zijn, en zoo zich er gedwee mee laat bedriegen; zooals ik verscheidene menschen hun jeneverdrinken hoorde verdedigen met de woorden: 'waarom is de jenever er anders?'[104]

---

[99] It is hard to tell whether this was, at the time, a position that Brouwer shared with many other mathematicians, or that it was a solitary one. In 1912, the journal *L'Enseignement mathématique* published the results of a questionnaire on the working methods of mathematicians. Answers had been sent in by more than one hundred mathematicians from various parts of the world. One of the questions put to them was which internal images or which forms of interior speech (*parole intérieure*) one used. Unfortunately, only 26 mathematicians answered this specific question, which, according to the organizers, was excusable since it was the last question in a list of thirty. The answers that *were* given do not reveal much about general attitudes held towards this issue by mathematicians, since they vary quite a lot. Some mathematicians describe mathematics as pure thought, whereas the most popular internal image mentioned is a visual one. [Fehr, Flournoy & Claparède 1912, pp. 119–120]

[100] [Brouwer 1992, p. 21]

[101] See 2.2.4.

[102] [Brouwer 1907, pp. 166–170]

[103] Letter from Brouwer to Korteweg, 23/01/1907; [Van Dalen 1981, p. 23]

[104] 'Theoretical logic does not teach anything in the present world, and one knows this, at least sensible people do; it only serves lawyers and public leaders, not to instruct other people, but to

Brouwer applies his position of the separation between mathematics and mathematical language to four areas, all of which are criticised for not conforming to this idea: the axiomatic foundations of mathematics, Cantor's theory of transfinite numbers, Peano's and Russell's logistics, and Hilbert's logical foundations of mathematics.[105] Before discussing the parts that are most relevant to us, I turn to what Brouwer wrote on what were to become the two central themes of the foundational debate: the principle of the excluded middle and the question of mathematical existence.

**The principle of the excluded middle**  While discussing the relationship between mathematics and logic, Brouwer touches upon the status of the principle of the excluded middle.[106] His exposition is rather brief. The theorem 'either a function is differentiable or it is not differentiable', Brouwer maintains, does not express anything; it is equivalent to 'if a function is not differentiable, it is not differentiable',[107] which is a tautology. Although Brouwer thus denies any value to this logical principle, he does not reject it in his dissertation.

It is not clear where this early conclusion came from. Van Dalen's conjecture that Brouwer obtained it from Bellaar-Spruyt, an Amsterdam philosophy lecturer, is the best answer available.[108] Later, Brouwer was to push his criticism of the principle of the excluded middle further, and claim that it could not be used unrestrictedly in mathematics.[109]

On a different level, Brouwer is more critical of the principle of the excluded middle. Countering a statement of Hilbert, Brouwer maintains that it is not certain whether every mathematical problem either has a solution, or can be proved to be unsolvable.[110] In his famous address at the 1900 International Mathematicians' Conference in Paris, Hilbert had claimed that every mathematician was convinced of the truth of this disjunction,[111] now known as Hilbert's Dogma.[112] Later, Brouwer was to use this observation in the construction of his counterexamples against the principle of the excluded middle. At the time, however, he apparently saw no link between this statement and the status of the principle of the excluded middle.

---

deceive them. And this is possible because the mob unconsciously reasons: that language with its logical figures exists, hence it will probably be useful, and in this way docilely lets itself be deceived; just as I heard several people defend their drinking gin with the words: 'why else would gin exist?' '

[105] [Brouwer 1907, p. 172]

[106] Generally, this principle is referred to as the *law* of the excluded middle. As far as I know, Brouwer was the first to label it a *principle*.

[107] [Brouwer 1907, pp. 170–171]

[108] [Van Dalen 1999A, pp. 106–107]

[109] See 2.3.2.

[110] [Brouwer 1907, p. 181]

[111] [Hilbert 1901, p. 297]

[112] [Van Dalen 1999A, p. 105]; Hilbert called it the 'axiom of the solvability of every problem' (*'Axiom von der Lösbarkeit eines jeden Problems'*).

## 2.3. THE FIRST ACT OF INTUITIONISM

**Mathematical existence** The other subject that was to become dominant in the foundational debate, the question of mathematical existence, figures in the dissertation, too. However, Brouwer does not put forward his view on pure existence statements prominently. In discussing the axiomatic foundations of mathematics, he remarks that[113]

> (...) nergens bewezen [wordt, DH], dat als een eindig getal aan een stelsel voorwaarden moet voldoen, waarvan bewezen kan worden, dat ze niet contradictoir zijn, dat dan dat getal ook bestaat.[114]

Although Brouwer did not specify the concept of the non-contradictoricity of a set of conditions (containing a free variable), one can try to formulate his assertion in modern terminology. As Van Dalen pointed out to me, translating the assertion into formal predicate calculus leads to the following. If we label the conditions $C_i$ (for $i \in \{1, \ldots, n\}$), denote $C_1(x) \wedge C_2(x) \wedge \ldots \wedge C_n(x)$ by $C(x)$ and use $\neg \varphi$ as a shorthand for $\varphi \to \bot$, the assumption that the conditions can be proved non-contradictory can be stated as $\nvdash \forall(x)\neg C(x)$. We can then describe what Brouwer questions as whether the conclusion

$$\nvdash \forall(x)\neg C(x) \Rightarrow\ \vdash \exists x C(x)$$

is legitimate. In reading this, we should keep in mind that Brouwer had just before defined the 'interpretation' of an axiomatic system as the construction of a mathematical 'edifice' satisfying the demands of the system.[115] Although he does not state so explicitly, it seems reasonable to assume that he had the same 'constructivistic' idea about the existence of a mathematical object as about axiomatic systems.

This interpretation of Brouwer's statement is in accordance with a consequence one can draw from Gödel's second incompleteness theorem.[116]

Brouwer only describes his alternative in general terms: to exist in mathematics means to be constructed intuitively.[117]

Brouwer dismisses the idea that consistency proofs play an important role in mathematics. He does so by again emphasizing the separation between mathematics and the mathematical language:[118]

---

[113][Brouwer 1907, pp. 180–181]

[114]'(...) it [has, DH] nowhere been proved that if a finite number must satisfy a set of conditions, which can be proved to be non-contradictory, that then this number actually exists.'

[115]'vinden we vervolgens een wiskundige interpretatie voor de axioma's (die dan natuurlijk bestaat in den eisch, een wiskundig gebouw te construeeren met aan gegeven wiskundige relaties voldoende elementen)', [Brouwer 1907, p. 180]

[116]Take $Prov(x, [0 = 1])$ for $C(x)$. From the second incompleteness theorem follows that $\forall x \neg Prov(x, [0 = 1])$ is not provable in PA (Peano Arithmetic), since otherwise the consistency of PA would be provable in PA. Thus, $\nvdash_{PA} \forall x \neg Prov(x, [0 = 1])$. However, $\exists x Prov(x, [0 = 1])$ is not provable in PA. For Gödel's second incompleteness theorem, see 5.4.2.

[117]'*bestaan* in de wiskunde beteekent: *intuitief zijn opgebouwd*', [Brouwer 1907, p. 216]

[118][Brouwer 1907, p. 171]

## CHAPTER 2. THE GENESIS OF BROUWER'S INTUITIONISM

> En wanneer het gelukt *taal*gebouwen op te trekken, reeksen van volzinnen, die volgens de wetten der logica op elkaar volgen, uitgaande van taalbeelden, die voor werkelijke wiskundige gebouwen, wiskundige grondwaarheden zouden kunnen accompagneeren, en het blijkt dat die taalgebouwen nooit het taalbeeld van een contradictie zullen kunnen vertoonen, dan zijn ze toch alleen wiskunde als taalgebouw en hebben met wiskunde van buiten dat gebouw, bijv. met de gewone rekenkunde of meetkunde niets te maken.[119]

Note that this is the same position that Borel was to take a few years later.[120]

The reader is reminded of the words Brouwer wrote in *Leven, kunst en mystiek* on the subject of consistency. There, he had claimed that 'to be able to continue speaking for a while, without being caught in contradictions, is definitely an art, but to be appreciated only as one appreciates an acrobat.'[121] Now, Brouwer applies this idea to mathematics. Even if it is proved that no contradiction will occur in a mathematical linguistic system that could accompany mathematical constructions, the proof would be of limited value. It would only tell us something about the mathematical language, looked upon in a mathematical way, but not about mathematics proper. Brouwer uses the same argumentation to protest against the central role which Hilbert ascribes to a consistency proof.[122]

In this case, too, Brouwer turns the argumentation round. Not only is a consistency proof of limited value to the mathematician, in the cases that matter it is not even needed. If we have a set of logical axioms, Brouwer argues, and we can point out a mathematical system of which the logical axioms can be considered to express properties, we know that no contradiction can occur, because a constructed mathematical system cannot contain a contradiction. In Brouwer's view, Hilbert's idea of securing the foundations of mathematics has to fail. Hilbert denies value to intuitive mathematics, and he considers a consistency proof to be the only way to secure mathematical existence. Besides the fact that in Brouwer's view a consistency proof is not sufficient, he notes that Hilbert is now caught in an infinite regress. For he again has to secure the mathematical existence of the system in which he proved the consistency, which, following his own criterion, can only be done by giving a consistency proof, etc.[123] At that time, Hilbert had not yet introduced the meta-mathematical level in which the consistency proof had to be given.

---

[119]'And if one succeeds in constructing *linguistic* buildings, series of sentences that follow each other according to the laws of logic, originating from linguistic images, that could accompany fundamental mathematical truths in case of real mathematical constructions, and these linguistic buildings turn out never to show the linguistic image of a contradiction, then still these are only mathematics as a linguistic building and have nothing to do with mathematics outside the building, such as normal arithmetic or geometry.'

[120]See 1.3.2.

[121]See 2.2.4.

[122][Brouwer 1907, pp. 210–211]

[123][Brouwer 1907, p. 176]

## 2.3. THE FIRST ACT OF INTUITIONISM

**Cantorian set theory**  Regarding Cantorian set theory, Brouwer remarks that we can indeed posit new ordinal numbers after the (standard) natural numbers and in this way obtain arbitrarily large denumerable ordinals.[124] But by then defining the second number class, as Cantor does, as the totality of all ordinal numbers thus made, one loses mathematical ground. Brouwer claims that a totality made by an 'etcetera' can only be thought of mathematically if the 'etcetera' applies to an order type $\omega_0$[125] of the same procedure that is iterated. In the case of the second number class, there are two procedures that are iterated.[126] In Brouwer's opinion, the only thing Cantor could have done was to introduce all transfinite numbers as a purely logical concept, devoid of mathematical contents. But then he should have given a consistency proof, which he did not do either.[127] Note that Brouwer does recognize the value of consistency proofs in the case someone would like to act inside a purely logical system.[128] Only, Brouwer does not consider this interesting, since he wants to have mathematical constructions.

Brouwer reproaches Cantor for having introduced meaningless objects. Cantor and his followers postulated ever bigger ordinal numbers, and in the end they defined the totality of all ordinal numbers. Then, they discovered the Burali-Forti paradox.[129] This should not come as a surprise, Brouwer claims, for the Cantorians had already long before left the domain of mathematics. If one only makes logical constructions without mathematics, the construction made could a priori just as well be consistent as contradictory.[130]

The only cardinalities Brouwer accepts are the finite ones; the denumerable infinite; the denumerable unfinished infinite; and the continuum. By 'denumerable unfinished' Brouwer means a set of which only a denumerable part is well-defined, to which part always new elements which belong to the set can be added following

---

[124] For an explanation of Cantor's transfinite numbers, see 1.1.1.

[125] Brouwer uses '$\omega$' where we speak about '$\omega_0$'.

[126] Cantor used two so-called principles of generation in defining the transfinite ordinals: adding a unit to produce a new number and postulating a new number bigger than all numbers in a sequence with no biggest element, [Cantor 1883, pp. 104–105]; see 1.1.1.

[127] [Brouwer 1907, pp. 183–185]

[128] It is worth noting that Hilbert had criticised Cantor with the very same argumentation: '(...) System aller Mächtigkeiten überhaupt oder auch aller Cantorschen Alephs, für welches, wie sich zeigen läßt, ein widerspruchsloses System von Axiomen in meinem Sinne nicht aufgestellt werden kann, und welches daher nach meiner Bezeichnungsweise ein mathematisch nicht existierender Begriff ist.' ('(...) the system of all cardinal numbers or of all Cantor's alephs, for which, as may be shown, a system of axioms, consistent in my sense, cannot be set up. Either of these systems is, therefore, according to my terminology, mathematically non-existent.' English translation cited from [Ewald 1996, vol. II, p. 1105]).

[129] The Burali-Forti paradox comes into being if one allows all ordinals to be joined in a set. Since the set of all ordinals $\Omega$ is a well-ordered set, $\Omega$ has an ordinal number $o$, which by definition is bigger than all the ordinals in $\Omega$. However, since $\Omega$ is the set of all ordinals, $o$ has to be in $\Omega$, whereby $o$ is bigger than itself.

The paradox was named after Burali-Forti and for decades ascribed to him. However, Burali-Forti did not conclude that there is a paradox, but that the set of all ordinal numbers is not well-ordered, cf. [Burali-Forti 1897].

[130] [Brouwer 1907, pp. 186–191]

a fixed procedure.[131]

It is interesting to note that Brouwer does accept the actual infinite, albeit not in Cantor's form. In Brouwer's view, one should restrict oneself to the intuitively constructable, and not attempt at enlarging this by logical combinations. Finally, Brouwer joins Borel in rejecting Zermelo's axiom of choice.[132]

**Logistics**  Brouwer uses the same argumentation to criticise logistics. Here, too, people work with word systems, using such terms as 'propositional function', 'all', etcetera, and apply logical principles, without caring about whether there is a mathematical system underlying these words or not. Again, the fact that they find contradictions in their system need not come as a surprise. Brouwer attacks Russell's suggestion that one should reject the notion of 'all things', while retaining 'any arbitrary thing'. The reason Russell gave was that at least the logical principles are valid for any arbitrary thing. That logical principles should hold for any arbitrary thing, Brouwer maintains, is exactly where logistics goes wrong: logical principles only hold for words that have mathematical meaning.[133]

**Hilbert's foundations of mathematics**  In the final part of chapter three, Brouwer directs his criticism towards Hilbert's view on the foundations of mathematics. In doing so, Brouwer bases his opinion mainly on Hilbert's views as expressed in his 1904 Heidelberg lecture. In this lecture, Hilbert had sketched his axiomatic method, with the consistency proof as one of its characteristics.[134] Brouwer again asserts that nothing of mathematical value is obtained if a consistency proof is given. But he sees Hilbert's method as even worse than the logistic one. In order to clarify this, Brouwer analyses in various stages what happens in doing axiomatic mathematics. Brouwer sees this process as stages of 'linguistic engineering',[135] the first of which are:

1. pure construction of intuitive, languageless mathematical systems;

2. the introduction of a language that accompanies the pure mathematical construction;

3. looking at that language in a mathematical way and observing certain regularities, so-called logical principles;

4. ignoring the contents of the elements of the language and using the thus obtained empty terms to reconstruct languagelessly the logical figures that occurred in phase two, by means of a new mathematical system (of the

---

[131][Brouwer 1907, pp. 187–188]
[132][Brouwer 1907, p. 215; pp. 191–192]; the discussion on Zermelo's axiom of choice is treated in 1.3.1.
[133][Brouwer 1907, pp. 198–202]
[134][Hilbert 1905A, p. 273]. Hilbert's early view on mathematical existence is discussed in 4.2.2.
[135][Van Stigt 1998, p. 10]

## 2.3. THE FIRST ACT OF INTUITIONISM

second order); as soon as one starts generalising at this level, contradictions become possible;[136]

5. the language that accompanies the second construction;

6. looking at that language in a mathematical way.[137]

For Brouwer, only the first level constitutes real mathematics, whereas the second cannot be missed for practical purposes. All the other levels lie outside the domain of mathematics.[138] Brouwer's scheme is clearer than Hilbert's 1904 lecture, where Hilbert had not even indicated what one could use in order to give a consistency proof; Hilbert's own examples were proved in a more or less informal way. Furthermore, it clarifies the substantial difference between Brouwer's view on mathematics and a purely formalistic one, even before the foundational debate had started.

### Summary

The summary of the thesis only deals with foundational issues. Brouwer starts by stating that mathematics is a free creation, developed from the primordial intuition and independent of experience. In Brouwer's view, mathematical systems are projected on experience, where one system may turn out to be more useful than another. The primordial intuition may be described as permanence in change or unity in multitude.[139]

Brouwer's moral point of view asserts itself when he treats the question of mathematical definitions. In Brouwer's opinion, mathematical definitions should not be looked upon in a mathematical way, but should only be used as a support for our memory. Basic concepts, such as 'continuous', 'once again', 'etcetera', have to be irreducible.

A logical construction of mathematics, Brouwer concludes, is impossible — the only thing one can obtain is this way is a linguistic construction, which is always separated from real mathematics. What is needed is the mathematical primordial intuition.[140]

### The rejected parts

Although the parts of Brouwer's dissertation rejected by Korteweg were mainly outside the domain of mathematics, they do offer important information on the way in which Brouwer looked at mathematics. In *Leven, kunst en mystiek* Brouwer

---

[136] This is where, in modern terms, we would say a formalistic conception of mathematics starts.
[137] This is the level later called meta-mathematics.
[138] [Brouwer 1907, pp. 208–214]
[139] On this basis, Heyerman conjectured that Brouwer's mathematical experience has its place in the process of turning into oneself, between the experience of the multitude of the outer world and the mystical experience of unity, [Heyerman 1981, p. 40].
[140] [Brouwer 1907, pp. 217–218]

only treated the subject of mathematics marginally. He placed mathematics in the intellect, which he generally regarded with distaste.[141] In the parts intended for the dissertation, Brouwer is more specific.

Here, Brouwer maintains that one can focus intellectual observation[142] on sequences as such, independently of feelings of lust or fear. This opens up new possibilities, since Brouwer disliked the means the intellect provided for fulfilling lusts and thus the obstacle it meant for turning inward. Now, Brouwer states that mathematics, if done for its own sake, can obtain the same harmony as music and architecture. Only when mathematics is applied to the outer world, it appears as something inferior, far removed from wisdom or religion.[143]

**The defence** When Brouwer defended his dissertation, seconded by his 'paranimfs' Adama van Scheltema and Rudolf Mauve,[144] Mannoury and Barrau carried out the opposition from the floor. Both of them were later to play a role in the foundational debate as well. Especially Barrau's criticism is interesting. He suggested that, since Brouwer only wanted to work with discrete sets, he should drop the continuum intuition and accept that, from his own point of view, the whole of mathematics should be constructed from the basic idea of 'two distinct points'.[145] Later, Brouwer indeed dropped the continuum intuition, when he executed the so-called second act of intuitionism.[146]

The quality of Brouwer's dissertation was recognized by awarding it the degree *cum laude*.[147]

### 2.3.2 The unreliability of the logical principles

The year after he obtained his Ph.D., Brouwer published a short paper in a Dutch philosophical journal on the 'unreliability of the logical principles'. In this paper, Brouwer expands upon his earlier thoughts on the relationship between mathematics and logic, and draws a far more radical conclusion concerning the principle of the excluded middle.

The paper was not accepted for publication easily. Most members of the editorial board of the *Tijdschrift voor wijsbegeerte* ('Journal of Philosophy') stated that they 'hardly understood anything'[148] of the article, and they could only be convinced to accept the paper by the promise that Brouwer would explain his views in a series of papers in a more readable way later.[149]

---

[141] See 2.2.4.
[142] 'aanschouwing', [Van Dalen 1981, p. 26]
[143] [Van Dalen 1981, pp. 30–34]; cf. [Heyerman 1981, pp. 38–39]
[144] [Van Dalen 2001, p. 97]
[145] Letter from Brouwer to Korteweg, 16/2/1907; [Van Dalen 1981, p. 24]
[146] See 2.6.
[147] [Van Stigt 1990, p. 44]
[148] 'zoo goed als niets', letter from Kohnstamm to Brouwer, 3/1/1908, [MI Brouwer, CB.AKO.1]
[149] As far as I know, Brouwer never published such a series. The fact that Brouwer and Mannoury had, upon request of Maas & Van Suchtelen, the publisher of Brouwer's dissertation, tried to set

## 2.3. THE FIRST ACT OF INTUITIONISM

After a philosophical introduction which, for the uninitiated reader, rather obscures than clarifies his views, and in which he frequently refers to *Leven, kunst en mystiek*, Brouwer asks whether logic is secure for mathematical systems. He repeats the viewpoint expressed in his dissertation, namely that by applying logical principles one focuses on the mathematical language, whereas it is mathematics as constructed from the primordial intuition that matters. The paradoxes which have been constructed recently, Brouwer maintains, come into being[150]

> waar regelmatigheid in de taal, die wiskunde begeleidt, wordt uitgebreid over een taal van wiskundige woorden, die geen wiskunde begeleidt (...).[151]

Then Brouwer asks where exactly the mistake comes from. When is it allowed to move in the linguistic building, using only the logical principles, relying on the assumption that mathematical constructions can be made to which the linguistic forms will fit? Brouwer's answer is that this is allowed for the principles of syllogism and contradiction, but not for the principle of the excluded middle.[152]

**Rejection of the principle of the excluded middle**  Brouwer's argumentation for rejecting the principle of the excluded middle goes as follows. In Brouwer's interpretation, the principe of the excluded middle is equivalent to asserting that, as he calls it, every supposed fitting into each other of systems in a certain way can either be finished or be proved impossible. Thus, it expresses that there are no unsolvable problems. This interpretation is in accordance with Brouwer's idealistic philosophy, in which the human mind is the central feature of the world, thus also of truth. As Brouwer had remarked in his dissertation, there is not even a hint of a proof for the solvability of every mathematical problem. Therefore, the principle of the excluded middle is as yet not reliable. It can only be used as a reliable tool in finite systems, where every possibility can, at least in principle, be checked by means of a finite process.[153]

However, Brouwer continues, by using the principle of the excluded middle in infinite systems one will never find out that the arguments used are not correct by getting caught in a contradiction. For if we assume that the principle of the excluded middle leads to a contradiction, this means that neither a certain proposition nor its negation holds. But then the negation of this proposition and its double negation would hold, which is ruled out by the principle of contradiction.[154]

---

up a philosophical journal and that the *Tijdschrift voor Wijsbegeerte* was founded by professional philosophers as a reaction to that, may have played a role in the negative judgement as well; [Van Dalen 1999A, p. 108].

[150] [Brouwer 1908, p. 256]
[151] 'where regularities in the language which accompanies mathematics, are extended over a language of mathematical words which does not accompany mathematics (...).'
[152] [Brouwer 1908, pp. 255–256]
[153] [Brouwer 1908, p. 257]
[154] [Brouwer 1908, p. 258]. In modern notation: assume $(\varphi \vee \neg\varphi) \to \bot$, then $\neg(\varphi \vee \neg\varphi)$, which, by De Morgan's law, is equivalent to $\neg\varphi \wedge \neg\neg\varphi$.

Brouwer finishes the paper by mentioning a mathematical theorem that, in his view, has to be regarded as not proved, because it illegitimately relies on the principle of the excluded middle: the statement that every ordinal number[155] is either finite or infinite. Brouwer argues that this theorem states that for every number $\gamma$ either a mapping can be constructed from $\gamma$ onto a finite part of the natural numbers, or onto the whole sequence of natural numbers. It is clear that we do not possess such a general procedure. As a consequence, one cannot be sure whether questions such as 'Is there an infinite number of pairs of successive equal digits in the decimal development of $\pi$?' have a solution.[156]

The reader may have noticed that there is another novelty in Brouwer's paper, which, however, is more hidden: the way he interprets the negation. Opposed to the classical idea, where the negation of a proposition is seen as the mere impossibility of proving the proposition, Brouwer posits the notion that the negation of a proposition means that an actual contradiction can be derived from the proposition. This interpretation of negation would later be called the strong negation.

Thus, Brouwer's objections against the universal validity of the principle of the excluded middle as expressed in the paper are mostly a matter of principles. Brouwer interprets the logical principle in a constructive way, and then concludes that it cannot hold generally in infinite systems. Later, Brouwer was to work out more concrete mathematical counter-examples against the principle of the excluded middle.[157]

1908 was also the year of Brouwer's first international conference. Brouwer lectured on Hilbert's fifth problem (the treatment of Lie groups without the differentiability conditions) and on cardinal numbers in constructive mathematics at the International Conference for Mathematicians in Rome. Most mathematicians who were to play a role in Brouwer's later career were present there, amongst others Borel, Bernstein, Blumenthal, Carathéodory, De Donder, Hahn, Hardy, Levi-Civita, Koebe, Hadamard, Hilbert, Poincaré and Zermelo.[158]

## 2.4 Topology

In 1909, Brouwer was accepted as *privaat-docent* at the university of Amsterdam. In his first public lecture, entitled *Het wezen der meetkunde* ('The nature of geometry'), Brouwer combines philosophical concerns with topological interests. He argues that geometry has no right to the a priori status it had in the past; the only a priori in mathematics is the intuition of time. Even though geometry is not singled out by an a priori position, there is a mathematical way of distinguishing

---

[155] Or possibly cardinal number; Brouwer only speaks about a number in transfinite number theory.
[156] [Brouwer 1908, pp. 258–259]
[157] See 2.6.2 and 2.7.1.
[158] [Van Dalen 1999A, p. 204]

## 2.4. TOPOLOGY

geometry from other areas of mathematics. One of the characteristics of geometry, Brouwer maintains, is that one classifies spaces according to their transformation groups, as stated in Klein's Erlangen program.[159] Brouwer sees a role for 'topology' in the foundation of parts of mathematics:[160]

> Zoo zullen (...) ook uit andere theorieën, als het gelukt ze op de analysis situs[161] te grondvesten, coördinaten en formules niet geheel verbannen hoeven te worden, maar de formulelooze, de 'meetkundige' behandeling zal het uitgangspunt vormen, de analytische wordt een ontbeerlijk hulpmiddel.[162]

One can again note Brouwer's preference for languageless, or at least formulaless, mathematics, this time expressed in the claim that by using topological methods one can dispense with analytical formulas. Seen in this light, Brouwer's preference for topology was not far removed from his philosophical inclination.

In the summer of 1909, Hilbert and Brouwer met in the dunes of the Dutch beach resort Scheveningen.[163] During one of their meetings that summer, Brouwer told Hilbert about the difference between what he called first-order and second-order mathematics;[164] Hilbert later was to label the latter meta-mathematics.

In the first years as a *privaat-docent*, Brouwer worked out the problems mentioned in his public lecture. The period 1908–1912 was one of extreme activity, during which he published over 40 papers in topology. These papers, some of which were of a 'startling originality', 'completely transformed' the new area of topology and showed Brouwer's 'almost prophetic insight'.[165] His most important results include the example of a curve that divides the plane into three parts in such a way that every point on the curve is a boundary point of all three parts;[166] a proof of a

---

[159][Brouwer 1909, pp. 9–15]. In general, a transformation group $(G, S)$ consists of a group G and a set S such that the elements of $G$ act on $S$ and preserve the structure of $S$, [Hazewinkel 1988–1993, vol. 9, p. 242]. In topology, one usually works with continuous transformations that preserve the topological structure of the space they act on. In fact, Brouwer later moved away from such a transformational viewpoint in topology, [Johnson 1987, p. 74].
[160][Brouwer 1909, p. 23]
[161]In those days, 'analysis situs' was the name used for what we now call 'topology'.
[162]'In this way (...) coordinates and formulas will not have to be banned completely from other theories either, if one succeeds in grounding them in analysis situs. But the formula-free, 'geometric' treatment will form the starting-point, the analytical one becomes a dispensable tool.'
[163]Letter from Brouwer to Adama van Scheltema, 9/11/1909; in: [Van Dalen 1984A, p. 100]
[164][Brouwer 1928, p. 410]
[165][Kreisel & Newman 1969, pp. 46–53]
[166][Brouwer 1910C, p. 355; p. 359]

generalisation of Jordan's theorem;[167] the invariance of dimension;[168] and several fixed-point theorems.[169] In doing so, Brouwer almost singlehandedly developed new methods and concepts that were to determine the course of topology, such as the definition of an $n$-dimensional manifold, the concept of degree of a continuous mapping, and the simplicial method.[170]

**Unity or disunity in Brouwer's work** Brouwer's topological results differ markedly from the more philosophical declarations he had made in earlier papers, mainly regarding the permissibility of the use of the principle of the excluded middle. The standard interpretation has it that Brouwer's scientific work falls into two separate parts: his intuitionistic work and his work in topology. Thus, in the introduction to Brouwer's Collected Works, Heyting speaks about two 'almost disjoint parts' in Brouwer's work.[171] However, in his joint paper with Freudenthal on Brouwer's life, Heyting conjectures that it was 'the same mental disposition, that drove him [Brouwer, DH] on the one hand into constructive methods [in topology, DH], on the other hand into a constructivist philosophy'.[172]

When one reads Brouwer's topological papers, one notices characteristics such as the frequent use of drawings and the description of procedures that can actually be carried out. At the same time, there is a certain tendency towards intuitive and constructive mathematics in a broad sense. However, Brouwer did use the principle of the excluded middle also in infinite sets, which is strange if we keep in mind that he had already published *De onbetrouwbaarheid der logische principes*,[173] in which he rejected its unlimited use. Thus, there are elements which seem to divide Brouwer's topological work from the intuitionistic part, while they also seem to share certain characteristics.

Regarding this contrary evidence, Koetsier and Van Mill argued that the main dividing line in Brouwer's work does not run through his topological work on the one hand and his intuitionistic work on the other, but rather separates his

---

[167][Brouwer 1910E]. Jordan's theorem states that any plane simple closed curve decomposes the plane $^{-2}$ into two connected components and is their common boundary, where a closed curve is defined as the image of a continuous map from the circle. Brouwer generalised the theorem to arbitrary dimension $n$, [Hazewinkel 1988-1993, vol. 5, pp. 241-242].

[168][Brouwer 1911C]. In this paper, Brouwer proved that it is impossible to construct a one-to-one and continuous mapping (i.e., a homeomorphism) between an $m$-dimensional manifold and an $m + h$-dimensional manifold, for $h > 0$.

[169]Brouwer's most famous fixed-point theorem states, in modern terms, that for any continuous map $f$ from the full circle $C$ to itself there is a point $(x, y) \in C$ such that $f((x, y)) = (x, y)$, [Kahn 1975, p. 139]. The theorem for the sphere is in [Brouwer 1910F, pp. 247-248].

[170][Van Stigt 1990, pp. 51-56], [Freudenthal & Heyting 1967, p. 337], [Johnson 1987, p. 88]. Brouwer's ideas in topology were further developed by mathematicians such as Alexander, Schmidt, Hopf and Alexandroff. A more detailed account of Brouwer's topological work can be found in [Van Dalen 1999A, pp. 122-196] and, for the early period, in [Johnson 1987].

[171][Heyting 1975, p. XIII]

[172][Freudenthal & Heyting 1967, p. 339]. Dubucs argued, though not convincingly, that Brouwer attempted to obtain results which showed the importance of combinatorial notions in topology, [Dubucs 1988, p. 143].

[173]'The unreliability of the logical principles', see 2.3.2.

## 2.4. TOPOLOGY

pre-1917 work, whether topological or intuitionistic, from his post-1917 achievements.[174] Van Dalen goes further and claims that Brouwer's experience in topology paved the way for his idea of dealing with the infinite constructively by means of choice sequences. A person without Brouwer's penchant for topology, Van Dalen conjectures, would probably have embraced a more combinatorial or finitist constructivism, like Kronecker's.[175]

Koetsier and Van Mill's argumentation about the basic unity in Brouwer's pre-1917 work rests on two arguments. Firstly, Brouwer's dissertation was not purely foundational in the philosophical sense — it contained many subjects which are more topological. For example, Brouwer defined arithmetical operations on the (measurable) continuum by means of group theory and topological notions, and he defined different geometries by placing topological restrictions on the group of one to one continuous mappings on the measurable continuum. Thus, already in his thesis Brouwer used topology in order to provide a foundation for mathematics. Brouwer was to work out these ideas in his topological papers.[176] Secondly, by the time Brouwer published his topological work, his intuitionism was only halfway developed. The 'second act of intuitionism', the introduction of constructive set theory based on choice sequences, was not conceived until 1917.[177] The later intuitionism, Koetsier and Van Mill maintain, was definitely in contrast with Brouwer's topological work — as it was with his dissertation.[178]

I agree with Koetsier and Van Mill that there is more that links Brouwer's pre-1917 intuitionistic work to his topology than separates it. However, Van Dalen's conjecture about Brouwer's intuitionism being 'a topologist's constructivism' is one of the best explanations available of the genesis of Brouwer's intuitionism, notably regarding the inclusion of choice sequences after 1917. In this way, the unity in Brouwer's work is stressed even more.

Thus, Brouwer's dissertation and his topological work mostly follow naturally from the same basic principles. The only variance that can be found lies in his opinion on the principle of the excluded middle. This may be explained by Brouwer's then still quite pragmatic stance. His pragmatism is clear from what he wrote in 1917:[179]

> Inderdaad heeft men op het intuitionistische standpunt (...) het recht, zoodanige onderstellingen, als ter wille van de levensvatbaarheid der theorie wenschelijk zijn, in de constructieprincipes geïmpliceerd te denken.[180]

---

[174][Koetsier & Van Mill 1997, p. 160]
[175][Van Dalen 1999A, p. 383]
[176][Koetsier & Van Mill 1997, pp. 142–145]
[177]See 2.6.
[178][Koetsier & Van Mill 1997, pp. 145–146]
[179][Brouwer 1917A, p. 263]
[180]'Indeed, from the intuitionistic point of view (...) one has the right to consider presuppositions desirable for the viability of the theory implied in the construction principles.'

This is the main argument I found in the sources which could explain Brouwer's use of the excluded middle in topology while he had rejected it in one of his intuitionistic writings. It implies that, if Brouwer saw no other way for keeping topology as a viable theory than by using the principle of the excluded middle, this was his argument to justify its use. It is a weak argument for someone who adheres to a single philosophical conviction, but it is a most reasonable argument for the working mathematician who wants do develop the subject further.[181]

Brouwer's work in topology brought him international recognition. In 1911, he visited Göttingen, the mathematical Mecca, for the first time. He remained a regular visitor as one of the 'extra-territorial' members of the Göttingen group, often combining a visit to Göttingen with a few weeks in the nearby Harz mountains, where he later bought a house.[182]

## 2.5 Intuitionism and formalism

In 1911,[183] Brouwer introduced two terms that were to dominate the foundational debate: intuitionism and formalism. He did so in a review of Mannoury's book *Methodologisches und Philosophisches zur Elementar-Mathematik* ('Methodological and philosophical remarks on elementary mathematics'). Brouwer distinguishes between two visions presented in Mannoury's work: the formalistic one, defended by the author as well as by Russell, Hilbert and Zermelo, and the intuitionistic one of Poincaré and Borel. Brouwer also describes a proof he himself had given as intuitionistic. He characterises formalism by its identification of mathematics with mathematical language, a theme familiar from Brouwer's dissertation. Following Poincaré, Brouwer argues that formalists cannot do without the intuitive use of the principle of complete induction, whereby they strengthen intuitionism rather than formalism.[184]

Contrary to what the review may suggest,[185] Mannoury did not use the terms 'intuitionism' and 'formalism' in his book. Instead, he described the two groups as *Kantianismus* and *Symbolismus*.[186] Thus, the question arises how Brouwer came to employ these names. In general, it seems that the term 'formalism' had a negative connotation,[187] which obviously served Brouwer well. As to the choice of

---

[181] Later, Brouwer claimed that he had tried to derive only such results in his topological works as he could hope to keep within the intuitionistic framework, [Brouwer 1919D, p. 231].
[182] [Van Dalen 1999A, p. 213; 252]
[183] As pointed out by Mancosu, [Mancosu 1998, p. 180].
[184] [Brouwer 1911A, pp. 199–200]
[185] and to what Mancosu concludes, [Mancosu 1998, p. 180].
[186] [Mannoury 1909, pp. 139–149; 272]
[187] Cf., e.g., Rickert: *'mögen manche Philosophen auch noch so sehr auf den Formalismus schelten'*, ('even if many philosophers may denounce formalism strongly'), [Rickert 1921, p. 351]; in Göttingen, one argued against the appointment of Nelson to a chair in philosophy by pointing to his 'repulsive formalism' (*'abstoßenden Formalismus'*, [Dahms 1987B, p. 171]); others, however, used formalism in a more or less neutral sense, as e.g. Scheler in [Scheler 1913].
Frege, in his discussion with Thomae in the *Jahresbericht der deutschen Mathematiker-*

## 2.5. INTUITIONISM AND FORMALISM

the name 'intuitionism', I see three possible explanations for Brouwer's adoption. The first one is that Brouwer made the name up himself. Kant and Schopenhauer are authors who might have influenced him in this respect. Also Poincaré, who stressed the role of intuition in mathematics,[188] may have been influential. The second one is that he took the term from someone else. Felix Klein and Henri Bergson are the most natural persons to think of.[189] Van Stigt maintains that there is no doubt about Brouwer's familiarity with the latter's philosophical ideas.[190] The third possibility relies on Brouwer's interest for theories on morality. In a letter to his friend the Dutch poet Adama van Scheltema in 1905, Brouwer wrote that he was working on a book on morality.[191] In morality theory, the word 'intuitionism' had been used already. Clauberg and Dubislav's *Systematisches Wörterbuch der Philosophie* ('Systematic dictionary of philosophy') was published in 1923 and can therefore be considered as a reasonable knowledge base for the preceding two decades. In it, the term 'intuitionism' is used only in one context: ethics. There, not only intuitionism is described, but also 'formal' currents in ethics, including one called 'logicism'. The ethical definitions given there correspond quite well with the ones used in the foundations of mathematics. Especially in the case of a formal theory the resemblance is striking. Clauberg and Dubislav's description is:[192]

> Eine Lehre der Ethik heißt eine 'formale', der zufolge das ethisch Gesollte nur durch allgemeine Normen, nicht aber durch Bezeichnung der Gegenstände, welche begehrt werden sollen, bestimmt werden kann.[193]

This is exactly what Brouwer wanted to express with the name formalism: it proceeds only formally, without paying attention to the contents of the concepts involved.

One of the sub-categories of the class of formal theories in ethics is logicism, described as a doctrine for which consistency of motives is sufficient to ensure ethical value. In mathematics, this is normally seen as a characteristic of formalism in general, not specifically of logicism.

Finally, intuitionism in ethics is designated as a current according to which ethical norms are not acquired, but innate. This could be related to Brouwer's intuitionism by means of the primordial intuition, from which mathematics is constructed. The strong point in this explanation, therefore, is that it not only accounts for Brouwer's choice of the name 'intuitionism', but also of 'formal-

---

*Vereinigung*, argued strongly against Thomae's views, using words such as 'formal' and '*Formalarithmetiker*', but he did not coin Thomae's view 'formalism'; cf. [Thomae 1906A], [Frege 1906] and [Thomae 1906B]. A similar remark applies to Frege's criticism in the *Grundgesetze der Arithmetik* of Thomae's view; cf. [Frege 1893, pp. 96-139].

[188] See 1.4.2.
[189] See 6.3.2.
[190] [Van Stigt 1990, p. 114]
[191] Letter from Brouwer to Adama van Scheltema, 7/4/1905; in: [Van Dalen 1984A, pp. 58-59]
[192] [Clauberg & Dubislav 1923, p. 161]
[193] 'An ethical theory is called 'formal' if it considers that what ethics prescribes can only be determined by general norms, not by describing the objects that should be desired.'

ism'. However, I have not found any explicit reference to intuitionism in ethics in Brouwer's writings.

Whatever the origin of the names may be, it is a fact that in the 1920s people involved in the foundational debate saw the terms 'intuitionism' and 'formalism' as coined by Brouwer. Bernays, looking back at the debate from the 1970s, even claimed that all three terms—intuitionism, formalism and logicism—originated from Brouwer.[194]

In April 1912, after a successful campaign by Korteweg, Brouwer was elected member to the Dutch Academy of Scieces. Three months later, at the age of 31, he was offered a chair at the university of Amsterdam. At first, it was a position as associate professor (*extra-ordinarius*), but within half a year this was changed into a full professorship (*ordinarius*) under pressure of an offer to Brouwer from the university of Groningen. The official description of his chair was 'set theory, function theory and axiomatics'. Brouwer's fast rising to the position of full professor was established mainly because of the ongoing efforts of Korteweg, who recognized Brouwer's extraordinary talent. From 1909 on Korteweg had campaigned internationally for a position for Brouwer, collecting impressive support from world famous mathematicians such as Hilbert, Klein, Poincaré and Borel. In the end, Korteweg even vacated his own chair in favour of Brouwer.[196] On October 14, 1912, Brouwer delivered his inaugural address *Intuitionisme en formalisme* ('Intuitionism and formalism').[197]

In his oration, Brouwer mainly follows the line set out in his dissertation. He characterises science as the ordering of nature by means of causal sequences. He modifies Kant by dispensing with the spatial intuition, while keeping the intuition of time. He stresses that the correctness of a mathematical theory cannot be assured by a consistency proof.

Brouwer again presents his distinction between two parties, intuitionists and formalists, but this time he explains in more detail how he sees both currents. The former were mainly French, the latter mainly German. Brouwer traces intuitionism back to Kant; he calls the altered position, where the spatial intuition is abandoned, 'neo-intuitionism'.[198]

From the text one can infer that Brouwer considers Poincaré and Borel to be intuitionists, Cantor and Zermelo formalists. Hilbert is not mentioned explicitly, but the way in which Brouwer refers to the first of Hilbert's 23 problems makes

---

[194] [Bernays 1971, p. 171]. The claim for logicism seems to be incorrect; it probably originates from Carnap in 1929. [195] This indicates that Klein's Evanston lectures were little known; see 6.3.2.
[196] [Van Stigt 1990, pp. 58–60], [Van Dalen 1999A, pp. 216–226]. Korteweg retired in June 1913. Brouwer was to hold the chair until 1951.
[197] In publications from this period, one often finds the Dutch word *'intuitionisme'* spelled without the diaeresis.
[198] [Brouwer 1912, p. 7]. Brouwer had abandoned the spatial intuition in his dissertation, see 2.3.1.

## 2.5. INTUITIONISM AND FORMALISM

it quite clear that he is to be included in the formalists.[199] The basic difference between intuitionism and formalism is, in Brouwer's view, a matter of ontology:[200]

> Op de vraag, waar de wiskundige exactheid (...) bestaat, antwoorden beide partijen verschillend; de intuitionist zegt: in het menschelijk intellect, de formalist: op het papier.[201]

He further describes the formalistic position as follows:[202]

> Weliswaar leiden wij uit axiomatisch vooronderstelde relaties tusschen wiskundige entiteiten, volgens vaste wetten andere relaties af in de overtuiging, dat op die manier door logische redeneering uit waarheden andere waarheden worden afgeleid, doch deze onwiskundige waarheids- of echtheidsovertuiging mist elke exactheid en is niets anders dan een vaag lustgevoel, teweeggebracht door het bewustzijn der doelmatigheid van de projectie der genoemde relaties en redeneerwetten op de ervaringswereld. De wiskundige exactheid ligt dus voor den formalist uitsluitend in de wijze van ontwikkeling der relatieseriën, en is onafhankelijk van de beteekenis, die men aan de relaties of aan de daardoor verbonden entiteiten zou willen toekennen.[203]

Brouwer reproaches formalists for leaving to psychologists the question why we believe in certain logical systems and not in others.[204]

I have quoted Brouwer at length because, in describing the formalistic position as completely dispensing of contents in mathematical reasoning, he shows an early understanding of where formalistic ideas ultimately would lead to. It was to take others much longer to come to the same conclusion.

Brouwer's distinction between intuitionism and formalism was not completely new. In 1900, Hilbert had already described a similar division, although on a less fundamental level. In a paper for the *Jahresbericht der deutschen Mathematiker-Vereinigung* Hilbert distinguished between two currents: a genetic and an axiomatic one. When using the genetic method, mathematical concepts have to be

---

[199] [Brouwer 1912, p. 23]
[200] [Brouwer 1912, p. 7]
[201] 'The question where mathematical exactness (...) exists is answered differently by both parties; the intuitionist says: in the human intellect, the formalist: on paper.'
[202] [Brouwer 1912, p. 8]
[203] 'It is true that from certain relations among mathematical entities, which we assume as axioms, we deduce other relations according to fixed laws, in the conviction that in this way we derive truths from truths by logical reasoning; but this non-mathematical conviction of truth or reality has no exactness whatever and is nothing but a vague sensation of delight arising from the awareness of the efficacy of the projection onto the world of experiences of these relations and laws of reasoning. For the formalist, therefore, mathematical exactness consists merely in the method of developing the series of relations, and is independent of the significance one might want to give to the relations or the entities which they relate.' Translation based on [Brouwer 1913].
[204] [Brouwer 1912, pp. 9–10]

generated. Following to the axiomatic method, however, one postulates the existence of certain elements, then uses axioms in order to characterise the relations between these elements, and finally one needs proofs for consistency and completeness.[205] Hilbert added that, despite the high pedagogic and heuristic value of the genetic method, he preferred the axiomatic method in order to secure logically 'the contents of our knowledge'.[206] At the time, Hilbert probably still thought that he could handle meaningful mathematical concepts by the axiomatic method. In this respect, Brouwer saw clearer than Hilbert what the consequences of the latter's view on the foundations of mathematics were.[207]

The opposition between intuitionism and formalism is posited by Brouwer as a strong one. He speaks about two parties that are 'fighting each other', and about a 'fundamental matter of controversy, that divides the mathematical world'.[208] This seems somewhat exaggerated, if we consider the number of people actually involved in the discussion concerning, mainly, set theory and the status of logic.[209] Brouwer finishes his address with the prophetic words:[210]

> Tot beide partijen behooren geleerden van den allereersten rang, en de kans, dat men het in afzienbaren tijd eens zal worden, is vrijwel uitgesloten. Om met Poincaré te spreken: 'Les hommes ne s'entendent pas parce qu'ils ne parlent pas la même langue et qu'il y a des langues qui ne s'apprennent pas.'[211]

Brouwer should be given credit for having been the first person to clearly distinguish between two concepts of mathematics and describing them *in nuce*. The fact that he described the difference between the two as unsurpassable fits into the picture that, at least in the beginning of the debate, people observed a huge rift between them.[212]

**Brouwer's position**  I should like to point out that, strange as it may seem to us today, Brouwer does not explicitly take position in his address. To the listener

---

[205]It is interesting to note that, if one uses this definition, Cantor would have been on the genetic side.

[206]'*des Inhaltes unserer Erkenntnis*', [Hilbert 1900, p. 181]. Hilbert did not mention anything about a meta-mathematical level at which the consistency proof should be given.

[207]Mehrtens draws the same conclusion, [Mehrtens 1990, p. 188]. Hilbert's view on mathematical existence and its consequences for the contentual character of mathematics is discussed in 4.2.2.

[208]'elkaar bestrijdende theorieën', 'fundamenteele strijdvraag, die de wiskundige wereld verdeeld houdt', [Brouwer 1912, p. 5; p. 29]

[209]See 1.3.1 and 1.4.1.

[210][Brouwer 1919B, p. 29]

[211]'There are eminent scholars on both sides and the chance of reaching an agreement within the foreseeable future is practically excluded. To speak with Poincaré: 'People do not understand each other because they do not speak the same language, and there are languages that cannot be learned.' ' Translation, except for the French citation, based on [Brouwer 1913, p. 96].

[212]This did not stop later participants to the debate to see possibilities for a compromise, see 3.3.2.

## 2.5. INTUITIONISM AND FORMALISM

who did not know any of Brouwer's writings before, the oration must have sounded as a more or less neutral description of two currents in the foundations of mathematics.[213] Brouwer frequently splits mathematics into different parts, especially in the case of set theory. Parts of set theory may be important for the formalist, while the intuitionist regards them as devoid of meaning. Only the more critical attitude towards formalistic practices may have hinted at Brouwer's preference for intuitionism.

To the informed listener, on the other hand, it must have been clear that some of the characteristics Brouwer mentioned as neo-intuitionistic were actually his own. This applies, for example, to the modification of Kant's position and the rejection of consistency proofs as a means of assuring a mathematical theory. On the latter point, Brouwer explicitly disagrees with both Poincaré and Borel:[214]

> Nimmer mag dan ook de intuitionist de juistheid eener wiskundige theorie verzekerd achten door waarborgen als het bewijs harer niet-contradictoriteit, de definieerbaarheid harer begrippen door middel van een eindig aantal woorden, of de practische zekerheid, dat zij in de verstandhouding der menschen nooit tot misverstand aanleiding zal geven.[215]

However, the criticism of set theory and the intuitive acceptance of the principle of complete induction are characteristics which were defended by Borel and Poincaré respectively.[216] Thus, intuitionism as presented here by Brouwer is a mixture of his own ideas and those of mainly Borel and Poincaré.

With regard to terminology, it is worth noting that by now Brouwer has replaced the word *opbouwen* ('building up'), which appeared frequently in his dissertation, by *construeeren* ('to construct'), *constructief* ('constructive'), *constructie* ('construction'), etc. It is not clear why he did so. Possibly, Brouwer wanted to stay closer to the French terms, since he had described intuitionism as mainly French.

The two dominant themes of the foundational debate after 1921, the question of mathematical existence and the status of the principle of the excluded middle, play only a marginal role in the address. The former is not even mentioned; the latter figures only in a footnote in the very end of the oration. This makes it clear that the opposition between formalism and intuitionism was not defined in the terms that were to play a dominant role in the debate; in Brouwer's address, the opposition was more a matter of principle (the question of mathematical exact-

---

[213] Cf. Hölder, who on the basis of Brouwer's address maintained that he went further than Brouwer by not seeing formalism and intuitionism as equally justified views, claiming instead that formalism should be rejected as a foundation of mathematics, [Hölder 1924, p. 277].

[214] [Brouwer 1912, pp. 13–14]

[215] 'For this reason the intuitionist may never feel assured of the exactness of a mathematical theory by such guarantees as the proof of its being non-contradictory, the possibility of defining its concepts by a finite number of words, or the practical certainty that it will never lead to a misunderstanding in human relations.' Translation based on [Brouwer 1913, p. 86].

[216] See 1.4.2 and 1.3.2.

ness), and it was applied mainly to set theory. This was to change mainly because of Hermann Weyl's intervention.[217]

Finally, it should be mentioned that Brouwer again touches upon the subject of a 'choice sequence'. Interestingly, he places choice sequences at the formalist side of mathematics. In his view, real numbers in the intuitionistic sense can only be constructed by means of finite generating laws. Formalistic real numbers, however, can be determined by fundamental sequences of free choices of numbers, forming a sequence of decimals.[218] In 1917, he was to present a modification that placed choice sequences at the intuitionistic side.[219]

Brouwer's inaugural address was his first foundational contribution that was translated. It appeared in 1913 in English in the Bulletin of the American Mathematical Society.[220]

Brouwer's professorship not only brought him recognition and a decent salary, but also extra tasks. He complained about the teaching load and other academic demands, which, together with the refereeing work he did for the *Mathematische Annalen*, left him little time for doing research. Since Brouwer's professional contacts were to a large extent with Göttingen, this problem was more or less solved automatically by the outbreak of the war in 1914.[221]

## 2.6 The second act of intuitionism

During the 'Great War' (1914-1918), international communication became much more difficult and mathematical activities at the German centres diminished. Visa regulations severely restricted international travelling. Brouwer shifted his attention to significs, the establishment of an International Academy for Philosophy, university politics, and the development of his intuitionistic thoughts starting from set theory.

In 1914, Brouwer published a review of Schoenflies and Hahn's *Die Entwickelung der Mengenlehre und ihrer Anwendungen* ('The development of set theory and its applications') in the *Jahresbericht der deutschen Mathematiker-Vereinigung*, which he used mainly to put forward the intuitionistic view on set theory. Brouwer 'reminds'[222] the reader that for an intuitionist, only well-constructed infinite sets[223]

---

[217] See 4.2.1 and 5.2.1.
[218] [Brouwer 1912, pp. 22–24]
[219] See 2.6.
[220] [Brouwer 1913]
[221] [Van Dalen 1999A, pp. 227–229]
[222] '*erinnere ich daran*', [Brouwer 1914, p. 79]; Brouwer uses this word despite the fact that I found no indication that he had ever before published the statement he made here.
[223] Brouwer wrote 'species', because he used 'species' for what we nowadays call 'set', and 'set' for what is now called a 'spread'. Since it would be confusing to take over Brouwer's terminology,

## 2.6. THE SECOND ACT OF INTUITIONISM

exist, which can be put together from two parts. The first of these is a fundamental sequence;[224] the second makes use of 'a sequence of choices among the elements of a finite set or of a fundamental sequence'.[225] Apparently, Brouwer had changed his idea about choice sequences, since he now places them on the side of intuitionism. Brouwer again states, but now with more stress than in his inaugural address, that for an intuitionist the principle of the excluded middle cannot be accepted. Also, he points out that several parts of set theory as treated in Schoenflies' book are meaningless for the intuitionist.[226]

In the same year, Felix Klein offered Brouwer a position at the editorial board of the *Mathematische Annalen*. This was probably the most prestigious mathematical journal at the time; Brouwer accepted.[227]

In the Netherlands, too, Brouwer strengthened his position. He became president of the Dutch *Wiskundig Genootschap* ('Mathematical Society'), and he used an offer for a chair in Leiden to get his friend Mannoury appointed in Amsterdam.

During this period, Brouwer was mainly interested in what was called significs. This was a socio-philosophical enterprise that pleaded for investigations into the different uses of language and for linguistic changes for the benefit of society. Thus, Brouwer argued against the 'anarchy in the formation of words' in order to eradicate the injustice defended by words-of-power such as 'fatherland'.[228] Apart from Brouwer, the people involved were the author and first Dutch psychiatrist Frederik van Eeden, H. Bloemers, a sociologist, H. Borel, a sinologist, the physicist Leonard Ornstein and the poet and lawyer Jacob Israel de Haan. This Signific Circle published a manifesto, with the idea that the movement should become an international one. Peano was one of the people who was invited to join the group. However, the Circle was never really to expand beyond a small group of interested academics in the Netherlands, and was finally dissolved in 1926.[229]

Partly overlapping in time and persons was the attempt to establish an International Academy for Philosophy. From September 1915 on, Brouwer headed a committee with the task of founding such an academy. Van Eeden, Henri Borel and Bloemers also took part in the enterprise, along with people from outside

---

I as a rule adhere to the modern terms. These concepts are explained in 2.6.

[224] Brouwer did not explain what he meant by a 'fundamental sequence'. In most cases, as presumably here, he meant a lawlike sequence; cf. [Brouwer 1918, p. 161].

[225] 'eine Folge von Auswahlen unter den Elementen einer endlichen Menge oder einer Fundamentalreihe', [Brouwer 1914, p. 140]

[226] [Brouwer 1914, pp. 141–142]

[227] [Van Stigt 1990, p. 63]

[228] 'anarchie in de woordvorming', 'vaderland', [Brouwer 1917D, p. 110]

[229] [Van Stigt 1990, pp. 65–68; pp. 77–78], [Van Dalen 1999A, pp. 243–250; 372]; Brouwer's role in the signific movement, and significs in the Netherlands in general, is discussed in [Schmitz 1990].

the signific circle. By December 1916, the group around Brouwer and Van Eeden had split from the others and set up statutes for their own academy. Brouwer got Mannoury to join their group, who subsequently took organisational matters in hand. The goal of the Academy was the 'renewal of the valuation of the elements of life of the individual and society'. The International Institute for Philosophy, as it was by then called, was finally established in 1918. It functioned for some time, though never internationally and never as flourishing as its founders had hoped, and was dissolved in 1922.[230]

Around 1914, Cor Jongejan, a fellow-student of Lize's daughter Louise at the domestic science school,[231] joined the Brouwer household. She became Brouwer's assistant and an intimate friend, often accompanying him on his travels.[232]

### 2.6.1 Intuitionistic set theory

In 1918, Brouwer published a paper which he later considered to mark the 'second act of intuitionism'.[233] The paper was published as *Begründung der Mengenlehre unabhängig vom logischen Satz vom ausgeschlossenen Dritten* ('Foundation of set theory independent of the logical theorem of the excluded middle') in the Dutch Academy's *KNAW Verhandelingen*. Contrary to what the title might suggest, it contained no polemical attack against the principle of the excluded middle, nor any reference to Cantorian set theory, as Brouwer had done in earlier publications. In fact, both these subjects were not even mentioned in the main text of the paper. Instead, Brouwer mainly gave definitions of his own set theoretical concepts.

The opening line of the paper is puzzling. In it, Brouwer maintains that the basis of set theory is formed by an 'unbounded sequence of signs'.[234] One wonders why Brouwer used the term 'signs' to indicate the basis of set theory, where he had before taken the view that signs are nothing but a linguistic description of mathematics, at best. Brouwer himself later commented on this issue that '[b]ecause mathematics is independent of language, in this definition the word *sign* (*Zeichen*) (...) must be understood in the sense of *mental signs*, consisting in previously obtained mathematical concepts.'[235] Perhaps it was also an attempt to keep all polemics out and conform to mathematical practice with the purpose of getting his new theory accepted.[236] A similar remark can be made for the term 'there exists', which Brouwer uses various times in the classical sense.

---

[230][Van Dalen 1999A, pp. 258–270; 334]
[231]*huishoudschool*
[232][Van Dalen 1999A, pp. 250–252]
[233][Brouwer 1992, p. 23]
[234]'eine unbegrenzte Folge von Zeichen', [Brouwer 1918, p. 150]
[235]'[w]egens de taalloosheid der wiskunde behoort in de genoemde definitie bij het woord *teeken* (*Zeichen*) (...) gedacht te worden aan *gedachtenteekens*, bestaande in reeds verkregen mathematische denkbaarheden.', [Brouwer 1947B, p. 339]; English translation based on the translation in [Brouwer 1975–1976, vol. 1, p. 477].
[236]I found no instances in the secondary literature where this point is discussed. The only person during the debate who criticised Brouwer on his 'quibbling with words' was Bentley, cf. [Bentley 1932, pp. 153–155].

## 2.6. THE SECOND ACT OF INTUITIONISM

Brouwer introduces the main novelty in the second paragraph. In order to recreate the impression it might have made on the reader at the time, the paragraph is cited here in full:[237]

> Eine *Menge* ist ein *Gesetz*, auf Grund dessen, wenn immer wieder ein willkürlicher Ziffernkomplex der Folge $\zeta$ [N, DH] gewählt wird, jede dieser Wahlen entweder ein bestimmtes Zeichen, oder nichts erzeugt, oder aber die Hemmung des Prozesses und die definitive Vernichtung seines Resultates herbeiführt, wobei für jedes $n$ nach jeder ungehemmter Folge von $n-1$ Wahlen wenigstens ein Ziffernkomplex angegeben werden kann, der, wenn er als $n$-ter Ziffernkomplex gewählt wird, *nicht* die Hemmung des Prozesses herbeiführt. Jede in dieser Weise von der Menge erzeugte Zeichenfolge (welche also im allgemeinen nicht fertig darstellbar ist) heisst ein *Element der Menge*. Die gemeinsame Entstehungsart der Elemente einer Menge $M$ werden wir ebenfalls kurz als *die Menge M* bezeichnen.[238]

The description would not win the prize for didactical clarity. Apart from the style, which makes it difficult to read, there are some other aspects that do not foster understanding. In the first place, Brouwer uses a term, 'set', to define something quite different from what was then (and now) called a 'set'. It would have helped if he had chosen a new name for his new concept. Secondly, another concept is hidden in the definition, one that Brouwer had before placed at the side of formalism: that of a choice sequence. In the sentence following the citation just given, Brouwer speaks about a *'Wahlfolge'*, but he does not give a definition.

Indeed, many people complained about Brouwer's writing style. The German geometer Study commented on this definition as follows:[239]

> Alles, was ich hiervon zu begreifen vermag, ist, dass 'ein Sack Kartoffeln', da er bestimmt kein Gesetz ist, nach Brouwer keine 'Menge' von Kartoffeln sein kan.[240]

The historian of mathematics and science Dijksterhuis later asked Heyting regarding this paper of Brouwer's:[241]

---

[237][Brouwer 1918, p. 150]
[238]'A *spread* [literary: set, DH] is a *law* on the basis of which, if an arbitrary digit complex is chosen from the sequence $\zeta$ [¯, DH] over and over again, each of these choices generates either a specific sign, or nothing, or causes the blocking of the process and the definite destruction of its result, where for every $n$ after each non-blocked sequence of $n-1$ choices at least one digit complex can be indicated which, if chosen as the $n$-th digit complex, does *not* lead to the blocking of the process. Each sequence of signs generated in this way by the spread (which therefore in general cannot be presented as finished) is called an *element of the spread*. We will also denote the common mode of generation of the elements of a spread $M$ as *the spread M* for short.' English translation based on the translation in [Van Stigt 1998, p. 24].
[239]*Prolegomena zu einer Philosophie der Mathematik* ('Prolegomena of a philosophy of mathematics'), unpublished manuscript, p. 18, [BMIM Study]; on the *Prolegomena*, see 6.2.1.
[240]'All that I can understand of this is that 'a sack of potatoes', as it definitely is not a law, cannot be a 'set' of potatoes according to Brouwer.'
[241]Letter from Dijksterhuis to Heyting, 21/12/1927; [TLI Heyting, B dijk1-271221]

## 62    CHAPTER 2. THE GENESIS OF BROUWER'S INTUITIONISM

> (...) schrijft Brouwer werkelijk zoo hoogst onduidelijk of lijkt dit alleen maar zoo, zoolang men niet voldoende intuitionistisch denkt?[242]

Whereupon Heyting answered diplomatically:[243]

> Hoewel het tegen Prof. Brouwer herhaaldelijk gerichte verwijt van onduidelijkheid niet geheel gegrond is, daar ieder van zijn zinnen zijn bedoeling volkomen weergeeft, zijn zijn geschriften door den uiterst gedrongen zinsbouw, die bij het lezen groote concentratie op ieder woord eischt, als inleiding minder geschikt.[244]

Van der Waerden voiced the same complaint about the 'unhealthy concentration of attention'[245] required for reading Brouwer's papers.

Although Du Bois-Reymond and Borel had discussed choice sequences before, Brouwer was the first to actually make use of them in mathematics. He probably got this most powerful insight during the lecture course he gave in 1915–1916, notes of which have survived.[246] In order to clarify Brouwer's spread definition, I give a modern explanation of choice sequences.[247]

One of the easiest ways to understand choice sequences is by reflecting upon what kind of sequences are used in classical (i.e., non-intuitionistic) mathematics. In classical mathematics, a sequence is usually defined by a law. The law determines the elements of the sequence *ad indefinitum*. Sequences of this kind are hereafter called *lawlike sequences*. If one would like to broaden the concept of a mathematical sequence, the opposite concept is a sequence which is not bound by any law at all. In intuitionistic mathematics, this means that we allow for sequences the elements of which are determined by the mathematician who is constructing the sequence mentally. This kind of sequence is called a *lawless sequence*.[248] A lawless sequence is thus never completely determined; however, one always knows an initial segment and, if necessary, the segment can be extended. This means that mathematical operations can be carried out on the known part of the sequence, which suffices for treating them as mathematical concepts. For example, the standard way of defining the sum of two (lawlike) sequences is by means of the sum

---

[242]'(...) does Brouwer really write so extremely unclear, or does it only seem that way as long as one does not think intuitionistically enough?'

[243]Letter from Heyting to Dijksterhuis, 10/1/1928; [TLI Heyting, B dijk1-280110*]

[244]'Although the repeatedly uttered reproach of unclarity directed against Prof. Brouwer is not completely grounded, since each of his sentences represents his intention completely, his writings are less suited as an introduction, because of the extremely terse sentence structure, which demands great concentration on every word when reading.'

[245]'ongezonde aandachtskoncentratie', letter from Van der Waerden to Heyting, 28/5/1925; [TLI Heyting, B wae-250528]

[246][Van Dalen 1999A, pp. 240–241]

[247]The explanation was based on [Heyting 1934], [Troelstra 1969], [Troelstra 1977] and [Troelstra & Van Dalen 1988]. A justification for Brouwer's choice sequences based on Husserl's transcendental phenomenology is given in [Van Atten 1999].

[248]The term 'lawless' was introduced in print only in the 1960s by Kreisel, following a suggestion by Gödel; [Kreisel 1968], [Van Atten 1999, p. 37].

## 2.6. THE SECOND ACT OF INTUITIONISM

of its elements, starting from the first one: let $a = (a_i)$, $b = (b_i)$ be two arbitrary sequences, then the sum $c = a + b$ of these sequences is defined as $c = (c_i)$ with $c_i = a_i + b_i$ for all $i$.[249] This means that if we want to know the sequence $c$, we have to add $a_1 + b_1$ to find $c_1$, $a_2 + b_2$ to find $c_2$, etcetera. That is, we work on initial segments of the sequences. The same can be done with lawless sequences.

The concept of a choice sequence covers the whole area from lawlike to lawless sequences. That is to say: both of these appear as special cases of the general notion of a choice sequence. It can be defined as follows.

**Definition 1** *A choice sequence is a sequence of mathematical objects, in which each time the next element in the sequence is decided upon by the free human subject who is generating the choice sequence. The choice of the next element may be limited by a law, which can depend on the choices made earlier. At every moment in the process, there must be at least one object that can be chosen.*

The claim just made that lawlike and lawless sequences are special cases of a choice sequence can be substantiated by varying the limiting law mentioned in the definition. If the limiting law is taken so restrictive as to allow for only one object to be chosen each time, the choice sequence becomes a lawlike sequence. If the limiting law disappears altogether, the choice sequence becomes a lawless sequence.

If we look back at the citation from Brouwer's paper given above, we can only recognize the making of choices as an indication of the concept of a choice sequence. In later lectures and publications, too, Brouwer almost invariably either did not mention choice sequences separately, or only in direct connection with spreads. Weyl and Heyting were more explicit about it and treated them separately. The late professor De Iongh (1915–1999), a student of Brouwer, stressed that for Brouwer it was not the concept of a choice sequence that mattered, but the possibility of speaking about 'all choice sequences in a spread'.[250] Indeed, in most of Brouwer's published works he only speaks of choice sequences in the context of intuitionistic set theory.

Still, the introduction of choice sequences in intuitionistic mathematics marks a fundamental change in Brouwer's views. Before, he had identified infinite sequences with algorithms or laws of generation. Now, Brouwer had come to accept also infinite sequences as processes of generation in progress.[251]

I now continue my explanation of intuitionistic set theory in modern terms. The basic concept introduced by Brouwer is a 'spread'.[252]

For the purpose of intuitionistic set theory, it suffices to consider choice sequences of natural numbers. It turns out to be helpful to look at these choice

---
[249] Cf. [Kreisel 1968].
[250] Personal communication to the author, fall 1994.
[251] [Van Stigt 1998, p. 12]
[252] The explanation given here was based on [Heyting 1956].

sequences as branches in the universal tree $\mathbb{N}^{<\mathbb{N}}$, consisting of all finite sequences of natural numbers, as is shown in figure 2.1. It should be noted that this heuristic is one of classical mathematics, not of intuitionistic mathematics.

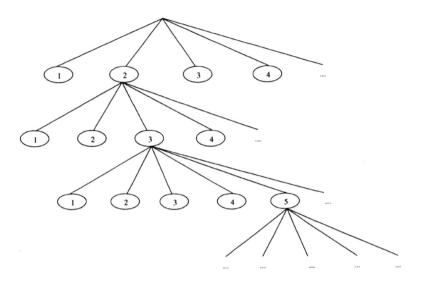

Figure 2.1: *The choice sequence $\langle 2, 3, 5, \ldots \rangle$ as a branch in the universal tree*

A spread can be defined as follows.[253]

**Definition 2** *A spread $M$ consists of two laws, a spread law $\Lambda_M$ and a complementary law $\Gamma_M$.*
*A spread law $\Lambda_M$ is a lawlike characteristic function on $\mathbb{N}^{<\mathbb{N}}$, the elements of which are denoted by $\langle a_1, \ldots, a_n \rangle$, such that:*

1. *For every natural number $k$, $\Lambda_M(\langle k \rangle)$ is decidable;*

2. *If $\Lambda_M(\langle a_1, \ldots, a_n, a_{n+1}\rangle) = 1$, then $\Lambda_M(\langle a_1, \ldots, a_n\rangle) = 1$;*

3. *If $\Lambda_M(\langle a_1, \ldots, a_n\rangle) = 1$, then for every natural number $k$, $\Lambda_M(\langle a_1, \ldots, a_n, k\rangle)$ is decidable;*

4. *If $\Lambda_M(\langle a_1, \ldots, a_n\rangle) = 1$, then at least one natural number $k$ can be found such that $\Lambda_M(\langle a_1, \ldots, a_n, k\rangle) = 1$.*

---

[253]Following [Heyting 1956, pp. 34–35], with modernised terminology suggested by Van Dalen.

## 2.6. THE SECOND ACT OF INTUITIONISM

A complementary law $\Gamma_M$ of a spread $M$ is a rule which assigns a definite mathematical entity to any finite sequence $\langle a_1, \ldots, a_n \rangle$ for which $\Lambda_M(\langle a_1, \ldots, a_n \rangle) = 1$.
A choice sequence $\{a_n\}$ in which, for every $n$, $\Lambda_M(\langle a_1, \ldots, a_n \rangle) = 1$, is called an admissible choice sequence *for the spread $M$*.
The choice sequence which we get by applying the complementary law $\Gamma_M$ to the sequences $\langle a_1 \rangle, \langle a_1, a_2 \rangle, \ldots$ of an admissible choice sequence $\{a_n\}$ of $M$ is called an element *of the spread $M$*.

So, speaking in terms of trees, a spread consists of a method for determining a subtree (without terminating branches) of the universal tree, together with a method for assigning mathematical objects to each initial part of the subtree. Brouwer gives as an example the spread law which allows all sequences composed of positive integers, together with the complementary law which assigns to the sequence $\langle a_1, a_2, \ldots, a_n \rangle$ the element $\frac{1}{2^{a_1}} + \frac{1}{2^{a_1+a_2}} + \ldots + \frac{1}{2^{a_1+a_2+\ldots+a_n}}$, as shown in figure 2.2. In this way, one can generate all real numbers between 0 and 1 (including 1, excluding 0).[254]

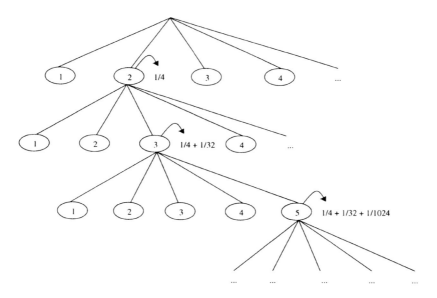

Figure 2.2: *The spread with complementary law* $\Gamma_M(\langle a_1, a_2, \ldots, a_n \rangle) = \frac{1}{2^{a_1}} + \frac{1}{2^{a_1+a_2}} + \ldots + \frac{1}{2^{a_1+a_2+\ldots+a_n}}$

---

[254][Brouwer 1918, p. 156]

66     CHAPTER 2. THE GENESIS OF BROUWER'S INTUITIONISM

Because of the uniqueness of every path in the universal tree, one can depict the mathematical object assigned to the initial part of the subtree from the top to a certain node $a$ in the node $a$. The choice sequences which we get by going through the determined subtree and taking the assigned mathematical objects instead of the natural numbers are the elements of the spread.[255] The spread could be called the intuitionistic counterpart of the classical set: it does not collect mathematical objects that may or may not have been created before, but instead gives a common mode of generation for its elements. This finishes the explanation in modern terms; I now return to Brouwer's paper.

The next concept Brouwer introduces is one which does collect mathematical objects made beforehand: a species. A *species* (of the first order) is a property which only mathematical entities can be supposed to possess, in which case these entities are called a *member* of the species.[256]

The paper continues with a multitude of further definitions, many of them refinements of classical mathematical concepts. For example, Brouwer distinguishes between species which are *abzählbar, zählbar, auszählbar, durchzählbar* and *aufzählbar*, where some of the distinctions are related to the question of decidability.[257]

One of the applications of Brouwer's spread definition concerns the continuum. By giving a truly constructive definition of the continuum, Brouwer succeeded in dropping the continuum intuition he had introduced in his dissertation. This is one in a series of several changes Brouwer made to intuitionistic mathematics while developing it – something that was later sometimes used against him, when some of his opponents claimed that even Brouwer did not know for sure what was intuitively clear. Though such changes may seem a weak point, this need not have disturbed Brouwer. Since Brouwer was a solipsist, the only thing that mattered to him, philosophically speaking, was his present self. Therefore, he could dismiss his old ideas and beliefs without straining his own philosophy too much.

Brouwer presented the intuitionistic continuum in various forms, to various degrees of explicitness and at various occasions.[258] Since the presentation of the real numbers given in the *Begründung* paper was improved in Brouwer's *Mathematische Annalen* paper, I treat the intuitionistic continuum there.[259]

In the second paper on the foundation of set theory independent of the principle of the excluded middle, Brouwer further elaborated his set theory in the area

---

[255][Troelstra 1982, p. 471]
[256][Brouwer 1918, pp. 150–151]
[257][Brouwer 1918, p. 154]; since these terms have not become commonplace in mathematics, no English translations are available.
[258]Cf., for example, [Brouwer 1918, p. 156], [Brouwer 1921A, p. 236], [Brouwer 1925A, p. 251], [Brouwer 1992, p. 28] and [Brouwer 1930, p. 433]. Bishop, who developed constructivism further, remarked on this special interest of Brouwer's: 'In Brouwer's case there seems to have been a nagging suspicion that unless he personally intervened to prevent it the continuum would turn out to be discrete.', [Bishop 1967, p. 6].
[259]See 2.6.2.

## 2.6.2 Further development of intuitionistic mathematics

The years after Brouwer's second act were his most productive ones for intuitionistic mathematics. In the period 1923–1928 alone, Brouwer published 27 papers on intuitionism (translations not included), most of which contained original contributions.[261]

Brouwer's development of intuitionism would deserve an extensive treatment by itself. However, I am here dealing with the genesis of his intuitionism as far as it is relevant to its reception. Since most reactions did not go deeply into intuitionistic mathematics, neither shall I.[262] Instead, I restrict myself to giving an impression of how Brouwer developed intuitionism by treating the main ideas of some of his papers. Being aware that only the top of the intuitionistic iceberg is presented here should suffice to appreciate the gap between intuitionism as Brouwer developed it, and the reactions to it.

Brouwer's papers appeared mostly in KNAW[263] publications, in the *Jahresberichte der deutschen Mathematiker-Vereinigung* and in the *Mathematische Annalen*. About two-thirds of his publications were in German, the other third in Dutch; many papers appeared in both languages. In his titles, Brouwer frequently used the term 'intuitionistic'. Until 1923, the principle of the excluded middle figured prominently in many titles; after that year not anymore, even though Brouwer did not change his position on the excluded middle. It is not clear why Brouwer changed this presentation policy.

In 1919, Brouwer was offered a chair at the university of Berlin and that of Göttingen almost simultaneously. The Berlin call, for Carathéodory's chair, motivated the choice for Brouwer by reference to his topological work, and stressed Brouwer's originality:[264]

> In der Originalität seiner Methoden wird Brouwer von keinem Mathematiker der jüngeren Generation erreicht.[265]

Although Brouwer valued these universities highly, university salaries in post-war Germany were very low. Furthermore, there was a lot of political and social unrest in Berlin. In the end, Brouwer declined both offers. He did buy a house in Berlin

---
[260][Van Stigt 1990, p. 75]
[261]Still, his interest was not one-sided: for example, in the period 1919–1921 he published some 14 papers on the topology of surfaces. Furthermore, a lot of his energy was taken up by the so-called Denjoy conflict from the end of 1920 to the beginning of 1923, [Van Dalen 1999A, pp. 292; 344–356; 367].
[262]Brouwer's intuitionistic work is treated in more detail in [Van Dalen 1999A].
[263]*Koninklijke Nederlandse Akademie van Wetenschappen*, 'Royal Dutch Academy of Sciences'
[264]*Vorschlag* 19/12/1919; cited from: [Biermann 1988, p. 192]
[265]'No mathematician of the younger generation reaches the same originality of methods as Brouwer.'

and remained a regular visitor. Also, he used the offers to obtain extra money from the university of Amsterdam for books and journals.[266] He furthermore campaigned for the establishment of a mathematical centre in Amsterdam, which he called 'my Göttingen'.[267] One of the subjects to be taught would be 'epistemological mathematics': set theory, analysis situs, axiomatics, and the mathematical foundation of natural philosophy.[268] One again notes that for Brouwer, topology and foundations were not far removed from each other.

In a 1919 re-edition of his 'Unreliability of the logical principles' paper, Brouwer maintains that he had not changed his point of view, even though there were few people who supported it.[269]

In the same year, Brouwer published his first intuitionistic paper in a non-Dutch journal after the second act of intuitionism. In the paper *Intuitionistische Mengenlehre* ('Intuitionistic set theory'), which appeared in the *Jahresbericht der deutschen Mathematiker-Vereinigung*, Brouwer gave an introduction to his earlier papers on the foundations of set theory independent of the principle of the excluded middle. An important novelty lies in the use of labels: here, Brouwer for the first time explicitly calls his own point of view 'intuitionistic', and refers to his set concept as a 'constructive set definition'.[270] Van Dalen conjectured that the late use of a label to identify his own view on mathematics stemmed from the desire to reform all of mathematics, instead of claiming a corner of the mathematical empire for himself.[271]

The two main principles his work rests on, Brouwer claims, are the rejection of the axiom of comprehension as a basis for mathematics,[272] and the rejection of the principle of the excluded middle as a mathematical proof tool.[273] He explicates the reasoning already implicit in *De onbetrouwbaarheid der logische principes*[274] that classical logic was abstracted from finite systems, and that it has illegitimately been extended to infinite ones. In infinite systems, the principle of the excluded middle loses its universal validity, since there is no means of checking all possibilities. Brouwer mentions that some classical theorems, such as Cantor's main theorem,[275] are no longer true in their original form in intuitionistic mathematics.[276]

---

[266] [Van Dalen 1999A, pp. 300–304]
[267] [Van Dalen 2001, p. 228]
[268] [Van Stigt 1990, pp. 68–69; 80–84]
[269] [Brouwer 1919B]
[270] '*konstruktive[n] Mengendefinition*', [Brouwer 1919D, p. 230]
[271] [Van Dalen 1999A, p. 281]
[272] The axiom of comprehension states that all things which have a certain property can be united into a set.
[273] Brouwer mentions it as strongly as it is expressed here, not merely that the principle is not universally valid.
[274] See 2.3.2.
[275] Cantor's main theorem holds that, for any set $A$, $2^A > A$.
[276] [Brouwer 1919D, pp. 230–233]

## 2.6. THE SECOND ACT OF INTUITIONISM

**Brouwerian counter-examples** In September 1920, mathematicians, physicists and medical researchers met at the Bad Nauheim *Naturforscherversammlung* ('Meeting of natural scientists'). The meeting was the German counterpart of the Allied-dominated mathematical conference organised at exactly the same time in Strasbourg. The vast majority of the speakers was German; three came from neutral countries, and none from the Allied powers. Some 2500 scientists participated in the conference. The list of participating mathematicians was impressive and included Bernstein, Bieberbach, Brouwer, Hausdorff, Koebe, Landau, E. Noether, Pólya, Schoenflies, Schur, Weitzenböck, and Weyl.[277]

Brouwer lectured on the intriguing question whether every real number has a decimal expansion.[278] Not surprisingly, Brouwer's answer is negative. He maintains that one can only claim that a number has a decimal development if one actually has a method for deciding what the decimal development is. The counter-example he gives against the positive statement is based on our limited knowledge of the decimal development of $\pi$.[279] Since this knowledge can be extended over time, Brouwer remarks that it is possible that these specific counter-examples 'are disposed of'[280] one day. In that case, however, one can always replace them by others.[281] The time-dependency of mathematics, which is a further elaboration of Brouwer's earlier, more philosophical remarks on mathematics as a human, mental construction, is totally opposed to the dominant Platonistic philosophy of mathematics and must have seemed strange to a major part of his audience.[282]

In the same paper, Brouwer gives a presentation of the intuitionistic continuum based on nested intervals, as a one-dimensional specification of the nested quadrats described in his 1919 paper.[283] The intuitionistic continuum figures prominently, at the beginning of the paper. Brouwer uses closed intervals $\lambda_n$ with dual fractions $\frac{a}{2^n}$ and $\frac{a+2}{2^n}$ as end points.[284] Brouwer then defines a point of the continuum as an unlimited sequence of intervals $\lambda_n$, where each time the next interval in the sequence is nested in the former.[285]

As an explanation, I would like to add the following. Firstly, it should be noted that, since Brouwer speaks about dual fractions, $a \in \mathbb{Z}$. Secondly, one can put the sequence generated here more in explicit accordance with Brouwer's definition of an element of a spread.[286] Thus, the choice consists of choosing a pair of numbers $(a, n)$, with $a \in \mathbb{Z}$, $n \in \mathbb{N}$. The spread law then generates the interval $[\frac{a}{2^n}, \frac{a+2}{2^n}]$, if

---

[277] [Van Dalen 1999A, p. 325; 342]
[278] The contribution was published the following year in the *Mathematische Annalen*, [Brouwer 1921A].
[279] See below for more details on Brouwerian counter-examples.
[280] '*hinfällig werden*', [Brouwer 1921A, p. 245]
[281] [Brouwer 1921A, pp. 244–245]
[282] Even so, the remark only played a minor role in the arguments used against Brouwer's intuitionism.
[283] [Brouwer 1919A, p. 191]
[284] In the paper, the $n$th power in the first fraction is missing, presumably due to a misprint.
[285] [Brouwer 1921A, p. 236]. Brouwer remarks that the word 'real number' used in the title of the paper is used in a wider sense here than in his 1918 *Begründung* paper.
[286] See 2.6.1.

this interval is nested in the previous interval in the sequence; if not, the process is blocked. In this way, Brouwer gave a constructive procedure for generating points in the continuum.

At the same Bad Nauheim conference, Brouwer approached Weyl with an offer for a chair at the University of Amsterdam. This was part of Brouwer's plan to build a mathematical institute in Amsterdam comparable to that in Göttingen. Weyl, however, decided to stay in Zürich, where the ETH met all his wishes.[287]

In 1923, Brouwer lectured in Antwerpen at the *Vlaamsch Natuur- en Geneeskundig Congres* ('Flemish Physical and Medical Congress') and in Marburg at the *Jahresversammlung der deutschen Mathematiker-Vereinigung* ('Annual meeting of the German Mathematical Society') on the role of the principle of the excluded middle in mathematics, especially in function theory. Both versions were published, the German one in the *Journal für die reine und angewandte Mathematik*.[288] The idea of using the classical laws of logic to prove the consistency of certain axiomatic systems is without value, Brouwer maintains:[289]

> (...) een door geen weerleggende contradictie te stuiten onjuiste theorie is daarom niet minder onjuist, zooals een door geen reprimeerend gerecht te stuiten misdadige politiek daarom niet minder misdadig is.[290]

Here we see Brouwer's idea re-appear, already expressed in *Leven, kunst en mystiek*, that the sole consistency of a linguistic system does not guarantee any value.[291]

In order to substantiate his claim against the universal validity of the principle of the excluded middle, Brouwer presents a counter-example which hits mathematics more in the heart than his earlier one. It is an example against the trichotomy law, which implies that every real number is either greater than zero or less than zero or equal to zero.[292] In a somewhat more elaborated form, it runs as follows.

Brouwer argues that if the principle of the excluded middle holds, then for any pair of real numbers $a$ and $b$, $a < b$ 'is *either* valid *or* impossible',[293] and the same goes for $a > b$. Thus, for all $a, b \in \mathbb{R}$, either $a < b$, or $a = b$, or $a > b$. Note that there exists a procedure to calculate the decimal expansion of $\pi$.[294] Let $d_m$ be

---

[287] [Van Dalen 1999A, pp. 306–311]
[288] In what follows I cite from the Dutch version, since this was Brouwer's mother tongue.
[289] [Brouwer 1923, p. 3]
[290] '(...) an incorrect theory, even if it cannot be checked by any contradiction that would refute it, is none the less incorrect, just as a criminal policy is none the less criminal even if it cannot be checked by any court that would curb it.' Translation based on [Van Heijenoort 1967, p. 336].
[291] See 2.2.4.
[292] Interestingly, Cantor seems to have agreed with Brouwer on this particular subject. Cantor had admitted that the trichotomy law of cardinals was not proved, and he intentionally refrained from using it, [Moore 1980, p. 102].
[293] 'òf geldt òf onmogelijk is', [Brouwer 1923, p. 3]
[294] As Brouwer had shown in [Brouwer 1921A, pp. 242–243].

## 2.6. THE SECOND ACT OF INTUITIONISM

the $m$th digit in this expansion. One assigns the value $m$ to the variable $k$ if, at $d_m$, it happens for the first time that the sequence from 0 to 9 appears in the decimal expansion of $\pi$, that is, that $d_m d_{m+1} \ldots d_{m+9} = 0123456789$.[295] At the time, such a sequence had not been found, nor was there a proof for the impossibility of its existence.[296] Define

$$c_i = \begin{cases} (-1/2)^k & \text{if } i \geq k, \\ (-1/2)^i & \text{otherwise.} \end{cases}$$

Note that one can, for any value of $i$, decide whether $i \geq k$ by checking if at $d_i$ the sequence 0123456789 does or does not appear (on the assumption that smaller values of $k$ have been checked earlier in the process). Then the infinite sequence $\langle c_1, c_2, c_3, \ldots \rangle$ (as a Cauchy sequence) defines a real number $r$, for which neither $r < 0$, nor $r = 0$, nor $r > 0$ is valid. Thus the principle of the excluded middle is not valid.

This kind of examples, of which Brouwer made more in the years to come,[297] are called weak counterexamples, because they rest upon our inability to prove something (in this example: to prove one of the claims $r < 0$, $r = 0$ or $r > 0$, which are normally considered to hold for all real numbers). The inability, in turn, stems from an unsolved (but not proved unsolvable) problem (in this example: the problem 'does there exist a sequence $01\ldots 9$ in the decimal expansion of $\pi$?'). Since these problems might still prove to be solvable, these specific counterexamples might disappear. But, since Brouwer did not believe that every mathematical problem is solvable, it would always be possible to construct new ones. Later, Brouwer was to make strong counterexamples against the principle of the excluded middle, too.[298]

Brouwer devoted the remaining part of the lecture to pointing out that certain classical theorems, such as the Bolzano-Weierstrass theorem and the Heine-Borel covering theorem, have to be dropped in intuitionism.[299]

---

[295] The definition is somewhat strange, since it is not clear if $k$ exists. In modern terms, one could express this by using the existence predicate $E$, indicating that the object to which it applies exists. Then, Brouwer's definition would read: $E(k) \to k = m$ (cf. [Troelstra & Van Dalen 1988, vol. 1, pp. 50–56]).
Brouwer actually uses $k_n$, to denote the $n$th time that the sequence $01\ldots 9$ appears in the decimal expansion of $\pi$, but only $k_1$ is needed in this example.

[296] One has recently found that the sequence does occur in the decimal development of $\pi$. We now know that we have to assign the value 17,387,594,880 to $k$ ([Borwein 1998]). Thus, since $k$ is even, the $r$ defined below has become bigger than zero, and this is no longer a counter-example. Of course, one can always make a new Brouwerian counter-example by choosing a sequence of which we do not yet know whether it appears in the decimal development of $\pi$ or not.

[297] Brouwer later gave the names 'pendulum number' (*Pendelzahl*), 'fleeing property' (*fliehende Eigenschaft*) and 'solution number' (*Lösungszahl*) to $c$, 'having (for the first time) the sequence 0-9 as the digits $n - n + 9$ in the decimal expansion of $\pi$' and $k_1$, respectively, [Brouwer 1929, p. 161]. These counter-examples apply to what Brouwer later called the reduced continuum, made up of lawlike real numbers; see 2.7.1.

[298] See 2.7.1.

[299] The Bolzano-Weierstrass theorem holds that every infinite, bounded set has a limit point; the Heine-Borel theorem states that all closed and bounded subsets of ¯ are compact, i.e., if a closed and bounded subset $I$ of ¯ is covered by a collection $C$ of open sets, then there is a finite subset of $C$ that covers $I$.

In the same year, Brouwer lectured at a meeting of the Amsterdam Academy of Sciences on the 'intuitionistic splitting of fundamental mathematical concepts', which was published as a short paper. There, Brouwer specifies his criticism of the principle of the excluded middle. Intuitionism not only rejects its general use, Brouwer claims, but also the more specific form of the principle of reciprocity.[300] In this form, the principle of the excluded middle allows one to conclude a statement by the negation of its negation. In a footnote, Brouwer remarks that Bernays pointed out to him that the principle of the excluded middle is equivalent to the principle of reciprocity.[301]

As a counterexample against the principle of the excluded middle, Brouwer again uses the number $r$ defined above by means of the decimal expansion of $\pi$. He asks whether $r$ is rational. Note that, if we find a value for $k$, the sequence $\langle c_1, c_2, c_3, \ldots \rangle$ stops at a rational value. Therefore, we cannot claim that $r$ is not rational, since this would mean that $k$ does not exist, in which case $r$ would be equal to zero and thus rational. However, we cannot claim that $r$ is rational either, since we cannot compute two integers $p$ and $q$ of which $r$ is the quotient. Therefore, we cannot conclude that $r$ is rational from the fact that it is impossible that $r$ is not rational, and thus the principle of reciprocity fails.

The fact that in intuitionism a double negation does not cancel, does not mean, however, that one is left with infinite sequences of negations. As Brouwer points out in the paper, the following theorem, which is given in his formulation, holds intuitionistically:[302]

**Theorem 1** *Absurdity of absurdity of absurdity is equivalent to absurdity.*

As Brouwer points out, the theorem is easily proved in the following way:
$\Rightarrow$ If a proposition $\psi$ follows from a proposition $\varphi$, then the absurdity of $\varphi$ follows from the absurdity of $\psi$. Thus, since absurdity of absurdity follows from correctness, absurdity of absurdity of absurdity implies absurdity.
$\Leftarrow$ Since the correctness of any statement implies the absurdity of its absurdity, especially the correctness of the absurdity of a statement, i.e., its absurdity, implies the absurdity of the absurdity of its absurdity.

Compared to present-day standards, the absence of logical symbolism in Brouwer's writings is striking. At the time, symbolic language for logic was available, for example in the works of Russell and Whitehead. But Brouwer remained true to his preference for formula-free writing.[303]

---

[300]This is also called *reductio ad absurdum*. In the formal language of propositional logic it reads: $(\neg \varphi \rightarrow \bot) \rightarrow \varphi$.
[301][Brouwer 1925E, p. 276]
[302][Brouwer 1925E, p. 277]
[303]See 2.4.

## 2.6. THE SECOND ACT OF INTUITIONISM

**New Results** The year 1924 meant an important breakthrough in the development of intuitionistic mathematics.[304] In this year, Brouwer was able to harvest results of the new concepts he had elaborated the years before. In a short paper published in the KNAW Proceedings,[305] Brouwer proved a positive result that classical mathematics could impossibly prove: that every full function is uniformly continuous.[306]

In the proof, Brouwer makes use of what is now called a bar.[307] In order to understand what a bar is, one should note that a function $f : \mathbb{N} \to \mathbb{N}$ can be seen as a path in the universal tree by considering the function as a sequence $\langle f(0), f(1), \ldots \rangle$. A bar $B$ can then be defined as a set of nodes in the universal tree of all finite sequences of natural numbers such that for each function from $\mathbb{N}$ to $\mathbb{N}$ there is an initial segment of the function that lies in $B$. Brouwer used two fundamental insights in the proof. The first is that for every function $F$ from the set of all choice sequences of natural numbers to $\mathbb{N}$ there exists a bar $B$ such that $F$ is determined by the initial segments of the choice sequences in $B$. The second is that every thin bar, i.e., every bar of which only the 'highest' node in the same branch is taken, can be defined inductively.[308]

Despite the extraordinary power of the uniform continuity result, it is to be doubted how many people actually were aware of it. Brouwer not only worked within a set theory that very few people knew about, he also made only limited use of formalisation and used no drawings at all, features which did not facilitate understanding the paper.

In the same paper, Brouwer treats another very important theorem, which he later called the 'fundamental theorem of finite sets' (spreads). The theorem, which became known as the 'fan theorem' after Brouwer had re-labelled it in his post-World War II lectures,[309] is, as the name indicates, a statement about finite spreads or, if we look at sets as trees, about finitary branching trees. It states that if a natural number $n(e)$ is assigned to every element $e$ of a finite spread, then one can determine a natural number $z$ such that $n(e)$ is determined by the first $z$ choices of $e$. In others words: if there is a bar $B$ such that every path of the tree is cut off by $B$ at $n(e)$, then there is a $z \in \mathbb{N}$ such that for all paths in the tree there is a natural number $x \leq z$ such that the path is cut off by $B$ at place $x$.[310] The fan theorem is a direct corollary of the bar theorem, and plays a fundamental role in the development of intuitionistic mathematics. Apparently, Brouwer was not satisfied with its proof. Van Stigt pointed out that all of Brouwer's attempts at a systematic intuitionistic reconstruction of analysis abruptly end at this point.[311]

Also in 1924, Brouwer lectured for the Göttingen Society on the consequences

---

[304] Cf. [Van Dalen 1999A, p. 376]
[305] [Brouwer 1924D2]
[306] A full function is a function that is defined everywhere on its domain.
[307] In the original paper, Brouwer does not introduce a name for the concept.
[308] A more detailed explanation, in modern notation, is given in [Van Dalen 1999A, pp. 377–381].
[309] 'fan' means 'finitary spread'
[310] [Brouwer 1924D2, p. 289], [Van Dalen 1999A, pp. 381–382]
[311] [Van Stigt 1990, p. 93]

## 74   CHAPTER 2. THE GENESIS OF BROUWER'S INTUITIONISM

of the intuitionistic point of view in mathematics. After the lecture, Hilbert is reported to have stood up and proclaimed that the goal is to obtain more, not less theorems.[312]

In 1925, Brouwer started a series of expository papers on intuitionism in the *Mathematische Annalen*, which, because of the status of the journal, contributed substantially to the respectability of intuitionistic mathematics. The first of these, *Zur Begründung der intuitionistischen Mathematik. I* ('On the foundation of intuitionistic mathematics. I'), was an updated version of his earlier *Begründung* paper. The most striking differences between the two papers lie in the terminology. The difference in the titles is clear: the paper is now presented as 'intuitionistic', whereas before stress was laid on the independence of the principle of the excluded middle. In the paper itself, many of the statements that had before been formulated in terms of 'there exists' are now stated as 'if one is certain' or 'can be pointed out'. Presumably, Brouwer also wanted to adapt the classical mathematical terms to his intuitionistic ideas. This reminds one of Mannoury's more human-oriented style.[313]

As one of the examples of intuitionistic set theory Brouwer presents the intuitionistic continuum. As a spread law,[314] Brouwer gives the law which generates the real number $\frac{1}{2^{a_1}} + \frac{1}{2^{a_1+a_2}} + \frac{1}{2^{a_1+a_2+a_3}} + \ldots$, once the natural numbers $a_1, a_2, a_3, \ldots$ are chosen. Thus, by choosing each next natural number one further specifies the real number. In this way, Brouwer claims, one obtains a generating tool for all dually developable real numbers $> 0$ and $\leq 1$.[315] In the 1918 version, Brouwer had used the same example as generating *all* real numbers $> 0$ and $\leq 1$.[316] Since he has widened his concept of real number,[317] however, this example has now come to indicate only part of the real numbers.

In papers not treated here, Brouwer further elaborated the intuitionistic reconstruction of mathematics. Among other things, he gave new definitions of common concepts in analysis and algebra, new proofs of classical theorems, which sometimes had to be re-formulated, and he developed new ordering systems for the intuitionistic continuum. Brouwer also gave an intuitionistic version of the dimension concept he had earlier developed in classical topology.[318] Finally, Brouwer published an intuitionistic proof of the fundamental theorem of algebra together with his Ph.D. student De Loor.[319]

---

[312] [Van Dalen 2001, p. 305]

[313] See 2.2.3.

[314] See 2.6.1.

[315] [Brouwer 1925A, p. 251]. A real number $a$ is called dually developable if for every finite dual fraction $e$ the relation $a > e$, i.e., $a$ is separated from $e$ by a finite dual fraction $e_1 > e$, is decidable.

[316] [Brouwer 1918, p. 156]

[317] See 2.6.2.

[318] Cf. [Van Dalen 1999A, pp. 388–392]

[319] [Brouwer & De Loor 1924]. The fundamental theorem of algebra holds that every complex

## 2.7 The Brouwer lectures

In the end of the 1920s, Brouwer gave a number of influential lectures on intuitionism.

### 2.7.1 Berlin

In 1927, Brouwer was invited to give a series of lectures on intuitionism in Berlin. By that time, Berlin was not only the official capital of Weimar Germany, but it had also developed into its cultural capital, one of the liveliest and most modern cities in Europe.[320] Brouwer's lectures started in January and lasted until halfway March. They show the extent to which Brouwer, almost single-handedly, had developed intuitionism.[321] Publisher Walter de Gruyter had agreed to publish the lectures, but although Brouwer worked on the manuscript for many years, the idea never materialised.[322]

The different versions of the manuscript that have been left reflect the changing way Brouwer looked at intuitionism. In the original version, the opening words are:[323]

> Bevor ich Ihnen über einige Gegenstände der intuitiven oder direkten Mathematik vortrage, will ich einiges über die Vorgeschichte des Intuitionismus auseinandersetzen. Sie ist identisch mit der Geschichte der Anschauungen über den Ursprung der Exaktheit der Mathematik; nämlich einerseits über die Existenz oder nicht-Existenz einer sei es objektiven, sei es intuitiven oder aprioristischen Grundlage derselben, — andererseits über die Rolle, welche die menschliche Sprache und die Logik in bezug auf den exakten Charakter der Mathematik spielen.[324]

In the last version of the manuscript that Brouwer worked on, these words have been changed into:[325]

---

polynomial equation has a solution in ¯. Weyl gave an intuitionistic proof of it in the same year, [Weyl 1924], independently of Brouwer and De Loor, as did Skolem, [Skolem 1924].
    The peculiar way in which De Loor obtained his Ph.D. is recounted in [Van Dalen 1999A, p. 404].

[320][Gay 1969, pp. 134–139]

[321]The survey presented here was based on the posthumously published version of the Berlin Lectures, [Brouwer 1992]. This version supposedly reflects the original lectures to a large extent, since most of the changes Brouwer made in the manuscript were stylistic.

[322][Van Dalen 1992, pp. 7–8]

[323][Brouwer 1992, p. 59]

[324]'Before I lecture to you about some objects of intuitive or direct mathematics, I would like to explain something about the history of intuitionism. It is identical with the history of the views on the origin of exactness in mathematics, namely on the one hand about the existence or non-existence of a beit objective, beit intuitive or a priori foundation of it, — on the other hand on the role which human language and logic play with respect to the exact character of mathematics.'

[325][Brouwer 1992, p. 19]

76    CHAPTER 2. THE GENESIS OF BROUWER'S INTUITIONISM

> Der Intuitionismus hat seine historische Stellung im Rahmen der Geschichte der Anschauung *erstens* über den Ursprung der mathematischen Exaktheid; *zweitens* über die Umgrenzung der als sinnvoll zu betrachtende Mathematik.[326]

The second version never reached the public during Brouwer's life-time,[327] but it does reflect a shift in the presentation of intuitionism that Brouwer had in mind. In both versions, Brouwer singles out two characteristics in which intuitionism deviates from classical mathematics. In the original version, these are the view on mathematical existence and on logic — exactly the two main themes of the foundational debate. In the later version, Brouwer replaced these by two more abstract characteristics: the view on mathematical exactness and on the meaning of mathematics. Indeed, the contentual character of mathematics was a theme which became more important in the course of the debate.[328] It is not clear whether Brouwer, in changing his characterisation, was following the shifting emphasis of the foundational debate.

Brouwer starts his lectures with a historical survey of intuitionism, mentioning Kronecker, Poincaré, Borel and Lebesgue as 'pre-intuitionists'. For intuitionism proper, he distinguishes two main 'acts'.[329] The first act of intuitionism was the separation between mathematics and mathematical language.[330] Brouwer characterises mathematics as 'a languageless construction carried out by the human mind'.[331] Since there is no mathematics outside the human mind, there is no mathematical truth outside it either. The second act of intuitionism is its construction of sets.[332]

After this, Brouwer repeats some of his earlier set theoretical works, again with formulations such as 'if one is sure that' and 'if we are in the possession of a means to'. He defines the mathematical continuum in a way analogous to his Bad Nauheim paper,[333] distinguishing between the (full) continuum made of real numbers based on choice sequences and the reduced continuum made of lawlike real numbers,[334] and presents different orderings on it.

Brouwer again explains the fundamental theorem of finitary spreads, and uses

---

[326]'The historical position of intuitionism lies in the framework of the history of the views *in the first place* on the origin of mathematical exactness; *in the second place* on the definition of what mathematics is to be considered meaningful.'

[327]It bears some resemblance, however, to the later, again changed presentation which Brouwer gave in lectures delivered in South Africa in 1952, where he singled out as the two main factors *'die oorsprong van die matematiese sekerheid'* ('the origin of mathematical certainty') and *'die begrensing van die objek van die matematiese wetenskap'* ('the delimitation of the object of mathematical science'), [Brouwer 1956, p. 186]; English translation cited from [Brouwer 1952, p. 508].

[328]See 3.3.1.

[329]*'Handlungen'*, [Brouwer 1992, pp. 21–23]

[330]This step was carried out in Brouwer's dissertation; see 2.3.1.

[331]*'eine vom menschlichen Geiste vollzogen sprachlose Konstruktion'*, [Brouwer 1992, p. 21]

[332]See 2.6.

[333]See 2.6.2.

[334][Brouwer 1992, p. 28]

## 2.7. THE BROUWER LECTURES

it to prove the indecomposability of the continuum, i.e., the theorem that states that if the continuum is split into two sets, one of them has to be empty, while the other is the full continuum.[335] He applies this to do something new: to construct a strong counter-example against the generalised principle of the excluded middle. This principle says, in particular, that every real number[336] is either rational or irrational. Brouwer does not further elaborate this point, but it is clear that if every number is either rational or irrational, we can split the continuum into two, which contradicts its indecomposability. Here, for the first time, Brouwer does not give a weak counter-example, based on our incomplete knowledge, but a strong one. He explicitly mentions the 'contradictoricity'[337] of this consequence of the principle of the excluded middle, which proves that, for intuitionistic mathematics, this principle is actually false.[338]

Brouwer finishes his lectures with some elementary theorems that do not hold in intuitionistic mathematics. Among these are the theorem of the existence of a maximum for every full function on the unit interval; the Bolzano-Weierstrass theorem; the fundamental theorem of algebra; and Brouwer's own topological fixed-point theorem.[339]

### 2.7.2 Amsterdam

In December 1927, Brouwer lectured before the *Koninklijke Nederlandse Akademie van Wetenschappen* ('Royal Dutch Academy of Sciences') about 'intuitionistic reflections on formalism'.[340] In the lecture, which was published both in German and, in a summarized form, in Dutch,[341] Brouwer tries to build a bridge to formalism:[342]

> Die Richtigkeitsdifferenzen zwischen der formalistischen Neubegründung und dem intuitionistischen Neubau der Mathematik werden beseitigt sein, und die Wahl zwischen beiden Beschäftigungen sich auf eine Geschmacksangelegenheit reduzieren, sobald die folgenden in erster Linie auf den Formalismus bezüglichen, aber in der intuitionistischen Literatur zuerst formulierten Einsichten allgemein durchgedrungen sein werden. Dieses Durchdringen ist deshalb nur eine Zeitfrage, weil es sich um reine Besinnungsergebnisse handelt, die kein diskutables Element

---

[335] Van Dalen later proved the stronger result that even a 'perforated' continuum is indecomposable, cf. [Van Dalen 1999B].
[336] Brouwer speaks about every 'infinite dual fraction'.
[337] '*Kontradiktorität*'
[338] [Brouwer 1992, pp. 49–50]
[339] [Brouwer 1992, pp. 53–57]
[340] '*Intuitionistische Betrachtungen über den Formalismus*'
[341] The Dutch summary appeared in the *KNAW Verslagen* 36 (1927), p. 1189. I cite from the integral German version.
[342] [Brouwer 1928, p. 375]

enthalten und zu denen jederman der sie einmal verstanden hat, sich bekennen muss.[343]

The statement is rather bold, reminding one of Schopenhauer, one of Brouwer's favourite philosophers. He had used a similar reasoning, when he claimed that *''Die Welt ist meine Vorstellung' — ist (...) ein Satz, den Jeder als wahr erkennen muß, sobald er ihn versteht'*,[344] whereas philosophers had been in dispute over the question for centuries. Likewise, it is hard to imagine that Brouwer meant these words seriously, since the four 'insights' he mentioned were precisely points that had been under discussion during the foundational controversy. If he meant them seriously, it shows that he had little compassion with his opponents, which is not impossible.

Note the distinction Brouwer makes between formalism, which tries to provide new foundations for existing mathematics, and intuitionism, which aims at reconstructing mathematics, implying that the latter leaves its foundations untouched.

**The four insights** The four insights which, in Brouwer's view, would suffice to end the foundational debate, are the following.[345] Note that, in Brouwer's presentation, the insights are as much about priority claims as they are about content.

1. In the first place, the recognition that formalism divides mathematics into an inventory of mathematical formulas and an intuitive or meaningful theory about these formulas, where the intuitionistic theory of natural numbers is indispensable for the latter.[346] Brouwer claims priority for this 'insight', referring to his dissertation, where he had spoken about mathematical language and mathematics of the second order. In addition, Brouwer states that he had informed Hilbert about this idea orally as early as 1909. The literature Brouwer refers to, which contains only writings of himself and of Hilbert, clearly gives priority to Brouwer for the concept. There is no further evidence of the oral transmission from Brouwer to Hilbert, and Hilbert has never denied it. Thus, it seems that Brouwer justly claimed priority for the first insight.

2. Secondly, Brouwer mentions the insight that one should reject the thoughtless use of the principle of the excluded middle, and that its validity in intuitive mathematics does not extend beyond finite systems. According to Brouwer, this was recognized for the first time in the formalistic literature in 1922 in Hilbert's

---

[343]'The disagreement over which is correct, the formalistic way of founding mathematics anew or the intuitionistic way of reconstructing it, will vanish, and the choice between the two activities be reduced to a matter of taste, as soon as the following insights, which pertain primarily to formalism but were first formulated in the intuitionistic literature, are generally accepted. The acceptance of these insights is only a question of time, since they are the results of pure reflection and hence contain no disputable element, so that anyone who has once understood them must accept them.' Translation from [Van Heijenoort 1967, p. 490].

[344]' 'The world is my representation' — is (...) a proposition that everybody has to recognize as true, as soon as he understands it', [Schopenhauer 1844, p. 10]

[345][Brouwer 1928, pp. 375–377]

[346]Indeed, later modifications of Hilbert's proof theory not only included the intuitionistic theory of natural numbers on the meta-mathematical level, but even more; see 3.3.2 and 4.4.3.

## 2.7. THE BROUWER LECTURES

*Die logischen Grundlagen der Mathematik.*[347] Since Brouwer had already rejected it in 1908 in *De onbetrouwbaarheid der logische principes*, he claims priority for this as well. Again, the priority claim is correct. Whether Hilbert had taken over this point from intuitionism is another question.[348] Note that this is not what Brouwer claims; he only maintains that it was formulated first in the intuitionistic literature.

The first two insights, Brouwer rightly claims, have been taken over by formalism. The following two have not.

3. The third insight states that the principle of the excluded middle and the principle of the solvability of every mathematical problem are one and the same.

4. Finally, Brouwer mentions the recognition that the justification of formalistic mathematics by means of a consistency proof contains a vicious circle, since it rests on the assumption that the correctness of a proposition follows from its consistency, which is based on the (contentual) correctness of the principle of the excluded middle.

Brouwer finishes the list with an appeal to formalism:[349]

> Nach dem Vorstehenden hat der Formalismus vom Intuitionismus nur Wohltaten empfangen und weitere Wohltaten zu erwarten. Dementsprechend sollte die formalistische Schule dem Intuitionismus einige Anerkennung zollen, statt gegen denselben in höhnischem Ton zu polemisieren und dabei nicht einmal die richtige Erwähnung der Autorschaft einzuhalten.[350]

However, that was not the way the story would end.

### 2.7.3 Vienna

In March 1928, on Menger's recommendation, Brouwer was invited by the *Komitee zur Veranstaltung von Gastvorträgen ausländischer Gelehrter der exakten Wissenschaften* ('Committee for the organisation of invited lectures by foreign scientists') to lecture in Vienna. He delivered two lectures, the first of which was

---

[347] See 5.2.2.

[348] Regarding the more general question of Brouwer's influence on Hilbert's foundational views, Sieg argued that 'Hilbert's program was not created at the beginning of the twenties solely to counteract Brouwer's intuitionism, but rather emerged out of broad philosophical reflections on the foundations of mathematics and out of detailed logical work', [Sieg 1999, p. 1]. The parts of Hilbert's unpublished foundational lectures that Sieg uses in his argumentation indeed show a larger concern on Hilbert's side regarding the foundations of mathematics and a more open attitude than what is known from his published works. The forthcoming publication of a large selection of Hilbert's *Nachgelassene Schriften* on the foundations of mathematics and natural sciences should throw more light onto this question.

[349] [Brouwer 1928, p. 377]

[350] 'According to what precedes, formalism has received nothing but benefits from intuitionism and may expect further benefactions. The formalistic school should therefore accord some recognition to intuitionism, instead of polemicizing against it in sneering tones while not even observing proper mention of authorship.' Translation based on [Van Heijenoort 1967, p. 492]

called *Mathematik, Wissenschaft und Sprache* ('Mathematics, science and language'). The address was Brouwer's first more philosophical one after a series of mathematical-intuitionistic lectures and papers. Brouwer does not put forward many new ideas, but gives a refined and more coherent exposition of intuitionism. He repeats the views on time and causality from his dissertation, he explains the primordial intuition, he expresses some rudimentary psychological and sociological ideas, and he restates his objections against classical logic as mentioned before in *De onbetrouwbaarheid der logische principes*. His judgement of formalism is rather mild: he describes its view on mathematics as one in which 'the contents of mathematical concepts and of the relations between these concepts are not further discussed'.[351]

Intuitionism is introduced as putting forward 'the extra-linguistic existence of pure mathematics'.[352] Next, Brouwer gives some weak counter-examples to the principle of the excluded middle. He declares that if intuitionism gains ground, substantial parts of mathematics will have to be given up while other parts have to be constructed in a different way.[353]

Towards the end of the lecture, Brouwer relaxes his attitude towards formalism even more. He acknowledges that, on the basis of the intuitionistic insights, one can not only develop correct theories, without using the principle of the excluded middle, but also non-contradictory ones, in which the principle of the excluded middle can be used. In order to do so, one should mechanise the language of the 'intuitionistic-non-contradictory mathematics'. However, since there are strong counter-examples against the unlimited use of the principle of the excluded middle, one should be cautious in doing so.[354] These remarks build on the line set out in *Intuitionistische Betrachtungen über den Formalismus*, where Brouwer had already recognised two forms of mathematics next to each other.[355] In the Vienna lecture, however, he sees a bigger role for intuitionism in the 'non-contradictory' mathematics. This marks that Brouwer had moved further away from his original position, when he saw the difference between intuitionism and formalism as almost unsurpassable.[356]

Brouwer's second Vienna lecture, *Die Struktur des Kontinuums* ('The structure of the continuum'), delivered four days later, was somewhat more technical. In this lecture, Brouwer first presents the formalistic and 'old-intuitionistic' (Poincaré, Borel) views on the continuum, before moving on to intuitionism. It is worth noting that, in doing so, Brouwer speaks about the 'discovery' of non-Euclidean geometry.[357] The use of this term contradicts his view, which he held during all his life, that mathematics is a human construction. Brouwer was usually quite precise in the choice of terms; this may have been a slip of the pen.

---

[351]'*Der Sinn der mathematischen Begriffe und Begriffsverknüpfungen wird dabei nicht näher erörtert*', [Brouwer 1929, p. 424]
[352]'*die außersprachliche Existenz der reinen Mathematik*', [Brouwer 1929, p. 424]
[353][Brouwer 1929, p. 427]
[354][Brouwer 1929, p. 428]
[355]See 2.7.2.
[356]See 2.5.
[357][Brouwer 1930, p. 430]

In intuitionism, Brouwer explains, one should distinguish between the reduced continuum and the full continuum. In the definition of the full continuum Brouwer does not use the word 'choice sequence'. He claims that the intuitionistic continuum does justice to the old claim by Kant and Schopenhauer that the continuum is pure intuition a priori.[358] Brouwer finishes the lecture by mentioning some results on the ordering of the continuum and its inseparability.[359]

Both lectures, thus, contained hardly anything new. However, their influence on such important thinkers as Wittgenstein and Gödel should not be underestimated.[360]

## 2.8 The *Mathematische Annalen* and afterwards

**The Bologna Congress** The development of Brouwer's intuitionism was dealt a definitive blow in 1928 in what Einstein called the *Frosch-Mäusekrieg*.[361] In several lectures, starting in 1921, Hilbert had opposed Brouwer's views in most polemic terms.[362] This was in sharp contrast to the period before, when Brouwer and Hilbert had had a cordial relationship. Until 1919, Brouwer wrote Hilbert about the nice walks they had had together in the Dutch dunes, the knowledge he had obtained from Hilbert, and he assured him of his friendship.[363] In that year, Brouwer, together with Hk. de Vries and J. de Vries, put forward Hilbert's candidacy for the Dutch Academy of Sciences, describing his mathematical theories as adding 'monuments of crystalline simplicity to the spiritual property of humanity.'[364]

The prelude for the final clash between Brouwer and Hilbert lay in the International Congress of Mathematicians at Bologna, held in September 1928. In order to understand the situation, it should be pointed out that after the First World War an international scientific organisation had been set up which excluded scientists of certain nationalities. This was the *Conseil International de Recherches* ('International Council of Research'), set up by the Central Powers as a branch

---

[358]Posy put forward an interesting Kantian framework for Brouwer's intuitionism, in which the continuum takes a central place, cf. [Posy 1998, pp. 308–314].
[359][Brouwer 1930, pp. 432–439]
[360]See 4.4.2 and 5.4.2.
[361]Letter from Einstein to Born, 27/11/1928; [JNUL Einstein, 8-184]; [Van Dalen 1990, p. 26]. 'The frog and mice battle' was a Greek parody of Homer's Iliad. The first public appearance of this term in the debate is in [Study 1929, p. 4]. I do not know whether Study got the term from Einstein; they had been in contact since 1918. A full account of the events, based on correspondence of the people involved, can be found in [Van Dalen 1990]. The report given here was based on Van Dalen's paper, unless stated otherwise.
[362]See the following chapters.
[363]Letters from Brouwer to Hilbert, 28/10/1909, 28/8/1918 and 20/8/1919, [MI Brouwer, CB.DHI.6, CB.DHI.23, CB.DHI.25]
[364]'*monumenten van kristallijnen eenvoud aan het geestelijk bezit der menschheid heeft toegevoegd*', nomination proposal David Hilbert for '*Buitenlandsch Lidmaatschap der Akademie*' ('Foreign Membership of the Academy'), [Van Dalen 1999A, p. 305]

of the League of Nations. The organisation had a statutory boycott of German, Austrian, Bulgarian and Hungarian scientists.[365] Thus, Germany was excluded from international conferences, there was a ban on the German language in scientific discourse, and institutes were reallocated to countries of the allied part of the world.[366] Brouwer had campaigned against this situation at several occasions. In 1926, Germany was admitted as a member of the League of Nations. The Bologna congress was the first post-war international mathematical conference to which Germans were admitted. Although they were allowed to give lectures and the like, their status was that of observers, not of full members. Brouwer had lobbied hard to obtain full membership for them, but the lobby had failed. In the spring of 1928, he wrote a circular letter to mathematicians planning to attend the Bologna Congress, calling upon them to reconsider their participation.[367]

The argumentation Brouwer uses in the letter is based upon the organising institutions behind the Congress and the intolerant attitude that some mathematicians associated with those institutions had manifested. The 1928 International Congress of Mathematicians, with Benito Mussolini as one of the presidents of the honorary committee,[368] was organised by the *Union Mathématique Internationale*, the mathematical part of the *Conseil International de Recherches*. However, at the time there was some unclarity about this situation. As Brouwer states in his letter, the *Union Mathématique Internationale* was mentioned in earlier announcements of the Congress, but left out in later ones. Brouwer reminds his colleagues of the words Painlevé, a leading mathematician and president of the French *Académie des Sciences*, had used to introduce the *Conseil International de Recherches* in 1919. At that occasion, Painlevé had spoken about science on the other side of the Rhine as 'a gigantic enterprise where a whole people, with a patient servility, was trying to the utmost to produce the most formidable killing machine that had ever existed'.[369] And these words, Brouwer underlines, have not been withdrawn. Brouwer finishes his letter with an appeal to the individual responsability of each mathematician whether or not to attend the Congress.

Bieberbach supported Brouwer in an open letter.[370] Some German and Austrian mathematicians followed Brouwer's example and stayed away from the conference, and there was no official delegation from Berlin.[371]

---

[365] [Van Stigt 1990, p. 85], [Zanichelli 1928, p. 5]
[366] [Van Dalen 1999A, p. 340]
[367] The matter will be dealt with in more detail in the second part of Van Dalen's Brouwer biography.
[368] [Zanichelli 1928, p. 23]
[369] 'une gigantesque entreprise où tout un peuple, avec une patiente servilité, s'acharnait à fabriquer la plus formidable machine à tuer qui ait jamais existé', letter privately printed by Brouwer; [MI Brouwer]
[370] [Siegmund-Schultze 1993, p. 51]
[371] At this point, the old Berlin–Göttingen rivalry may have played a role. As an indication of the powerful position Berlin had had, it may be pointed out that in the 1870s almost all the mathematical professorships in Prussia were held by graduates of the Berlin school ([Rowe 2000, p. 56]). By the end of the 1920s, however, Göttingen had achieved fame not only nationally but also internationally, and a mathematical institute was being constructed, supported by the Rockefeller foundation.

## 2.8. THE MATHEMATISCHE ANNALEN AND AFTERWARDS

Hilbert strongly opposed the boycott. As he wrote in a concept for his circular Bologna letter, he saw the boycott as a form of political oppression, headed by Brouwer:[372]

> In Deutschl[and] ist ein polit[isches] Erpressertum schlimmster Sorte entstanden[:] Du bist kein Deutscher, der deutsch[en] Geburt unwürdig, wenn Du nicht sprichst und handelst, was ich Dir jetzt vorschreibe. Es ist sehr leicht, diese Erpresser loszuwerden. Man braucht sie nur zu fragen, wie lange sie im deutschen Schützengraben gelegen haben. Leider sind aber deutsche Math[ematiker] diesem Erpresserthum zum Opfer gefallen[,] z.B. Bieberbach. Brouwer hat es verstanden diesen Zustand d[er] Deutschen sich zu Nutze zu machen u[nd] ohne (selbst?) sich im deutsch[en] (Schützengr[aben]??) sich [sic] zu betätigen, desto mehr zum Aufhetzen u[nd] zum Zwiespalt der Deutschen zu sorgen[,] um sich zum Herrn über d[ie] deutsch[en] Math[ematiker] aufzuspielen. Mit vollem Erfolg. Zum zweiten Mal[373] wird es ihm nicht gelingen.[374]

In the end, a substantial German delegation followed Hilbert to Bologna, and in total the German scientists came second, outnumbered only by the Italians.[375]

That, however, was not the end of the affair, but rather led to a series of events that was to finish the Brouwer-Hilbert-controversy by power politics. These events centered around the editorial board of the *Mathematische Annalen*.

**The *Mathematische Annalen*** Since 1915, Brouwer had been a member of the editorial board of the *Mathematische Annalen*, at the time the most prestigious mathematical journal inside and outside Germany. Brouwer was one of the 'associate editors', the group of people mentioned on the cover under the heading *unter Mitwirkung von* ('with cooperation of'). The 'chief editors', those mentioned under *Gegenwärtig herausgegeben von* ('presently published by'), were Hilbert, Einstein,

---

[372] cited from [Schappacher & Kneser 1990, p. 57] with small adaptations as indicated by Schappacher afterwards (electronic communication from Schappacher to the author, 22/10/2001; orginal in [NSUB Hilbert, Cod. Ms. Hilbert 494, 18/1-2, Anlage zu seinem Rundbrief in Sachen Bologna vom 29.6.1928]

[373] The first time, to which Hilbert implicitly refers, is probably, as Van Dalen pointed out to me, the Riemann affair, in which Brouwer succeeded in not having any French authors included in the *Mathematische Annalen Riemann Festschrift*.

[374] 'In Germany, a political blackmail of the worst sort has come into being. You are not German, German birth unworthy, if you do not speak and act as I now prescribe you. It is very easy to get rid of this blackmail. One need only ask them how long they have laid in a German trench. Unfortunately, German mathematicians have fallen victim to this blackmail, e.g., Bieberbach. Brouwer has understood how to make use of this situation of the Germans, and without himself giving a hand in German trenches, causing all the more incitement and discord among the Germans, in order to pretend being the master of German mathematics. With all success. He will not succeed a second time.'

[375] [Zanichelli 1928, p. 28; p. 63]

Blumenthal and Carathéodory. Brouwer was an active collaborator to the *Annalen*, and did his refereeing work in a most detailed way.[376]

When Hilbert returned from the Bologna Congress, he wrote a letter to Brouwer (with copies to the other chief editors) with the short message that, 'given the incompatibility of our views on fundamental matters',[377] Brouwer was dismissed as a member of the editorial board. However, Hilbert had not discussed this with the other editors, let alone that they had consented to his action. Carathéodory, who was a friend of Brouwer, was sent to Laren and arrived there before Brouwer, who had been warned by telegram, had opened Hilbert's letter. Even though Carathéodory disagreed with Hilbert's plan, he had to save the situation and he informed Brouwer about the intention of the chief editors to remove him from the editorial board. The main argument he gave was that Hilbert wished so, and that Hilbert's illness required giving in to him. By that time, Hilbert was indeed seriously ill again.[378]

The reasons Hilbert gave for dismissing Brouwer can be read in more detail in a letter he sent to Einstein, asking for his support. Hilbert states three objections to Brouwer's continued presence in the editorial board: Hilbert felt insulted by Brouwer, especially by his circular 'Bologna' letter; Hilbert thought that Brouwer held a hostile position against sympathetic foreign mathematicians; and Hilbert feared that Brouwer would take over the *Mathematische Annalen* after Hilbert had left, whereas he wanted to keep Göttingen as its main base. However, seen in the light of Hilbert's spontaneous letter to Brouwer, one may ask whether these are not rationalisations afterwards.[379]

The battle continued with both personal and legal arguments. Einstein refused to give in to Hilbert's wish and pleaded for tolerance: *'Sire, geben Sie ihm Narrenfreiheit!'*[380] Brouwer wrote a letter to Carathéodory telling him that he

---

[376] Brouwer did not hesitate to state his critical remarks clearly. In a dispute over a paper submitted by Schouten, who was world famous for his contributions to differential geometry and the tensor calculus, Brouwer wrote that the main work in his evaluation of Schouten's paper had been 'to trace among the great mass of trivialities, the few essential theorems, and finally to find out at which places, not cited by the author, those theorems, in so far as they are correct, have appeared earlier in the literature.', (*'unter der grossen Menge dabei herausgekommener Trivialitäten die wenigen wesentlichen Sätze heraussucht, und schliesslich ausfindig macht, an welchen vom Verfasser nicht zitierten Stellen diese Sätze, soweit sie richtig sind, schon früher in der Literatur auftraten.'*), letter from Brouwer to Klein, 7/8/1920, [Van Dalen 1999A, p. 298].

[377] *'bei der Unvereinbarkeit unserer Auffassungen in grundlegenden Fragen'*, letter from Hilbert to Brouwer, 25/10/1928, [JNUL Einstein, 13.144]

[378] In the fall of 1925, it was discovered that Hilbert suffered from pernicious anemia, a blood disease generally considered fatal. He was saved by a treatment based on raw liver flown over from the United States. He recovered very well, but had a relapse in the fall of 1928, [Reid 1970, pp. 179–188].

[379] I did not find the slightest evidence for Fraenkel's later claim that the reason for Hilbert's removal of Brouwer was that Brouwer would have protested against the large number of Jewish authors from Eastern European countries contributing to the *Mathematische Annalen*, [Fraenkel 1967, p. 161].

[380] 'Sire, give him a jester's freedom!', letter from Einstein to Hilbert, 19/10/1928, [JNUL Einstein, 13.141]; permission granted by the Albert Einstein Archives, The Jewish National & University Library, The Hebrew University of Jerusalem, Israel.

## 2.8. THE MATHEMATISCHE ANNALEN AND AFTERWARDS

would only comply with his request if Hilbert's physician would state in writing that Hilbert was 'of unsound mind'.[381] This letter did not strengthen Brouwer's position; it actually lost him the support of his friend Blumenthal. Brouwer then wrote a circular letter to the publisher and the editors, arguing against his dismissal. But the conflict was beyond reason, and the only question was how to solve the painful situation as smoothly as possible. In the end, a solution was found in the dissolution of the entire editorial board. A new board was set up with Hilbert, Blumenthal and Hecke as the sole editors; Einstein and Carathéodory refused to join it.

For Brouwer, his dismissal from the editorial board of the *Mathematische Annalen* by means of a 'coup d'état',[382] even carried out as a re-organisation, meant a serious insult. In the end, the mountains did not make way. Van Dalen put forward the dismissal from the *Mathematische Annalen* as the main motivation for Brouwer's withdrawal from the foundational debate.

**After the *Mathematische Annalen*** In the following years, Brouwer remained silent. His only public performance in the period covered here was an address delivered in Amsterdam in December 1932. The lecture was similar to *Mathematik, Wissenschaft und Sprache*, except for one thing. In explaining intuitionism, Brouwer now adds reflections on an idealised person. A human mind with unlimited memory, Brouwer maintains, would be able to do exact pure mathematics in solitude and without using linguistic signs. However, once human communication is involved, the exactness will disappear.[383] Brouwer once more admits that intuitionistic mathematics makes life harder.[384]

> Doch de sferen der waarheid zijn nu eenmaal minder permeabel, dan die der illusie.[385]

In 1934, Brouwer was again offered a chair in Göttingen, which he refused. By then, he mostly spent his time on such things as local government, school organisation and the fight for duped shareholders of a Hungarian spa enterprise. He did, at the invitation of Gonseth, lecture on intuitionism in Genève.[386] Finally, Brouwer founded the mathematical journal *Compositio Mathematica* in 1934. Brouwer himself took little part in managing the journal, leaving this mainly to his assistant Hans Freudenthal.[387]

---

[381] Brouwer must have been well aware of the seriousness of pernicious anemia, since his wife Lize had had the same illness, [Van Dalen 1999A, p. 375]. However, it is not clear if Brouwer knew of which disease Hilbert suffered.
[382] '*staatsgreep*'; the term comes from Heyting; letter from Heyting to Brouwer, 7/10/1929; [TLI Heyting, B bro-291007*]
[383] [Brouwer 1933, p. 189]
[384] [Brouwer 1933, p. 190]
[385] 'But the spheres of truth happen to be less permeable than those of illusion.'
[386] [Van Dalen 2001, p. 382]
[387] [Van Stigt 1990, pp. 105–106]

86     CHAPTER 2. THE GENESIS OF BROUWER'S INTUITIONISM

With this, the development of Brouwer's intuitionism for the period until 1933 (and somewhat afterwards) is finished.[388] However sad the end may be, Brouwer's results are impressive. The following chapters are devoted to the way the academic world reacted to his creation. Before turning to the reception, however, I touch upon a factor that may also have been of importance in the debate, namely Brouwer's character.

## 2.9  Brouwer's personality

There were two characteristics of Brouwer about which friend and foe would agree. The first was that Brouwer was a great mathematician. Thus, in 1921 Planck noted that Brouwer was 'the first mathematician in the Netherlands and together with Hilbert (...) the first of our time.'[389] Weyl said that '[o]f all mathematicians I have met, Brouwer more than anybody else with the exception of Hilbert, impressed me as a man of genius.'[390] Hilbert himself described Brouwer as 'a scholar of extraordinary talent (...) with a rare acumen.'[391] And in Brouwer's obituary, Heyting and Freudenthal, who had both worked with Brouwer as his assistant, characterised Brouwer as follows:

> Zijn lange, magere maar gespierde gestalte, zijn scherp ascetische gelaatstrekken, zijn hoewel niet krachtige, maar desalniettemin besliste stem, zijn zeer persoonlijk handschrift bleven tot in zijn hoge ouderdom het indrukwekkende fysieke correlaat van een genie, dat uitmuntte in originaliteit, diepte en geestelijke zelftucht en dat geen grenzen van vak en discipline kende.[392]

The second characteristic universally recognized was that Brouwer was an outstanding personality. Freudenthal considered Brouwer to be 'the most impressive person I have ever met'.[393] Husserl, too, was impressed by Brouwer and

---

[388] After 1933, Brouwer published some thirty more papers on intuitionism, some of which were technical contributions to intuitionistic mathematics, others of a more expository nature. Compared to his ideas before 1930, no spectacular new insights were presented in the later papers. Brouwer died in 1966.

[389] As reported in [Siegmund-Schultze 1993, p. 39].

[390] Draft for a lecture at the Bicentennial conference, 1946, [ETH Weyl, HS 91a:17]

[391] 'einen Gelehrten von ungewöhnlichen Begabung, (...) und von seltenen Scharfsinn.', draft for an answer to a letter from Korteweg to Hilbert, 6/2/1911; [NSUB Hilbert, 464-1]

[392] 'His tall, lean and muscular figure, his sharp ascetic features, his soft but firm voice and his very typical handwriting remained even in his later years the physical reflection of a genius which excelled in originality, depth and self-discipline and which was not restricted by the narrow bounds of one subject and its methods.' English version based on [Brouwer 1975–1976, vol. 2, p. XV].

[393] [Freudenthal 1981, p. 253]. Freudenthal even modelled the principal character of his first Dutch novel after Brouwer. Although Freudenthal himself always refused to admit so (oral communication by De Iongh to the author), this becomes quite clear from the novel itself. Even the name of the principal character, Von Blowitz, bears some resemblance to L.E.J. Brouwer. (Freudenthal liked name games. He wrote the book under the pseudonym 'V. Sirolf', the ret-

## 2.9. BROUWER'S PERSONALITY

described him as a 'completely original, radically honest, real and very modern man'.[394] Einstein recognized Brouwer's genius, but saw some dark sides, too. In a letter to Hilbert, he described Brouwer as *'ein unfreiwilliger Verfechter von Lombrosos Theorie der nahen Verbindung von Genie und Wahnsinn'*.[395] Nowadays, in terms of the Myers-Briggs Type Indicator, one would say that Brouwer was an INTP bordering on INFP.[396]

Brouwer's views on intuitionism were firm and unshakable. In his opinion, intuitionism constituted the 'real' mathematics. He explicitly stated so in an unpublished review:[397]

> Ein Titel: 'Int[uitionistische] Einführung des Dimensionsbegriffes' heisst dan auch bei mir 'Richtige Einführung des Dimensionsbegriffes' und sagt implizite aus, dass alle frühere Einführungen dieses Begriffes (in erster Linie meine eigenen aus dem Jahre 1913) falsch sind.[398]

Brouwer has, not without reason, been described as an 'expert at nurturing grudges'.[399] He was an emotional person, who had a strong sense of justice and could defend his ideas fiercely.[400] This caused the list of fellow mathematicians

---

rograde of 'Floris V'.) One can imagine Freudenthal's refusal to admit that Von Blowitz was modelled after Brouwer, since the character is not always very positive. Von Blowitz is characterised as an emotional and suspicious, almost paranoid person, who easily thinks that he is personally attacked and who attaches great weight to his own point of view. (A good example of such behaviour can be found in the letter Brouwer wrote to Heyting on the occasion of his 70th birthday, where he listed all the wrongs the mathematical community had done him; letter from Brouwer to Heyting, 23/2/1951; [TLI Heyting, B bro-510223].) Von Blowitz is not afraid of confronting the person he sees as his enemy, as long as the fight is honest. Maybe the clearest hint at the resemblance between Brouwer and Von Blowitz is when Von Blowitz echoes an argument Brouwer had used in the foundational debate, by saying that 'it remained a piece of villainy, even if the major signed ten treaties.' (*'het bleef een schurkenstreek, al ondertekende de majoor nog tien verdragen'*, [Sirolf 1947, p. 165]. One of the famous quotes that rang through the foundational debate was Brouwer's allegation that an incorrect theory, even if it cannot be inhibited by any contradiction that would refute it, is none the less incorrect, just as a criminal policy is none the less criminal even if it cannot be inhibited by any court that would curb it. The original citation, which was published in [Brouwer 1923, p. 3], is given in 2.6.2.)

[394]'eines völlig originellen, radikal aufrichtigen, echten, ganz modernen Menschen', letter from Husserl to Heidegger, 9/5/1928; cited from: [Husserl 1994, Band IV, p. 156]

[395]'an involuntary advocate of Lombroso's theory of the close relationship between genius and madness', letter from Einstein to Hilbert, 19/10/1928, [JNUL Einstein, 13.141]; permission granted by the Albert Einstein Archives, The Jewish National & University Library, The Hebrew University of Jerusalem, Israel.

[396]'INTP' stands for 'Introversion, Intuition, Thinking, Perceiving'; in 'INFP', the 'Thinking' has been swapped for 'Feeling'. The classical INTP type is the logician, the mathematician, the philosopher; the INFP is the idealist; cf. [Keirsey & Bates 1984, pp. 176-178; 186-188].

[397]From Brouwer's notes with respect to a review of [Fraenkel 1927A], undated, cited from [Van Dalen 2000, p. 309].

[398]'A title 'Int[uitionistic] introduction of the dimension concept' means to me 'Correct introduction of the dimension concept' and implies that all earlier introductions, first of all my own from 1913, are wrong.'

[399][Grattan-Guinness 1981, p. 501]

[400][Heyting 1967, p. 674]

he had disputes with, e.g. about priority claims, to become larger and larger. It included, among others, Jahnke, Lebesgue, Koebe, Kohnstamm, Denjoy, Menger, Schouten, and, of course, Hilbert.[401]

The description given in this chapter on Brouwer's intuitionistic activities might suggest that Brouwer was single-mindedly interested in the foundations of mathematics. That, however, was definitely not the case. Not only was Brouwer also involved in topology and significs, he furthermore led campaigns at various occasions in favour of world peace, against the inclusion of philosophy in science-degree courses,[402] for the introduction of associate-professorships, against the *Union Mathématique Internationale*, and for 'photogrammetry' to be used in defence matters.[403] What is true is Van Dalen's remark that Brouwer preferred revolutionary innovations to routine mathematical work. His work, whether it be inside or outside foundations, contains few, if any, run-of-the-mill papers.[404]

In theory, Brouwer disliked people. This attitude can be found in his profession of faith:[405]

> Mijn God (...) heeft mij ook het streven gegeven om mijn leven, d.w.z. mijn voorstellingen zo schoon mogelijk te maken; daaruit vloeit voort dat ik in de, deel van mij zijnde, mij omgevende wereld, getroffen word door walgelijkheden, en die wil trachten weg te nemen, ook wat betreft de menschenwereld. Liefde voor mijn naaste zal ik dit ternauwernood kunnen noemen, immers aan de meeste menschen heb ik het land, bijna nergens vind ik mijn eigen gedachten en zieleleven terug: de menschenschimmen om mij heen zijn voor mij het leelijkste deel van mijn voorstellingenwereld.[406]

In practice, however, he enjoyed company – be it on his own terms.[407]

---

[401][Van Stigt 1990, pp. 28–57]. A number of these conflicts is treated in [Van Dalen 1999A]. Van Dalen reported the anecdote of the Dutch topologist De Groot, who once drove Brouwer home after a meeting. In the car, Brouwer summed up all colleagues with whom he had a bad relationship at the time. Suddenly he said: 'But you and I still have an unfinished quarrel, too!' De Groot suggested Brouwer not to pursue the matter further, [Van Dalen 1984A, p. 170].
[402]Cf. [Brouwer 1921H] and [Brouwer 1922].
[403][Van Stigt 1990, pp. 78–85]
[404][Van Dalen 1999A, p. 177]
[405][Van Stigt 1990, p. 390]
[406]'My God (...) has also given me the ambition to make my life, i.e., my images, as beautiful as possible. From this follows that I am struck by loathsomeness in the world surrounding me, which is part of me, and that I want to try to remove this, also in relation to the human world. I can hardly call this love for my fellow-man, for I cannot stand most people; I hardly recognise my own thoughts and inner life anywhere. I find the human shades surrounding me the ugliest part of my world of images.'
[407]Cf. [Van Dalen 1984A, p. 25; 44]

Taking together the different evidence we have of Brouwer's personality, the description of Brouwer that, I think, fits best comes from Heyting:[408]

> Brouwer was an individualist who in social life as well as in his philosophy took little notice of generally accepted norms and ideas.

Brouwer was a non-conformist in every aspect of his life. The reactions to his intuitionism would show that not everybody appreciated this.

## 2.10 Conclusion

Brouwer developed his intuitionistic view on mathematics from 1907 on. In his dissertation, published that year, he carried out what he later identified as the first act of intuitionism: the separation between mathematics and mathematical language. In doing so, Brouwer set forth a line of thought already present in his mystical work *Leven, kunst en mystiek* from 1905, where he had played down the value of language. In Brouwer's view, mathematics is a languageless creation of the human mind, constructed from the primordial intuition. In this view, one recognizes both Mannoury's influence, who had stressed the human character of mathematics, and Brouwer's own philosophical view, which was idealistic. The following year, Brouwer drew the radical consequence that the principle of the excluded middle is not universally valid. In Brouwer's view, logic is nothing but a description of regularities appearing in the mathematical language. In infinite totalities, there is no ground for asserting the principle of the excluded middle.

In 1912, Brouwer became professor at the university of Amsterdam. His inaugural lecture *Intuitionisme en formalisme* was devoted to the foundations of mathematics. The lecture is important since it introduced the names for the two currents which were seen as the main opponents in the foundational debate. Brouwer himself did not explicitly take position, nor did he explicitly identify Hilbert as a formalist.

After his work in topology, which brought him world fame, and after the first World War, Brouwer launched the second act of intuitionism: the introduction of constructive set theory, based on choice sequences. In the following years, Brouwer almost singlehandedly developed intuitionistic set theory and function theory. He also made weak counter-examples against the principle of the excluded middle, based on unsolved, but not proved unsolvable, problems. Instead of 'negation', Brouwer used 'absurdity', and he proved that a triple absurdity equalled a single one. Brouwer opposed the idea that consistency was enough to justify a mathematical system, claiming that 'an incorrect theory, even if it cannot be checked by any contradiction that would refute it, is none the less incorrect, just as a criminal policy is none the less criminal even if it cannot be checked by any court that would curb it.' Brouwer published a series of expository papers on intuitionistic mathematics in the *Mathematische Annalen*, which, however, were hard to read.

---
[408][Heyting 1978, p. 7]

In 1927–1928, Brouwer was at the top of his intuitionistic height. He lectured in Berlin and Vienna, significantly influencing such important persons as Wittgenstein and Gödel. In the same period, Brouwer formulated what the main differences were between the intuitionistic and the formalistic point of view. If formalists recognised that the principle of the excluded middle was to be identified with the solvability of every problem, and that the justification of formalistic mathematics by means of a consistency proof rested on a vicious circle, Brouwer maintained, the choice between the two would be reduced to a matter of taste.

However, the story ended differently. Hilbert, upon returning from the 1928 Bologna conference of which Brouwer had called for a boycott, dismissed Brouwer from the editorial board of the *Mathematische Annalen*. This is presumably the reason why Brouwer withdrew from the foundational debate.

Thus, Brouwer differed from mathematicians such as Kronecker, Borel and Poincaré, who are often seen as his predecessors, in that he not only criticised classical mathematics, but also offered an alternative. Furthermore, Brouwer's criticism was more radical, since it extended to the unlimited use of the principle of the excluded middle. Although Brouwer sometimes used polemics, most of his works on intuitionism were purely mathematical. Brouwer's intuitionistic mathematics is coherent and follows logically once one, as a mathematician, accepts the idea that mathematics is a human construction. However, Brouwer was not the most didactically gifted person, which made it hard for others to penetrate his intuitionistic ideas. Moreover, Brouwer was a strong personality with a strict sense of right and wrong, and such a dogmatic position did not help in the spreading of intuitionism, either.

# Chapter 3

# Overview of the foundational debate

> (...) ich halte diese Alternative [between intuitionism and formalism, DH] überhaupt für eine logisch unzulässige Anwendung des 'Tertium non datur' (...).[1]
>
> Ernst Zermelo[2]

## 3.1 Introduction

In this chapter, I give an overview of the foundational crisis in mathematics in the beginning of the 20th century. In order to determine the temporal extension and other characteristics of the foundational crisis, we have to look at four factors. There has to be a group of influential mathematicians who (1) reflect on the mathematical process, (2) express doubts concerning the validity of certain methods or results, and (3) request changes in the mathematical process. Furthermore, there has to be (4) a sense of crisis among the participants to the debate.[3] The first three criteria have been met from the end of the 19th century on, when Kronecker and others criticised existing mathematical practices and results.[4] The fourth criterion, however, was only met once intuitionism was presented as a full-blown alternative to classical mathematics, threatening its very existence. From that moment on,

---

[1] '(...) I take this alternative [between intuitionism and formalism, DH] to be a logically inadmissable application of the 'tertium non datur' (...).'

[2] Report to the *Notgemeinschaft der Deutschen Wissenschaft* ('Emergency Society of German Science'), written sometime between 1930 and 1933; published in [Moore 1980, pp. 130–134] (citation on page 131).

[3] See 0.1.

[4] See 1.2.

the debate on the foundations of mathematics became more emotional and polemical, as seen among other things in the use of metaphors.[5] Therefore, I have taken the reactions to Brouwer's intuitionism as a main indicator of the course of the foundational debate. The overview of the foundational crisis presented below is thus limited to those characteristics which can be derived from the reactions to intuitionism.

The description of the foundational crisis can be divided into a quantitative and a qualitative part. The quantitative survey was based on numbers taken from two sources. The first is the *Jahrbuch über die Fortschritte der Mathematik*, the main mathematical review journal of those days; the second consists of all public reactions to intuitionism that I could find. The advantage of the first method is that all contributions originated from one single source, so that, assuming that the *Fortschritte* had a constant coverage of mathematical journals, changes in the reception of Brouwer's intuitionism could be measured more objectively. The advantage of the second is that more reactions could be included. Since I am dealing with the reception of Brouwer's intuitionism, Brouwer's own works were not included in the numbers. Neither were reviews of intuitionistic works, since reviews are almost automatically produced once a work is published, and thus are not part of a more or less spontaneous development of the debate.

The qualitative view focuses on themes that played an important role in the foundational debate, on the 'tone' of the reactions, on the different school and currents, and on the social background of the participants.[6]

Combining these two parts leads to a characterisation of the debate both in time and in contents. The delimitation thus found is applied in the remainder of the book.

In order to start the process, a first chronological demarcation is needed. I chose to start as early as possible, with the first reactions to Brouwer's intuitionism I could find. This was Mannoury's *Methodologisches und Philosophisches zur Elementar-Mathematik* ('Philosophical and methodological remarks on elementary mathematics'), which appeared in 1909.[7] Thus, the period covered here starts in 1909.

As a preliminary end point, I took 1933. There are two main reasons for doing so. The first is Hitler's rise to power in Germany, which had a significant impact on the mathematical world, too.[8] The second reason is that, by 1934, references to Brouwer and intuitionism have become so common that they do not constitute anything special anymore, and furthermore it has become hardly possible to cover them all.

---

[5]The use of metaphors and its consequence for delimiting the foundational crisis is discussed in 6.2.

[6]Characteristics which can be treated in a more or less isolated way, such as the number of female participants or the languages that were used, are dealt with in this chapter. The more general relationship between the foundational debate and its cultural context is treated in chapter 6.

[7][Mannoury 1909]

[8]See 6.6.

## 3.2 Quantitative inquiry

### 3.2.1 The *Fortschritte*

The German journal *Jahrbuch über die Fortschritte der Mathematik* or, for short, the *Fortschritte*, was the main mathematical review journal until the beginning of the 1930s.[9] Therefore, it stands out as the most natural choice as a means to measure the development of the foundational debate, by taking reactions which were reviewed in the *Fortschritte*. Before turning to the numbers that can be taken from it, however, a preliminary remark should be made on the reliability of information drawn from the *Fortschritte*.

The *Fortschritte* suffered serious delays in publication. For example, volume 48 covering the years 1921–22 only appeared between 1925 and 1928; volume 52, dealing with 1926, came out in 1935. This is likely to have been caused by post-war inflation problems and the boycott of German science.[10] Furthermore, the order in which the *Fortschritte* appeared was not strictly chronological. For example, in 1927 volume 48 (1921–22) had not yet been completed, but one had 'already' started publishing parts of volume 49 (1923). Because of these delays, the picture obtained from an analysis of the *Fortschritte* volumes may be coloured by the fact that its editors could work with hindsight. In this way, it is possible that they projected later knowledge on earlier periods. Whether this happened or not depends at least partially on the way in which papers to be treated in the *Fortschritte* were chosen. Unfortunately, I have no information on this issue.

Since not all publications reviewed in the *Fortschritte* of a certain year were published in the same year, I had to do some rearranging.[11] Thus, I did not count the publications dealing with Brouwer's intuitionism by volume of the *Fortschritte*, but according to the year of publication.[12]

Figure 3.1 shows the number of publications cited in the *Fortschritte* from 1914 to 1933 that at least mention intuitionism or Brouwer (not in the field of topology).[13]

---

[9] In 1931, the rival *Zentralblatt für Mathematik und ihre Grenzgebiete* was founded by Göttingen mathematicians, [Siegmund-Schultze 1993, p. 200].

[10] Cf. [Siegmund-Schultze 1997, p. 142].

[11] Sometimes, even papers from year $y + 1$ were included in the Fortschritte of year $y$.

[12] The first volumes after 1933 do not include reviews of reactions to intuitionism published before 1933.

[13] The numbers given here are minimum numbers, in the sense that only those publications were included about which it was almost completely sure, either by the publication itself, or by its title, or by the review, that it deals with Brouwer's intuitionism. In case of doubt about the year of publication, the year of the *Fortschritte* in which it appeared was taken. It remains possible that some reviews of publications do not mention references to intuitionism that are present in the work reviewed, and that I therefore missed some relevant contributions.

Figure 3.1: Reactions to intuitionism mentioned in the *Fortschritte*, 1914–1933

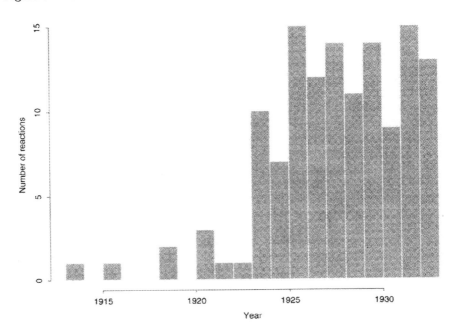

As one can see from the histogram, the number of reactions to intuitionism grew in 1921, when Weyl's paper was published,[14] and reached a peak in 1926. From that year on, the debate remained at its high level, wavering up and down until the end of the period covered here. This leaves us with the picture of a debate with a clear beginning, but with no clear end. I return to this problem later.

Apart from these numbers, there are some other characteristics of the *Fortschritte* which reveal interesting aspects of the foundational debate. The *Fortschritte* is divided into several categories. The categories in which foundational papers were placed changed over the years covered here. In the beginning of the 1920s, publications dealing with Brouwer's intuitionism could be found either in the section 'Arithmetic and Algebra', subsection 'Foundations'; or under the same subsection of the section 'Analysis'; or in the section 'Set Theory'; or, finally, in the section 'History, philosophy and didactics'. Over time, almost all contributions to the foundational debate were placed under the latter section. The subsection 'Foundations' under Analysis was restricted to general analysis books, and 'Set theory' to strictly technical set-theoretical works. From volume 51 on (1925, published in 1932) the subsection 'Foundations' under 'Arithmetic and Algebra' disappeared. In volume

---
[14][Weyl 1921]

## 3.2. QUANTITATIVE INQUIRY

61 (1935, published in 1936), the philosophy part was renamed 'foundations' and joined with abstract set theory rather than history and didactics.

These developments indicate that people came to regard foundational questions more as a whole, and not separated by the specific mathematical disciplines they were applied to, such as analysis or algebra. This reveals that, in the process of the foundational debate, foundational questions were established as a separate field of mathematical research.

Finally, it should be pointed out that the publications mentioned here represent but a fraction of the total number of publications reviewed in a single issue of the *Fortschritte*. For example, a rough estimate tells us that the 1926 volume of the *Fortschritte* mentions almost 5,000 publications.[15] Therefore, it is clear that even in such a top year of the discussion, only a small part of the mathematical production was related to the foundational crisis, namely about 0.3%.

### 3.2.2 'All' public reactions to intuitionism

In this subsection, rather than presenting all data from a single source, such as the *Fortschritte*, I included all public reactions to intuitionism known to me.[16] This means that I included all published books and papers dealing with intuitionism, all lectures given (or at least announced) on intuitionism, and all reviews which mention intuitionism while discussing publications that do not. In this case, lectures were placed in the year they were given rather than the year of publication, if any. This differs from the method followed for numbers obtained from the *Fortschritte*, where all materials were published and thus the year of publication was used for lectures, too. Furthermore, the method followed in this section means that also non-mathematical reactions were taken into account.[17]

The advantage of including all known contributions is that the amount of material covered is more than twice as big as that from the *Fortschritte*. The

---

[15] The estimate was based on countings made from the name register in the back, leaving out cross-references, and a check by countings made from random pages in the body itself.

[16] The only other relevant source that seemed possible was the number of lectures mentioned in the *Jahresbericht der deutschen Mathematiker-Vereinigung*. If other journals had been taken as a measure, one would only get a somewhat arbitrary sample of numbers that had mostly already been included in the *Fortschritte*. Since announcements and reports of lectures were not included in the *Fortschritte*, the choice of the *Jahresberichte* would provide us with independent numbers on the development of the debate. However, this material would lead to a too large degree of uncertainty. Firstly, from most of the lectures only the title is given. This makes it hard to judge whether or not Brouwer's intuitionism was discussed. Secondly, the mentioning of lectures in the *Jahresbericht* was not always done regularly. For example, in some volumes more universities are mentioned than in others. Taken together, these restraints mean that not more than ten talks would survive, a number far too small to provide relevant information on the development of the debate.

[17] The full list of all public reactions to Brouwer's intuitionism until 1933 is given in appendix B.

disadvantage is that the numbers are more likely to have been distorted by my own personal interest and capacities. In particular, most of my attention was drawn to the beginning of the debate, which means that the numbers from after 1930 could well be too low.

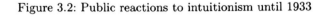

Figure 3.2: Public reactions to intuitionism until 1933

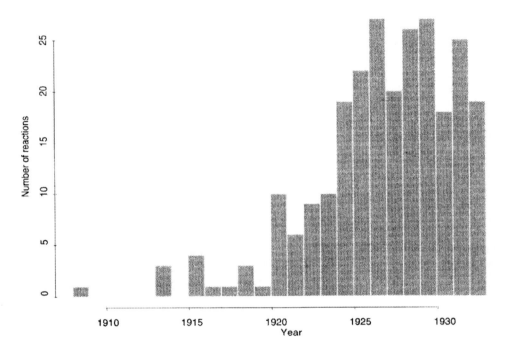

Figure 3.2 gives a somewhat different image of the development of the reactions to Brouwer's intuitionism than the first picture. What remains unaffected is the beginning of the debate: 1921 still clearly stands out as the first year in which a considerable number of people reacted, also more than the first years afterwards. However, the figure we now have gives a smoother image of the debate. Between 1922 and 1927, there is a monotone increase in the number of reactions, reflecting a significantly growing interest in intuitionism. Also, it has now become possible to identify more clearly the zenith of the debate, which lies in 1927.[18] After that year,

---

[18]Thus, Mehrtens' claim that the foundational crisis is to be situated primarily between 1920 and 1925 and that from halfway the 1920s the foundational crisis turned into a specific *discours*

## 3.3. QUALITATIVE INQUIRY

the number of reactions fluctuates. With this picture, I do not see any clear relation between the development of the foundational crisis and the general situation in Weimar Germany.[19]

An analysis of the differences between the picture obtained from the *Fortschritte* material and the one from all known contributions reveals that the different shapes were not so much caused by adding more contributions, but by re-ordering the given lectures. This means that, if we accept that lectures should be classified according to the year in which they were given rather than according to the year in which they were published, we get a reasonably smooth picture of the development of the debate. Classifying lectures by the year in which they were given seems most reasonable, since they then count in the year in which the reaction was first expressed (though possibly not for the largest audience).

## 3.3 Qualitative inquiry

### 3.3.1 Themes

In order to structure the contents of the debate, I identified its dominant themes. The selection of the themes was done by analysing which subjects were mentioned in the written sources that I used as a basis for this study. This means that the categories according to which I set up this chapter (and others which were based on this one) were formed, at least in principle, by following the categories as they were used by the historical actors. However, some actors, such as Fraenkel and Weyl, were much more explicit in identifying themes in the debate than others. Especially Weyls' *Grundlagenkrise* paper played a dominant role in setting the agenda. Therefore, the selection presented here was carried out by looking at themes that were recognised by some historical actors as themes in the debate, and next checking whether these issues occured in contributions of others as well, even if only as side remarks.

Following this procedure, the main themes which were discussed in contributions that can be classified as reactions to Brouwer's intuitionism are the following:

- Mathematical existence and constructivity;

---

among foundational researchers is not correct, [Mehrtens 1990, pp. 294–295]. The top of the crisis clearly came later, and most of the contributors to the foundational crisis after 1925 were *not* specialists in the foundations of mathematics, if only because the number of foundational researchers was still very small. This substantially weakens Mehrtens' claim that, with the consolidation of the Weimar republic, the foundational crisis faded away, [Mehrtens 1990, p. 294].

[19]The history of Weimar Germany is normally divided into three periods: the first from the founding of the republic in 1918 until 1923, a period in which the republic was still threatened in its existence; the second from the end of 1923 until the end of 1929, during which the economy recovered and there was (relative) political consolidation; the third from the world economic crisis in October 1929 until 1933, during which both economic and democratic circumstances worsened and which finally meant the end of the Weimar republic, [Müller, H.M. 1996, pp. 226–227].

- The principle of the excluded middle and logic in general;[20]
- The contents of mathematics and formalisation;
- The roles of intuition and philosophy in mathematics;
- The foundational crisis and intuitionism in general.

I will now clarify what should be understood by these categories by mentioning some of the questions that were addressed in the discussion of these themes.

The first category, on mathematical existence and constructivity, can be characterised by questions like: What does 'existence' mean in mathematics? Does consistency suffice for mathematical existence? Or is constructivity required? What objects exist constructively in mathematics? Is mathematics, or a certain current within the foundations of mathematics, constructive? Should mathematics be constructive? Can mathematics be constructive? What does 'constructive' mean?

The dispute about the principle of the excluded middle, the second item, can be seen as a special case of divergent opinions on the status of logic. Questions which are treated under this header are: Is mathematics dependent on logic, is it the other way round, or should both be developed simultaneously? What logical principles may be used in mathematics? Is the principle of the excluded middle always reliable? Does the principle of the excluded middle apply even if the alternatives in the statement transcend our present state of knowledge? Can one do mathematics without the principle of the excluded middle? Should one do so?

The third category treats the issue in what sense formal mathematics is or can be meaningful, if at all, and whether it is allowed to transcend contents by doing purely formal mathematics.

The final two categories are less coherent than the first three. The one on intuition and philosophy deals with extra-mathematical factors: Can or should intuition be used within mathematics, and if so, in what way? Can mathematics be founded on intuition? Can it be founded on philosophy? Should it be founded on philosophy? Or is mathematics independent of intuition and philosophy?

The issue of the 'foundational crisis and intuitionism in general' is a rest category for matters still linked to intuitionism and the foundations of mathematics, such as the crisis in the foundations of mathematics, finitism, the infinite, set theory and the antinomies, subjectivism in mathematics, etc.

Note that many of the questions mentioned here are in the 'should'-form: they contain value judgements about mathematics or concern the question of delimitating the field of mathematics.

It should be pointed out that the dominance of these themes, above all the questions of mathematical existence and the principle of the excluded middle, means that in the debate most attention was paid to critical aspects of intuitionism, where certain parts of classical mathematics were rejected. Particularly

---

[20]Sometimes, the principle of the excluded middle is referred to as the law of the excluded middle; I prefer the former name.

## 3.3. QUALITATIVE INQUIRY

Hilbert specialised in focusing on the negative aspects of intuitionism. Brouwer's alternative, based on choice sequences and a constructive set theory, received little attention, especially from mathematicians. Philosophers tended to be more positive about this aspect. Thus, the foundational debate focused on themes that gave intuitionism a 'destructive' image.[21]

In the course of the debate, the focus on the main themes remained remarkably constant. This becomes clear by analysing not only all public contributions to the debate, such as lectures and papers, but also private notes, correspondence, and unpublished manuscripts. In this way, we find that throughout the debate the two themes that are discussed most are mathematical existence and constructivity, and the principle of the excluded middle and logic. The only significant change that takes place in this respect is that the two subjects become recognised more widely as the central ones in the foundational crisis. In the beginning of the 1920s, it is still fairly common to find contributions to the debate which treat only one of these issues. Around 1930 this has changed, and reactions to Brouwer's intuitionism would typically at least mention both.

The other thing that can be remarked regarding a changing thematic focus during the debate is that there is a shifting attention regarding the third and fourth theme. Up to about 1928, there are significantly more reactions discussing the question of intuition and philosophy in mathematics than those treating the issue of the contents of mathematics and formalisation. After that year, the mentioning of questions of philosophy and intuition drops, whereas the question of the contents of mathematics and its formalisation becomes about as important as the two dominant themes.[22]

Finally, the focus on what kind of intuitionism is discussed changed around 1930. No longer was Brouwer's intuitionism and Weyl's presentation thereof, between which most people did not differentiate, the only issue discussed. From then on, Heyting's formalisation of intuitionistic logic as published in 1930 became one of the main themes.[23] Especially people like Barzin and Errera discussed Heyting's work extensively.[24] This means that the foundational debate continued in a substantially modified form.

### 3.3.2 Tone

In the previous section I identified the main themes in the debate and I pointed out some shifts that took place over the years. A more important change than this one, but harder to pinpoint, is the change in tone of the debate.

---

[21]Thiel draws the same conclusion, [Thiel 1972, p. 112].
[22]Thus, it seems that this development in mathematics ran contrary to the more general debate among German academics, where intuition was used *more* after about halfway the 1920s, [Ringer 1969, p. 403].
[23]See 5.4.1.
[24]See 5.4.3.

# 100  CHAPTER 3. OVERVIEW OF THE FOUNDATIONAL DEBATE

One of the first contributions in which the changed tone can be found is Fraenkel's lecture at the first *Tagung für Erkenntnislehre der exakten Wissenschaften* ('Conference on epistemology of the exact sciences') in Prague in 1929. In the lecture, Fraenkel points out the polemics involved in the debate and tries to find a synthesis between intuitionism and formalism:[25]

> Wie mir scheint, hat Brouwer den größten Erfolg für seine Anschauungen dadurch erzielt, daß er als Anhänger seiner Ausgangsposition – Hilbert gewonnen hat! Nur die Schärfe der Polemik zwischen beiden Forschern und ihren Schülern auf der einen Seite, die völlige Entgegengesetztheit der Schlußfolgerungen auf der anderen Seite konnte es, glaube ich, verschleiern, daß Hilbert tatsächlich die Forderung der Konstruktivität und die Ablehnung eines einsichtigen Grundes für die Anwendung der Aristotelischen Logik auf unendliche Gesamtheiten übernommen hat.[26]

In the end of the lecture, Fraenkel appears conciliatory, stating that he believes that none of the foundational currents is completely wrong; they are just looking at the same problem from different angles.[27]

The most striking examples of the changed tone can be found in the 1930 Königsberg conference on the foundations of mathematics, where both formalism, intuitionism and logicism were presented.[28] One may compare the following contributions, made there, to earlier ones.

In the introduction of his lecture Von Neumann, who represented formalism, states the following on the three currents:[29]

> Die scharfe Formulierung der Mißstände in der klassischen Mathematik durch *Brouwer*, die exakte und erschöpfende Beschreibung ihrer Methoden (der guten und der bösen) durch *Russell*, und die von *Hilbert* geschaffenen Ansätze zur mathematisch-kombinatorischen Untersuchung dieser Methoden und ihrer Zusammenhänge — diese drie wichtigen Vorstöße im Gebiete der mathematischen Logik haben es mit sich gebracht, daß heute in den Grundlagenfragen immer mehr eindeutige Fragestellungen und nicht Geschmacksunterschiede zu untersuchen sind.[30]

---

[25] [Fraenkel 1930A, p. 294]
[26] 'As I see it, Brouwer has reached the biggest success for his point of view by winning as an advocate of his starting position – Hilbert! Only the sharpness of the polemics between both researchers and their students on the one hand, the complete opposition of the consequences on the other hand could, I think, veil that Hilbert indeed has taken over the demand of constructivity and the rejection of a contentual ground for the application of Aristotelian logic on infinite totalities.'
[27] [Fraenkel 1930A, p. 302]
[28] See 4.4.1.
[29] [Von Neumann 1931, p. 116]
[30] '*Brouwer*'s sharp formulation of the serious deficiencies of classical mathematics; *Russell*'s exact and exhaustive description of its methods (the good and the bad ones); and *Hilbert*'s

## 3.3. QUALITATIVE INQUIRY

Carnap, who spoke on logicism, had already in the announcement of his lecture hinted at a possible rapprochement between the different currents.[31] He points out that logicism has certain correspondences with both intuitionism and formalism. With intuitionism it shares, in Carnap's words, the constructive tendency in making concepts; with formalism, the idea of working purely formally within an established logico-mathematical system.[32] And finally Heyting, the speaker on intuitionism, takes up a pluralist position, too. He describes the question of taking a starting position in how to found mathematics as a matter of choice. Only once one has chosen the position that mathematics should consist of constructions in the human mind, intuitionism provides the only possible way of doing so.[33]

Dawson rightly describes the tone of the discussion as 'conciliatory'.[34] One may compare these quotes to citations from the earlier period: Weyl, who called Brouwer 'the revolution'; Hilbert, who described Brouwer and Weyl's intuitionism as a 'dictatorship of forbiddings', and who reproached them of 'treachery to our science'; Ramsey, who spoke of the 'Bolshevik menace of Brouwer and Weyl'; or Brodén, who reproached intuitionism of 'extreme cowardice'.[35] The vehemence of the terms used in describing the other party indicates that more was at stake, in the participants' views, before 1930 than afterwards.[36]

Now Von Neumann, Carnap and Heyting had never been the radicals of the debate. In Von Neumann's earlier publications dealing with the foundational debate, he had already argued in favour of intuitionism and had tried to link intuitionism and formalism by identifying the formalistic meta-mathematical level with intuitionistic mathematics.[37] Carnap, too, had treated intuitionism as a current in its own right in an earlier publication,[38] and he had no problems in working with concepts for which the principle of the excluded middle did not hold.[39] Heyting, finally, had earlier complimented formalists such as Bernays for possessing a true understanding of intuitionism.[40]

---

initiation of a mathematical-combinatorial inquiry into these methods and their relations — these three important advances in the field of mathematical logic have resulted in there nowadays being more and more unambiguous problems to be investigated in foundational questions and not matters of taste.'
[31] [Tagung für exakte Erkenntnislehre 1930, p. 1068]
[32] [Carnap 1931, pp. 104–105]
[33] [Heyting 1931A, p. 115]
[34] [Dawson 1984, p. 114]
[35] [Weyl 1921, p. 226], [Hilbert 1922, p. 159], [Hilbert 1926, p. 170], [Ramsey 1926A, p. 380], and [Brodén 1925, p. 236] respectively. I should add that the tone used in Ramsey's paper is much more relaxed than for example in Hilbert's work.
[36] On the use of metaphors in the debate, see 6.2.
[37] [Von Neumann 1924, p. 239] and [Von Neumann 1927, pp. 2–3] respectively.
[38] [Carnap 1927, p. 363]. Carnap continued his conciliatory remarks, as, e.g., in [Carnap 1930, p. 308].
[39] [Carnap 1927, p. 364]
[40] [Heyting 1925, p. 2]

## 102   CHAPTER 3. OVERVIEW OF THE FOUNDATIONAL DEBATE

The following year, both Carnap and Fraenkel openly stated that they saw a synthesis between intuitionism, formalism and logicism as achievable.[41]

The fact that people like Carnap, Heyting and Von Neumann moved to the center of the discussion means that something had changed. No longer Brouwer and Hilbert, two giants fighting for their sole right to mathematical truth, were dominating the foundational debate, but people more willing to see the other side of truth.[42] Even though the foundational debate continued in 1930 and beyond, the foundational crisis was by and large over.[43]

Another difference in tone can be noted if we vary not time but space. This becomes clear by looking at the reactions that Brouwer's intuitionism evoked in the US. First of all, these reactions were both small in number and late. Even though Frizell reacted to the translation of Brouwer's inaugural address in the Bulletin of the American Mathematical Society as early as 1914,[44] it remained silent on the other side of the Atlantic for years afterwards. In 1924 Arnold Dresden, a Dutch-born mathematician who had taken his Ph.D. at the University of Chicago and who was the translator of Brouwer's inaugural address, published a paper on Brouwer's contributions to the foundations of mathematics.[45] Still no public discussion arose. In 1927, Dresden lectured on intuitionism at a joint meeting of the American Mathematical Society and the Mathematical Association of America, and so did Pierpont.[46] The following year, Church published a paper on the law of the excluded middle.[47] The number of American reactions does not compare to the large number of publications on intuitionism in Europe.

More telling even than these numbers is the tone in which these contributions were made. The calm way in which Americans drew their conclusions is maybe best illustrated by the following words of Church:[48]

> (...) we may accept a system of logic in which the law of the excluded middle is assumed, a system in which the law of the excluded middle is omitted without making a contrary assumption, and a system which contains assumptions not in accord with the law of the excluded middle as all three equally admissible, unless one of them can be shown to lead to a contradiction. If we had to choose among these systems of logic, we could choose the one most serviceable for our purpose, and we might conceivably make different choices for different purposes.

---

[41][Carnap 1930, p. 310], [Fraenkel 1930B, p. 297]

[42]Brouwer was removed from the editorial board of the *Mathematische Annalen* in 1928 by Hilbert and left the foundational debate disillusioned, see 2.8; Hilbert suffered from the normally fatal blood disease pernicious anaemia and had to retire at the mandatory age of 68 in 1930, [Reid 1970, pp. 179; 190].

[43]In fact, the end of the foundational crisis is best put in 1928; see 6.2.

[44][Frizell 1914]

[45][Dresden 1924]

[46]See 4.3.2 and 5.3.1.

[47]See 5.3.1.

[48][Church 1928, p. 77]

## 3.3. QUALITATIVE INQUIRY

Church takes up a pragmatic position and does not seem to be worried at all by the fact that two of the world's leading mathematicians had attacked a logical principle that had been used by mathematicians for centuries. This makes it clear that, even though some elements of the foundational debate had crossed the ocean, the foundational crisis definitely had not.[49]

Finally, the synthesis was completed when Hilbert and Bernays, too, acknowledged the value of intuitionism. The important works to mention in this respect are the first volume of Hilbert and Bernays' *Grundlagen der Mathematik* ('Foundations of mathematics'), published in 1934, and Bernays' paper *Sur le platonisme dans les mathématiques* ('On Platonism in mathematics'), published the following year.[50] In the former, Hilbert and Bernays for the first time publicly acknowledge that Brouwer's remarks concerning the principle of the excluded middle were justified.[51] Having explained that in finitary mathematics existential and universal statements and their 'sharp' negations are not opposed as contradictories, they write:[52]

> Die komplizierte Situation, die wir hier in betreff der Verneinung von Urteilen beim finiten Standpunkt vorfinden, entspricht der These Brouwers von der Ungültigkeit des Satzes vom ausgeschlossenen Dritten für unendliche Gesamtheiten.[53]

In 1935, Bernays went even further by taking up a position that, in a way, meant a kind of reconciliation between intuitionism and formalism. He explicitly acknowledges the value of intuitionism, claiming that[54]

> (...) ces deux tendances, intuitionniste et platonicienne [formaliste, DH], sont nécessaires, elles se complètent et il faudrait se violenter pour renoncer à l'une d'elles.[55]

Further on in the paper, it becomes clear why Bernays is so positive about intuitionism. Part of the reason is one of Gödel's incompleteness theorems. The theorem states that, given a formal system $S$ which is at least as strong as the *Principia Mathematica* and assuming that $S$ is consistent, the consistency proof of $S$ cannot be expressed in $S$ itself.[56] Bernays realised that Gödel had destroyed Hilbert's

---
[49] See also 6.2.
[50] [Hilbert & Bernays 1934–1939, vol. I]; [Bernays 1935]
[51] It is not clear whether really both of them wrote this remark. By that time, Hilbert generally left most of the work to Bernays. In the light of what Hilbert had said before on intuitionism, it seems probably that Bernays was the one who inserted the remark.
[52] [Hilbert & Bernays 1934–1939, vol. I, p. 34]
[53] 'The complicated situation that we find here with respect to the negation of judgements from the finitary point of view corresponds to Brouwer's thesis of the invalidity of the principle of the excluded middle for infinite totalities.'
[54] [Bernays 1935, p. 66]
[55] '(...) these two tendencies, intuitionistic and platonic [formalistic, DH], are both necessary; they complement each other and it would be doing oneself violence to renounce one of them.'
[56] See 5.4.2.

original programme. Now, he was looking for a way to modify it.[57] Bernays's solution was to use intuitionistic mathematics as the meta-theory in which to prove the consistency of formal mathematics.[58]

By proposing this, Bernays showed himself to go even further than what Brouwer had claimed. In 1928, Brouwer had formulated four 'insights' which formalism should recognize in order to reduce the choice between intuitionism and formalism to a matter of taste. One of these was that for meaningful metamathematics the intuitionistic theory of the natural numbers was necessary.[59] Now, Bernays chose to include all of intuitionistic mathematics on the metamathematical level.

### 3.3.3 Currents and schools

**Hilbert-school** In terms of academic schools,[60] it is interesting to look at the reactions to intuitionism from the Hilbert-school: people who had written their dissertation under Hilbert. Since these people were all trained by Hilbert, at least in principle, in doing scientific research, one might expect a certain sympathy for Hilbert's point of view. A large number of these, eleven, took part in the foundational debate: Ackermann, Behmann, Bernstein, Courant, Grelling, Hamel, Hedrick, Hellinger, Lietzmann, Schmidt, and Weyl.[61]

Of one of these I do not know what opinion he held on the foundational debate, since he gave a lecture of which I only found the title. This was Hellinger, who lectured on Weyl's research into the foundations of mathematics in June 1921 in Frankfurt am Main.[62] Bernstein only mentioned Brouwer in a paper.[63] Ackermann mentioned some aspects of intuitionism, as well as of other currents in the foundations of mathematics, but remained strictly neutral.[64] Lietzmann simply described the formalistic and intuitionistic views on the foundations of mathematics as its respective adherents saw it.[65] Courant gave a joint lecture with Bernays on Weyl's and Brouwer's new arithmetical theories in February 1921, which was not published.[66] A 1927 lecture of Courant on the general meaning of mathematical thinking, which he gave for the conference of German philologists and pedagogues held in Göttingen, was published. Here, he pointed out that if intuitionism would

---

[57] Gödel maintained that there was no contradiction between his theorem and Hilbert's programme, because a finitary proof need not be formalisable in the system under consideration; see 5.4.2.
[58] [Bernays 1935, p. 69]
[59] [Brouwer 1928, p. 410]; see 2.7.2.
[60] The term 'academic school' is here taken in the loose meaning of people who share a specific, common scientific background.
[61] The [Verzeichnis Hilbert-Dissertationen] contains a list of people who wrote their dissertation under Hilbert. It is not clear if this list contains all Hilbert students.
[62] [Hellinger 1921]
[63] [Bernstein 1919, p. 70]
[64] [Ackermann 1927, pp. 450–451]
[65] [Lietzmann 1925]
[66] [Courant & Bernays 1921]

## 3.3. QUALITATIVE INQUIRY

win, a substantial part of mathematics would be lost and furthermore that, despite the genius of its advocate, intuitionism is untenable.[67] In the beginning of the debate, Behmann only replied negatively to a comment Brouwer made to his talk,[68] but he began to work on constructivity himself in 1930.[69] Hamel saw positive sides in Brouwer's intuitionism and tried to reconcile it with Hilbertian formalism.[70] Hedrick and Grelling were not only quite sympathetic to Brouwer's view, they also criticised Hilbert. Hedrick sympathised with Brouwer especially regarding the principle of the excluded middle. Countering Hilbert's argument that large portions of mathematical knowledge would disappear if mathematicians all became intuitionists, Hedrick concludes his lecture as follows:[71]

> If we are to deny new developments whenever they require relinquishment of the ideas of the past, we shall be serving not truth but only our vanity.

Grelling found the axiomatic method 'unsatisfactory' with respect to set theory, and he shared Brouwer's criticism of the principle of the excluded middle (though with a different argumentation).[72] In a later publication, Grelling not only gave a clear and appropriate characterisation of intuitionism,[73] but he also seemed to criticise the way in which Hilbert defended classical mathematics:[74]

> If intuitionists have been characterized with a certain property as revolutionists who overturned the *ancien regime*, Hilbert might be compared with a Napoleon who, without regard to considerations of legitimacy, established, through a brilliant political stroke, a new order whose success is the substitute for legitimacy.

It is not clear to me whether Grelling put stress on the loss of legitimacy in formalistic mathematics, or that for him the success of this type of mathematics was more important. On the whole, Grelling remained mostly neutral in his paper. Schmidt, in his rector's address at the *Friedrich-Wilhelms-Universität* Berlin in 1929, was critical of intuitionism.[75] Finally, Weyl started as a staunch supporter of Brouwer and remained sympathetic to intuitionism throughout his life.

Two conclusions can be drawn. Of all these Hilbert students, only Weyl, and possibly Grelling and Lietzmann, can be seen as really involved in the foundational debate; the others made not more than one public contribution each. Furthermore, Weyl was clearly on Brouwer's side, at least in the beginning; Behmann seemed

---

[67] [Courant 1928, p. 92]
[68] [Behmann 1924, p. 67]
[69] See 4.4.3.
[70] [Hamel 1928, pp. 11–15]
[71] [Hedrick 1933, p. 343]
[72] '*hat (...) etwas Unbefriedigendes*', [Grelling 1924, pp. 46–47]
[73] [Grelling 1928, pp. 99–101]
[74] [Grelling 1928, p. 103]
[75] [Schmidt 1929, pp. 60–63]

sympathetic to intuitionism, at least later; Grelling and Hedrick supported some of Brouwer's views and criticised Hilbert's position; Bernstein, Hamel and Lietzmann were more or less neutral; Hellinger's position is unknown; and only Courant and Schmidt openly sided with Hilbert. This means that within his own school, Hilbert never had strong public support, and his criticism of intuitionism was not taken over in similarly vehement terms.

Finally, it is worth noting that Hilbert's assistant in the foundations of mathematics, Paul Bernays, made not a single direct public reference to intuitionism between 1922 and 1929, the period during which the debate raved most actively.[76] This is all the more remarkable since presumably Bernays was the person most *au courant* with the details of the formalistic foundational programme.

**Husserl-school** Another person with influence on a number of participants to the foundational debate was the philosopher Edmund Husserl (1859–1938). Although Husserl worked as a philosopher, his interest in the foundations of mathematics goes back to his originial training as a mathematician in Berlin under Kummer, Kronecker and Weierstrass.[77] The ways in which people involved in the foundational crisis were affected by Husserl varied. Becker wrote his *Habilitationsschrift* with Husserl and worked as his assistant; Carnap took part in Husserl's seminars in 1924–25; Weyl attended Husserl's lectures in Göttingen; Schmidt did the same, and he was a personal friend of the Husserl family; Kaufmann had an intense personal contact with Husserl; Lipps and Mahnke studied philosophy in Göttingen with Husserl; and finally Heidegger was Husserl's assistant at the philosophical seminar from 1919 to 1922.[78] It should be remarked that Husserl did not get to know Brouwer personally until 1928, when the former lectured in Amsterdam.[79]

There seems to be no consistent influence of Husserl as to what opinion to hold vis à vis intuitionism, since these people took different positions. Lipps, Mahnke and Schmidt were critical of Brouwer's achievements; however, they mostly devoted only some side remarks to intuitionism.[80] Heidegger was neutral, or at least he did not express his opinion on the issue.[81] Carnap was a logicist, who however was open to implement some of Brouwer's ideas, albeit in a modified form.[82] Kaufmann agreed with Brouwer on some points, but disagreed with him on others. Finally,

---

[76] In the papers Bernays published on foundations during this period, he either made technical comments, as in [Bernays 1927A], or gave a more subtle presentation of the formalistic position than Hilbert used to do, as in [Bernays 1927B, p. 15] on the status of logic or in [Bernays 1928, p. 202] on the question of mathematical existence.
[77] [Van Atten 1999, p. 24]
[78] [Sepp 1988, pp. 424–442], [Schuhmann 1977, pp. 68, 158, 269 and 281]
[79] [Schuhmann 1977, p. 330]
[80] [Lipps 1925, p. 71], [Lipps 1927–1928, vol. II, pp. 98–102], [Mahnke 1927, p. 286], and [Schmidt 1929, pp. 60–63]. Lipps' critical attitude may be related to the fact that he worked in the Göttingen mathematical-physical faculty.
[81] [Heidegger 1927, p. 9]
[82] See 4.4.3.

## 3.3. QUALITATIVE INQUIRY

the two persons of this group who by far contributed most to the debate, Weyl and Becker, were positive about Brouwer's intuitionism, at least in the beginning. Halfway the 1920s, however, Weyl became much more positive about the need for symbolic mathematics, and their views diverged substantially.[83]

**Polish logic** One might expect that there was some relation between the development of intuitionistic logic and the school of Polish logicians, who developed many-valued logic in the same period. Lukasiewicz put forward a system of three-valued logic as early as 1920.[84] However, the first time intuitionism and many-valued logic were mentioned together was in Lukasiewicz' 1930 paper *Philosophische Bemerkungen zu mehrwertigen Systemen des Aussagenkalküls* ('Philosophical remarks to many-valued systems of propositional logic'). Lukasiewicz remarked that Brouwer's ideas had until then not lead to a system, and that the construction of such a system was still 'completely unclear'.[85] Furthermore, they were treated jointly in Zawirski's paper *Les logiques nouvelles et le champ de leur application* ('The new logics and their field of application') in 1932, and at the 1934 *Prager Vorkonferenz des Ersten Internationalen Kongresses für Einheit der Wissenschaft* ('Prague Pre-conference of the First International Congress for Unity of Science'), where both Ajdukiewicz and Reichenbach placed intuitionistic logic and many-valued logic in a non-classical framework.[86]

**Logicism** Nowadays, the image most people have of the foundational debate in the 1920s and early 1930s is that it was a discussion between three currents: formalism, intuitionism and logicism. Such a conception is fostered by source books such as Benacerraf and Putnam's well-known *Philosophy of mathematics*, which opens with English translations of the three Königsberg contributions by Carnap, Heyting and Von Neumann on logicism, intuitionism and formalism respectively.[87] However, the debate was primarily seen as one between intuitionism and formalism by contemporaries, and logicism only played a marginal role.[88] During the

---

[83]Weyl and Becker corresponded on the relationship between mathematics and phenomenology; their mutual understanding deteriorated when Weyl had reacted in strongly negative terms to a paper that Becker had submitted for publication in *Symposion*. An analysis of the situation, including the two letters dealing with the foundations of mathematics in full, will be publlished in [Mancosu & Ryckman 2002].

[84][Lukasiewicz 1930, p. 65]. It is worth noting that Lukasiewicz' argumentation for the acceptance of a non-two-valued logic was contentual rather than formal.

[85]'völlig im Unklaren', [Lukasiewicz 1930, p. 75]

[86][Zawirski 1932] and [Ajdukiewicz 1935, p. 155], [Reichenbach, H. 1935, p. 37], resp.

[87][Benacerraf & Putnam 1964, pp. 31–54]. Dawson describes these translations as 'widely read and discussed', [Dawson 1984, p. 112].

[88]Mehrtens draws the same conclusion, [Mehrtens 1990, p. 291]. Dawson's remark that by 1920 'the logicist philosophy propounded by Whitehead and Russell had become the dominant philosophy of mathematics' refers more to the views of logicians than of mathematicians ([Dawson 1984, p. 112]; electronic communication to the author, 4/11/1998).

Interbellum, logicism fell substantially in reputation.[89] Its representatives were few, most notably Ramsey and Carnap. Russell, one of the fathers of logicism, to my knowledge never reacted to intuitionism publicly during the foundational debate,[90] while Wittgenstein only did so after he had heard Brouwer speak in Vienna in 1928, when he broke with his *Tractatus* position.[91] Furthermore, Carnap only entered the debate in 1927,[92] whereas Ramsey was relatively isolated in the United Kingdom. From the reactions to intuitionism during the foundational crisis (1921–1927), only some ten percent included remarks about logicism as well.[93] It was not without reason that Hahn claimed in 1929 that Germany had only heard about the fight between intuitionism and formalism.[94] Fraenkel did mention logicism in his *Einleitung in die Mengenlehre* from the second edition in 1923 on, but it did not figure as prominently as intuitionism and formalism. Thus, it seems that the 1930 Königsberg discussion between all three currents was the starting point of the image of a tripartite debate, which was then projected backwards onto the earlier period of the debate.

**Platonism** Presumably, there was a different current which did play a role, and maybe even an important one. For it is rather plausible that the philosophy most mathematicians adhered to was some kind of platonism. Thus, a belief in a platonic world of mathematical ideas may have been the background of the silent majority, of the working mathematicians who kept on working while others occupied themselves with foundational questions. However, since the silent majority remained silent, it is hard to say anything on its views.

There are two other, more philosophical currents that deserve being mentioned: the *Wiener Kreis* ('Vienna Circle') and the Berlin *Gesellschaft für empirische Philosophie* ('Association for empirical philosophy').

**Wiener Kreis** The *Wiener Kreis*[95] (Vienna Circle) started in 1924 under the direction of the philosopher Moritz Schlick. Its original members included the mathematician Hans Hahn, the sociologist-economist Otto Neurath, and the philosophers

---

[89] Cf. [Grattan-Guinness 1981, p. 497]. An overview of the development of logicism between the two World Wars can be found in [Grattan-Guinness 1984].

[90] Russell did touch upon intuitionism in the introduction to the second edition of his 'Principles of Mathematics', which appeared in 1937. There, he described the formalist as 'a watchmaker who is so absorbed in making his watches look pretty that he has forgotten their purpose of telling the time, and has therefore omitted to insert any works.' Intuitionism, in Russell's view, was 'a more serious matter', [Russell 1937, p. vi].

[91] See 4.4.2.

[92] [Carnap 1927]

[93] Of course, this does not exclude the possibility that there were contributions which discussed logicism in its own right, without reacting to intuitionism.

[94] [Hahn 1930, p. 101]. Carnap made a similar remark in [Carnap 1930, p. 298].

[95] The name stems from a manifesto published in 1929 by Carnap, Neurath and Hahn, [Dawson 1997, p. 26].

## 3.3. QUALITATIVE INQUIRY

Herbert Feigl and Friedrich Waismann. The philosopher Felix Kaufmann, too, frequented meetings of the Circle, but he did not consider himself a member. Soon, they were joined by the young German mathematician Kurt Reidemeister, who had been appointed in Vienna in 1922. One of the characteristics of the Circle was that all of its members had a first-hand acquaintance with some field of science.

Reidemeister suggested that the *Wiener Kreis* should read Wittgenstein's *Logisch-philosophische Abhandlung*, later known under the title of its English translation, *Tractatus Logico-Philosophicus*. Hahn enabled the members to discuss the work by explaining to them the ideas of Whitehead and Russell's *Principia Mathematica*. In 1926, the young philosopher Rudolf Carnap was accepted as *Privatdozent* at the university of Vienna; Reidemeister had by that time left Vienna for Königsberg. Hahn saw Carnap, who joined the Circle in the same year, as the person who would carry out Russell's program of using symbolic logic for doing philosophy in an exact way. The *Wiener Kreis* read the *Tractatus* a second time. Although Wittgenstein lived in Vienna from 1927 to 1929 and there were personal contacts between him and some of the *Wiener Kreis* members, he never joined the Circle. His main interest at the time was in architecture.

From 1927 on, the Circle attracted foreign visitors, like the Polish logician Alfred Tarski. In the late 1920s, the mathematician Karl Menger, who had succeeded Reidemeister, and the logician Kurt Gödel attended meetings of the Circle. In 1929, the *Wiener Kreis* published a manifesto, the tone of which Ringer characterised as 'that of outsiders, men who were fed up with the 'growth of metaphysical and theologizing tendencies' in the philosophy of the German academic establishment.'[96] Carnap and Neurath saw the Circle's philosophy as an expression of the *neue Sachlichkeit*, which was propagated by the *Bauhaus*. In 1931, the name 'logical positivism' was introduced to describe the philosophy of the *Wiener Kreis*.[97]

Whether intuitionism was discussed in the Circle discussions is not clear. In any case since Reidemeister joined the Circle, the members could have been aware of the fact that intuitionism existed. Some of its most important members, most notably Carnap, were close to Russell's logicism. Many members of or persons close to the *Wiener Kreis* reacted to Brouwer's intuitionism at one time or the other, including Kaufmann, Reidemeister, Carnap, Gödel, and Menger.

---

[96] [Ringer 1969, p. 308]
[97] [Feigl 1969, pp. 630–641], [Carnap 1963, pp. 20–30]

***Gesellschaft für empirische Philosophie*** The Berlin section of the *Internationale Gesellschaft für empirische Philosophie*[98] was founded in 1927.[99] It was set up to foster the development of philosophy on the basis of experiences in the different sciences. To this end, the group organised public lectures. The first of these was Joseph Petzoldt's *Rationales und empirisches Denken*, delivered in May 1927, in which he touched upon Brouwer's criticism of the principle of the excluded middle.[100] Reichenbach's lecture on the philosophical foundations of mathematics, delivered before the Association in November 1927, also treated Brouwer's intuitionism.[101] In the list of lectures that Danneberg and Schernus reconstructed, there are some by Dubislav which might have included a reaction to intuitionism, too.[102] However, it is not sure whether all these lectures were delivered, and the precise contents of them is unknown as well.[103] Such lectures were visited by some 100–300 persons on average, mostly academics, with an over-representation of medical professions. The local press, like the *Vossische Zeitung*, showed interest in the lectures, too.[104] Furthermore, the *Gesellschaft* could use substantial parts of the journal *Annalen der Philosophie* to promote its ideas.[105] There was a lively interest in such an enterprise, for by the end of 1927 there were already more than one hundred members. These included the mathematician Richard von Mises (originally a Viennese), the philosophers Walter Dubislav and Joseph Petzoldt, and the couple Alexander and Lily Herzberg. Apart from Von Mises, all of these publicly reacted to intuitionism. Hans Reichenbach, who is nowadays associated most with the Berlin *Gesellschaft*, only joined in October 1928.[106] John von Neumann and the physicist Fritz London, who contributed to the foundational debate, were in more or less close contact with the Association.[107]

---

[98]Apart from the Berlin section, hardly anything seems to be known about the International Association to which the section refers; cf. [Danneberg & Schernus 1994, pp. 394–396]. In 1931, it changed its name into *Gesellschaft für wissenschaftliche Philosophie* ('Association for scientific philosophy'), following a suggestion by Hilbert ([Hoffmann 1994, p. 29]). There also was a *Berliner Gruppe* or *Berliner Kreis*, which overlapped to a large degree with the Berlin section in terms of members. Some people, however, like Kurt Grelling and Carl Gustav Hempel, were a member of the *Berliner Gruppe* but not of the *Gesellschaft*. The *Berliner Gruppe* was organised in a more informal way. Details on the exact relationship between the two are not known yet, cf. [Danneberg & Schernus 1994].

[99]Carnap incorrectly gives 1928 (or later) as the year of society's origin, [Carnap 1963, p. 29] (cf. [Hoffmann 1994, p. 22]). This may be connected to Carnap's likewise mistaken claim that Reichenbach was appointed in Berlin in 1928; he actually started there two years earlier, [Biermann 1988, p. 226].

[100]See 5.3.1 and 6.4.2.

[101][Herzberg, L. 1928], a short report on the lecture.

[102]Like Dubislav 12/12/1927 *Konventionelle und moderne Logik* ('Conventional and modern logic'), Dubislav 13/1/1931 *Die Grundlagenkrise der Mathematik* ('The foundational crisis in mathematics'), and Dubislav 13/12/1932 *Das Unendlichtkeitsproblem in Logik und Mathematik* ('The problem of the infinite in logic and mathematics').

[103][Danneberg & Schernus 1994, pp. 425–428]

[104][Danneberg & Schernus 1994, pp. 405–407]

[105][Hoffmann 1994, p. 22]

[106][Danneberg & Schernus 1994, p. 393]

[107][Hoffmann 1994, pp. 26–27]

## 3.3. QUALITATIVE INQUIRY

There were strong connections between the Berlin and Vienna group, also because of the good personal relationship between Carnap and Reichenbach, who had been in contact since 1920.[108] From 1930 on, Carnap and Reichenbach jointly edited the journal *Erkenntnis* as the home journal of the *Wiener Kreis* and the Berlin *Gesellschaft*. In 1930, the two groups also joined forces to organise the Königsberg conference on the foundations of mathematics, at which intuitionism was one of the main currents discussed.[109]

### 3.3.4 People

**General** The character of the study carried out in this book implies that I did not only look at the 'great names' of a period, such as Brouwer, Hilbert and Weyl, but that I also included works of numerous lesser-known mathematicians and philosophers. Altogether, some 120 people were involved in the discussion. The historical actors discussed here have one characteristic in common, which is that most of them held a university position. The foundational debate was a highly academic topic.

Some basic data may serve to further characterise the participants to the foundational debate.

Within this group of academics, the number of mathematicians greatly exceeded that of philosophers.[110] Also, the male predominance was almost complete. In fact, I only found three women who reacted to intuitionism: Lily Herzberg, Marie Deutschbein and Alice Ambrose. Herzberg contributed by writing a one-and-a-half-page report of a lecture Reichenbach had given on the foundational crisis.[111] Deutschbein published two contributions to the foundational debate. One was a paper on the philosophical importance of mathematics for *Bildung*, in which she devoted one page to intuitionism.[112] The other was a higher education textbook on the philosophical foundations of mathematics, written together with Walther Brand, in which the foundational debate was treated in a subsection.[113] Ambrose, a student of Wittgenstein in Cambridge in the 1930s, published a paper on 'a controversy in the logic of mathematics' in 1933.[114] I found no data about ethnic minorities, homosexuals, etc., which might have been interesting in terms of revealing existing networks. On the whole, the debate seems to have been a discussion between white, male, western academics – like presumably most mathematicians in those days.

---

[108] [Reichenbach, M. 1994, p. 13]
[109] See 4.4.1.
[110] The importance of such a distinction should not be exaggerated. As is still the case nowadays, people doing mathematical logic can be found in both mathematical and philosophical departments, and they may have studied either mathematics or philosophy.
[111] [Herzberg, L. 1928]
[112] [Deutschbein 1929, p. 332]
[113] [Brand & Deutschbein 1929, pp. 45–50]
[114] [Ambrose 1933]

## 112 CHAPTER 3. OVERVIEW OF THE FOUNDATIONAL DEBATE

**Age** As to the age of the participants, the generation most involved in the debate was clearly Brouwer's, not Hilbert's. When the debate started in 1921, Hilbert was 59 years old, Brouwer 40. The average age of the participants was, measured in 1921, about 35.

The persons from Hilbert's generation[115] who were involved were: Brodén, Hadamard, Hölder, Korselt, Mannoury, Marcus, Petzoldt, Pierpont, Schoenflies, Study, and Voss. Of these, Brodén, Korselt, Marcus and Petzoldt were clearly negative about intuitionism;[116] Hölder, on the other hand, preferred intuitionism to formalism, and also Pierpont judged intuitionism positively;[117]; Mannoury disagreed with many of the ideas of his personal friend Brouwer, but did not see formalism as an alternative;[118] Hadamard and Study disliked the whole controversy;[119] whereas Schoenflies remained neutral and Voss did not express his opinion on the issue.[120] Thus, on average those involved from Hilbert's generation seemed to be somewhat negative about intuitionism.

The youngest generation involved consisted of those born after 1900: Ambrose, Church, Freudenthal, Gödel, Heiss, Herbrand, Hirsch, Kolmogorov, Menger, Von Neumann, and Ramsey. Of these, the philosopher Heiss was negative about Brouwer's conclusions;[121] Church and Freudenthal remained neutral;[122] so did the philosophers Ambrose,[123] a student of Wittgenstein in Cambridge, and Hirsch, who devoted his dissertation to the foundational crisis[124] and mostly published newspaper articles of an expository nature on the debate; Herbrand looked for a kind of synthesis between intuitionism and formalism;[125] Menger was at first interested in intuitionism, but he became more critical over the years;[126] Von Neumann saw positive sides to intuitionism;[127] Ramsey first resisted intuitionism, but later made some positive contributions to it, which were however not published during his lifetime;[128] and Gödel and Kolmogorov made several positive contributions to

---

[115] i.e., who were at most five years younger than Hilbert
[116] [Brodén 1925, pp. 235–236]; [Korselt 1916, p. 135]; [Marcus 1928]; [Petzoldt 1927, pp. 157–158]
[117] [Hölder 1924, p. 277], [Pierpont 1928, p. 53]
[118] [Mannoury 1924, p. 40]
[119] [Hadamard 1926, p. VI]; [Study 1929]
[120] [Schoenflies 1922, p. 102]; Voss only mentioned Brouwer in [Voss 1914, p. 146].
[121] [Heiss 1928, p. 405]
[122] [Church 1928, pp. 76–77]; [Freudenthal 1932, p. 98]
[123] [Ambrose 1933, p. 611]
[124] [Hirsch 1933]. I have not seen the dissertation; there is a short description of its contents in [Gruenberg 1988, pp. 350–351].
[125] [Herbrand 1930A, p. 164]
[126] Menger was still positive about intuitionism in [Menger 1926, p. 115], but he was more critical in [Menger 1928A, p. 225] and [Menger 1928C, p. 303]; see 4.4.3.
[127] See 3.3.2.
[128] Ramsey's criticism of intuitionism can be found in [Ramsey 1926A, p. 339] and, in a more subtle form, in [Ramsey 1926B, pp. 216–217]; Ramsey's notes on intuitionism are now available as [Ramsey 1991B] and [Ramsey 1991C]. In the introduction to the book which contains the latter items, Galavotti, the editor, mentions Ramsey's 'conversion to intuitionism by the end of his life' and refers to an earlier edition of some of Ramsey's published and unpublished

## 3.3. QUALITATIVE INQUIRY

the development of formalised intuitionistic logic.[129] Thus, the youngest generation involved was on the whole not negative about Brouwer's intuitionism, and some actually contributed to it positively.[130]

### 3.3.5 Languages and media

**Languages** More insight in the spreading of the debate can be obtained by looking at the languages in which contributions were made. If we split all public contributions to the debate according to language, the German predominance is clear. More than half of the contributions was made in the German language, French coming second with less than one fifth. After that, English and Dutch were the languages most used respectively. The odd publication was made in, e.g., Norwegian, Greek or Catalan.[131]

After the publication of Weyl's *Grundlagenkrise* paper in 1921, the debate started as a German affair, with some Dutch contributions. In 1923, the first Italian and Hebrew contributions appeared, but this had no lasting influence on the debate.[132] The following year, Wavre and Dresden made the debate accessible to the French and English reading readers respectively.[133] From 1933 on the German language lost its position as being used more often than all other languages together, and the use of French and English became more frequent.

**Media** The media which the persons involved in the foundational debate chose to make their contribution reveal some characteristics of the debate, too. Of the public contributions that were available for this research, about one fourth were first delivered as a lecture and published later. The other three fourths consisted of papers and books that were published directly. The majority of the contributions consisted of more or less popular presentations, that were either meant for a large audience or could be understood without any specific knowledge. Hilbert was the champion of talks. The five papers he published on the foundational debate were all

---

works edited by Braithwaite, [Galavotti 1991, p. 22]. Braithwaite, in his introduction, tell us that Ramsey was 'converted to a finitist view which rejects the existence of any actual infinite aggregate' [Braithwaite 1931, p. xii]. It seems probable that Braithwaite, who wrote the introduction in 1930, did not distinguish between intuitionism and finitism, just like Ramsey himself (cf. [Ramsey 1991B, p. 201]). Braithwaite does not mention what makes him believe that Ramsey was converted to a finitist view. Having read the above-mentioned posthumously published notes, I would say that Ramsey became more sympathetic towards some aspects of intuitionism, but remained critical of others, like Brouwer's theorem of the equivalence between a triple negation and a single one, [Ramsey 1991B, p. 202], [Ramsey 1991C, pp. 215–217]. Majer argued that Ramsey was inspired by Weyl's intuitionism, [Majer 1989].

[129] See 5.3.1 and 5.4.2.
[130] I have not been able to find biographical data on all the people involved, so it is possible that I missed some persons in these categories.
[131] [Skolem 1926], [Oikonomou 1926] and [Garcia 1933] respectively.
[132] The Italian paper was [Levi 1923]; the Hebrew one was [Fraenkel 1923B], which was published in a multi-lingual journal, German being the language of the other version of this paper.
[133] [Wavre 1924] and [Dresden 1924]

reports of lectures he had given earlier, and mostly quite popular presentations. Nevertheless, Hilbert managed to get most of them published in the normally strictly mathematical *Mathematische Annalen*, of which he himself was one of the editors-in-chief.[134] The journals used were in vast majority scientific ones, although besides mathematical journals also philosophical ones played a considerable role.

However, the trends sketched here are not conclusive, since they are to a large degree dependent on the way in which this study was carried out. In the first place, almost all unpublished lectures have been lost, or at least they could not be traced. Secondly, there were newspaper articles in which the foundational debate was discussed, but these are much harder to find. I mainly used personal archives for finding them, which means that the search process was not carried out systematically. Finally, the character of this work automatically has a bias towards scientific journals. Since this is a study in the history of mathematics, I tend to focus on mathematical journals in the first place. Therefore, what was presented may be slightly biased.

It took some time before journals devoted specific attention to the foundational debate. In 1930, the *Blätter für Deutsche Philosophie* devoted an issue to the philosophical foundations of mathematics.[135] This included papers by Scholz, Fraenkel, Carnap, Menger, and Bernays on the foundational debate. The journal *Erkenntnis* followed the following year with a special on the foundational crisis, including reports from lectures given at the 1930 Königsberg conference.

The foundational crisis entered a high school book in 1927. In that year, Lietzmann published his *Aufbau und Grundlage der Mathematik* ('Construction and foundation of mathematics'), meant for the higher classes of secondary school. He motivated the publication of the book by reference to the new Prussian directives, cited by him as saying:[136]

> Logik und Erkenntnistheorie finden einen Platz in der Mathematik. Auch die psychologischen Grundlagen des mathematischen Denkens soll der Unterricht berühren.[137]

Lietzmann only makes a short remark about intuitionistic mathematicians who do not accept the universal validity of the principle of the excluded middle.[138]

In the 1930s, a new medium was introduced in the debate: the radio. However, it was used scarcely, with only Fraenkel and Scholz delivering radio lectures on the foundational debate.[139]

---

[134] [Hilbert 1922], [Hilbert 1923], [Hilbert 1926], [Hilbert 1927], [Hilbert 1930A]
[135] Issue 3 of the 1930–1931 volume.
[136] [Lietzmann 1927, p. III]
[137] 'Logic and epistemology will find a place in mathematics. The education has to touch upon the psychological foundations of mathematical thinking, too.'
[138] [Lietzmann 1927, p. 13]
[139] [Fraenkel 1932] and [Scholz, H. 1933]

## 3.4 Conclusion

The features of the foundational debate just described lead to the following image. The foundational crisis started in 1921. Before that time, there were some separate reactions to Brouwer's intuitionism, but nothing like a debate. This means that it was not in the first place the mathematical material itself that evoked the reactions, since that had been available since 1918.[140] In 1921, after the publication of Weyl's paper on the foundational crisis, a large number of reactions appeared. Among the people who reacted were Hilbert and Bernays, seen as the most prominent adherents of formalism. From 1922 on, the debate grew monotonously, until it reached its top around 1927. After 1927, the number of reactions to Brouwer's intuitionism fluctuated.

Around 1930, the tone of the discussion changed markedly. Before that year, when Brouwer, Hilbert and Weyl were the main actors, words like 'revolution', 'betrayal' and 'Bolshevik menace' could apparently be used legitimately within the debate, although such contributions did not dominate numerically. From 1930 on, important contributors to the debate like Von Neumann, Carnap and Heyting defended their own position less vehemently and were more open to the advantages of other currents in the foundations of mathematics. The search for a synthesis had begun.

In general, the participants to the foundational debate were academics, most of them mathematicians, some philosophers. The dominant generation was of Brouwer's age: about 35 years old when the debate started, and male. The crisis was heaviest in Europe and was expressed mostly in the German language. From 1933 on, the influence of the changed political climate in Germany becomes evident in mathematics, too, and the number of German contributions to the foundational debate dropped markedly.

The earliest clear point found here at which the foundational crisis is definitely over is the 1930 Königsberg discussion. However, there are other reasons to put the end of the debate earlier, namely in 1928.[141] Therefore, in the remainder of this thesis the foundational debate is discussed at length only for the period 1921–1927.[142]

During the period 1909–1933, the two themes that were discussed most remained unchanged: the question of mathematical existence and constructivity, and the status of the principle of the excluded middle and logic. The only change that took

---

[140] See 2.6.

[141] These reasons, connected to the use of metaphors, are treated in 6.2.

[142] I thus disagree with Fraenkel, who sticked to a much broader demarcation of the foundational crisis. In his view, the crisis started with the reactions to Zermelo's well-ordering proof in 1904 and lasted until after the Second World War ([Fraenkel 1947, p. 19] and [Fraenkel 1951, p. 7]). I think the episode about the axiom of choice should only be seen as a prelude to the foundational crisis, since the number of contributions before World War I does not compare to that afterwards. As to the end of the debate, I find the changed tone in the contributions to be decisive in taking 1928 as marking the end of crisis.

place was that, over time, more people recognised that these were the main themes. Furthermore, in the beginning of the debate the third main theme was the role of intuition and philosophy in mathematics; after about 1928, this was replaced by the question of the contents of mathematics and the legitimation of formalising mathematics. Additionally, after 1930 a substantial part of the reactions to intuitionism focused on Heyting's formalisation of intuitionistic logic instead of on Brouwer's work or Weyl's presentation thereof.

From these themes, I shall treat the first two in detail in the following chapters and leave out the others. The main reason for this is that these themes are the dominant ones throughout the whole debate. Furthermore, the subject of philosophy and intuition was, I think, for most people involved still a side issue, whereas the question of contents and formalization only started to play a considerable role after 1928, when the foundational crisis was already over.

# Chapter 4

# Reactions: existence and constructivity

Die Wolfskehl-Stiftung veranstaltete einen Vortragskreis von Planck über Quantentheorie. Ein Ausflug wird angesagt: Versammlung vor dem Hause des andern Planck, (...) Hainholzweg 44. Stimme aus dem Publikum: 'Wie findet man da hin?' [Hilbert:] 'Aber das ist doch ganz einfach (sprich[t] ostpreußisch!), da gehn Sie nach Hainholzweg 42, und dann noch ein Haus weiter, dann sind Sie da.'[1]

Walter Lietzmann[2]

## 4.1 Introduction

### 4.1.1 Mathematical existence

In mathematics one deals with various objects. Some theorems are about numbers, others about functions or groups, to name but a few. But even though mathematicians are able to work with these mathematical objects, their ontological status usually cannot be inferred.[3] What precisely *are* these objects mathematicians speak about? Are they pre-existing entities that we discover? Are they creations of the human mind? Or are they nothing more than signs written down on paper?

---

[1]'The Wolfskehl Foundation organised a sequence of lectures by Planck on quantum theory. An excursion is announced: assembling in front of the other Planck's house, (...) Hainholzweg 44. Voice from the public: 'How do we get there?' [Hilbert:] 'But that is very easy (speak[s] east-Prussian!), you go to Hainholzweg 42, then one house further, and then you are there.''

[2][Lietzmann 1960, p. 26]

[3]Ontology, as explained in the glossary, is the study of being.

## 118  CHAPTER 4. REACTIONS: EXISTENCE AND CONSTRUCTIVITY

Does, for instance, the number '2' have an existence independently of us? And what about a function like $f : x \mapsto x + 1$?

Some mathematicians might argue that such existence questions are too philosophical to be part of mathematics. But they are linked to other problems which come closer to the core of mathematical argumentation. It is certainly reasonable to ask when objects are mathematical enough to be included in a mathematical system. In other words: the question what we need to know about objects in order to consider them as legitimate objects of mathematical study is one which is of importance to mathematics proper. The answer to this question may vary according to which position is taken with respect to the issue of mathematical existence. For a metaphysical realist, there is no need to create mathematical objects, since they already exist. Therefore, a realist will generally have fewer demands than an idealist.[4]

One may distinguish between three questions of mathematical existence. The first is the most philosophical one: it is the question about the ontology of mathematical objects. If we take the example of non-Euclidean geometry, the ontological existence question would be: do planes, lines and triangles of non-Euclidean geometry exist in any way, for example in a Platonic world of ideas? The second one is about the existence of mathematical systems. In this example, the appropriate existence question would be: on what grounds could or should we include the whole system of non-Euclidean geometry into the study of mathematics? Finally, the third question is about existence statements within mathematics. This would ask how we could prove the existence of, for instance, a certain line in non-Euclidean geometry.[5]

---

[4] In philosophy proper, two main currents are distinguished concerning both metaphysical and epistemological questions. Metaphysical realism, following Plato, states, put roughly, that there exists an outer reality which in its existence is independent of our experience of the world. Opposed to this are various views joined together under the name anti-realism, with idealism as one of its main representatives. Metaphysical idealism holds that the mind is the only bearer of reality, [Van Dooren 1983, p. 119]. An analogous distinction can be made within epistemology. Epistemological realism takes the objects to play a primary role in the process of knowing, and leaves the knowing subject with a mostly receptive role, [Gethmann 1995, p. 500]. Therefore, it is possible for an epistemological realist that propositions are true, even though we do not and maybe even cannot recognize so. Epistemological idealism, following Kant, maintains that all conceivable propositions deal with human experiences, [Dancy 1989, pp. 136–137]. In other words: there is no evidence-transcendent truth. This leaves us with three possible combinations: one can be a metaphysical and an epistemological realist (Plato's position), one can be a metaphysical realist and an epistemological idealist (Kant's position), or one can be a metaphysical idealist and therefore also an epistemological idealist (Fichte's position). Brouwer's intuitionism is an idealistic philosophy of mathematics in the metaphysical sense (see 2.2.2).

In the primary and secondary literature on the foundational debate in mathematics, however, there is some confusion of terminology, which can be misleading. For example, Ridder labels intuitionists 'realists', but describes their point of view correctly, [Ridder 1931, p. 12]; similarly Wavre calls formalists 'idealists', [Wavre 1924, pp. 243–245]. I will throughout this work stick to the realist-idealist terminology as just explained.

[5] The phenomenologist Moritz Geiger made a different distinction between existence questions, for which Von Freytag introduced a terminology ([Geiger 1928, p. 403]; I used [Freytag 1937] as

## 4.1. INTRODUCTION

Even those who find that the ontological question lies outside the domain of mathematics proper will have to answer the questions when something can be called a mathematical system, and when we can state the existence of a mathematical object within a mathematical system. Usually, people who leave out ontological considerations see mathematics as a formal system, in which consistency suffices for existence and in which existence statements are derived formally.

In the foundational debate in the 1920s, all three existence questions were discussed. Brouwer focused on the existence of mathematical systems, with a clear link to his view on the ontological status of mathematical objects.[6] However, most discussions were about mathematical systems or objects, and many participants to the debate dismissed the ontological question as an illegitimate invasion of philosophy into mathematics. Becker and Heyting did treat the ontological existence question, but this aroused few reactions.[7]

As can be seen from the history of mathematics, the various existence questions are of all times. They were asked frequently in the past, especially when new mathematical objects were introduced. For instance, for a long time square roots of negative numbers were considered mysterious. The Renaissance scholar Cardano, who lectured in Bologna and Milan, was probably the first to voice this opinion. In his major work *Ars magna*, which was published in 1545 and which is often taken to mark the birth of modern mathematics, he described these roots as 'sophisticated' and 'as subtle as useless'.[8] Around 1700, the question had still not been settled, and Leibniz called the imaginary numbers amphibians between being and non-being.[9] Later, in the 18th century, the British metaphysician Berkeley attacked Newton's method of fluxions. He called the supposed existence of a finite ratio between absolutely vanishing terms absurd.[10] And in the decades before the foundational debate the issue of mathematical existence, both of systems and of existence statements, was raised by mathematicians like Kronecker, Poincaré and

---

cited by [Bockstaele 1949, pp. 16–19]). Following this characterisation, there are two categories into which existence questions can be divided. The so-called big existence problem asks what the ontological status of mathematical objects is. This should be distinguished from the small existence problem of which criterion mathematical objects have to fulfill in order to be admissible within a mathematical system. The latter characterisation, however, is rather vague. If it is taken literally, one may ask how we could have obtained a mathematical system without having mathematical objects. If it is taken to apply to existence statements within a mathematical system, it contains the big existence question in disguise. For, as Van Dalen pointed out to me, if we have proved a statement of the form $\neg \forall (x) \neg f(x) = 0$, and we ask ourselves if there exists a number that satisfies $f(x) = 0$, then we are not asking for the existence of a number. Rather, we are looking for a number $n$, which already existed, which satisfies the given equation. That is, we are looking for a proof $p$ such that $p$ proves $f(n) = 0$. But this is a question about the ontology of the proof $p$.

[6]Cf., e.g., [Brouwer 1907, p. 81; pp. 217–218].
[7]See 4.2.3, 4.3.2 and 4.3.1.
[8][Cardano 1545, p. 287]
[9][Becker, O. 1964, p. 214]
[10][Cajori 1980, p. 219]; Berkeley spoke about 'the ghosts of departed quantities'.

# 120    CHAPTER 4. REACTIONS: EXISTENCE AND CONSTRUCTIVITY

the French semi-intuitionists.[11]

The same kind of questions are still being discussed now, and they can always re-appear in the future.[12] Putting the question is nothing special. However, at some times the importance ascribed to it seems, to the mathematicians involved, much bigger than at others. The foundational debate in the 1920s definitely belongs to a period when great significance was attached to these issues. Existential questions were among the main items discussed.[13]

## 4.1.2   A short history of constructivism

**Antiquity**    Related to the existence question is the assertion that all mathematics should be constructive. This subject, too, has a long history. A famous thesis by the Danish mathematician-historian Zeuthen, formulated by the end of the 19th century, holds that already in Antiquity constructions were needed in geometry in order to ensure the existence of the mathematical object under investigation. Thus, Euclid would only use an object such as the middle of a line segment after he had proved by a construction that the object in fact existed.[14] Presumably relying on Zeuthen's thesis, Kline argued that Gauss' proof of the fundamental theorem of algebra 'inaugurated a new approach to the entire question of mathematical existence', since it did not yield a method for computing the desired roots.[15] However, Zeuthen's argumentation is far from conclusive, and modern commentators have argued against it.[16] Also, it should be pointed out that it is not clear which constructions were ranked as admissible by the ancient Greeks. Although Euclid restricted constructions to ruler-and-compass ones in his *Elements*, it is clear that in early Greek geometry more general constructions were admitted.[17]

**Kant**    'The grandfather of mathematical constructivism',[18] Kant, put forward constructions as the decisive characteristic of mathematics. As he formulated it in the *Kritik der reinen Vernunft*:[19]

---

[11]See 1.2.2,1.4.2 and 1.3.1 respectively.

[12]Thiel even maintains that one can just as well read the literature from the 1920s on philosophical-foundational matters as the current literature, [Thiel 1972, p. 159].

[13]Perhaps not incidentally, the discussion on mathematical existence runs parallel to the rise of existentialism in German philosophy, as developed primarily by Jaspers and Heidegger. Where *Lebensphilosphie* stressed life, existentialism placed existence in a prominent position, [Lukács 1960, p. 430]. For *Lebensphilosophie*, see 6.3.1.

[14][Zeuthen 1896, p. 223], [Zeuthen 1912, pp. 40–46]

[15][Kline 1972, p. 599]. Kline contrasts Gauss' proof to the Greeks' criterion for existence, constructibility. The non-constructive proof indeed became one of the targets of intuitionism, and both Brouwer & De Loor and Weyl published (independently of each other) constructive versions of the proof, [Brouwer & De Loor 1924] and [Weyl 1924], as did Skolem ([Skolem 1924]; see also 4.3.

[16]Cf., e.g., [Mueller 1981, p. 15] and [Knorr 1983, p. 140].

[17]Cf. [Fowler 1987, pp. 287–291] for the example of the so-called *neusis*-construction that was freely used by, among others, Hippocrates and Archimedes.

[18][Posy 1998, p. 315]

[19][Kant 1781/1787, A p. 741/B p. 485]

## 4.1. INTRODUCTION

> Die *philosophische* Erkenntnis ist die *Vernunfterkenntnis* aus *Begriffen*, die mathematische aus der *Construction* der Begriffe.[20]

A modern interpretation hereof would read as follows. In Kant's view, mathematical objects have to be constructed in intuition. Intuition is seen as a direct apprehension of appearances by means of forms, which enable one to order the appearances. The only pure forms in intuition which Kant recognizes are space and time.[21] Furthermore, according to Kant mathematics is not analytic, but synthetic.[22] Thus, by construction Kant meant a synthetical procedure to perform judgements in space and time of intuition.

In the beginning of this century, Kant was probably the most valued philosopher at German universities, and few German academics could escape noting his views. In this way, the Kantian stress on constructions in mathematics may have played its role in the foundational debate more or less regardless of its interpretation.[23]

Fichte went further and claimed that constructibility was a demand for all of science.[24] In this light, it is not surprising that Weyl mentioned Kant and Fichte as two of his main philosophical influences.[25]

**Intuitionistic constructivism**   As far as I know, the use of the term 'constructive' in its present, intuitionism-inspired meaning can be traced back to Lebesgue, who used it in his 1904 *Leçons sur l'intégration et la recherche des fonctions primitives* ('Lessons on integration and the research into primitive functions').[26] There, he distinguished between descriptive definitions, in which characteristic properties are given, and constructive ones. On the latter he wrote:[27]

> Dans les définitions *constructives*, on énonce quelles opérations il faut faire pour obtenir l'être que l'on veut définir.[28]

This is basically the same notion as the one used by Brouwer and later intuitionists: constructions are mathematical operations which are completely performable in

---

[20]'*Philosophical* cognition is *rational cognition* from *concepts*, mathematical cognition that from the *construction* of concepts.' English translation cited from [Kant 1998, p. 630].

[21][Kant 1781/1787, A pp. 34–36/B pp. 56–57], [Caygill 1995, p. 265]

[22][Kant 1781/1787, A p. 14/B p. 42]. An analytic judgement is a judgement that is necessarily true on purely logical grounds, because the meaning is already implicit in the subject; a synthetical judgement gets its meaning from non-logical sources as, e.g., experience.

[23]Opinions expressed during the 1920s tended to agree on the fact that mathematics should be constructive, but to disagree on what this should mean; see, e.g., 4.2.2, 4.2.2 and 4.4.1. Bernays was the only person I found who explicitly referred to the Kantian terminology when speaking about intuitive constructions in mathematics, [Bernays 1923, p. 162].

[24][Ritter & Grunder 1971–1995, vol. 4, pp. 1011–1113]

[25][Weyl 1954, pp. 632–637]

[26]Kronecker seemed to prefer the wording 'to arithmetize', cf. [Kronecker 1887, p. 253].

[27][Lebesgue 1904, p. 99]

[28]'In *constructive* definitions one formulates which operations one has to carry out in order to obtain the object one wants to define.'

principle.[29] The same interpretation of existential statements was given in Russell and Whitehead's monumental *Principia Mathematica*:[30]

> An asserted proposition of the form '$(Ex).fx$' expresses an 'existence theorem', namely 'there exists an $x$ for which $f(x)$ is true.' The above proposition gives what is in practice the only way of proving existence-theorems: we always have to find some particular $y$ for which $fy$ holds, and thence to infer '$(Ex).fx$'.

The opposed, classical view of identifying an existence statement with a statement of the form 'not for every not' dates back to Frege and Peano.[31]

One of the first prominent pure existence proofs[32] came from Hilbert. Gordan had worked long to prove his conjecture of the existence of a finite basis in invariant space theory, and he had only found some proofs under special circumstances. These proofs were algebraic and constructive in nature. Hilbert, however, proved the whole conjecture — but in a purely existential way. Gordan refused to accept the proof; he called it 'not mathematics, but theology'.[33] In 1920, the what we would nowadays call non-constructive character of classical existence statements was pointed out by Skolem. Introducing what is now known as Skolem-functions,[34] Skolem remarks that these can only be thought of by using the axiom of choice.[35] Although the paper in which this concept was established was one of those that 'at once' made Skolem one of the foremost among logicians,[36] it does not seem to have influenced the debate on mathematical existence at all. Why this is the case remains unclear.

Hermann Weyl stated his view on mathematical existence most manifestly for a large audience, stressing the issue of existence statements within a mathematical

---

[29][McCarty 1983, p. 108]

[30][Whitehead & Russell 1910, p. 20]

[31][Moore 1980, p. 97]

[32]A pure existence proof, as explained in the glossary, is a proof in which the existence of a mathematical object is proved in a non-constructive way, i.e., the proof gives no method for actually constructing the object.

[33]Cf. [Hilbert 1923, p. 188]. In fact, Hilbert seems to have taken Gordan's criticism to heart, cf. [Rowe 2000, p. 60].

[34]In modern notation, a Skolem-function is a function $f_\varphi$ associated with a formula $\varphi$ such that the so-called Skolem axiom

$$\forall x_1 x_2 \ldots x_n (\exists y\, \varphi(x_1, x_2, \ldots, x_n, y) \rightarrow \varphi(x_1, x_2, \ldots, x_n, f_\varphi(x_1, x_2, \ldots, x_n)))$$

holds, where $n+1$ indicates the number of free variables of $\varphi$. In words: if, for all $x_1, x_2, \ldots, x_n$, there exists an element $y$ such that formula $\varphi(x_1, x_2, \ldots, x_n, y)$ holds, then the Skolem-function $f_\varphi$ picks out this element $y$. The definition given here is a generalisation of the one introduced by Skolem in his 1920 paper. It can be proved that for each theory $T$ there exists a Skolem expansion $T^*$ obtained by extending $T$ with the Skolem axioms and the original language $L$ with the Skolem functions, such that $T^*$ preserves all the original theorems. Cf. [Van Dalen 1997, p. 139].

[35][Skolem 1920, p. 107]; on the axiom of choice, see 1.3.1.

[36][Fenstad 1970, p. 11]

## 4.1. INTRODUCTION

system. He clearly linked the subjects of mathematical existence and constructivity. Through Weyl, both subjects entered into the foundational debate.[37]

In the 1920s, there appears to have been no generally accepted division between constructive and non-constructive mathematics. Brouwer used words such as *opbouwen* ('to construct') from the start, and Weyl gave mathematical constructions a central place in his exposition of intuitionism in 1921.[38] Nevertheless, constructive mathematics was not seen by all involved in the debate as being solely the terrain of the intuitionists. Although the term 'constructive' appeared regularly in the debate, it was not until the beginning of the 1930s that it got a mathematically more specified meaning based on the concept of computability, by the work of Turing and others.[39]

**Modern constructivism** After the publication of Weyl's paper, the demand that mathematics be 'constructive' was continued by Heyting. The latter followed Brouwer by defining intuitionistic mathematics as coming into being by means of a constructive ability of our mind.[40] In was only in 1967, however, that the term 'constructive mathematics' came to serve as a general label. This was one result of Errett Bishop's research program, which started with the book 'Foundations of constructive analysis'.[41] He explicitly mentioned Brouwer as the one who started 'constructivising' mathematics and termed him a 'constructivist'. In Bishop's view, the goal of constructive mathematics is to give numerical meaning to as much as possible of classical mathematics.[42] He clarified this by showing how constructive mathematics is done by proving theorems in a constructive way. Bishop paid little attention to philosophical discussions and focused on the practice of constructive mathematics.

Nowadays, the term 'constructivism' is used to describe a collection of views on mathematics placing certain restrictions on the kinds of arguments seen as admissible.[43] These include various branches of mathematics, not only Bishop's constructive mathematics, but also finitism, Markov's constructive recursive mathematics, and intuitionism.[44] In this way, constructivism has become a branch of mathematics that can be distinguished quite clearly from pure formalism.

The present-day intuitionistic view on existence statements is part of the Brouwer-Heyting-Kolmogorov interpretation of logic, in which an argument is ac-

---

[37] See 4.2.1.
[38] See 2.3.1 and 4.2.1 respectively.
[39] See 4.4.3.
[40] [Heyting 1934, p. 2]
[41] Lorenzen's 'constructive' mathematics is of a very different kind, since he deals with symbolic sign constructions, [Lorenzen 1950, p. 163].
[42] [Bishop 1967, p. ix]
[43] [Detlefsen 1994, Vol. 1, p. 656]. Note that this is a negative way of characterising constructive mathematics, which does not attempt to give meaning to the 'constructive' part in this ism.
[44] [Troelstra & Van Dalen 1988, vol. 1, pp. 1–5]. More information on the developments of constructive mathematics after Brouwer can be found in this book.

cepted as a proof for $\exists x A(x)$ if the argument provides a $d \in D$, where $D$ is the domain, together with a proof of $A(d)$.[45] Within intuitionism, the tradition of working with 'constructions' continues until today. It was stimulated by Kreisel's interpretation of intuitionistic logic, where 'construction' and 'constructive proof' are the basic notions.[46] Troelstra, a student of Heyting and one of the main modern intuitionists, described intuitionism as 'a theory of mental constructions'.[47]

## 4.2 The beginning of the debate

The debate on mathematical existence and constructivity developed in several phases. First, there was the 'inner circle' of people involved: Brouwer, Weyl, Hilbert and Bernays. This was also the period in which it was not clear to Hilbert and Bernays what the intuitionists Brouwer and Weyl meant by existence and constructivity. From 1924 on, the differences became clearer and the debate centered on what 'existence' and 'constructive' should mean within mathematics. At the same time, these questions moved to a more central position in the foundational debate. Finally, around 1930 the question was settled in so far that mathematicians kept on using existence in the classical way (but with a modified meaning),[48] whereas intuitionism won the battle about the predicate 'constructive'.

In the remainder of the chapter, I present my interpretation of the debate on mathematical existence. More specifically, I defend the following theses. In the first place, the debate on mathematical existence was more one of clarification than of real discussion. The philosophical part of the debate took place at the outskirts; at the core of the debate, participants stated *what* they stood for rather than *why*. Secondly, the concept of constructivity played a major role in the debate on mathematical existence. Claimed first by all sides involved, it was later generally associated with intuitionism and consequently used as a means of distinguishing between the two kinds of mathematical existence involved. Nobody denied that the difference existed, and most people were perfectly willing to accept Brouwer's analysis of the situation—but not his conclusion that non-constructive existence statements should not be accepted. Hence, the intuitionistic notion of mathematical existence was incorporated into mathematics by a specification of mathematical language: pure existence was distinguished from constructive existence. In the 1930s, the concept of constructivity was further developed by Turing, Gödel and others in terms of computability, and it was also accepted by formalists like Bernays. I support these theses by giving an overview of the debate in a roughly chronological order.

---

[45][Troelstra & Van Dalen 1988, vol. 1, p. 9]
[46][Sundholm 1983, p. 152]; Kreisel's interpretation is in [Kreisel 1962].
[47][Troelstra 1983, p. 199]
[48]See 4.4.1.

## 4.2. THE BEGINNING OF THE DEBATE

**Pre-Weyl reactions**  Before Hermann Weyl published his 1921 paper, there were few reactions to Brouwer's intuitionism in general and to his view on mathematical existence in particular. The first to touch upon the issue was Ludwig **Bieberbach** in his inaugural lecture *Über die Grundlagen der modernen Mathematik* ('On the foundations of modern mathematics') as a professor of mathematics in Basel in 1914.[49] Bieberbach's early reaction may be explained by the fact that he had met Brouwer personally at a *DMV* congress in 1912.[50] In the lecture, Bieberbach treats the question of mathematical existence, defined by him as: given a set of axioms, do objects of thought exist to which these axioms apply consistently? In his view, such a question is meaningless to intuitionists. Bieberbach himself supports the formalistic point of view, which, he maintains, is the only one that does justice to the actual state of mathematics.[51] The intuitionistic demand that mathematical objects be constructed was mentioned correctly by **Vollenhoven** in his theological dissertation published in 1918, but only as a side remark.[52] Only De Haan referred to Vollenhoven's contribution during the debate, and he considered it 'powerless' compared to Brouwer's work.[53] Felix **Bernstein** was somewhat more explicit about what he called the finitist view on mathematics. Bernstein, who had written his dissertation under Hilbert, had founded the statistical institute of the Göttingen university in 1918.[54] In a paper published the following year in the *Jahresberichte der deutschen Mathematiker-Vereinigung*, he explains that finitists consider no other constructions thinkable than those based on natural numbers. However, Bernstein considers such various mathematicians as Poincaré, Richard, Borel, Lindelöf and Brouwer as constituting the finitist movement, without always clearly distinguishing between their views.[55] In this way, it does not seem probable that Bernstein reacted specifically to Brouwer's view on mathematical existence. The paper was mentioned in the second edition of Fraenkel's *Einleitung in die Mengenlehre*, where Bernstein was criticised for not distinguishing between the different forms of finitism.[56] In 1920, finally, Hugo **Dingler** wrote Brouwer a letter to tell him how pleased he was to see that Brouwer demanded set theory to be 'constructive'.[57]

### 4.2.1  Weyl's *Grundlagenkrise*

Hermann Weyl's 1921 paper *Über die neue Grundlagenkrise der Mathematik* ('On the new foundational crisis in mathematics') put the themes of mathematical existence and constructivity on the agenda at once. Weyl had studied mathemat-

---

[49] Bieberbach was appointed in 1913. He held the inaugural lecture a year later.
[50] [Van Dalen 2001, p. 341]
[51] [Bieberbach 1914, p. 901]. Later, Bieberbach radically changed his position, see 5.3.1.
[52] [Vollenhoven 1918, p. 234]
[53] 'machteloos', [De Haan 1919, p. 30]
[54] [Gottwald, Ilgauds & Schlote 1990, p. 53]
[55] [Bernstein 1919, pp. 64–70]
[56] [Fraenkel 1923A, p. 173]
[57] Letter from Dingler to Brouwer, 26/7/1920; [MI Brouwer]

## 126  CHAPTER 4. REACTIONS: EXISTENCE AND CONSTRUCTIVITY

ics and physics in Göttingen and München from 1904 to 1908.[58] He had chosen Göttingen for his studies primarily because the principal of his lyceum was a cousin of Hilbert's and had written him a letter of recommendation. Soon, however, Weyl was gripped by Hilbert's mathematical work.[59] He also took lectures from Husserl. Weyl obtained his degree from Hilbert in 1908 on the subject of integral and differential equations. In 1910, he handed in his *Habilitationsschrift*[60] and became *Privatdozent* in Göttingen. At that time, Weyl held a Cantorian view on the foundations of mathematics.[61]

In 1913, Weyl published his first book, on the subject of Riemann surfaces,[62] which established his name in mathematical circles. The book was inspired by Brouwer's topological work, which Weyl later considered 'the outstanding topological event of my life'.[63] In the same year, he became full professor of geometry at the *Eidgenössische Technische Hochschule* in Zürich. In September he married Helene Joseph, a student of Husserl's. Weyl served as a soldier for one year in the German army, until the Swiss government managed to get him back to lecturing. Weyl was relieved. With the publication of his book *Raum · Zeit · Materie* ('Space · Time · Matter') on relativity theory in 1918, which saw five printings in five years, Weyl's name was also set outside the domain of mathematics.[64] In *Das Kontinuum*, published in the same year, Weyl presented his own attempt to improve the foundations of mathematics. The continuum should avoid the impredicative[65] definitions which appear in Dedekind's approach. Arithmetic should be founded independently of set theory and be based on the primordial intuition of iterating.[66] Grattan-Guinness conjectured that Weyl's attack on impredicativity may have been influenced by Poincaré, who had come to lecture on exactly that topic in Göttingen when Weyl was still studying there.[67] Soon, however, Weyl again moved one step further away from his original, Cantorian view on the foundations of mathematics. He abandoned his own project and joined Brouwer as an 'apostle of his intuitionism'.[68]

---

[58] Weyl's mathematical work, other than intuitionistic, is not treated here. An appreciation thereof may be found in [Chevalley & Weil 1957].

[59] [Chevalley & Weil 1957, p. 158]

[60] In Germany, academics had to publish a so-called *Habilitationsschrift* in order to become a university lecturer.

[61] In his *Habilitationsvortrag*, Weyl maintained that the natural numbers were founded on set theory and that therefore 'nowadays set theory appears in logical respect as the actual foundation of the mathematical sciences', [Weyl 1910, p. 302] *('(...) erscheint uns denn die Mengenlehre heutzutage in logischer Hinsicht als die eigentliche Grundlage der mathematischen Wissenschaften')*.

[62] [Weyl 1913]

[63] Draft for a lecture at the Bicentennial conference, [ETH Weyl, HS 91a:17]

[64] [Chevalley & Weil 1957, p. 159]

[65] The term 'impredicative' is explained in the glossary.

[66] A modern appreciation of *Das Kontinuum* may be found in [Feferman 1988].

[67] Personal communication from Grattan-Guinness to the author, June 2000

[68] Draft for a lecture at the Bicentennial conference, 1946, [ETH Weyl, HS 91a:17]

## 4.2. THE BEGINNING OF THE DEBATE

**Weyl's conversion**  Weyl was one of the earliest converts to Brouwer's intuitionism, and certainly the most important one for its promotion. He had been in contact with Brouwer at least from 1911 on.[69] Weyl had turned to intuitionism because of the talks Brouwer and he had when they spent a short Summer holiday together in Engadin in 1919.[70] In January 1920, Weyl described this event to Bernays in the following way:[71]

> Eine Zusammenkunft mit Brouwer im Sommer hat der Sache neuen Impuls gegeben; ich modifiziere meinen Standpunkt wesentlich. Brouwer ist ein Mordskerl und ein wunderbar intuitiver Mensch. Ich war durch die paar Stunden Zusammensein mit ihm ganz beglückt.[72]

It is striking to note how easily Weyl changes from writing about Brouwer's influence on his view on foundations to Brouwer as a person.[73] Indeed, Weyl later admitted that '[p]ersonal relationships were often a contributing cause for my attention' and he described the summer 1919 events by saying that he 'fell under the spell of Brouwer's personality and ideas'[74] – note the order. This may have nourished Hilbert's later complaint that 'even in circles of mathematicians, the suggestive force of a single temperamental and penetrating man can have the most unlikely and eccentric effects.'[75]

However, Brouwer's personality was certainly not the only thing that appealed to Weyl. Brouwer's neo-Kantian view on the foundations of mathematics came close to Weyl's own phenomenological conviction. Weyl did not explicitly express this in his 1921 paper, but it is clear from his physical writings. Thus, in his 1918 *Raum · Zeit · Materie* he wrote:[76]

> In prinzipieller Allgemeinheit: die wirkliche Welt, jedes ihrer Bestandsstücke und alle Bestimmungen an ihnen, sind und können nur gegeben sein als intentionale Objekte von Bewußtseinsakten. Das schlechthin Gegebene sind die Bewußtseinserlebnisse, die ich habe – so wie ich sie habe.[77]

---

[69] Letter from Brouwer to Hilbert, 31/3/1911, [MI Brouwer, CB.DH1.9], [NSUB Hilbert, 49-9]
[70] Letter from Brouwer to Fraenkel, 28/1/1927; published in [Van Dalen 2000, p. 303].
[71] Letter from Weyl to Bernays, 9/1/1920, [ETH Weyl, HS 91:10]
[72] 'A meeting with Brouwer this summer has given the matter a new impulse; I modify my point of view substantially. Brouwer is a hell of a guy and wonderfully intuitive person. I was made completely happy by the few hours we spent together.'
[73] In the fall of 1920, Weyl visited Brouwer in the Netherlands, and his feelings for Brouwer were as strong as before: '*Brouwer ist ein Mensch, den ich von ganzer Seele lieb habe.*' ('Brouwer is a man whom I love with all my heart.') Letter from Weyl to Klein, 15/11/1920, [NSUB Klein, 12-296]
[74] Draft lecture for the Bicentennial conference, 1946, [ETH Weyl, HS 91a:17]
[75] '*auch im Kreise der Mathematiker die Suggestivkraft eines einzelnen temperamentvollen und geistreichen Mannes die unwahrscheinlichsten und exzentrischsten Wirkungen auszuüben vermag.*', [Hilbert 1927, p. 81]. Courant echoed the complaint in [Courant 1928, p. 92].
[76] [Weyl 1918B, p. 3]
[77] 'In principal generality: the real world, all of its components and all determinations about it, are and can only be given as intentional objects of acts of consciousness. What is simply given are experiences of consciousness, which I have – as I have them.'

Furthermore, Weyl had already acknowledged time as 'the primordial form of the flowing of consciousness'.[78] In view of these belief of Weyl's, it is not surprising that he was attracted to Brouwer's intuitionistic view of treating mathematics as and only as a meaningful substance, based on the primordial intuition of time.

Weyl and Brouwer stayed in touch during these years. In 1920, Brouwer offered Weyl a chair in Amsterdam, which Weyl declined in the end.[79]

Whatever the relative influence of Brouwer's personality and the common philosophy may have been, Weyl's paper *Uber die neue Grundlagenkrise der Mathematik* certainly put the cat among the pigeons.[80]

**Lectures and paper**  Weyl's 1921 paper was based on three lectures he had given in Fueter's seminar in Zürich. The lectures were given on December 2, 9 and 16, 1919. In the Gonseth archive in Lausanne, I found notes which the philosopher Ferdinand Gonseth took when he attended Weyl's lectures.[81] From the notes, it becomes clear that all the essential parts of the paper, like the rejection of pure existence statements in favour of constructions, the criticism of the principle of the excluded middle, the introduction of choice sequences and the continuum as a medium of free becoming, were already present in the lectures. The main difference seems to be that none of the polemical terms of the paper appear in the lectures. Since the notes only comprise three pages, it is very well possible that Gonseth left them out and only noted the essentials. It is equally well possible, however, that Weyl added the polemics later on. The change in the title from 'On the foundations of mathematics' in the lectures to 'On the new foundational crisis in mathematics' in the paper supports the latter assumption.

Pólya was one of the persons who attended the lectures, and Gonseth included parts of the discussions between Pólya and Weyl after the lectures.[82] The most interesting one took place after the third lecture and runs as follows:[83]

---

[78]'*Die Urform des Bewußtseinstromes*', [Weyl 1918B, p. 5]

[79]Letters from Brouwer to Weyl, 7/9/1920 and 1/1/1921; [ETH Weyl, HS91-492] and [ETH Weyl, HS 91-493]

[80]Weyl's first publication in which he treats intuitionism, though without mentioning so explicitly, is [Weyl 1920]; see 6.4.2.

[81]'Ueber die Grundlagen der Analysis', [BCUL Gonseth, IS 4323/8/30/17]

[82]Pólya and Weyl had discussed foundational matters before. There is a famous bet between them, which was made before Weyl was converted to intuitionism, in February 1918. In it, Weyl predicted, among other things, that within twenty years the majority of the leading mathematicians would admit that concepts such as 'number' and 'set' are completely vague and that one can say as little about the truth of propositions containing these concepts as about those of Hegelian philosophy. Some years after the bet expired, Weyl admitted that he had lost 'by 49 to 51 percent'. The bet was later published as [Pólya 1972].

[83]'Ueber die Grundlagen der Analysis', [BCUL Gonseth, IS 4323/8/30/17]

## 4.2. THE BEGINNING OF THE DEBATE

Polya: Sie sagen: die mathematische Sätze sollen nicht nur wahr, sondern auch sinnvoll sein. Was heisst sinnvoll?

Weyl: Das ist eine Sache der Ehrlichkeit.

Polya: Es ist eine Verirrung, philosophische Sätze in die Wissenschaft zu mengen. (Polya nennt Weyl's Kontinuum-Auffassung Gefühl.)

Weyl: Was Polya Gefühl und Rhetorik nennt, dass nenne ich Einsicht und Wahrheit; was er Wissenschaft nennt, nenne ich Buchstabenreiterei. Polya's Verteidigung der Mengenlehre (man könne diesen Formulierungen vielleicht mal einen Sinn unterschieben) ist Mystik. – Abscheidung der Mathematik als formal aus dem Geistesleben tötet sie, macht sie zur Schale. Zu sagen, nur das Schachspiel ist Wissenschaft, und die Einsicht ist keine, das ist Einschränkung. (Polya hatte gesagt, man dürfte die Forschung der Mengenlehre nicht einschränken.)[84]

Note that, if Gonseth's notes are correct and if Pólya paraphrased Weyl's position correctly, it means that Weyl believed in the truth of statements which are not contentual. Besides that, the notes show that Weyl used arguments to support intuitionism which fitted very well into *Lebensphilosophie*.[85]

After his Zürich lectures, Weyl lectured on the continuum in the Göttingen *Mathematische Gesellschaft* on May 11, 1920. Furthermore, he delivered a series of lectures on the foundations of mathematics in the mathematical seminar in Hamburg on July 28, 29 and 30, 1920.[86] I do not know what the contents of these lectures was, but one may assume that they were similar to the ones Weyl gave in Zürich.

---

[84] 'Polya: You say: mathematical theorems should not only be true, but also meaningful. What do you mean by meaningful?
Weyl: That is a matter of honesty.
Polya: It is a mistake to mix philosophical statements into science. (Polya calls Weyl's view of the continuum sentiment.)
Weyl: What Polya calls sentiment and rhetoric, I call insight and truth; what he calls science, I call letter pedantry. Polya's defence of set theory (one could maybe give meaning to those formulations one day) is mysticism. – Separating mathematics as formal from spiritual life kills it, turns it into a shell. To say that only the chess game is science, and insight not, that is curtailment. (Polya had said that one should not curtail the research of set theory.)'

[85] See 6.3.1.

[86] *Jahresbericht der deutschen Mathematiker-Vereinigung* 29 (1920), p. *33, 54*

130     CHAPTER 4. REACTIONS: EXISTENCE AND CONSTRUCTIVITY

**Contents of the paper**  In one of the key passages of the 1921 paper,[87] Weyl analyses the status of existence statements in mathematics.[88] He does not treat existence statements as a separate subject, but discusses them in the context of the intuitionistic continuum and the question of the acceptability of the principle of the excluded middle. His conclusion is that they are not real propositions, as for instance statements about a definite natural number are, but mere proposition-abstracts:[89]

> *Ein Existentialsatz* —etwa 'es gibt eine gerade Zahl'— *ist überhaupt kein Urteil*[90] *im eigentlichen Sinn, das einen Sachverhalt behauptet*; Existential-Sachverhalte sind eine leere Erfindung der Logiker. '2 ist eine gerade Zahl': das ist ein wirkliches, einem Sachverhalt Ausdruck gebendes Urteil; 'es gibt eine gerade Zahl' ist nur ein aus diesem Urteil gewonnenes *Urteilsabstrakt*.[91]

The citation does not make it clear how one can obtain existence statements. Further on in the paper, Weyl clarifies this point. He states that it is 'meaningless'[92] to negate universal statements. Furthermore, he maintains that an existence statement by itself is nothing; if the proposition from which it was derived gets lost, only the incentive remains to look for the proper proposition again.[93] Thus, it seems that Weyl envisaged that existence statements could only be derived from an instantiation one already had. More polemically, he expressed this point in the following way:[94]

> Bezeichne ich Erkenntnis als einen wertvollen Schatz, so ist das Urteilsabstrakt ein Papier, welches das Vorhandensein eines Schatzes anzeigt, ohne jedoch zu verraten, an welchem Ort. Sein einziger Wert kann darin liegen, dass es mich antreibt, nach dem Schatze zu suchen.[95]

---

[87]Weyl's remarks on the principle of the excluded middle are treated in 5.2.1. The 'crisis' and 'revolution' metaphors Weyls used in the paper are discussed in 6.2.1.

[88]Weyl had touched upon the theme before. In his 1910 paper *Über die Definitionen der mathematischen Grundbegriffe* ('On the definitions of basic mathematical notions'), he remarked that the method of implicit definitions, which is a non-constructive way of defining concepts, is always but a temporary one. What we want in the end are 'explicitly defined' (*explizit definierter*) concepts, [Weyl 1910, p. 301].

[89][Weyl 1921, p. 224]

[90]As Van Dalen pointed out, Weyl's use of *'Urteil'* does not coincide with what we at present call 'judgement'. Rather, it is probably better translated as 'proposition', as can be seen from [Weyl 1918A, p. 1]; [Van Dalen 1995, p. 157].

[91]*'An existence statement* —say, 'there is an even number'— *is not at all a proposition in the strict sense, which expresses a state of affairs*; existential states of affairs are empty inventions of logicians. '2 is an even number': this is a real proposition, expressing a state of affairs; 'there is an even number' is merely a *proposition abstract*, obtained from this proposition.' Translation based on [Weyl 1998, p. 97]; for the translation of 'Urteil', see the preceding footnote.

[92]'sinnlos', [Weyl 1921, p. 226]

[93][Weyl 1921, p. 226]

[94][Weyl 1921, pp. 224–225]

[95]'If I designate knowledge as a precious treasure, then the proposition abstract is a piece of

## 4.2. THE BEGINNING OF THE DEBATE

The metaphor was to echo through the foundational debate.

Referring to the inflatory situation of the German economy, Weyl summarizes his view on mathematical existence with the following metaphor:[96]

> (...) die Mathematik [erscheint, DH] als eine ungeheure 'Papierwirtschaft'. Realen Wert, den Lebensmitteln in der Volkswirtschaft vergleichbar, hat nur das Unmittelbare, das schlechthin Singuläre; alles Generelle und alle Existenzaussagen nehmen nur mittelbar daran teil. Und doch denken wir als Mathematiker gar selten an die Einlösung dieses 'Papiergeldes'! Nicht das Existenztheorem ist das Wertvolle, sondern die im Beweise geführte Konstruktion.[97]

At the time of publication, the inflation of the German mark had only just begun.[98] By time, the metaphor was to win in power, especially among academics, who as public employees and as people who (in general) owed quite some savings were among the groups that were hardest hit by the inflation.[99]

Regarding Weyl's description of an existence statement, it should be remarked, as Sundholm pointed out to me, that Weyl's interpretation is stricter than the one Heyting later adhered to. Weyl explicitly states that the mere possibility of a construction does not suffice:[100]

> Hier ist also von der *Möglichkeit* der Konstruktion gar nicht die Rede, sondern nur im Hinblick auf die *gelungene Konstruktion*, den *geführten Beweis* stellen wir eine derartige Existential-Behauptung auf.[101]

Heyting was to allow such a possibility.[102]

Earlier in the article, Weyl had explained what he meant by mathematical constructions.[103] In Weyl's view, there are six 'definition principles' which do not

---

paper indicating the presence of a treasure, yet without revealing at which place. Its only value can lie in stimulating me to look for the treasure.' Translation based on [Weyl 1998, pp. 97–98].

[96][Weyl 1921, p. 225]

[97]'(...) mathematics [appears, DH] as a monstrous 'paper economy'. Only the immediate, the quitessentially singular has real value, comparable to foods in economics; everything general and all existence statements partake in it only indirectly. And yet we, as mathematicians, very seldom consider the redemption of this 'paper money'! The valuable thing is not the existence theorem, but the construction carried out in the proof.' Translation based on [Weyl 1998, p. 98].

[98]The German mark rose from 8.9 marks to the US dollar in January 1919 to 192 in January 1922 to 4200 billion in the fall of 1923, [Ringer 1969, p. 62].

[99][Ringer 1969, pp. 62–63]

[100][Weyl 1921, p. 222]

[101]'We are thus not at all talking of the *possibility* of the construction; rather, we only form such an existential claim in view of the *succeeded construction*, of the *proof carried out*.' Translation based on [Weyl 1998, p. 95].

[102]See 4.4.3; see also 5.4.3, where Heyting argues against Lévy that the intuitionistic affirmation of a proposition $p$ should be understood as 'one can prove $p$' (not as 'one has proved $p$').

[103]In his 1919 *Der circulus vitiosus in der heutigen Begründung der Analysis* ('The vicious circle in the present foundation of mathematics'), Weyl had already spoken about fixing the principles of logical constructions. He then claimed that the essence of mathematical-physical knowledge lies in the method of theoretical (*begrifflich*) construction — without, however, explaining what was to be understood by 'constructions', [Weyl 1919, p. 88].

# 132    CHAPTER 4. REACTIONS: EXISTENCE AND CONSTRUCTIVITY

contain a vicious circle. He had already mentioned the principles in *Das Kontinuum*.[104] They are, in his formulation:[105]

1. identification of variables: from $P(x,y)$, $P(x,x)$ comes into being;

2. negation: from $P(x)$, $\neg P(x)$ comes into being;

3. combination by 'and': from $P(x)$ and $Q(y)$, $P(x)$ and $Q(y)$ comes into being;

4. combination by 'or': from $P(x)$ and $Q(y)$, $P(x)$ or $Q(y)$ comes into being;

5. replacing a variable by a constant: from $P(x)$, $P(5)$ comes into being (assuming we are in the domain of the natural numbers);

6. replacing a variable by an existence statement: from $P(x)$, there is an $x$ such that $P(x)$ comes into being.

These principles suffice, Weyl maintains, to build new properties and relations throughout mathematics — but excluding set theory. As soon as the latter enters, a theory of types is needed and restrictions should be placed on the principles 5 and 6 as to their application to concepts of different types.

I should like to point out that Weyl does *not* mention anything about formalism or Hilbert in the paper. He only states that 'the old order' cannot be maintained, and that only two alternatives are known, namely his own *Das Kontinuum* and Brouwer's intuitionism.[106] The old order Weyl refers to is not that of formalism, for formalism still had to be developed as a coherent foundation of mathematics, but that of contentual, classical mathematics. It is that view from which Weyl distances himself.

Weyl's *Grundlagenkrise* paper appeared in print on April 13, 1921.[107] At that time, Weyl was seen as one of the brightest stars rising at the mathematical firmament. Whereas Hilbert was the major mathematician alive, Weyl was Hilbert's main student.[108] Finally, the choice for publication of the paper in the *Mathematische Zeitschrift* instead of in the main rival mathematical journal, the *Mathematische Annalen*, where Hilbert was an editor, suggests that Weyl wanted to speak freely.[109] Hilbert could hardly avoid reacting.

---

[104][Weyl 1918A, pp. 4–6]. For a further analysis of the use of the construction principles in Weyl's work, cf. [Leupold 1961, pp. 72–85].

[105][Weyl 1921, p. 215]

[106][Weyl 1921, p. 211]

[107]It was received by the *Mathematische Zeitschrift* on May 5, 1920.

[108]Incidentally, Weyl's revolt against Hilbert perfectly matches the main Expressionist theme of that period, the revolt of the son against the father; cf. [Gay 1969, p. 119].

[109]Husserl claimed that Weyl had promised to publish the paper in his *Jahrbuch für Philosophie und phänomenologische Forschung* (letter from Husserl to Weyl, 22/4/1922, published in [Van Dalen 1984B, pp. 6–7]). I do not know if Husserl's claim is correct, and, if so, why Weyl changed his plans.

## 4.2.2 Hilbert's first reactions

David Hilbert was born in 1862 in Königsberg, eastern Prussia (nowadays Kaliningrad, Russia). He studied mathematics in the same city, and finished his dissertation in 1885. He then spent a semester in Leipzig with Klein, and the next one in Paris. After that, he returned to Königsberg for a six-year period as *Privatdozent*. In 1892 he was appointed *außerordentlicher Professor*, and the following year ordinary professor at the university of Königsberg. In the whole period of nine years Hilbert worked at the Königsberg faculty, he never (with one exception) lectured on the same subject twice.[110] In 1895, upon an offer from Klein, Hilbert left Königsberg for Göttingen, where he stayed until his superannuation in 1930. During those years, he became one of the, if not the, leading mathematicians of his generation. He frequently changed his specialisation, producing substantial contributions to such different fields as invariant theory, number theory, foundations of geometry, integral equations, theoretical physics, and foundations of mathematics.[111] As Rowe pointed out, many of these contributions, especially those before 1900, have little to do with axiomatics per se, contrary to Hilbert's present-day image.[112] In 1900, Hilbert delivered a lecture before the Second International Mathematical Conference in Paris, in which he formulated 23 major unsolved mathematical problems. The list played an important role in the further development of mathematical research. The first six problems dealt with foundational issues; the first two addressed the status of the continuum. Problem one asked for a (preferably direct) proof of Cantor's continuum hypothesis[113] and of the well-ordering of the continuum.[114] Problem two concerned the consistency of the axiomatic system of the real numbers.[115]

During the period 1905–1917, Hilbert continued working on foundations, as Sieg pointed out, beit behind the scenes. He delivered a number of lectures on principles and foundations of mathematics, none of which was pusblished.[116] The year 1917 marks Hilbert's (published) return to foundational matters, with his Zürich lecture *Axiomatisches Denken* ('Axiomatic thinking'). There, he stressed the interdependence of mathematics and physics and the importance of the axiomatic method for both.[117] Important problems he mentioned in the lecture were the solvability in principle of every mathematical question, and the relationship between content and formalism in mathematics and logic. In Hilbert's view, the axiomatisation of logic would form the crowning of the axiomatic method.[118]

During the Summer semester of 1920, Hilbert lectured on mathematical logic

---

[110] [Rowe 2000, p. 58]
[111] [Weyl 1944, p. 617]
[112] [Rowe 2000, p. 72]
[113] See 1.1.1.
[114] The concept 'well-ordered' is explained in the glossary.
[115] [Hilbert 1901, pp. 298–301], [Rowe 2000, pp. 72–75]
[116] [Sieg 1999, p. 8]
[117] [Rowe 2000, pp. 83–84]
[118] [Hilbert 1918, p. 153]

## 134 CHAPTER 4. REACTIONS: EXISTENCE AND CONSTRUCTIVITY

in Göttingen. He argued against the 'dictatorial' tendencies of Kronecker and Poincaré, accusing them of throwing out the baby with the bathwater. He specifically mentioned items such as propositions involving the infinite and the *tertium non datur* as falling prey to their restrictve behaviour.[119]

Hilbert had missed Weyl's Göttingen lecture on the continuum in May 1920, since he did not know that Weyl was going to speak. By that time, Hilbert was still positive about Weyl's research. In a letter to Weyl written in May 1920, Hilbert stated that, although their basic tendencies seemed to differ, many of Weyl's ideas were similar to the ones he had developed over the last years.[120] It is not clear if Hilbert by that time was aware of Weyl's conversion to intuitionism.

Hilbert's lecture notes from the same period, as partially published by Sieg, show a similarity between Hilbert's and Weyl's ideas that goes much further than what is known from the published sources at the time. In the lecture *Logik-Kalkül* ('Logical calculus'), from the winter term of 1920, Hilbert puts forward as his opinion that:[121]

> (...) erscheint es als der geeignete Weg, dass man die mathematischen Konstruktionen an das Konkret aufweisbare anknüpft und die mathematischen Schlussmethoden so interpretiert, dass man immer im Bereiche des Kontrollierbaren bleibt.[122]

Sieg describes the view put forward by Hilbert at that occasion as 'strict finitist number theory'.[123] The main question which these new sources raise is why Hilbert, who apparently was open to intuitionist-constructivist views already at an early stage of the foundational debate, opposed them so strongly in public.

In 1921, Hilbert responded for the first time publicly to the intuitionistic challenge. He did so in a series of lectures, first in Copenhagen for the university and the polytechnic, then in Hamburg for the mathematical seminar, where Weyl had spoken the year before. Before going into the contents of these talks, let us first consider if there was any special reason for choosing these places. In order to answer this question, some attention should be drawn to the situation of the Hamburg university.

**Hamburg University** One of the most important developments in the field of higher learning during the Weimar period was the establishment of three new urban universities, in Frankfurt, Hamburg and Cologne.[124] The Hamburg university was

---
[119][Ewald 1996, vol. II, pp. 943–946]
[120]Letter from Hilbert to Weyl, 16/5/1920; [NSUB Hilbert, HS 91:606]
[121][Sieg 1999, p. 24]
[122]'(...) it seems appropriate to connect the mathematical constructions to what can be concretely exhibited and to interpret the mathematical inference methods in such a way that one always stays within the domain of what can be checked.', English translation based on [Sieg 1999, p. 24].
[123][Sieg 1999, p. 24]
[124][Ringer 1969, p. 75]

## 4.2. THE BEGINNING OF THE DEBATE

founded in May 1919. Within the year, the *Universität Hamburg* employed three mathematical lecturers: Blaschke, who until then had held a position in Tübingen, Hecke, who had left his professorship in Göttingen for Hamburg, and Radon, from the *Technische Hochschule* in Vienna. This quick enlargement of the mathematical staff was facilitated by the fact that the Hamburg university was seen as something completely new (which it *was* compared to the German universities in small provincial cities with great academic traditions) and by an active attitude of the city state government. Furthermore, Hamburg was the first German university to create real assistant positions also for theoretical professors, including those in mathematics; i.e., positions to be filled not by students but by employees.[125] This reinforced the image of Hamburg as a well-equipped university.[126]

The concept of the *Mathematische Seminar* in Hamburg as a place where mathematicians came to give a lecture and meet was quite common at that time: it had been organised even during the Great War in Berlin, Göttingen, Frankfurt am Main and Dresden, and it continued after the war had ended.[127] However, in post-war, poverty-struck Germany opportunities for mathematicians from different universities to meet professionally were few. In Hamburg, the former mayor Von Melle was active in obtaining funds for the local scientific foundation, in order to invite speakers. These were the early years of inflation: the German mark rose from 45 to the US dollar in January 1921, to 60 in spring and summer, to 100 in September and ended the year at 120 to the dollar.[128] Under these circumstances, the fact of having enough money available put forward Hamburg as one of the few universities that could afford to invite prominent speakers from other universities and even from abroad.[129]

This was a general feature that made Hamburg an attractive place to speak. On top of that, there were many personal contacts between Hamburg and Göttingen. Hecke had written his dissertation under Hilbert and had later become his colleague in Göttingen, and Blaschke had studied there for a short period. Furthermore, in 1920 Kurt Reidemeister, whose interests included philosophy, and the Ukrainian mathematician Alexander Ostrowski came from Göttingen to Hamburg as Hecke's assistants.[130]

This, taken together, may well explain why Hilbert reacted to the intuitionistic challenge in Hamburg: money, contacts, and people interested in foundational matters.

---

[125] Paid assistant positions were known before; e.g., from the beginning of the century on these existed in Göttingen, albeit low-paid, [Reid 1970, p. 108; p. 130].
[126] [Behnke 1978, pp. 43–44]
[127] As can be seen from the then volumes of the *Jahresbericht der deutschen Mathematiker-Vereinigung*.
[128] [Gay 1969, p. 160]; the general figures are even more dramatic: in July 1914 a dollar had costed 4.20 DM; in January 1923 it costed 17,972 DM, and in November 1923, 4.2 billion marks, [Müller, H.M. 1996, p. 242].
[129] [Behnke 1978, p. 44–45]
[130] [Artzy 1972, p. 96], [Behnke 1978, p. 46]

As for Copenhagen, I know of no special reason why Hilbert would have spoken there on intuitionism.

**Hilbert's 1921 lectures** Hilbert's Copenhagen lectures were, according to a German newspaper article, very popular. Hilbert gave a series of five lectures: first a general one on *'Natur und mathematisches Erkennen'* ('Nature and mathematical cognition'), then four more specific ones for a mathematical audience. The big newspapers devoted articles to Hilbert's visit, and he received a honorary doctorate from *Københavns Universitet*.[131]

Hilbert's Hamburg lectures drew a large audience, too.[132] Hilbert lectured with joy, and the lectures were followed by lively discussions.[133] In his lectures, Hilbert outlined his ideas on how to provide mathematics with axiomatic foundations. In Hilbert's view, the only demand for mathematical systems is consistency,[134] which should be proved intuitively.[135]

Alexander Ostrowski (1893-1986) wrote about Hilbert's lectures in a Hamburg newspaper. Although the exact date of the newspaper article is not clear, it seems to be a preview, since it states the time and place of Hilbert's lectures.[136] The Ukrainian Ostrowski had come to Marburg to study with Hensel and had been a civil prisoner there during the War.[137] He had next come to Göttingen in 1918 to write his dissertation under Hilbert and Landau,[138] so it is reasonable to assume that he must have had some knowledge of Hilbert's foundational views.

---

[131]'Geheimrat Hilbert in Kopenhagen', anonymous newspaper article in an unknown newspaper, [NSUB Hilbert, 751]

[132]Although the lecture *Neubegründung der Mathematik* ('New foundation of mathematics') which Hilbert gave in Hamburg is well-known in its published form, there is some confusion about the year in which the event took place. Both Hilbert's biographer Reid ([Reid 1970, p. 155]) and Hilbert himself (afterwards, [Hilbert 1927, p. 65]) give 1922 as the year in which he gave his first lecture on the new foundations of mathematics. Other authors claim the same, such as Benz in his small history of the beginnings of the *Mathematische Seminar* in Hamburg ([Benz 1983, p. 283]) and Behnke, at the time a student of Hecke's in Hamburg, in his autobiography ([Behnke 1978, p. 47]). It is clear, however, that Hilbert spoke in Hamburg on the foundations of mathematics on July 25-27, 1921. Reidemeister explicitly states so in a report on Hilbert's lectures in the *Jahresbericht der deutschen Mathematiker-Vereinigung* of 1921 ([Reidemeister 1921A, p. 106]). Therefore, Reid c.s. must be mistaken. (Hilbert himself admitted that he had a poor memory. For a memorable description hereof, see a newspaper article reprinted from the *Frankfurter Zeitung* in [Reidemeister 1971, pp. 85-86].)

The mistake probably stems from the fact that in the published version of Hilbert's lecture ([Hilbert 1922]) it is stated that the content of the paper is basically the same as that of the lectures given in 'the summer of this year' in Hamburg. Also, the completed version of volume one of the *Abhandlungen aus dem Mathematischen Seminar der Hamburgischen Universität*, in which the paper was published, appeared in 1922.

[133]'Hilbert über die Grundlagen der Mathematik', three newspaper articles in the *Hamburgische Correspondent* by 'R-r.' (presumably Reidemeister), each on one of the lectures, [NSUB Hilbert, 751].

[134]German: *Widerspruchsfreiheit*

[135][Reidemeister 1921A]

[136]In that case, one wonders where Ostrowski based his article on.

[137][Fraenkel 1967, p. 111], [Reid 1970, p. 145]

[138][Jeltsch-Fricker 1988, p. 34]

## 4.2. THE BEGINNING OF THE DEBATE

He picked out Hilbert's concept of securing mathematical existence by means of a consistency proof as one of the main issues, and wrote:[139]

> In diesem neuen, philosophisch bedeutsamen Begriff der Existenz— der für den modernen Mathematiker allein Maßgebend ist—liegt wohl die schönste Erkenntnis die der Hilbertschen Axiomatik entsprungen ist.[140]

The use of the word *'allein'* is somewhat ambiguous here. One could interpret Ostrowski's words as saying that only 'modern' mathematicians accept the securing of existence by means of a consistency proof. However, maybe his claim was stronger, and he meant that Hilbert's new interpretation of mathematical existence was the only important thing to modern mathematicians. In both cases, Ostrowski tried to discredit Brouwer and Weyl by excluding them from the rank of 'modern' mathematicians.

**Hilbert's view on mathematical existence** Hilbert himself had never been very clear on what exactly he considered mathematically existent. His famous *Grundlagen der Geometrie* ('Foundations of geometry'), published in 1899, which is usually taken to mark the birth of an axiomatic view on mathematics,[141] opens with the following words:[142]

> *Erklärung.* Wir denken drei verschiedene Systeme von Dingen: die Dinge des *ersten* Systems nennen wir *Punkte* (...); die Dinge des *zweiten* Systems nennen wir *Gerade* (...); die Dinge des *dritten* Systems nennen wir *Ebenen* (...).[143]

In this description, Hilbert speaks about things we think about, thus about objects of thought. However, Blumenthal told the anecdote that Hilbert, influenced by a lecture by Wiener, had maintained that 'one should always be able to say 'tables, chairs, beer-glasses' instead of 'points, lines, planes'.'[144] This descriptions points to a purely formal point of view, where contents is completely dispensed of.

---

[139] Ostrowski, A., 'David Hilbert, Zu seinen Vorträgen über die Grundlagen der Mathematik', newspaper article, July 1921; [NSUB Hilbert, 751]; the article presumably stems from a Hamburg newspaper.

[140] 'The most beautiful knowledge that has emanated from Hilbert's axiomatics is this new, philosophically important notion of existence — which is the only decisive one for the modern mathematician.'

[141] See, however, [Toepell 1986] for various sources from which Hilbert drew his inspiration.

[142] [Hilbert 1899, p. 2]

[143] '*Elucidation.* We think three different systems of things: we call the things of the *first* system *points* (...); the things of the *second* system *lines* (...); the things of the *third* system *planes* (...).'

[144] 'Man muß jederzeit an Stelle von 'Punkte, Geraden, Ebenen' 'Tische, Stühle, Bierseidel' sagen können', [Blumenthal 1935, p. 403]

## 138  CHAPTER 4. REACTIONS: EXISTENCE AND CONSTRUCTIVITY

Hilbert maintains that he had held the belief that consistency and existence coincide from when he started thinking about these questions. As he wrote in a letter to Frege[145] in December 1899:[146]

> Sie schreiben: '(...) Aus der Wahrheit der Axiome folgt, dass sie einander nicht widersprechen.'[147] Es hat mich sehr interessiert, gerade diesen Satz bei Ihnen zu lesen, da ich nähmlich, solange ich über solche Dinge denke, schreibe und vortrage, immer gerade umgekehrt sage: Wenn sich die willkürlich gesetzten Axiome nicht einander widersprechen mit sämtlichen Folgen, so sind wie wahr, so existieren die durch die Axiome definirten Dinge. Das ist für mich das Criterium der Wahrheit und der Existenz.[148]

Thus, Hilbert makes the existence of mathematical objects relative to the axiomatic system in which they are defined.[149] Note that Hilbert speaks about the truth of axioms. Unfortunately, the citation does not elucidate what kind of existence Hilbert was thinking of, once a collection of axioms was given.[150] More questions remain. Should the axioms define the object(s) uniquely, i.e., should the axiom system be categorical?[151] Should the system under investigation be decidable?[152] But these questions were not so clearly formulated at the end of the 19th

---

[145] Frege and Hilbert strongly disagreed on the value of the axiomatic method, as their correspondence shows. Frege challenged Hilbert to publish their correspondence, but the latter refused. Consequently, Frege decided to publish a paper in the 1903 *Jahresbericht der deutschen Mathematiker-Vereinigung* attacking Hilbert's views. It is interesting to note that, at the time, this did *not* lead to a sustained public discussion on axiomatics ([Rowe 2000, p. 76]).

[146] [Frege 1976, p. 66]; the text cited here was not taken from the original letter, which seems to have been lost, but from a partial copy made by Frege. We can be reasonably sure of its wording, however, since it differs only marginally from both Hilbert's concept or excerpt version, which still exists, and from Husserl's version of the letter published in [Husserl 1970, p. 449].

[147] It is interesting to note that, on this issue, Cantor agreed with Frege (and thus later with Brouwer, too) and disagreed with Hilbert. In a letter to Dedekind written in 1899, Cantor writes: '(...) sogar für endliche Vielheiten ist ein 'Beweis' für ihre 'Konsistenz' nicht zu führen. Mit anderen Worten: Die Tatsache der 'Konsistenz' endlicher Vielheiten ist eine einfache, unbeweisbare Wahrheit (...).' ('(...) even for finite multitudes, a 'proof' for their 'consistency' can *not* be given. In other words: the fact of the 'consistency' of finite multitudes is a simple, unprovable truth (...).') [Cantor 1932, p. 447]

[148] 'You write: '(...) It follows from the truth of the axioms that they do not contradict each other.' I found it very interesting to read exactly this sentence in your letter, since as long as I have been thinking, writing and lecturing on such things, I have always said exactly the opposite: if the arbitrarily posited axioms do not contradict each other with all their consequences, then they are true, then the things defined by the axioms exist. For me, that is the criterion of truth and existence.' English translation based on the translation in [Moore 1987, pp. 109–110].

[149] [Hallett 1995, p. 42]

[150] In 1930, Gödel in a sense proved Hilbert's article of faith for first-order logic, by his completeness theorem stating that an axiom system $S$ has a model if and only if $S$ is consistent, cf. [Moore 1987, p. 111].

[151] An axiom system is called categorical if any two models of the axiom system are isomorphic; see the glossary.

[152] A theory $T$ is called decidable if there is an algorithm that, for each proposition $\varphi$, checks if $T \vdash \varphi$; see the glossary.

## 4.2. THE BEGINNING OF THE DEBATE

century.[153]

In his 1904 Heidelberg lecture, Hilbert held on to the line of the *Grundlagen*, declaring that 'an object of our thought is called a *thought-thing* or, for short, a *thing* and is designated by a sign'.[154] Thus, at that time signs, in Hilbert's view, referred to something outside the mathematical text, inside the human mind. Furthermore, Hilbert argued for the existence of an infinite set which was proved consistent 'since it now receives a definite meaning and a contents that is always to be applied on later occasions.'[155] Thus, he seems to think of mathematical existence as what we would now call a semantical concept, but proved by syntactical means. It is not impossible that a consistency proof implies semantical existence (what we now call the model existence lemma), but the matter is certainly not trivial.[156] Furthermore, Hilbert also mentions mathematical concepts that are 'consistently existent',[157] an expression which seems to imply that he also considered it possible to exist without being consistent.

In these works Hilbert spoke about the truth of mathematical axioms, the contents of mathematical concepts, and about signs used in mathematics which designate objects of thought. Thus, he did not take up a purely formalistic position.

**Hilbert's 1922 paper**  Many people asked for the publication of Hilbert's 1921 lectures,[158] which indeed took place in the *Abhandlungen aus dem mathematischen Seminar der Hamburgischen Universität*.[159] Hilbert indeed stresses the importance of a consistency proof, although he does not link it to mathematical existence. Instead, he speaks of mathematical systems which are 'thinkable',[160] and maintains that, if the consistency proof succeeds,[161]

---

[153]The concept of categoricity was developed in the work of Huntington and Veblen in 1902–1904, [Moore 1987, p. 113].

[154]'*Ein Gegenstand unseres Denken heiße ein* Gedankending *oder kurz ein* Ding, *und werde durch ein Zeichen benannt*', [Hilbert 1905A, p. 266]

[155]'denn sie erhält jetzt eine bestimmte Bedeutung und einen später stets anzuwendenden Inhalt', [Hilbert 1905A, p. 273]

[156]It took some time before the distinction between syntax and semantics became clear. In his 1917/18 lecture notes *Prinzipien der Mathematik* ('Principles of mathematics'), Hilbert provides a general description of how the two should be related. Semantical considerations should be involved in obtaining the premises and in interpreting the results of formal operations in an axiomatic system. In the same notes, however, one still finds syntactic notions interwoven with semantic concepts, [Sieg 1999, pp. 14–18]. Zach argued the case that Bernays was the first to completely distinguish between syntax and semantics, in his *Habilitationsschrift* of 1918, [Zach 1999, p. 342].

[157]'widerspruchsfrei existierend', [Hilbert 1905A, p. 273]

[158]Letter from Bernays to Hilbert, 17/10/1921 [NSUB Hilbert, 21/1]

[159]It is hard to trace the precise relation between the lectures Hilbert gave and the published paper. In the paper, Hilbert claims that the paper contains 'the essential contents' of the lectures given in Copenhagen and Hamburg, [Hilbert 1922, p. 157]. In Copenhagen, Hilbert delivered five lectures; in Hamburg, he lectured three times two hours. This is definitely more than the contents of the paper.

[160]'denkbar', [Hilbert 1922, p. 159]

[161][Hilbert 1922, p. 162]

> so stellen wir damit fest, daß die mathematische Aussagen in der Tat unanfechtbare und endgültige Wahrheiten sind (...).[162]

Thus, at the time Hilbert believed in the truth of all mathematical propositions in a system which was proved consistent.

Concerning the question what mathematics consists of, Hilbert claims that[163]

> die Gegenstände der Zahlentheorie [sind mir, DH] die Zeichen selbst. (...) Hierin liegt die feste philosophische Einstellung, die ich zur Begründung der reinen Mathematik (...) für erforderlich halte: *am Anfang* - so heißt es hier- *ist das Zeichen*.[164]

He describes these signs in the following way:[165]

> Diese Zahlzeichen, die Zahlen sind und die Zahlen vollständig ausmachen, sind selbst Gegenstand unserer Betrachtung, haben aber sonst keinerlei *Bedeutung*.[166]

Thus, Hilbert did not see the formal system of number signs as completely separated from their interpretation. Again, what he depicts is what we would nowadays describe as the hope that the formal system in question has a unique model, a belief which Skolem proved wrong in the beginning of the 1930s.[167] Bernays later apparently did see the problem and added the demand that, in order for consistency and existence to coincide, the formal system under consideration has to be complete (in the sense of: $\forall \varphi$, either $\Gamma \vdash \varphi$ or $\Gamma \vdash \neg\varphi$).[168] Hilbert's belief is in accordance with the fact that he had in 1899 still left geometrical concepts a place in the human mind, even though he had freed them from their usual contents. Only in 1925 he was explicitly to move more towards the formalistic conception of mathematics as described by Brouwer.[169]

---

[162]'then we can say that mathematical statements are in fact incontestable and ultimate truths (...).' English translation from [Ewald 1996, vol. II, p. 1121].

[163][Hilbert 1922, p. 163]

[164]'the objects of number theory [are for me, DH] the signs themselves. (...) The firm philosophical attitude which I think is required for the grounding of pure mathematics (...) is this: *in the beginning* – so is said here – *was the sign*.'

[165][Hilbert 1922, p. 163]

[166]'These number-signs, which are numbers and which completely determine the numbers, are themselves the object of our consideration, but otherwise they have no *meaning* at all.' English translation based on the translation in [Ewald 1996, vol. II, p. 1122].

[167]Skolem proved the non-categoricity of the Peano axioms for the natural numbers. Fraenkel reported that this came as a big surprise to the mathematical world, thus indicating that presumably more people shared Hilbert's belief; [Fraenkel 1959, p. 356].

[168][Bernays 1928, p. 202]. Bernays demanded what Hilbert believed to be true in the first place: if a formal system is complete in the sense described above, then it has a unique model, and vice versa.

[169]See 4.3.1.

## 4.2. THE BEGINNING OF THE DEBATE

Towards the end of the paper, Hilbert presents what he calls his proof theory.[170] He explains that all 'actual mathematics'[171] has to be formalised, so that it becomes a stock of provable formulas. To this, a new mathematics called meta-mathematics is added, in which meaningful argumentations are used. The meta-mathematical level is the level in which the consistency proof for formalised mathematics has to be given.

In the paper, Hilbert claims the term 'constructive' for his own view. The way he does so is illuminating. He reproaches Weyl for his 'artificial'[172] argumentation. For, so Hilbert maintains, after Weyl had defined his constructive principle and had next reached a vicious circle, he should have concluded that his point of view and therefore also his constructive principle could not be applied to mathematics.[173] Note the clear indication that for Hilbert mathematics as it existed came first, only afterwards followed by reflections of whatever kind. Furthermore, he maintains, his own method is the truly constructive one:[174]

> Erst der hier in Verfolgung der Axiomatik eingeschlagene Weg wird, wie ich glaube, den konstruktiven Tendenzen, soweit sie natürlich sind, völlig gerecht.[175]

Note the in-built restriction in the 'as far as they are natural'; Hilbert did not go on to explain what he meant by that.

**Bernays in 1921** Before Hilbert's address was published, his assistant Paul Bernays gave a lecture at the *Mathematikertagung* in Jena in September 1921, in which he presented Hilbert's thoughts on the foundations of arithmetic. Bernays (1888–1977) had written his dissertation under Landau in Göttingen in 1912, had then moved to Zürich, and had returned to Göttingen as Hilbert's assistant in 1917. Bernays discussed foundational questions with Hilbert, helped with the preparation of lectures and made lecture notes.[176] He was deeply interested in philosophy and had published papers in Nelson's *Abhandlungen der Fries'schen Schule*. Bernays' talk was later published in the *Jahresbericht*.

Bernays starts by explaining that Hilbert's new programme is a clarification of the old one as put forward in Heidelberg in 1904.[177] Turning to intuitionism, he

---
[170]'*Beweistheorie*', [Hilbert 1922, p. 174]
[171]'*eigentliche Mathematik*', [Hilbert 1922, p. 174]
[172]'*künstlich*', [Hilbert 1922, p. 158]
[173][Hilbert 1922, p. 158]; the paper referred to is [Weyl 1919].
[174][Hilbert 1922, p. 160]
[175]'In my opinion, only the path taken here, following axiomatics, will do full justice to the constructive tendencies, as far as they are natural.' Translation based on the English translation in [Ewald 1996, vol. II, p. 1119].
[176][Zach 1999, p. 345; 361]
[177]At the International Congress of Mathematicians in Heidelberg in 1904, Hilbert had put forward his first consistency proofs. Although the examples he used were simple, they showed that it was possible to prove the consistency of a theory without the construction of a model, that is, he used syntactical rather than semantical means. A further explanation is given in [Smoryński 1988, pp. 9–11]; the original Hilbert paper is [Hilbert 1905A].

admits that the goal of reaching a purely constructive formation of arithmetic is an attractive one. In that case, mathematics can build its own structure and one does not have to rely upon the presupposition of a certain system of things. However, the way Bernays presents the construction of mathematics along intuitionistic lines is rather strange. In Bernays' words, Brouwer and Weyl want to replace existence presuppositions by construction postulates. He compares this with similar practices in geometry, where one for example can replace the axiom which says that every two points determine a line by the postulation that a line can be constructed between any two given points.[178]

Bernays' description is peculiar, because it suggests that all intuitionism aims at is a replacement of one kind of assumption by another. That, however, is not the case. What is at stake for intuitionism are not assumptions, but certain construction principles that are acknowledged as valid before starting to work inside a mathematical theory.

On the other hand, the examples Bernays gives of Weyl's and Brouwer's refusal to accept certain classical results of mathematics because they are not constructive are correct.

Bernays concludes that all Brouwer and Weyl have shown is that by replacing existence axioms by construction postulates, no consistency proof is obtained. The conclusion marks what could be seen as his formalistic view on Brouwer and Weyl, for they never aimed at a consistency proof.

Bernays does not leave 'constructive' mathematics to intuitionism. Returning to *Herr Geheimrat*, he maintains that Hilbert is far from giving up the constructive tendency, which stems from the independence of mathematics. On the contrary: Hilbert's aim, so Bernays claims, is to bring out the constructive direction in its strongest possible form. This can be done by not restricting oneself to the prejudice that every construction should be a number construction, as the intuitionists do. Note that Bernays's presentation of the intuitionistic idea of constructions is in accordance with Kronecker's view, but not with Brouwer's. Constructions can also be obtained, Bernays continues, by formalising mathematics and working with signs. In this way, a rigorously constructive structure can be achieved, not only of arithmetic, but also of higher mathematics.[179]

From a letter Bernays wrote to Hilbert in 1921, it becomes clearer what Bernays meant with such constructions. Discussing what Hilbert should treat in his coming lectures, Bernays proposed that he, among other things, should consider 'constructive arithmetic' and 'the broader idea of constructive thought' as possible subjects for the lectures. In Bernays' sense, this means the construction of proofs, presumably within Hilbert's proof theory. The aim of these 'constructions' is to make formalisation possible and the problem of consistency easier to understand.[180]

---

[178][Bernays 1922A, p. 13]; Bernays does not use the term 'intuitionism' in the paper.
[179][Bernays 1922A, pp. 15–16]
[180]'*konstruktive Arithmetik*', '*Die weitere Fassung des konstruktiven Gedankens*'; letter from Bernays to Hilbert, 17/10/1921 ,[NSUB Hilbert, 21/1]

## 4.2. THE BEGINNING OF THE DEBATE

These remarks show, I think, two important things. Firstly, the concept of 'constructive' mathematics did not, at that time, have an unambiguous meaning. Where the intuitionists put forward the idea of mental constructions and maintained that linguistic constructions do not necessarily imply mathematical constructions, Bernays (and, one may suppose, Hilbert) advocated the view of sign constructions, i.e., of an effective syntax. The split between mental and sign constructions continues until today, when both Brouwer's intuitionistic mathematics and Hilbert's proof theory are regarded as parts of constructive mathematics.[181] Secondly and more importantly, they show the importance *both* currents attached to the concept of constructive mathematics. The reason why this was thought significant seems clear in Brouwer's case, since the idea of constructing mathematics in the mathematician's mind was a basic aspect of his philosophy of mathematics. For Hilbert and Bernays, I find it harder to understand their attachment to the term 'constructive'. A somewhat speculative explanation might be that mathematics has traditionally worked with constructions, especially in geometry, or that the popularity of the 'construction' concept in society at large played a role, for example in the rise of Constructivism in art.[182]

**Hilbert's 1922 lecture** Even though the debate on the foundations of mathematics had only spread to a limited group, Hilbert again spoke on the topic in 1922. This time it was in a lecture for the *Deutsche Naturforscher-Gesellschaft* ('German Naturalist Society') delivered in September 1922 in Leipzig.[183]

In the lecture, Hilbert starts by again explaining that he has split mathematics into a formalised part, expressing classical mathematics, and a meaningful meta-mathematical part, required in order to prove the consistency of the formalised part. He clarifies the formalised part of proof theory in the following way:[184]

> Die Axiome und beweisbaren Sätze, d.h. die Formeln, die in diesem Wechselspiel entstehen, sind die Abbilder der Gedanken, die das übliche Verfahren der bisherigen Mathematik ausmachen, aber sie sind nicht selbst die Wahrheiten im absoluten Sinne.[185]

It should be pointed out that, although Hilbert works with formalised mathematics, he does *not* take up a purely formalistic position. He does not put forward the

---

[181][Kushner 1988, p. 356]; only in propositional and predicate logic is 'constructive' seen as synonymous with 'intuitionistic', cf. [Mints 1988, p. 355], [Dragalin 1988, p. 361].

[182]See 6.5.1. The same phenomenon may have occurred in other sciences, too. Thus, Reichenbach labelled his way of setting up an axiomatic system for the theory of relativity *'konstruktive Axiomatik'* ('constructive axiomatics'), as opposed to deductive axiomatics, [Reichenbach, H. 1924, pp. 2–3].

[183]Published a year later in the *Mathematische Annalen*, [Hilbert 1923].

[184][Hilbert 1923, p. 180]

[185]'The axioms and provable theorems, i.e., the formulas which arise in this continuous variation, are the representations of the thoughts which constitute the usual method of traditional mathematics; but they are not themselves the truths in an absolute sense.'

axiomatic system as an arbitrary formal system, but adheres to a link between the axioms and the thoughts of which mathematics until then consisted. The truths in an absolute sense, however, are now presented differently. Whereas Hilbert in his 1922 paper had described all propositions of a consistent system as true, he now reserves the term 'truth' for proofs regarding provability and consistency.

Hilbert next gives some axioms which suffice for formalising the theory of the natural numbers, and claims that by using finitary logic in a 'purely intuitive'[186] way, one can obtain elementary number theory from these axioms. He clarifies the meaning of 'finitary logic in a purely intuitive way' by mentioning that this includes recursion and 'intuitive' induction on finite, given collections.[187]

Hilbert then remarks that, in his proof theory, he wants to transcend finitary logic by adding axioms in order to prove transfinite theorems of classical mathematics, and he asks himself when mathematics for the first time exceeds the concrete intuitive and finite. His answer is illuminating: in using the terms 'for all' and 'there is'.[188] As for the latter notion, Hilbert describes an existence statement as a (possibly infinite) disjunction,[189] and points out that there is a difference between 'there is' and 'there is available':[190]

> Bei endlichen Gesamtheiten sind *'es gibt'* und *'es liegt vor'* einander gleichbedeutend; bei unendlichen Gesamtheiten ist nur der letztere Begriff ohne weiteres deutlich.[191]

This is some combination of Brouwer's criticism of applying concepts which are valid in finite domains to infinite ones, and Weyl's criticism of pure existence statements.[192] As Becker was to remark a few years later about these words of Hilbert's: *'Diese Darlegungen könnte wörtlich ein Intuitionist geschrieben haben'*.[193]

---

[186]'*rein anschauliche*', [Hilbert 1923, p. 181]

[187][Hilbert 1923, pp. 179–181]

[188]In his 1904 Heidelberg lecture, Hilbert had already suggested that the principles according to which the 'laws of mathematical thinking' should be developed, should include replacing the 'for all' notion by 'any (arbitrary)', where the latter should be understood to apply only to those things that can be added consistently to the mathematical system under investigation, [Hilbert 1905A, p. 274]. Bernays later referred to this position when arguing that Hilbert had actually at some time taken a stricter stand on the meta-mathematical level than what later became finitism, and that finitism was already a kind of compromise to him, [Bernays 1954, p. 12]. Sieg supplemented this by pointing to notes of a lecture course on set theory given in the Summer of 1917, where Hilbert would have been so radical as to dismiss the use of any arbitrary letter to replace a concrete number, and would have claimed that Kronecker 'wasn't radical enough', [Sieg 1999, pp. 9–11]. It would be interesting to see the lecture notes in full, in order to appreciate all aspects of Hilbert's position at that time. What seems clear in any case is that Hilbert was far more occupied by foundational matters around the time of World War I than his publications suggest.

[189]This idea can be found with Peirce in 1885, [Moore 1980, pp. 96–98].

[190][Hilbert 1923, p. 182]

[191]'For finite totalities *'there is'* and *'there is available'* are synonymous; for infinite totalities, only the latter notion is clear without difficulty.'

[192]See 2.3.2 and 4.2.1.

[193]'These expositions could have literally been written by an intuitionist.', [Becker, O. 1927, p. 466]

## 4.2. THE BEGINNING OF THE DEBATE

Hilbert continues that one can overcome this difficulty by adding certain transfinite axioms, and proving that the system thus obtained is consistent.[194] He does not draw any conclusions as to what this implies for the contentual character of the mathematical objects involved. In 1925, he was to do so explicitly. It is hard to say whether Hilbert at the time of writing this paper had already seen all the consequences of his way of proceeding.

It is interesting to note that, whereas Hilbert had in his Hamburg address still favoured 'constructive' mathematics, now this label is nowhere mentioned.

**Brouwer's reaction** It is clear from Hilbert's lectures that he was occupied by the themes and problems which Brouwer and Weyl had put forward. However, when discussing these issues, Hilbert mostly followed his (proclaimed) own path, and only occasionally referred to Brouwer. Brouwer, so it seems, was not pleased with this attitude. As the Austrian Roland Weitzenböck, for whom Brouwer had been able to create a position in Amsterdam, reported in a letter to Weyl in April 1923:[195]

> Betreffs der Hilbert'schen Sachen zuckt Br[ouwer] vorläufig die Achseln. Ich glaube, er ist einigermassen darüber verstimmt, dass Hilbert seine Sachen so links liegen lässt.[196]

If Weitzenböck is right, Brouwer presumably was not satisfied with the fact that, even though Hilbert agreed with some of the intuitionistic views, he never recognized their contribution to the development of his foundations of mathematics.

### 4.2.3 Becker's phenomenology

Also from a more philosophical point of view, the first reaction to intuitionism did not take long. In January 1922, Oskar Becker handed in his *Habilitationsschrift* at the *Universität Freiburg*, which was published a year later as *Beiträge zur phänomenologischen Begründung der Geometrie und ihrer physikalischen Anwendungen* ('Contributions to the phenomenological foundation of geometry and its physical applications').[197] Becker had studied physics, chemistry and psychology in Oxford, then mathematics (under Hölder and Herglotz), physics, philosophy, archaeology and history of art in Leipzig. He had hoped to work for both Hilbert and Husserl, but when the latter left Göttingen for Freiburg this became impossible. He then decided, reportedly under the impression of the First World War,[198] to continue in philosophy, and worked under Husserl on the area between

---

[194][Hilbert 1923, pp. 183–184]
[195]Letter from Weitzenböck to Weyl, 16/04/1923, [ETH Weyl, HS 788]
[196]'Regarding the Hilbertian affairs Br[ouwer], for the time being, shrugs his shoulders. I think he is somewhat displeased that Hilbert ignores his things.'
[197][Becker, O. 1923]
[198][Pöggeler 1969, p. 304]. The reason Becker gave afterwards for this decision was that he did not want to specialise in a single scientific subject.

mathematics and philosophy.[199] Whereas his dissertation under Hölder had been within mathematics proper, his *Habilitation* was much more philosophical. It was written from a phenomenological point of view,[200] and applied some of Husserl's concepts to the mathematical continuum. Becker included both Brouwer's version and Weyl's 1921 version of the continuum, and opted for the latter.[201] As Fraenkel later wrote, this was the first time 'a philosopher who knows the matter well has decided for 'intuitionism' in its opposition with 'formalism' '.[202]

Whereas Weyl had given an explanation of 'constructions' within mathematics, Becker described what should be understood by 'constructive' in a more general sense. In the first chapter of his *Habilitationsschrift* he writes:[203]

> Das entscheidende Merkmal, das den rationalen Zusammenhang vor anderen auszeichnet, ist sein *konstruktiver* Charakter. Das besagt, daß er sich aus diskreten Konstruktionselementen in endlicher Zahl zusammensetzt und daß sich die Art der struktiven Verbindung zwischen ihnen lediglich logisch-formaler (...) Natur ist. Er hat, so wollen wir das ausdrücken, den Charakter eines *Algorithmus*.[204]

Note the emphasis put on constructivity here, too: it is, Becker claims, the *decisive* property that distinguishes rationality from irrationality. Furthermore, in Becker's view constructivity implies that the procedures used should be finite. The description just given does not, however, provide the reader with a clear criterion to decide whether formalism or intuitionism should be seen as properly constructive. On the one hand the 'logical-formal' character seems to point to formalism, whereas the algorithmic part is clearly intuitionistic.

Becker's main interest in his *Habilitationsschrift* lies in the philosophical foundations of geometry. Within that, he limits himself to the problems of the continuum and of non-Euclidean geometry. Therefore, he touches upon questions of existence and constructivity in mathematics only in relation to the continuum and set theory. For example, Becker maintains that one cannot consider an infinite set to exist *an sich*. He arrives at this conclusion firstly by referring to Weyl's warning that, if one has the characteristic property of an infinite set, one should not think of it as if one had the members of the set laying readily before one. Secondly, he uses the principle of transcendental idealism to support his claim. This principle

---

[199] [Pöggeler 1969, p. 299], [Pöggeler 1996, p. 27]
[200] Phenomenology or, more specifically, transcendental phenomenology is a sub-current of anti-realism which focuses on intentional objects by reflecting on consciousness. A further explanation is given in the glossary.
[201] [Becker, O. 1923, p. 410]
[202] 'zum erstenmal ein mit der Materie wohlvertrauter Philosoph in dem Gegensatz zwischen 'Intuitionismus' und 'Formalismus' für ersteren Partei nimmt', [Fraenkel 1927/28B, p. 391]
[203] [Becker, O. 1923, p. 402]
[204] 'The decisive feature which distinguishes rational coherence from other forms of coherence is its *constructive* character. This means that it is composed of a finite number of discrete construction elements and that the structural connection between them is only of a logical-formal (...) nature. We would like to express this by saying that it has the character of an *algorithm*.'

## 4.2. THE BEGINNING OF THE DEBATE

holds that an object only exists inasmuch as it can be determined in the intellect with the degree of evidence that is characteristic for that object.[205] This does not apply to an arbitrary infinite set, since one cannot think of such a set as being completely finished. Note that Becker is concerned with the ontological existence problem.

The way Becker analyses geometry is very much along Brouwer's line. He starts by pointing out that, since Husserl, two kinds of ontology are distinguished: the material-eidetic one and the formal one. In Becker's words, the former is concerned with the understanding of essences by means of so-called ideative[206] abstraction, while the latter deals with the laws of the 'empty something'.[207] I will try to clarify this in modern terms.

One can think of the term 'ideative', which stems from 'eidos' (plural: eidè), meaning 'essence', as synonymous to 'essential' or 'a priori'. The way in which the ideative abstraction involved in the material-eidetic ontology is used in order to obtain knowledge of eidè can be described as follows. First, some pre-philosophical and pre-scientific experience of the eidos involved is required. Phenomenologists accept this experience as a basis. The acceptance hereof does not presuppose too much, however, since all belief with regard to the actual existence of the objects involved is suspended while investigating eidè. The next step is to select an instance of the eidos under investigation. Since the objects involved are treated as if they were objects of pure imagination, the instance can be taken both from experience or from phantasy. The latter are necessary in order to avoid that one does not get further than empirical generalisations. Then, one uses phantasy to subject the chosen instance to arbitrary variations. Finally, attention is focused on the overlapping synthesis of identity during the process of variation. This yields the eidos one was looking for.[208] Opposed to this is formal ontology, which coincides with that part of logic that is concerned with the study of models of possible theories.[209]

For Becker, pure mathematics belongs to formal ontology, but geometry, which in his view deals with the essence of space, is part of the material-eidetic ontology. However, the material-eidetic character of geometry is only reflected in its axioms, since all deductions are done formally. The question Becker asks himself is how it is possible to start from material-eidetic axioms, proceed in a formal way, and still obtain a system of truths.[210] Note that, if we see through the 'material-eidetic' terminology, we get the same point Brouwer was critical of in his dissertation.[211]

---

[205] [Becker, O. 1923, pp. 394–405]
[206] *ideierende*; translations from Husserlian terms generally follow Cairns' guide [Cairns 1973].
[207] 'leeren Etwas', [Becker, O. 1923, p. 389]
[208] [Scanlon 1997, pp. 168–169]
[209] [Null 1997, p. 237]
[210] [Becker, O. 1923, pp. 389–390]
[211] See 2.3.1.

Becker finds the answer to this question in Husserl's characterisation of space as a definite manifold. A manifold, in Husserl's sense, should be read as a field of knowledge.[212] The definiteness of a field, Becker maintains after Husserl, is characterised by the property that there is a finite number of concepts and theorems in the field from which the totality of all possible formations (*Gestaltungen*) in the field are completely and uniquely determined in a purely formal-logical way.[213] Thus, in a definite manifold, the concepts 'true' and 'formal consequence of the axioms' coincide.[214] Or, in again other words: a definite manifold cannot be extended under preservation of its axioms.[215]

Becker next analyses Husserl's concept of definiteness. In Becker's view, this can be interpreted in three different ways: as elementary definiteness (*Elementardefinitheit*), following Cantor; as extent definiteness (*Umfangsdefinitheit*), inspired by Russell and Weyl 1918; or as decidability definiteness (*Entscheidungsdefinitheit*), following Brouwer and Weyl 1921. If we apply this distinction to sets, the first of these says that a set is definite if for every element it is determined whether the element belongs to the set or not. In other words: a definite set is determined by a decidable property. The second interpretation is inspired by Russell's criticism of the vicious circle principle and is designed to exclude impredicative sets. It adds to the demand of elementary definiteness the requirement that it should be determined whether there exist objects outside a certain closed domain which belong to the set. Becker remarks that sets which are extent definite are obtained by means of a construction, which ensures their 'closed' character. The interpretation as decidability definiteness, finally, holds that every question that can be asked about a definite set should be decidable.[216] Note that the latter characterisation does *not* apply to all Brouwerian sets, contrary to what Becker suggests, since one of the characteristics of intuitionistic sets based on choice sequences is exactly that certain properties of it are not determined (at a specific moment in time). It is not clear where Becker got his characterisation of an intuitionistic set from. He refers to Weyl's *Grundlagenkrise* paper, but Weyl treats intuitionistic sets based on choice sequences.

Having analysed the different forms of definiteness, Becker concludes that only decidability definiteness is the characteristic which satisfies Husserl's idea of a definite set. The way he supports this claim is of interest to us. Becker maintains that his conclusion is supported by the 'conclusive' philosophical argument of transcendental idealism, according to which[217]

> (...) man von keinem Sachverhalt sagen kann, daß er bestehe, wenn man nicht ein prinzipielles Mittel hat, zu entscheiden, ob er besteht oder nicht.[218]

---

[212][Null 1997, p. 237]
[213][Becker, O. 1923, p. 390]
[214][Husserl 1913, p. 152]
[215][Van Dalen 1984B, p. 10]
[216][Becker, O. 1923, pp. 403–409]
[217][Becker, O. 1923, p. 414]
[218]'(...) one cannot say of any state of affairs that it exists if one does not in principle have a means to decide whether it exists or not.'

## 4.2. THE BEGINNING OF THE DEBATE

If this is indeed taken as a conclusive argument, one need not be surprised that Becker sided with Brouwer and Weyl regarding the foundations of mathematics, even though his argumentation regarding the definiteness of Brouwerian sets was not adequate.

The problem for non-phenomenologists in reading Becker is that his discourse is loaded with phenomenological concepts, which often have a technical meaning. This probably did not work in Becker's favour in drawing attention from mathematicians.[219] Only those who were *au courant* with phenomenological research, such as Hermann Weyl, could understand his work.[220] Nevertheless, the work was referred to reasonably often. Weyl incorporated the work into his book *Philosophie der Mathematik und Naturwissenschaft* ('Philosophy of mathematics and natural science').[221] Fraenkel, who was very well-documented, mentioned Becker's *Habilitationsschrift* from 1923 on, and used it as an example to show that Brouwer's work was also appreciated in philosophical circles.[222] Bieberbach referred to it, too, remarking that he found the mathematical parts of Becker's work incomprehensible.[223] Furthermore, Becker's work was mentioned by the philosophers Petzoldt and Cassirer, and Carnap, who also had a philosophical background.[224]

### 4.2.4 Fraenkel's early commentaries

1923 saw the entrance of the man who was to become the most important commentator to the foundational debate while it raged, Adolf Fraenkel (1891–1965).[225] Fraenkel had studied mathematics in München, Marburg, Berlin and Breslau (Wrocław). He finished his Ph.D. summa cum laude in Marburg in 1914 on an algebraic subject. While serving in the German army, he wrote his *Habilitationsschrift* in the same field and obtained his *venia legendi* at the same university two years later.[226] Thinking about what to do next with his spare time in the army, Fraenkel remembered the lectures he had taken from Hellinger in 1911 on set theory. These had covered, among other things, Zermelo's axiomatic foundations of set theory. Fraenkel told his fellow soldiers about the subject at several occasions,

---

[219] The gap was worsened by the fact that Becker's work was published in the *Jahrbuch für Philosophie und phänomenologische Forschung*, which presumably few mathematicians read.
[220] Becker's work was brought to Weyl's attention by Husserl, who recommended the *Habilitationsschrift* as a synthesis between the views of Einstein, Weyl and himself (i.e., Husserl). Letter from Husserl to Weyl, 9/4/1922, [ETH Weyl, HS 91-620]; a transcription of the letter was published in [Van Dalen 1984A, pp. 6–7], a partial English translation can be found in [Tonietti 1988, pp. 369–370]. Weyl had attended Husserl's lectures while studying in Göttingen, and he was married to one of Husserl's students, Helene Joseph, [Weyl 1968B, p. 87].
[221] [Weyl 1927B, p. 44]
[222] [Fraenkel 1923A, p. 165], [Fraenkel 1927A, p. 35]
[223] [Bieberbach 1925, p. 30]
[224] [Petzoldt 1925, p. 350], [Cassirer 1929, p. 433], [Carnap 1927, p. 365]
[225] For obvious reasons, Fraenkel later changed his name into Abraham A. Fraenkel.
[226] In Germany, academics had to publish a so-called *Habilitationsschrift* in order to become a university lecturer.

and thus got the idea to publish a booklet about it. The book, *Einleitung in die Mengenlehre* ('Introduction to set theory'), was to grow into his *opus magnum*.[227]

As Fraenkel himself later stated, set theory was by that time only known to a limited group of people, so one could relatively easily become a specialist in the field.[228] Fraenkel did see the importance of set theory, since to him it constituted the foundations of mathematics. By specialising in set theory, Fraenkel more or less automatically ran into Brouwer's intuitionism. Until 1922, Fraenkel only made reference to Brouwer's works.[229] This changed in 1923. In the meantime, Fraenkel had become associate professor[230] at the university of Marburg, profiting from the more tolerant rules of the Weimar Republic towards, among others, Jews.[231]

*Einleitung in die Mengenlehre*  In 1923, the second edition of Fraenkel's popular *Einleitung in die Mengenlehre* appeared. The first edition had been published right after the war, in 1919, and therefore Fraenkel had missed work that had come out in the meantime.[232] This gap was filled by the new edition, which contained a substantial enlargement of questions of principle, including a treatment of intuitionism. The intuitionistic view on mathematical existence and constructivity is described correctly, but only as a means to explain the intuitionistic opinion on the principle of the excluded middle. It is interesting to note that Fraenkel uses the dividing line drawn by the intuitionists between pure existence and constructivity in order to argue in *favour* of the axiom of choice. For, so he claims, exactly because there is this difference, the axiom of choice holds — but one should realise that it only states the existence of a certain set, not that we could actually construct it. In Fraenkel's view, such a pure existence poses no problem to logic nor to intuition.[233]

Fraenkel and Brouwer had already met in person before Fraenkel finished the second edition. Their first contact had been established in 1920 or 1921, when Fraenkel, who had a Dutch wife, stayed with his parents-in-law in Amsterdam.[234] The fact that Fraenkel had learnt enough Dutch to be able to read Brouwer's early intuitionistic writings, and that he had attended Brouwer's lectures, made him more familiar with intuitionism than many of his contemporaries. When Fraenkel left Amsterdam, he wrote Brouwer a letter to thank him for his support. He further declared:[235]

---

[227][Fraenkel 1967, pp. 105–136]
[228][Fraenkel 1967, p. 149]
[229]As in [Fraenkel 1919, p. V], [Fraenkel 1922A, p. 230] and [Fraenkel 1922B].
[230]*Extraordinarius*
[231]In Wilhelminian Germany, it had been difficult for Jews to hold a high position at universities or in the civil service.
[232][Fraenkel 1919, pp. III–V]
[233][Fraenkel 1923A, p. 167; 210]
[234]Fraenkel came to his parents-in-law in Amsterdam regularly and met Brouwer there at several occasions, [Fraenkel 1967, p. 162].
[235]Letter from Fraenkel to Brouwer, 18/4/1923, [MI Brouwer, CB.AFR 3]

## 4.2. THE BEGINNING OF THE DEBATE

Es war mir u.a. auch sehr interessant, das frische Leben des mancherseits schon totgesagten Intuitionismus zu beobachten; in mir selbst gärt es noch in diesen Fragen.[236]

**Mathematikertagung** The personal visit by Fraekel to Brouwer was returned at the occasion of the *Deutsche Mathematikertagung* ('German Mathematical Conference') in Marburg in September 1923. At the conference, Fraenkel gave a lecture which was to be the first in a series of presentations of his on the foundational debate.[237] During the conference, Brouwer, who also gave a talk there, stayed with the Fraenkels in Marburg.[238]

In the lecture he gave at the German Mathematical Conference, Fraenkel distinguishes between two currents regarding the foundations of mathematics, to which he refers as the classical and the intuitionistic one. In Fraenkel's view, the main difference between the two lies in the acceptance or not of the principle of the excluded middle. The question of existence is only mentioned along the way. Fraenkel remarks that for an intuitionist, existence and truth do not coincide with consistency, so that the consistency proof Hilbert is striving for will not silence their criticism.[239] The term 'constructive' is used once, in the sense of opposed to pure existence.[240]

**Jewish journal** A third contribution of Fraenkel to the foundational debate in 1923 was a paper published in the *Scripta Universitatis atque Bibliothecae Hierosolymitanaryum*. The journal, published in Jerusalem, contained papers by such important Jewish authors as Landau, Harald Bohr, Hadamard, Einstein, Ornstein, and Levi-Civita. Each paper was published both in the original, European language and in a Hebrew translation.[241] Fraenkel's paper dealt with the axioms of set theory. He only touches upon intuitionism, remarking that one either has to apply the radical constraints of Brouwer and Weyl, or one has to abandon the idea of a 'constructive, genetic' foundation of set theory.[242] He then proceeds following the latter alternative, giving axiomatic foundations of set theory and using existence in a formalistic way.

---

[236] 'Among other things, it was very interesting for me to observe the fresh life of intuitionism, which has been pronounced dead from many corners. In my own mind, these questions are still fermenting.'
[237] The lecture was published in the *Jahresbericht* a year later, [Fraenkel 1924A].
[238] [Van Dalen 2000, pp. 285–286]. This means that Brouwer stayed at the upper floor of Hensel's house, which the Fraenkels rented since their marriage in 1920. It must already have been crowded there, with both the Hensels, both the Fraenkels and their two children all in one house, [Fraenkel 1967, pp. 142–143].
[239] [Fraenkel 1924A, p. 99]
[240] [Fraenkel 1924A, p. 103]
[241] [Fraenkel 1967, p. 83]
[242] [Fraenkel 1923B, p. V]

152   CHAPTER 4. REACTIONS: EXISTENCE AND CONSTRUCTIVITY

Since Fraenkel was the most important reporter on the foundational debate during the whole period, the paraphrase given here of Fraenkel's presentation of intuitionism will serve to point out later shifts he made in presenting intuitionism.[243]

### 4.2.5 Baldus' rector's address

On December 1, 1923, Richard Baldus (1885–1945) delivered his rector's address at the *Technische Hochschule Karlsruhe*. It was published the next year as *Formalismus und Intuitionismus in der Mathematik* ('Formalism and intuitionism in mathematics') in the series *Wissen und Wirken*, and it was one of the more influential publications in the foundational debate. Baldus had studied in Erlangen and at the time lectured in Karlsruhe as a full professor in geometry.[244]

Baldus introduces the subject of the foundations of mathematics by remarking that the question 'Is present mathematics logically compelling?' has played an ever bigger role in the past decades, especially at the post-war conferences of the *Deutsche Mathematikervereinigung*.[245] He introduces the opposition between intuitionism and formalism, and starts with a criticism of Hilbert's axiomatic point of view as put forward in the *Grundlagen der Geometrie*:[246]

> Es beginnt z.B. jemand das Hilbertsche Buch zu lesen ohne zu wissen was 'Punkt', 'Gerade' usf. ist. Er soll sich zunächst Dinge dreier Systeme denken, ohne daß ihm irgendwie gesagt wird, was für Dinge das sein sollen. Er denkt sich nun irgendwelche Dinge, kommt zum ersten Axiom und entdeckt, daß seine gedachten Dinge dieses Axiom nicht erfüllen. Vielleicht kann er sich nun andere Dingen denken, welche diesem Axiom genügen; er liest darauf weiter und findet, daß er jetzt an einem späteren Axiom scheitert. Bald wird er ein Prinzip suchen, nach dem er das Axiomensystem denken kann, und da er kein solches findet, wird er, wenn er nicht rein zufällig auf das Gesuchte kommt, die Frage aufwerfen, wie denn überhaupt dieses Axiomensystem gedanklich realisierbar ist.[247]

Hilbert's answer to this problem, Baldus points out, lies in the drawings, which suggest the standard interpretation of the concepts 'point', 'line' and 'space'. Baldus' criticism clearly marks the difficulty of the position Hilbert held at the end

---

[243] See 4.3.1.
[244] [Poggendorff 1936–1940, vol. VI, p. 116]
[245] [Baldus 1924, pp. 4–5]
[246] [Baldus 1924, p. 13]
[247] 'For example, someone starts reading Hilbert's book without knowing what 'point', 'line' etc. is. He first has to imagine things of three systems, without him being told in any way what kind of things these should be. He now imagines some things, comes to the first axiom and finds out that the things he imagined do not fulfil this axiom. Maybe he can now imagine other things which do satisfy this axiom; he reads on and finds out that he fails at a later axiom. Soon he will look for a principle according to which he can imagine the axiom system, and since he does not find one, he will, if what is sought does not strike him purely accidentally, raise the question how at all this axiom system can be realised in thought.'

## 4.2. THE BEGINNING OF THE DEBATE

of the 19th century, when he was loosening the bond with reality by not adhering to a definite meaning of mathematical concepts, but where he did not defend the purely formalistic point of view where every bond with the human mind is cut.

Baldus proceeds with a presentation of intuitionism. He characterises intuitionists as a group of mathematicians who are opposed to Cantorian set theory, starting with Kronecker, further developing by the French semi-intuitionists, and finding in Brouwer their most radical representative. The first characteristic of intuitionism which Baldus mentions is its view on existence. He mentions the set theoretical paradoxes as the immediate cause for the intuitionists to reconsider the logical means of proof employed in mathematics. Contrary to formalists, who take consistency as the only criterion for existence, Baldus explains that intuitionists consider certain objects to be intuitively given, like the natural numbers. Besides these, only those objects exist which can be constructed mentally or which can be expressed in a finite number of words. Pure existence statements do not constitute proper existence statements for intuitionists. Taking all characteristics on intuitionism into account, Baldus concludes that intuitionism has succeeded in singling out the logically compelling part of mathematics from consistent mathematics.[248]

Thus, Baldus essentially explains intuitionism well. The only problem is that he does not distinguish between the views of Kronecker, the French semi-intuitionists and Brouwer. Because of that, he overrates the role the set theoretical paradoxes played in (Brouwerian) intuitionism, and he uses two different descriptions of what mathematical existence means for intuitionists, one stemming from Brouwer, the other from Borel.[249]

**Hilbert as a formalist?** It should be pointed out that Baldus' address is the first contribution to the foundational crisis in which Hilbert is clearly characterised as a formalist. Brouwer had explicitly designated Hilbert as a formalist only in his review of Mannoury's book, and had implied so in his inaugural address *Intuitionisme en formalisme*.[250] Weyl, who started the whole debate, had never indicated Hilbert's position by this term. Hilbert himself, finally, used expressions such as 'axiomatics' and 'proof theory' to refer to his own views on the foundations of mathematics,[251] and even seemed to use the term 'formalistic' in a negative sense as well.[252] Bernays did use the word 'formalism' when describing Hilbert's view,

---

[248] [Baldus 1924, pp. 23–26; 34]
[249] On Borel, see 1.3.2.
[250] See 2.5.
[251] [Hilbert 1922, p. 160; 174]
[252] '*Während lange Zeit hindurch die symbolische Logik nichts anderes als eine formalistische äusserliche Weiterbildung der Aristotelischen Schlussfigurentheorie zu sein schien, (...).*' ('While symbolic logic for a long time seemed nothing more than a superficial formalistic development of Aristotle's theory of inference figures, (...).'), Hilbert manuscript on Behmann's dissertation, 1/2/1918, [Mancosu 1999B, p. 317; 327].

but in the sense of a 'formal system'.[253] And Fraenkel, the best-read commentator to the debate, described the foundational conflict as one between *classical* and intuitionistic mathematics in the second edition of his *Einleitung in die Mengenlehre* in 1923;[254] in the third edition, published in 1928, he was to use the denomination 'formalistic' for Hilbert only between quotation marks.[255]

It should also be noted that the description of formalism which Baldus presents differs from the one given by Brouwer. Although all the characteristics which Baldus mentions, such as consistency being sufficient for existence, and the unrestricted use of the principle of the excluded middle, do apply, the essential feature of looking at (ordinary) mathematics as a purely formal system without contents is notably absent. It seems that Baldus has some kind of contentual view on formalistic mathematics, possibly believing in a general validity of what later was proved as the model existence lemma. Of course, the strict distinction between syntax and semantics had only just been developed at the time.[256] Baldus claims that most mathematicians are formalists.[257] In his description of formalism, this was probably true. In Brouwer's, however, it most probably was not.

## 4.3 The debate widened

From 1924 onwards, the debate extended beyond the initial group of the directly involved Brouwer, Weyl, Hilbert, and Bernays, the commentator Fraenkel and the relative outsiders Becker and Baldus. Not only the number of people involved increased, but also the languages used, a fact that considerably widened the group of people that could participate in the discussion. For the first time after the publication of Weyl's paper, the debate was brought to the English reading public, by Dresden.[258]

The following overview shows who became involved in the discussion. It is worth noting that all but Von Neumann agree that the predicate 'constructive' belongs to the intuitionistic side. Also, whereas at least Weyl and Bernays had still tried to explain what 'constructive' should mean, this aspect has by now disappeared from the debate. Apparently, 'constructivity' had become a term which was judged more important as a label than for its contents. Finally, Weyl withdrew his full support for intuitionism as the only correct way to do mathematics.

**Weyl in 1924**  In a paper submitted in October 1923 and published the following year in the *Mathematische Zeitschrift*, Hermann Weyl substantially modified

---

[253][Bernays 1922A, p. 15], [Bernays 1922B, p. 98]
[254]See 4.2.4.
[255][Fraenkel 1928, p. 377]
[256]Zach argued the case that Bernays was the first to completely distinguish between syntax and semantics, in his *Habilitationsschrift* of 1918, [Zach 1999, p. 342].
[257][Baldus 1924, pp. 18–19]
[258][Dresden 1924]

## 4.3. THE DEBATE WIDENED

his initial position towards intuitionism. Futhermore, he clarified some of his intuitionistic ideas.

Weyl modifies his stance in that he now maintains that mathematics should not only be done in Brouwer's, but also Hilbert's way. For, Weyl maintains, the example of physics shows that it is not always necessary for every individual proposition to have meaning, but that one can use the system as a whole for meaningful purposes.[259] It is very well possible that Weyl's occupation with the space problem in the early 1920s contributed to his changed appreciation of intuitionism, as Leupold suggested.[260]

The clarification concerned the intuitionistic position on mathematical existence. Weyl takes up the fundamental theorem of algebra as an example of a theorem that had never been proved in a way satisfying Brouwer's demands. Applied to this theorem, the intuitionistic requirements read: given an approximation method of the coefficients $a_i$ of the equation

$$x^n + a_{n-1}x^{n-1} + \ldots + a_0 = 0,$$

determine the roots of the equation in such a way that, if the approximation of the coefficients becomes increasingly more precise, so must the determination of the roots. The crucial step in the classical treatment of the theorem, Weyl claims, is that it steps over the difficulty of explicitly giving such an approximation. This is due to its *Existentialabsolutismus*, which assumes that the coefficients have an existence anyway.[261]

Towards the end of the paper, Weyl expresses his earlier statement on the relation between constructions and proofs in a modified way. Formal argumentation, he maintains, should not be regarded too highly. For the actual difficulty mostly lies not so much in presenting a proof when a construction is given, but in finding the construction itself. It is just as in the case of the fundamental theorem of algebra: once the construction, i.e., the approximation of the roots has been given, the remaining proof that these are the roots of the equation is relatively simple. And this, Weyl claims, is almost always the case.[262]

**Dresden in 1924** In 1924, the first American contribution to the debate appeared. The Dutch-born Arnold Dresden, who had studied and written his dissertation at the university of Chicago,[263] was the first in this region to react to intuitionism. I do not know whether his Dutch background played a role in his interest for intuitionism, but his knowledge of the Dutch language (though not fluent) certainly did help. Dresden had translated Brouwer's inaugural lecture into

---

[259][Weyl 1924, pp. 149–150]
[260][Leupold 1961, pp. 28–29]
[261][Weyl 1924, pp. 142–143]
[262][Weyl 1924, p. 149]
[263][Poggendorff 1936–1940, p. 600]

## 156  CHAPTER 4. REACTIONS: EXISTENCE AND CONSTRUCTIVITY

English, so it is reasonable to assume that they had been in contact before.[264] In a paper presented to the American Mathematical Society in December 1923 and published the following year, Dresden treated Brouwer's contributions to the foundations of mathematics. Dresden had corresponded with Brouwer about the paper, and his objective was to clear the way for intuitionism in the United States.[265] He introduces intuitionism by stressing its constructive character:[266]

> Brouwer conceives of mathematical thinking as a process of construction, which builds its own universe, independent of the universe of our experience, somewhat as a free design, under the control of nothing but arbitrary choice, restricted only in so far as it is based upon the fundamental mathematical intuition.

This 'constructive' aspect of intuitionistic mathematics recurs time and again throughout the paper.

Dresden treats the issue of mathematical existence by quoting Brouwer, who held that it was never proved that once, e.g., a number satisfies a non-contradictory system of conditions, that it then also actually exists.[267] Although Dresden was the first to bring the foundational crisis to the English speaking world, there was no reaction. His paper was only mentioned in Heyting's dissertation, by Jørgensen, and, of course, by Fraenkel.[268]

**Doetsch in 1924** In January 1924, the Halle professor in geometry Gustav Doetsch gave philosophical lectures for the local sections of the *Kant-Gesellschaft* in Halle and Magdeburg on the meaning (*Sinn*) of pure mathematics and its applications. He included a short description of intuitionism, where he explicitly characterises intuitionism by its 'constructive' method, and opposes it to the formalistic existence absolutism. Having explained Hilbert's demand of consistency in order to secure the existence of a mathematical system and the problem of proving it, he turns to the intuitionistic alternative:[269]

> Nun kann man allerdings dem Problem der Widerspruchslosigkeit dadurch aus dem Wege zu gehen versuchen, daß man die axiomatische Methode, Begriffe durch ihre Relationen zu definieren, durch eine '*konstruktive*' Methode ersetzt: Man denkt die Gebilde durch schrittweise

---
[264] The translation was published in the Bulletin of the American Mathematical Society, [Brouwer 1913].
[265] Letter from Dresden to Brouwer, 17/9/1923, [MI Brouwer]. In the first of his letters to Brouwer, Dresden does not introduce himself, which indicates that there had indeed been contact at some earlier date.
[266] [Dresden 1924, p. 32]
[267] [Dresden 1924, p. 37]; Dresden refers to Brouwer's thesis, [Brouwer 1907, pp. 180–181]; see 2.3.1.
[268] [Heyting 1925, p. 93], [Jørgensen 1931, p. 55] and [Fraenkel 1927A, p. 36]
[269] [Doetsch 1924, p. 449]

## 4.3. THE DEBATE WIDENED

Konstruktion entstanden. (...) Annahmen über die Existenz von Begriffen kommen hier nicht vor, sondern alle Objekte werden regelrecht erzeugt.[270]

Doetsch's conclusion regarding intuitionism is implicitly negative, for he finishes the description by mentioning that by proceeding in this way, large parts of mathematics have to be given up, and then continues with the alternatives of Hilbert and Russell.

Doetsch's contribution appeared in the *Kantstudien*, the leading philosophical journal at the time.[271] Nevertheless, it did not evoke a single reaction.

**Fraenkel in 1924** In a lecture delivered for the *Gesellschaft zur Beförderung der gesamten Naturwissenschaften* ('Society for the advancement of all natural sciences') in Marburg in June 1924, Fraenkel spoke on the current crisis in the foundations of mathematics. He traces the method of constructive proofs to Kronecker, and mentions Brouwer's and Weyl's rejection of pure existence statements.[272] His treatment of mathematical existence and constructivity is marginal—this was to change a year later.

**Von Neumann in 1924** In the same year, a 21-year-old Hungarian mathematician made his first appearance in the debate. Janos (John) von Neumann at the time lived three different lives: formally enrolled as a mathematics student at the university of Budapest, at the same time studying chemistry at the *Eidgenössische Technische Hochschule* in Zürich, since his father wished that he should learn something practical, Von Neumann spent most of his time doing mathematical research in Zürich and Berlin, on his own or together with some of the local professors.[273] At the time, he frequented Hilbert's house in Göttingen, and he had also got to know Weyl and Fraenkel personally.[274] One of Von Neumann's first papers was published in the prestigeous *Journal für die reine und angewandte Mathematik*. Its main subject was the axiomatisation of set theory, but it also contained some small remarks which are of interest to us here.

In Von Neumann's paper, it is clear that he uses the term 'constructive' in Hilbert's sense. This is not surprising, since he wanted to contribute to Hilbert-style mathematics. For example, Von Neumann lists what he calls arithmetic and logical 'construction axioms', which contain nothing but pure existence statements.[275]

---

[270]'For the rest, one can try to evade the problem of consistency by replacing the axiomatic method of defining notions by means of their relations by a '*constructive*' method: one thinks of the buildings as come into being by gradual construction. (...) Here, assumptions about the existence of notions do not occur, but all objects are directly generated.'
[271]Virtually all German philosophers were member of the *Kant-Gesellschaft* and thus received the *Kantstudien*.
[272][Fraenkel 1924B, pp. 123–124], [Fraenkel 1925A, p. 253]. The lecture was published twice, in somewhat different forms.
[273][Fraenkel 1967, p. 168]
[274][Reid 1970, p. 172], [Ulam et al. 1969, p. 236], [Fraenkel 1967, pp. 168–169]
[275][Von Neumann 1924, pp. 224–225]

158     CHAPTER 4. REACTIONS: EXISTENCE AND CONSTRUCTIVITY

Having obtained a system of axioms for set theory, Von Neumann devotes a small paragraph to the question of possible models for his system. He notes that especially one axiom, which prescribes an upper bound on the admissible functions, gives rise to difficulties. And any axiom replacing the one causing the problem must, so he argues, be an impredicative one. The demands presented by the axiomatic system are so complicated, Von Neumann continues, that he cannot point out a corresponding model,[276]

> (...) obgleich die Konstruktion durchführbar sein muß, falls die Mengenlehre auf nicht-intuitionistischer Basis überhaupt möglich ist.[277]

Here, the word 'construction' is used against intuitionism, in line with the meaning Von Neumann attaches to the term.

### 4.3.1  Existence in a central position

Until halfway 1924, questions of mathematical existence were treated in the reactions to Brouwer's intuitionism, but as a rule not extensively. Apparently, most contributors to the debate did not see it as a central matter to intuitionism or to its opposition to formalism. No generally accepted image had developed of what the foundational crisis centered on.[278] It should be pointed out that all contributions but one to this part of the debate paid attention to the idea of constructions or of constructive mathematics, the exception being Hilbert's 1922 lecture.

From 1924 on, these aspects changed. Papers and books appeared which put mathematical existence in a more central position, notably Wavre's 1924 *Y a-t-il une crise des mathématiques?*, Fraenkel's 1927 *Zehn Vorlesungen über die Grundlegung der Mengenlehre* and Becker's 1927 *Mathematische Existenz*.[279] At the same time, more contributions were made in which existence statements were discussed without referring to their supposedly constructive character.

I first analyse the items which treated mathematical existence as a central question to the opposition between intuitionism and formalism, or in which a substantial contribution to the discussion on mathematical existence was made. In the next section, I consider the contributions between 1924 and 1928 which dealt with the subject of mathematical existence as a side issue.

**Wavre in 1924**  In a paper published in 1924, the Genève professor Rolin Wavre presented a survey of the foundational debate as it had developed until then. Wavre's own field was differential and integral calculus, and most of his publications were on mathematics applied to physics or astronomy. However, he also wrote three papers on foundational questions in the *Revue de Métaphysique et de*

---

[276] [Von Neumann 1924, p. 237]
[277] '(...) although the construction has to be feasible, if set theory on a non-intuitionistic basis is at all possible.'
[278] See 3.3.1.
[279] [Wavre 1924], [Fraenkel 1927A] and [Becker, O. 1927]

## 4.3. THE DEBATE WIDENED

*Morale*, of which this was the first one. He allegedly also defended the intuitionistic point of view at meetings of the French speaking Swiss mathematicians.[280] Wavre based his presentation on publications by Brouwer, Weyl, Hilbert and Bernays. The purpose of the paper was to[281]

> (...) faire entrevoir que l'opposition [entre formalisme et intuitionisme, DH] devient tout à fait nette 'à propos de la notion d'existence et d'une application suspecte du principe du tiers exclu'.[282]

Note that Wavre presents the subject of mathematical existence as one of the two issues where the opposition between intuitionism and formalism becomes clear.

Wavre starts by pointing out that Lebesgue had already earlier criticized the classical notion of existence,[283] a criticism which was pushed further by Brouwer and Weyl. He explains Weyl's view by means of the following example. Given a sequence $S$ of natural numbers, it is clear that the following two propositions are opposed to each other:

(o) All numbers in $S$ are odd;

(e) There is a number in $S$ which is even.

But the questions that matters is: are they opposed as contradictories or not?[284] As long as $S$ is a finite sequence, nobody will doubt this to be the case. But for an infinite sequence opinions may diverge.[285] The answer to the question will depend on the meaning one attaches to the existence statement in proposition (e). With respect to this issue, Wavre notes that[286]

> Le logisticien, le formaliste fait par définition de 'il existe', l'équivalent de 'non tous', c'est son droit; mais, en faisant cela, il introduit un nouvel axiome ou une nouvelle définition.[287]

Wavre's claim is clear: the intuitionists are the ones who adhere to the meaning of the word 'existence', while formalists and logicists change its meaning in order to save the principle of the excluded middle.[288] Note that whereas Weyl had only

---

[280] [Juvet 1927, p. 137]

[281] [Wavre 1924, p. 436]

[282] '(...) show that the opposition [between formalism and intuitionism, DH] becomes clear 'regarding the notion of existence and a dubious application of the principle of the excluded middle'.' The quotation marks indicate the subtitle of Wavre's paper.

[283] See 4.1.2.

[284] Aristoteles called two propositions 'opposed as contradictories' which cannot both be true and cannot both be false, [Kneale & Kneale 1964, pp. 54–67]; see 5.1.1.

[285] The importance of this matter for the debate on the validity of the principle of the excluded middle is discussed in 5.3.

[286] [Wavre 1924, p. 443]

[287] 'The logicist, the formalist by definition makes the 'there exists' equivalent to 'not all'. He may do so; but, in doing so, he introduces a new axiom or a new definition.'

[288] Wavre does not mention the changed interpretation of the 'or'. From the discussion he presented later in the paper, it is clear that he tacitly assumed a constructive interpretation: one has to be able to point out which part of the disjunction is true. Most contributors to the debate focused on the interpretation of the existential quantifier and paid little attention to the 'or'.

condemned the classical use of pure existence statements as meaningless, Wavre points out that they have a different meaning.[289]

The question arises whether Wavre's judgement about who adhered to the real meaning of mathematical existence was correct. This is hard to determine. As the debate on mathematical existence shows, the 'old', 'traditional' or 'true' meaning of 'there exists' in mathematics was more than mere consistency. Even those seen as formalists often argued that some kind of existence was implied *because* of the consistency of the system. Whether this means that existence was seen as constructivity, as the intuitionists argued it should be, or as some kind of Platonic existence, is harder to say. The important thing, however, is not which of the currents held on to the 'old' meaning of existence, but the fact that Wavre stressed the different meanings intuitionists and formalists attached to the same term. I return to this after the next example in Wavre's paper.

Wavre continues his exposition with the example of a number of the form

$$m_n = 2^{2^{n+8}} + 1 \qquad \text{with } n \in \mathbb{N},$$

and asks whether such a number $m_n$ exists which is composite (i.e., not prime).[290] He next introduces a fictional 'idealist' (formalist) and 'empiricist' (intuitionist),[291] who discuss the question in a way similar to Weyl's private discussion in his *Grundlagenkrise* paper.[292] The intuitionist argues that he can only prove the existence of such a number by actually presenting one. The formalist replies that they apparently have different conceptions of what 'existence' means, since he will be satisfied by a proof showing the existence of a composite $m_n$, also if the proof contains no construction of the required number. Whereupon the intuitionist elaborates his position: the construction itself is not necessary; the only thing he asks for is that, if he wished so, he could actually produce the required number in a finite number of steps. But, the formalist holds, this demand is superfluous; for if the number exists, it certainly can be found somewhere in the sequence of natural numbers after a finite number of steps. One simply starts with the numbers $1, 2, 3, \ldots$, and this number $m_n$ will either be found or it will not be found.[293] Whereupon the intuitionist replies: If you would only replace the 'the number will be found' by 'the number has been found', our opposition would disappear. For in your description

---

[289] Brouwer had made the same point before, but had not stated it prominently, see 2.6.2.

[290] The numbers Wavre uses are the Fermat-numbers $F_n = 2^{2^n} + 1$ from $n = 8$ on. Fermat had conjectured these numbers to be prime on the basis of this being so for $n = 1, \ldots, 4$. Later, Euler proved that $F_5$ is composite. Until now, all the other Fermat numbers that have been checked turned out to be composite. There is, however, no proof that $F_1$ to $F_4$ are the only primes.

[291] Wavre uses the terms which Paul Du Bois–Reymond employed in the end of the 19th century, [Du Bois-Reymond, P. 1882, p. 3]. For the sometimes confusing terminology, see 4.1.1.

[292] See 5.2.1.

[293] What Wavre's fictional formalist defends comes close to what is nowadays known as Markov's principle. In formalised form, this principle reads $[\forall(x)(A(x) \vee \neg A(x)) \wedge \neg\neg\exists(x)A(x)] \rightarrow \exists(x)A(x)$, where $A$ is an algorithmically decidable predicate. In Troelstra and Van Dalen's words, it can be paraphrased as holding that if there is an algorithm for testing $A$, and if we know by 'indirect means' that we cannot avoid encountering an $x$ such that $A(x)$ holds, then we can in fact *find* such an $x$, [Troelstra & Van Dalen 1988, Vol. 1, pp. 203–204].

## 4.3. THE DEBATE WIDENED

you would have to go through the infinite sequence of natural numbers in order to prove that you did not find such a number, which you will never do.[294]

Wavre used this discussion in order to analyse the arguments of his ideal-typical exponents of both currents. I think it is interesting, and I have presented it at some length, not only because it clarifies both positions, but also because it shows most accurately that when intuitionists and formalists engage in a debate, they are likely to talk at cross-purposes.[295] Although they often use the same terms, namely the ones mathematics has 'always' used ('there exists', etc.), the meaning of these terms may vary according to which current the speaker belongs to. Looked at in this way, the foundational debate between formalism and intuitionism can be seen as a battle about the meaning of classical mathematical terms.

The term related to existence, 'construction', does not appear prominently in Wavre's exposition. He mostly uses it in his description of the intuitionistic conception of set theory. In this view, Wavre claims, sets cannot be given by a characteristic property, but they should be constructed. 'To construct' in a narrow sense would mean to enumerate the members of the set one by one; but this would place too much restrictions on mathematics. Brouwer's way out, he explains, is to introduce the concept of a choice sequence.[296]

In his concluding remarks, Wavre returns to the question of mathematical existence:[297]

> L'existence idéale n'est pour l'empiriste [intuitionist, DH] qu'une fausse fenêtre pour la symétrie logique des propositions portant sur l'ensemble fini d'une part, sur l'infini de l'autre; fiction des logiciens imaginée non pour sauver la logique, elle n'est pas en danger, mais pour arrondir son royaume.[298]

Intuitionistic existence, Wavre maintains, is the 'representation' of an object, whereas the formalist is satisfied if consistency has been proved. In a less-than-clear comparison with the relativist school in physics, Wavre presents the intuitionists as the more prudent ones: their position is not to proclaim the existence of objects that cannot be defined. But because of the different meanings of the word 'existence' in formalism and intuitionism, Wavre concludes, there is not much hope that the dispute will be settled.[299]

---

[294] [Wavre 1924, pp. 443–445]
[295] Good examples of this behaviour are [Bernays 1922A] and the beginning of the discussion Wavre himself later had with Lévy, cf. [Wavre 1926A], [Lévy 1926A], [Wavre 1926B] and [Lévy 1926B].
[296] [Wavre 1924, p. 457]
[297] [Wavre 1924, p. 467]
[298] 'Ideal existence is for the empiricist [intuitionist, DH] nothing but a fake window for the logical symmetry of propositions bearing on finite sets on the one hand and on infinite sets on the other; logicians' fiction made up not to save logic, for it is not in danger, but to enlarge its kingdom.'
[299] [Wavre 1924, pp. 467–468]

162     CHAPTER 4. REACTIONS: EXISTENCE AND CONSTRUCTIVITY

Wavre's paper, which appeared in the *Revue de Métaphysique et de Morale*, was read more than most other papers on the foundational crisis. Various authors such as Heyting, Gonseth, Becker, Fraenkel, Dresden, and Jørgensen referred to it.[300] Heyting was the only one of them who commented negatively, by stating that the paper contained some errors and did not get to the core of the matter.

The importance of Wavre's contribution to the foundational debate lies in his demonstration of where the differences between intuitionism and formalism lay regarding the central issues of mathematical existence and the principle of the excluded middle, and of the different meanings they attach to the same terms. A coherent exposition of the intuitionistic point of view, however, was still lacking, and had to wait for Fraenkel's 1927 book.[301]

**Hilbert in 1925** In June 1925, the *Westfälische Mathematische Gesellschaft* ('Westphalian Mathematical Society') organised a meeting in Münster to commemorate Weierstraß.[302] One of the speakers was Hilbert, who delivered a lecture called *Über das Unendliche* ('On the infinite'). It was published the following year in the *Mathematische Annalen*.

The lecture witnesses one step more in the development of Hilbert's point of view, moving away from the contents of mathematical signs towards a more formal interpretation. Until then, Hilbert had held on to the idea that also the formalised part of mathematics was linked to contents, since it consisted of the representations of mathematical thoughts.[303] Now, he partially dropped this claim and moved more towards the purely formalistic position which Brouwer had depicted before. This time, Hilbert describes the formalised part of his proof theory as follows:[304]

> (...) so wird die Mathematik zu einem Bestande von Formeln, und zwar erstens solche, denen inhaltliche Mitteilungen finiter Aussagen entsprechen, und zweitens von weiteren Formeln, die nichts bedeuten und die *idealen Gebilde unserer Theorie* sind.[305]

Thus, Hilbert preserves his claim that formalised mathematics represents mathematical thoughts for finite propositions, but he drops it for infinite ones. The reason why he arrived at this conclusion is that he want to keep the simple rules of Aristotelian logic.[306]

---

[300][Heyting 1925, p. 93], [Gonseth 1926, p. 192], [Becker, O. 1927, p. 775], [Fraenkel 1927A, p. 36], [Dresden 1928, p. 440], [Jørgensen 1931, p. 51]
[301]See 4.3.1.
[302]Weierstraß was born in Westphalia on October 31, 1815; he had studied some months in Münster and taught there for a year. He died in 1897, [Biermann 1976, pp. 219–220]. The reason why such a commemoration should be held in June 1925 is not clear to me. Maybe it had been planned in 1915, but was postponed because of the war.
[303]See 4.2.2.
[304][Hilbert 1926, p. 175]
[305]'(...) in this way mathematics becomes a stock of formulas, namely in the first place of such formulas which correspond to contentual communications of finite judgements, and in the second place of further formulas, which signify nothing and which are the *ideal product of our theory*.'
[306]See 5.3.1.

## 4.3. THE DEBATE WIDENED

If we take into account Hilbert's lecture notes from which Sieg published citations, it seems that Hilbert had already moved towards a more purely formalistic point of view earlier. In the 1922/23 winter term notes we read:[307]

> (...) das Problem der Widerspruchsfreiheit gewinnt nunmehr eine ganz bestimmte, greifbare Form: es handelt sich nicht mehr darum, ein System von unendlich vielen Dingen mit gegebenen Verknüpfungs-Eigenschaften als logisch möglich zu erweisen, sondern es kommt nur darauf an, einzusehen, dass es unmöglich ist, aus den in Formeln vorliegenden Axiomen nach den Regeln des logischen Kalküls ein Paar von Formeln wie $A$ und $\neg A$ abzuleiten.[308]

This means that already by the end of 1922, beginning of 1923 Hilbert was no longer after a proof of the logical possibility of an axiomatic system, but only after a formal proof of its non-contradictority. Thus, from then on Hilbert drew the same conclusion regarding consistency as the intuitionists: a consistency proof does not prove anything other than that no contradiction can occur. In particular, it does not tell us anything about existence.

As to the question of mathematical existence, in the 1925 lecture Hilbert understands that a mere consistency proof would not convince his intuitionistic opponents. He adds a historical argument to why his view on mathematical existence is the correct one:[309]

> Auch alte Einwendungen, die man längst abgetan glaubte, treten in neuem Gewande wieder auf. So wird neuerdings etwa dies aufgeführt: Wenn auch die Einführung eines Begriffes ohne Gefahr d.h., ohne Widersprüche zu erhalten, möglich sei und dies erwiesen werden könne, so stehe damit noch nicht ihre Berechtigung fest.[310] Ist dies nicht genau der Einwand, den man seinerzeit gegen die komplex-imaginären Zahlen geltend machte, indem man sagte: freilich könne man zwar durch sie keine Widersprüche erhalten; aber ihre Einführung sei dennoch nicht berechtigd; denn die imaginären Größen existierten doch nicht?[311]

---

[307][Sieg 1999, p. 40]

[308]'(...) the consistency problem from now on acquires a most specific, concrete form: it is no longer a question of proving that a system of infinitely many things is logically possible, but only of recognizing that it is impossible to derive a pair of formulas like $A$ and $\neg A$ from the axioms extant in the formulas according to the rules of the logical calculus.'

[309][Hilbert 1926, pp. 162–162]

[310]Hilbert speaks about the 'justification' of introducing new concepts. This is further away from the more philosophical discussion of whether certain objects exist. However, I am not sure whether much weight should be attained to this, since I do not have the impression that Hilbert was always very careful in the choice of the terms he used in the discussion.

[311]'Even old objections that have long been regarded as settled reappear in a new guise. In this way, we in recent times come upon statements like this: even if we could introduce a notion safely (that is, without generating contradictions) and if this were demonstrated, we would still not have established that we are justified in introducing the notion. Is this not precisely the

## 164   CHAPTER 4. REACTIONS: EXISTENCE AND CONSTRUCTIVITY

The historical correctness of Hilbert's reasoning is dubious. In the first place, it is not clear to whom Hilbert is referring. Furthermore and more importantly, at the time of the introduction of the complex numbers the notion of consistency did not appear in the discussion. Thus, it seems that Hilbert projected his own ideas about the existence of a mathematical system back onto the history of mathematics.

Hilbert continues his discourse by adding that the only criterion which one could accept on top of that of consistency is the success a mathematical theory has.[312]

The argumentation used by Hilbert would not satisfy an intuitionist. Indeed, in 1927 Becker wrote to Mahnke about Weyl:[313]

> Hilbert selbst habe er [Weyl, DH] ofters nach dem Sinn seiner (H.'s) 'idealen Aussagen' gefragt und H. habe 'regelmäßig den Kopf weggewendet und von etwas anderem gesprochen.'[314]

But Hilbert's realisation that for intuitionists a consistency proof does not suffice means that from now on there was, at least in principle, the possibility of exchanging arguments why to take a certain position, instead of talking at cross-purposes.

The description Hilbert gives in the lecture of an existence statement is rather obscure. He claims that a statement of the form 'there is a number with such and such a property' only obtains meaning as a 'partial judgement', i.e., as part of a more specified judgement, the contents of which is unnecessary for many applications.[315] Sieg found lecture notes of Hilbert to which this position seems to go back. In his 1921/22 lecture, Hilbert had said almost literally the same. Then he had continued:[316]

> Die Existenzbehauptung hat hier überhaupt nur einen Sinn als ein Hinweis auf ein Verfahren der Auffindung, welches man besitzt, das man aber für gewöhnlich nicht näher anzugeben braucht, weil es im allgemeinen genügt, zu wissen, dass man es besitzt.[317]

This looks rather like Weyl's *Urteilsabstrakt* in a watered-down version.[318]

---

same objection as the one formerly made against complex numbers, when it was said that one could not, to be sure, obtain a contradiction by means of them, but their introduction was nevertheless not justified, for, after all, imaginary magnitudes did not exist?' Translation from [Van Heijenoort 1967, p. 370]
[312] Weyl would later pick up this reasoning, see 4.3.2.
[313] Letter from Becker to Mahnke, 9/1927; quoted from [Mancosu & Ryckman 2002]
[314] 'He [Weyl, DH] had asked Hilbert himself several times about the meaning of his (H.'s) 'ideal statements', and H. had 'regularly turned his head away and changed the topic of discussion.'
[315] [Hilbert 1926, p. 173]
[316] [Sieg 1999, p. 28]
[317] 'The existence theorem only has meaning at all as a reference to a method of discovery which one possesses, but which, as a rule, one does not have to indicate further, because in general it suffices to known that one has it.'
[318] See 4.2.1.

## 4.3. THE DEBATE WIDENED

It should be noted that, again, Hilbert does not claim the term 'constructive' any more. In the same year, Bernays went one step further by recognizing constructive mathematics as distinct from axiomatic mathematics.[319] Thus, they had both dropped their earlier claim to 'constructivity'.[320]

**Fraenkel from 1925 to 1928**  In June 1925, Fraenkel gave a series of lectures for the *Kant-Gesellschaft* ('Kant Society') in Kiel on the foundations of set theory. This happened on invitation of Heinrich Scholz, a philosopher at the university of Kiel who later was to specialise in mathematical logic. Scholz had contacted Fraenkel after he had read the second edition of his *Einleitung in die Mengenlehre*, and a friendship for life developed.[321] The Kant Society was an important one, since virtually all German philosophers were a member of it. Fraenkel's talks were published in 1927 as *Zehn Vorlesungen über die Grundlegung der Mengenlehre* ('Ten lectures on the foundations of set theory').[322]

One of the lectures dealt with intuitionism. In his exposition, Fraenkel first refers to an unpublished formulation of the basic principles of intuitionism which Brouwer had written in a letter to Fraenkel.[323] In Fraenkel's words, these are the following: firstly, the principle of the independence of mathematics from mathematical language, and secondly, the principle of a constructive set definition as a basis for mathematics.[324] Note that these principles are far more general than the ones Fraenkel had used in 1923 in presenting intuitionism, e.g. the refusal to accept the universal validity of the principle of the excluded middle.[325]

Fraenkel next moves on to his *own* presentation of intuitionism. This may seem strange, but it is explained by what had happened between Brouwer and Fraenkel. The incomplete information we have about the episode is the following.[326] From a letter by Brouwer to Fraenkel in January 1927, we read that the latter had seen some behaviour of Brouwer's as a 'declaration of war'.[327] Apparently, Fraenkel was referring to remarks Brouwer had made while proof-reading the manuscript of Fraenkel's book. In his response, Brouwer denies this definitely, claiming that, quite on the contrary, the fact that he had convinced the publisher Teubner again to accept major changes in Fraenkel's text shows that he is try-

---

[319] Letter from Bernays to Finsler, 18/01/1925 [ETH Bernays, HS 648-4]
[320] See 4.2.2.
[321] [Fraenkel 1967, pp. 179–181]. Scholz later lobbied to have Fraenkel appointed at the university of Kiel, which indeed happened in 1928.
[322] [Fraenkel 1927A]; the published text is the one on which this analysis was based. It is likely to differ from the lectured one, since it took Fraenkel until December 1926 to finish completing the manuscript.
[323] [Van Dalen 2000, p. 289]
[324] [Fraenkel 1927A, p. 35]
[325] See 4.2.4.
[326] A more complete version of the episode can be found in [Van Dalen 2000].
[327] '*Kriegserklärung*', [BF Fraenkel, Letter from Brouwer to Fraenkel, 28/1/1927]; [Van Dalen 2000, p. 303]. In fact, Brouwer had also described Fraenkel's behaviour as a declaration of war, by the description of intuitionism Fraenkel intended to put forward in his book.

# 166  CHAPTER 4. REACTIONS: EXISTENCE AND CONSTRUCTIVITY

ing to avoid every conflict between him and Fraenkel. Brouwer's letter shows that the formulation of the basic principles of intuitionism used in the final version of Fraenkel's booklet, paraphrased above, was taken literally from Brouwer. In this light, it seems reasonable to assume that Fraenkel allowed for two presentations of intuitionism in his book in order to pacify Brouwer.[328]

Fraenkel starts his own exposition by remarking that he will only present some of the basic ideas of intuitionism and that he will reduce them to 'a few roots — maybe only one root'.[329] He then gives the basic thesis of intuitionism: mathematical existence = constructivity:[330]

> Die Grundanschauung, aus der sich alle die zum Teil so überraschend scheinenden Behauptungen der Intuitionisten mehr oder weniger konsequent ableiten lassen, betrifft einen (...) Punkt: *die scharfe Unterscheidung zwischen Konstruktionen und reinen Existenzaussagen und die alleinige Anerkennung der ersteren unter Verwerfung der letzteren.*[331]

Note that Fraenkel speaks about the *sharp* distinction between constructions and pure existence statements. The following year, Menger was the first to voice a protest against the in his view vague use of the word 'constructive' in intuitionism, and many were to support him.[332]

Fraenkel's formulation means a major improvement in the presentation of intuitionism. Finally there was a book, written by a well-known mathematician, available to everybody from 1927 on, which stated in clear terms a unification of all intuitionistic claims under one consistent point of view. This was exactly Fraenkel's purpose with his new presentation of intuitionism.[333]

Fraenkel next applies the intuitionistic demands to set theory. It is the well-ordering theorem that is hit hardest by the intuitionistic criticism. This theorem states that every set can be well-ordered, without telling us how to achieve this. The only value, Fraenkel repeats after Weyl, that intuitionists can attain to pure existence proofs is that they stimulate people to look for its constructive version. Fraenkel briefly explains that constructions should be seen as thought constructions, originating from some basic elements given intuitively.[334]

Fraenkel apparently was sufficiently satisfied with this presentation of intuitionism, since he repeated it almost *verbatim* in the third edition of his popular *Einleitung in die Mengenlehre*, which appeared in 1928.[335] This meant that the

---

[328] Unfortunately, the Fraenkel part of the correspondence is missing.

[329] '*einige wenige Wurzeln – vielleicht (...) eine einzelne Wurzel*', [Fraenkel 1927A, p. 36]

[330] [Fraenkel 1927A, p. 36]

[331] 'The principal idea from which all the in part seemingly so surprising assertions of the intuitionists can be derived concerns one (...) point: *the sharp distinction between constructions and pure existence statements and the sole recognition of the former under rejection of the latter.*'

[332] See 4.4.3.

[333] [Fraenkel 1927A, p. VI]

[334] [Fraenkel 1927A, pp. 37-38; 156]

[335] [Fraenkel 1928, pp. 226–228]

## 4.3. THE DEBATE WIDENED

reach of Fraenkel's view on the opposition between intuitionism and formalism was enlarged substantially.

The prevailing opinion, Fraenkel further remarks, which stems not only from Hilbert but also from Poincaré, is to demand only consistency for mathematical existence. However, philosophers and logicians, Fraenkel claims, hold that it is exactly the other way round: only if an object exists, it is consistent. Finally, there is Brouwer's view that existence requires a mental construction from intuitively given basic elements. Correct as this presentation is, Fraenkel steps over the fact that he is, in fact, comparing different categories. In the case of a consistency proof, one always considers an axiomatic system, and hence a number of mathematical objects. The view presented here as that of philosophers and logicians, on the other hand, and also Brouwer's view, is concerned in the first place with the existence of an individual mathematical object.

Fraenkel sees little hope for a reconciliation between adherents of Hilbert's view and those of Brouwer:[336]

> Der Gegensatz zwischen beiden Anschauungen ist von wesentlich dogmatischer Art und läßt wenig Hoffnung auf einen Ausgleich durch Überzeugung des Gegners (...).[337]

**Dingler in 1926**  In 1926, the philosopher Hugo Dingler (1881–1954) published the book *Der Zusammenbruch der Wissenschaft und der Primat der Philosophie* ('The collapse of science and the primate of philosophy'). Dingler was a university professor in Munich, who had among other things worked on the foundations of geometry. The section 'The chaos of opinions'[338] is devoted to mathematics and physics. Regarding the former subject, Dingler deals with the foundational crisis. He makes a very clear distinction:[339]

> Dies Wort [to exist, DH] heißt (wenn wir uns auf die Mathematik beschränken) zweierlei: 1. daß sich ein solches Gebilde, wie es da verlangt wird, *wirklich konstruieren* lasse, d.h. soweit logisch explizit durch Bestimmungen festlegen, daß es im mathematischen Sinn eindeutig bestimmt ist; 2. daß der gebildete Begriff *widerspruchslos* sei. Durch unbewußtes Hin- und Herpendeln zwischen diesen beiden Bedeutungen des Terminus 'existieren', sind nun die meiste Schwierigkeiten dieses Fragenkomplexes entstanden.[340]

---

[336][Fraenkel 1927A, pp. 155–156]
[337]'The opposition between both views is essentially dogmatic and leaves little hope for a settlement by convincing the opponent (...).'
[338]'*Das Chaos der Meinungen*', [Dingler 1926, pp. 75–124]
[339][Dingler 1926, p. 90]
[340]'This word [to exist, DH] means (if we restrict ourselves to mathematics) two things: 1. that such a building, as is being required, can be *actually constructed*, i.e., fastened down logically and explicitly by determinations to such an extent, that it is determined uniquely in a mathematical sense; 2. that the concept built is *consistent*. Most problems in this complex of questions came

Dingler here uses the word 'to exist' as referring to an existence statement about a mathematical object, not about a mathematical system. Therefore, his description of the second meaning of existence is somewhat imprecise. The point is not whether the object itself is consistent, but whether the assumption of the existence of the mathematical object in question leads to a contradiction in the mathematical system we are considering.

Dingler proposes to call the first meaning 'constructively existent', the second 'logically existent'. This is the first time someone came up with the remark that, since constructivity and consistency are two different concepts, one should not use the same word 'existence' for it, but different names. Later, Menger was to suggest a similar solution.[341]

Turning to the foundational debate, Dingler notes that formalists are inclined to see existence as logical existence, without, remarkably enough, saying anything about intuitionists and their constructive interpretation of existence. Dingler's commentary becomes even stranger when he claims, without further explanation, that 'actually, concerning problems of decidability, the intuitionists are formalists, but Hilbert and Bernays are intuitionists.'[342]

Dingler sees the foundational debate as part of a general sliding down of science into chaos, since there is no definite means for one of the currents to refute the other. The only solution, Dingler claims, is his own 'constructive methodology'[343] – note once more the popularity of the constructivity label.

Dingler only played a marginal role in the debate.

**Wavre vs. Lévy, 1926** In 1926, a discussion developed in the *Revue de Métaphysique et de Morale* between Rolin Wavre and the French mathematician Paul Lévy (1886–1971), a student of Hadamard and professor at the *École Polytechnique*. Lévy mostly worked in the fields of functional analysis and integral equations, and he had just started publishing important work in modern probability theory, of which he was later recognised as the co-founder (together with Kolmogorov). Besides this, Lévy was also interested in philosophical questions.[344] The starting point of the discussion was a paper by Wavre in which he, among other things, once more explained the intuitionistic view on mathematical existence.[345] He illustrated the intuitionistic stance with a counter-example against the principle of the excluded middle taken from Brouwer, based on the decimal expansion of $\pi$. Lévy reacted with an article under the telling title *Sur le principe du tiers exclu et sur les théorèmes non susceptibles de démonstration* ('On the principle of the excluded middle and on theorems not capable of demonstration').

---

into being by unconsciously wobbling to and fro between these two meanings of the terminus 'to exist'.'

[341] See 4.4.3.

[342] 'im echten Sinne genommen, was die Entscheidbarkeitsprobleme anlangt, die Intuitionisten Formalisten, Hilbert und Bernays aber Intuitionisten sind', [Dingler 1926, p. 92]

[343] 'konstruktive Methodologie', [Dingler 1926, p. 95]

[344] [Lévy 1970], [Dieudonné 1972]

[345] [Wavre 1926A]

## 4.3. THE DEBATE WIDENED

In the paper, Lévy claims that the discussion on the example taken from Brouwer by Wavre is caused by a misunderstanding due to the definitions used. For suppose, so he argues, that we would define the words 'even' and 'odd' in the following way: an integer $n$ is called even if *one can find* another integer of which it is the double; and odd if $n - 1$ is even. Then the first winning number in the national lottery from now on would be neither odd nor even, because we cannot find a number $p$ such that the winning number equals either $2p$ or $2p + 1$. This kind of definition, Lévy concludes, is nothing but a word game. His solution is the following:[346]

> Une définition des mots *pair* et *impair* qui ne donne pas lieu aux mêmes difficultés peut être exprimée de la manière suivante: un nombre entier $n$ est pair s'*il existe* un entier dont il soit le double; il est impair si $n - 1$ est pair.[347]

Wavre, in his reaction, does not hesitate to point out that Lévy's proclaimed solution exactly constitutes the problem. Switching from the parity of a number to its rationality, where the same kind of different definitions can be made, he maintains:[348]

> (...) il faut se défier des mots qui n'étreignent souvent que des ombres, de l'expression *il existe* dont le sens n'est pas toujours immédiat. Il ne servirait à rien de dire: un nombre est rationnel s'*il existe* une fraction égale à ce nombre, car il faudrait encore trouver cette fraction.[349]

In the final contribution to the discussion, Lévy recognizes the value of the differences intuitionism has brought about. But he still considers the intuitionistic demands concerning mathematical existence 'arbitrary'.[350]

In a sense, the discussion was quite fruitful. Neither of the contributors managed to convince the other, which was unlikely from the start since they departed from different conceptions of mathematics, but at least Lévy in the end recognized clearer than before the peculiarities of the intuitionistic point of view.

The discussion between Wavre and Lévy drew some attention. Borel sent a letter to the editor of the *Revue de Métaphysique et de Morale* reacting to it, which was published the same year. In the letter, Borel did not go into the intuitionistic

---

[346][Lévy 1926A, p. 254]

[347]'A definition of the words *even* and *odd* that does not bring about the same difficulties can be expressed in the following way: an integer $n$ is even if *there exists* an integer of which it is the double; it is odd if $n - 1$ is even.'

[348][Wavre 1926B, p. 426]

[349]'(...) one has to mistrust words that often only embrace shadows, the expression *there exists* the meaning of which is not always direct. It would be of no use to say: a number is rational if *there exists* a fraction which is equal to this number, since one would still have to find that fraction.'

[350]'arbitraires', [Lévy 1926B, pp. 549–550]

view which the discussion started about.[351] Besides that, the Wavre–Lévy discussion was mentioned by Becker, Barzin and Errera, Fraenkel, and Heyting.[352]

**Becker in 1927**  By far the most extensive single contribution to the question of mathematical existence at large was made by Oskar Becker, whose 1927 *Mathematische Existenz. Untersuchungen zur Logik und Ontologie mathematischer Phänomene* ('Mathematical existence. Research into the logic and ontology of mathematical phenomena') counted more than 350 pages. Becker had worked on the publication at least from 1924 onwards.[353] The year before, in 1923, he had taken over Heidegger's position as Husserl's assistant in Freiburg.[354] Even though Becker worked together with Husserl, Heidegger's influence is notable; already the words 'Existenz' and 'Ontologie' in the (sub)title refer to Heidegger.[355] *Mathematische Existenz* appeared jointly with Heidegger's *Sein und Zeit* ('Being and time') in the *Jahrbuch für Philosophie und phänomenologische Forschung*, founded by Husserl. To Husserl, this marked a special occasion. These publications were designed to show that phenomenology was applicable both to the social sciences and to mathematics and the sciences.[356] Nevertheless, Husserl was to read the second half of Becker's publication only ten years later.[357]

Becker treats the subject of mathematical existence in a much broader sense than the way it usually was discussed in the foundational crisis. In the first place, he devotes a lot of attention to philosophical and historical analysis that is not strictly connected to the theme of mathematical existence as I am treating it here. Secondly, Becker studies intuitionism at large, considering choice sequences, the intuitionistic theory of the continuum, the principle of the excluded middle and the like, subjects we are not concerned with here.[358] Finally, Becker deals with the general question of mathematical existence, whereas we are only concerned with the reception of the intuitionistic view on mathematical existence. For these reasons, the treatment of Becker's work here is not nearly as extensive as one might expect because of its volume alone. An overview of what remains is given below.

Becker states that his publication is directed towards the *Seinssinn der mathematischen Phänomene* ('being-sense of mathematical phenomena')— clearly a philosophical (and Heidegger-inspired) goal. This objective should be reached by methods of both Husserl's transcendental phenomenology and Heidegger's herme-

---

[351] [Borel 1926]
[352] [Becker, O. 1927, p. 775], [Barzin & Errera 1927, p. 71], [Fraenkel 1928, pp. 352–353], and [Heyting 1930D, pp. 957–959]
[353] Letter from Becker to Weyl, 10/10/1924, [ETH Weyl, HS 91-473]
[354] [Pöggeler 1996, p. 27]
[355] Husserl called Becker's work a 'direct application of Heidegger's ontology' ('Direkte Anwendung der Heid[eggerschen, DH] Ontologie'), letter from Husserl to Heidegger, 24/5/1927; cited from: [Husserl 1994, Band IV, p. 143].
[356] [Pöggeler 1996, p. 10]
[357] [Schuhmann 1977, p. 484]
[358] Becker's remarks on the principle of the excluded middle are treated in 5.3.1.

## 4.3. THE DEBATE WIDENED

neutic phenomenology.[359] The point of departure for Becker's investigation is the foundational debate between intuitionism and formalism.[360]

Regarding the question of mathematical existence, Becker's first definition is the following:[361]

> Mathematisch existent heißen Gegenständlichkeiten, die zum Thema einer mathematischen Theorie gemacht werden und in dieser Theorie **widerspruchsfrei fungieren** können.[362]

Becker points out that the use of 'to function' is a way of circumventing the problematic issue of whether we are dealing with objects or not. Opposed to the above definition is the view that only those objects can be seen as mathematically existent, which can be constructed from a fixed starting-point by fixed means. These two definitions of mathematical existence can be linked to formalism and intuitionism respectively. Becker describes the difference between the two currents as a difference on what the *Seinssinn* of mathematics should be: consistency or constructions. Note once more the clear linking of constructions with intuitionism. Becker even goes further and reproaches formalists the use of pseudo constructions.[363]

Becker also explains the contrast between intuitionism and formalism in other terms. Going back to Greek philosophy, he argues that one should distinguish between deduction and demonstration. Demonstrations are related to truth, whereas deductions have a hypothetical character and do not say anything about the truth or falsity of the deduced statement.[364] Thus the problem of consistency is to be linked with deductions, that of decidability to demonstrations.[365] The opposition between intuitionism and formalism can then be stated as demonstration versus deduction. Becker argues that the problems of consistency and decidability do not play such an important role in the opposite current. An intuitionist sees consistency as self-evident, once the necessary constructions have been carried out. And formalists do not put such stress on decidability, as for them mathematical entities do not have to be put forward in order to exist.[366]

---

[359] One of the main differences between Husserl's transcendental phenomenology and Heidegger's hermeneutic phenomenology lies in the former's acceptance of the possibility of intuitive access to mental life, whereas the latter considers all form of human awareness to be interpretations, [Nicholson 1997, p. 304]. Moreover, Heidegger's goal is to obtain knowledge about being as such, whereas Husserl focuses on consciousness.

[360] [Becker, O. 1927, pp. 441–442]; Pöggeler reports that it was Becker's general view that world interpretations are determined by opposites, [Pöggeler 1969, p. 302].

[361] [Becker, O. 1927, p. 469]

[362] 'Objects are called mathematically existent if they can be made the subject of a mathematical theory and if they can *function consistently* in this theory.'

[363] [Becker, O. 1927, pp. 469–471; 521]

[364] [Becker, O. 1927, pp. 511–514]

[365] It is not clear where the question of decidability (*Entscheidbarkeit*) comes from. It seems that Becker identifies the possibility of a construction with decidability.

[366] [Becker, O. 1927, pp. 624–625]

## 172  CHAPTER 4. REACTIONS: EXISTENCE AND CONSTRUCTIVITY

Thus, Becker's presentation of the formalistic and intuitionistic views on mathematical existence states that formalists look for deductions and consistency, whereas intuitionists seek demonstrations, decidability and constructions. Now Becker's work is, as he himself points out, among other things based on the what he calls the principle of transcendental phenomenology. In short, this principle holds that every entity (*Gegenständlichkeit*) can in principle be reached.[367] Since Hilbert's view on existence allows for mathematical objects that cannot be reached, it does not come as a surprise that Becker's conclusion is completely on the side of intuitionism:[368]

> *Damit entscheidet die phänomenologische Analyse als hermeneutische, d.h. als auslegende auf das Dasein hin, die Streitfrage der Definition der mathematischen Existenz zugunsten des Intuitionismus.* Denn die intuitionistische Forderung, jeder mathematisch existente Gegenstand müsse durch eine in concreto und de facto vollziehbare Konstruktion 'dargestellt' werden können (beinahe im Sinne der 'Darstellung' eines reinen Stoffes in der Chemie) enthält nichts anderes als das Postulat: alle mathematische Gegenstände sollen durch faktisch vollziehbare Synthesen erreicht werden können.[369]

Although Becker valued the philosophical value of intuitionism highly, he was not very optimistic about the chances of an intuitionistic breakthrough. In 1926, he noted the first sign of a public victory of Hilbert's side. In a letter to Weyl he states that he can understand the latter's sympathy for symbolic mathematics, because of the regret of seeing the mathematical buildings break down. Becker continues:[370]

> Dass in der öffentlichen Meinung der Mathematiker Hilbert, oder (...) irgend eine (...) Erneuerung des alten 'Existentialabsolutismus' siegen wird, scheint mir beinahe gewiss.[371]

---

[367][Becker, O. 1927, p. 502]. Geiger, who also was a student of Husserl, and who occupied himself with the philosophy of mathematics at the university of Göttingen, severely criticised the use of this principle in his review of Becker's book. He claimed that Becker had nowhere grounded the principle, and that he ignored phenomenological analyses which maintained that the principle was not valid, [Geiger 1928, p. 409], [Dahms 1987B, p. 172].

[368][Becker, O. 1927, p. 636]

[369]'*Therefore, the phenomenological analysis as hermeneutical, i.e., as interpreting towards being, decides the question at issue of the definition of mathematical existence in favour of intuitionism.* For the intuitionistic demand that every mathematically existent object has to be 'processable' by a construction which is in concreto and de facto executable (almost in the sense of the 'process' of a pure substance in chemistry) contains nothing else but the postulate: all mathematical objects have to be attainable by actually executable syntheses.'

[370]Letter from Becker to Weyl, 16/8/1926 [ETH Weyl, HS 91-475]

[371]'It seems almost sure to me that in the public opinion the mathematician Hilbert, or (...) some (...) renewal of the old 'existence absolutism' will gain the victory.'

## 4.3. THE DEBATE WIDENED

Since not many people expressed themselves on which current the academic community thought was winning the fight, it is hard to say if Becker's view represented the *communis opinio* in those days.[372]

Brouwer seems to have taken Becker's work seriously. In 1931, when a philosophical chair was vacant in Amsterdam, he considered Becker as one of the candidates and asked Husserl for advice.[373]

**Hilbert and Weyl, 1927–1928** In July 1927, Hilbert again lectured on the foundations of mathematics in Hamburg, the place where his public controversy with Brouwer and Weyl had begun. Halfway the talk he turns to intuitionism, the current to which, in Hilbert's view, most publications in the foundations of mathematics belonged. He opens his description of intuitionism with Brouwer's declaration that pure existence statements are meaningless. Hilbert counters this by referring to Gauss' declaration that, for analysis, one does not need more than the complex numbers. In order to prove this statement, Weierstraß and Dedekind had gone through all kinds of difficult argumentations and complicated computations. Hilbert points out that he was able to avoid all of this by using his logical $\epsilon$-function,[374] even though the proposition he had thus proved was a pure existence statement which 'by its nature'[375] could not be turned into a proposition about constructivity. This pure existence statement was the only way to avoid all the difficulties of Weierstraß and Dedekind. Furthermore, Hilbert claims that only in this way the inner reason for the validity of Weierstraß' and Dedekind's argumentations was shown.[376] He summarizes his position with the following words:[377]

> Das Wertvolle der reinen Existenzbeweise besteht gerade darin, daß durch sie die einzelne Konstruktion eliminiert wird und viele verschiedene Konstruktionen durch einen Grundgedanken zusammengefaßt werden, so daß allein das für den Beweis Wesentliche deutlich hervortritt: Abkürzung und Denkökonomie sind der Sinn der Existenzbeweise.[378]

At first sight, the quote seems to suggest that Hilbert claims that every pure existence statement could, if wished so, be replaced by a constructive one. This, however, is not in accordance with Hilbert's earlier statement about the pure existence proposition which, by its nature, could not be converted into a theorem of constructivity. Therefore, it seems more reasonable to interpret Hilbert as saying

---
[372]Becker expressed the same view in *Mathematische Existenz*, [Becker, O. 1927, p. 749].
[373]Letter from Husserl to Mahnke, 12/5/1931; in: [Husserl 1994, Band III, p. 478]
[374]The $\epsilon$-function is explained in 4.3.2.
[375]'*seiner Natur nach*', [Hilbert 1927, p. 77]
[376][Hilbert 1927, pp. 77–78]
[377][Hilbert 1927, p. 79]
[378]'The value of pure existence proofs consists precisely in that the individual construction is eliminated by them and that many different constructions are subsumed under one fundamental idea, so that only what is essential to the proof stands out clearly; brevity and economy of thought are the *raison d'être* of existence proofs.' English translation from [Van Heijenoort 1967, p. 475].

## 174  CHAPTER 4. REACTIONS: EXISTENCE AND CONSTRUCTIVITY

that once a pure existence proof has been given, there is no longer a necessity to provide a constructive version.

Again, Hilbert does not claim the predicate 'constructive' for his proof theoretical mathematics.

In a discussion remark in reaction to Hilbert's lecture, Weyl comes to the defense of intuitionism. He does not treat any specific differences between intuitionism and proof theory, but concentrates on the basic points of view to which all differences can be reduced.

Weyl opens by stating that, before Hilbert had developed his proof theory, mathematics was regarded as a system of meaningful truths. Brouwer, however, was the first to recognize that mathematics had by far exceeded the borders of meaningful thinking, and this was also the tendency against which Weyl himself had protested with his 1921 paper. Hilbert, Weyl maintains, respects these same borders on the meaningful level of meta-mathematics. Therefore, one cannot speak of arbitrary forbiddings. The level of support Brouwer received, Weyl continues, should not come as a surprise, since his conclusions follow logically from a thesis which was supported by all mathematicians before Hilbert had put forward his proof theoretical point of view. Hilbert only managed to save classical mathematics by radically changing its interpretation. This, Weyl maintains, was not so much a free decision, but it became necessary under pressure of the circumstances.[379]

Weyl notes with pleasure that, in the epistemological evaluation of the new situation, nothing separates him from Hilbert. He maintains that even the strictest intuitionist has to recognize that proving real propositions by means of the addition of ideal elements, as supported by Hilbert, is legitimate – something to which Brouwer, however, did not agree. Weyl finishes with a philosophical conclusion:[380]

> Setzt sich die Hilbertsche Auffassung, wie das allem Anschein nach der Fall ist, gegenüber dem Intuitionismus durch, *so erblicke ich darin eine entscheidende Niederlage der philosophischen Einstellung reiner Phänomenologie,* die damit schon auf dem primitivsten und der Evidenz noch am ehesten geöffneten Erkenntnisgebiet, in der Mathematik, sich als unzureichend für das Verständnis schöpferischer Wissenschaft erweist.[381]

By that time, Weyl had restored his relationship with Hilbert. As he wrote in a letter in February 1927, he was glad to be in a harmonious relationship with Hilbert again, since the spirit in which he did mathematics was Hilbert's.[382]

---

[379][Weyl 1927A, pp. 86–87]
[380][Weyl 1927A, p. 88]
[381]'If Hilbert's view wins ground against intuitionism, as in all likelihood is the case, *then I consider that as a decisive defeat of the philosophical mentality of pure phenomenology,* which by that already in the most primitive field of knowledge, which is the first one opened to evidence, in mathematics, turns out to be insufficient for the understanding of creative science.'
[382]Letter from Hermann Weyl, 21/2/1927; cited in: [König 1956, p. 243]

## 4.3. THE DEBATE WIDENED

In his famous address at the International Mathematical Conference in Bologna in September 1928, Hilbert only marginally touched upon the question of mathematical existence. The one remark he makes on this subject is that the difficult existence proofs have, over the last decades, attained the highest degree of 'simplicity' and 'clarity' by adoption of methods taken from the calculus of variations.[383] Maybe Hilbert was hinting at the famous Dirichlet principle, of which Hilbert was the first to give a proof, be it under certain restrictions. The so-called direct method Hilbert used in the proof was set up in such a way that it was also applicable to other problems in the calculus of variations.[384] In the 1905 paper in which the proof was presented, Hilbert had praised the arguments used with literally the same terms of *'Einfachkeit'* and *'Durchsichtigkeit'*. Furthermore, he had already at that time pointed out the way the Dirichlet principle helps us in finding existence proofs, and he had mentioned the advantage of his method being also applicable to, e.g., mathematical physics.[385]

### 4.3.2  Existence as a minor subject

In the contributions to the debate mentioned in the preceding section, mathematical existence was treated as a major subject, or an important contribution was made to the discussion on mathematical existence. Apart from these contributions, there were many papers and lectures in which the issue of mathematical existence was also discussed, but only as a minor subject. Most of those cases were separate contributions to the debate on mathematical existence.

What follows is an overview of these separate items.

**Various authors, 1925–1927**  The intuitionistic view on mathematical existence was spread further in expositions by several authors. These included the Norwegian mathematician Thoralf **Skolem**, who gave a lecture for the *Norsk Matematisk Forening* ('Norwegian Mathematical Society') in September 1925.[386] At about the same time, Torsten **Brodén** gave a very Hilbertian presentation of intuitionism at the Scandinavian Mathematical Conference in Copenhagen. Brodén describes intuitionism as a continuation of the Kroneckerian programme, restricting mathematics to objects that one has come across. A construction procedure is needed in order to consider the object mathematically existent. In Brodén's view, this intuitionistic demand is primarily motivated by fear for (possibly future) antinomies. Since Brodén thinks the antinomies can be solved in a different way, he reproaches intuitionism for 'extreme cowardice':[387]

---

[383]*'Einfachkeit'*, *'Durchsichtigkeit'*, [Hilbert 1930A, p. 1]
[384][Monna 1975, pp. 55–56]
[385][Hilbert 1905B, p. 11; 14]
[386][Skolem 1926, pp. 12–13]. Skolem devoted more attention to the discussion on the principle of the excluded middle, see 5.3.1.
[387]'extreme Feigherzigkeit', [Brodén 1925, p. 236]

176    CHAPTER 4. REACTIONS: EXISTENCE AND CONSTRUCTIVITY

> (...) sie [die Intuitionisten, DH] scheinen (...) den sehr guten Grundsatz ganz über Bord zu werfen, dass man *alle logische Möglichkeiten offen halten soll*. (...) Schon bei der geringsten Gefahr retten sie sich so zu sagen durch die Flucht.[388]

Nobody referred to Brodén's contribution to the debate.

The gymnasium mathematics teacher Walter **Lietzmann**, who also lectured at the University in Göttingen and who was president of the *Deutsche Verein für Förderung des mathematischen und naturwissenschaftlichen Unterrichts* ('German Society for the Fostering of mathematics and science education'),[389] explained the intuitionistic view on mathematical existence in an article for a German didactical journal in 1925.[390] The paper did not play any role in the foundational debate. **Fraenkel** once more presented existence in intuitionistic mathematics in the journal *Scientia* in the same year.[391] **Rivier** mentioned it in a paper in a theological-philosophical journal.[392]

**Bieberbach** touched upon the issue of mathematical existence in two lectures, one delivered in April 1925, the other in February 1926. In the former, Bieberbach merely describes the intuitionistic point of view, stating that only concepts which can be exhibited explicitly by constructive methods have the 'mathematical citizen's right'.[393] The following year, he has become a full supporter of the intuitionistic point of view. He characterises intuitionism as the current which takes into account that mathematics is done by human beings. Regarding mathematical existence, intuitionists maintain that only objects exist for which a procedure is known which enables one to construct them. Existence in the formalistic conception of consistency does not, Bieberbach explains, constitute knowledge to the intuitionist.[394]

In a paper published in 1926, Otto **Hölder** merely mentions the intuitionistic point of view on mathematical existence, as put forward by Weyl.[395]

In his *Fiktionen in der Mathematik*, published the same year, Christian **Betsch** mentions that both in intuitionism and in proof theory one works constructively, not asking himself what the meaning of the term in any of the cases could be.[396]

In a paper in the *Revue de théologie et de philosophie*, **Larguier des Bancels** explains the intuitionistic point of view on mathematical existence by means of

---

[388]'(...) they [the intuitionists, DH] **seem** to (...) completely throw overboard the very good principle that one *has to keep open all logical possibilities*. Already in the slightest danger they so to speak save themselves by fleeing.'
[389][Lietzmann 1960, p. 5]
[390][Lietzmann 1925]
[391][Fraenkel 1925B, pp. 209–218]
[392][Rivier 1925, p. 216]
[393]'das mathematische Bürgerrecht', [Bieberbach 1925, p. 398]
[394][Bieberbach 1926, p. 21]
[395][Hölder 1926, p. 248]
[396][Betsch 1926, pp. 339–348]

## 4.3. THE DEBATE WIDENED

the occurrence of a series of digits in the decimal development of $\pi$: one has to have found the series in order to be able to claim its existence.[397]

The following year, **Carnap** published a paper in which he, too, mentions the intuitionistic point of view on mathematical existence. He adds that as soon as a formal model for an axiomatic system is known, that is, in the intuitionistic point of view, is constructible, one can find a method to derive arbitrary many models for the system. Carnap does not give any proof for the statement.[398]

**Barzin** and **Errera** maintain that, for intuitionists, existence can only be proved by a construction starting from the natural numbers.[399]

The other contributions were more substantial and are treated below.

**Heyting's dissertation, 1925** In his dissertation on intuitionistic axiomatics of projective geometry, written under supervision of Brouwer and published in 1925, Heyting[400] linked the questions of the principle of the excluded middle and mathematical existence. In the introduction to his thesis he writes:[401]

> Het geloof in het 'principium tertii exclusi' berust meestal op den waan, dat het woord 'bestaan' in ontologischen zin zonder meer duidelijk is. Daardoor zien de meeste wiskundigen de mogelijkheid en noodzakelijkheid, een dergelijk 'bestaan' als grondslag voor abstracte wiskunde te wraken, niet onmiddellijk in. Zij komen dan licht tot de meening, dat de vraag, of een entiteit met gegeven eigenschappen 'bestaat', een van ons denken onafhankelijke beteekenis heeft.[402]

Heyting explicitly excludes formalists from such a naive conception, since they are willing to give up the meaningful character of classical mathematics, in order to save its classical form. He even claims that some of them, e.g. Bernays, can therefore truly understand what intuitionism stands for. Heyting is the first person after Becker who at least mentions the philosophical existence problem.

**Zariski in 1925** In 1925, the importance of the matter of mathematical existence was stressed in Italy in a paper on set theory by the algebraic geometer Oscar Zariski. Born in Poland, he was at the time working at his dissertation in Rome

---
[397][Larguier des Bancels 1926, pp. 120–121]
[398][Carnap 1927, p. 363]
[399][Barzin & Errera 1927, pp. 56–57]
[400]Biographical information on Heyting is given in 5.4.1.
[401][Heyting 1925, pp. 1–2]
[402]'The belief in the 'principium tertii exclusi' is mostly due to the delusion that the word 'existence' in the ontological sense is clear without due consideration. Therefore, most mathematicians do not immediately see the possibility or the necessity of denouncing such an 'existence' as a foundation for abstract mathematics. They then easily come to the view that the question if an entity with given properties 'exists' has a meaning which is independent of our thought.'

under the guidance of Castelnuovo. Zariski again used different names for intuitionism ('nominalism') and (naive) formalism ('realism'), but his characterisation of the difference between them was quite correct:[403]

> Ora, dal punto di vista di un matematico nominalista [intuitionistic, DH], dare una corrispondenza (o dire que una corrispondenza esiste) significa *costruirla* effettivamente in base ad una *legge* precisa. (...) Vediamo che il dissenso [between intuitionism and formalism, DH] nasce qui in tema del significato da attribuire nella matematica all'attributo d'esistenza.[404]

The only point where Zariski goes wrong is that his explanation of 'giving a correspondence' is too narrow for an intuitionist. For it excludes choice sequences, which by definition are not law-like. Zariski's paper did not play any role at all in the debate.

**Weyl, 1925–1927** Weyl's 1925 paper on the foundational crisis, *Die heutige Erkenntnislage in der Mathematik* ('The present knowledge condition in mathematics'), did not contain many new ideas on the subjects of mathematical existence and constructivity. Weyl traces the term 'constructive' back to Euclid, and repeats some of his ideas put forward earlier in the *neue Grundlagenkrise*. He adds the remark that the view he had put forward on mathematical existence in that paper was in fact not exactly Brouwer's, but was what seemed natural to Weyl when he was captivated by Brouwer's ideas.[405]

By 1926, Weyl took up an even more pragmatic stance in the foundational debate than before. Becker reported the following quote of Weyl in a letter to Mahnke:[406]

> Für mich hat der Kampf Brouwer–Hilbert freilich eine ganz prinzipielle Bedeutung. Ich bin zu sehr Mathematiker, um mich dem Eindruck verschliessen zu können, dass praktisch die Brouwersche Mathematik nicht das ist, was wir brauchen, und sich nicht durchsetzen wird. Ich bin in dieser Hinsicht geschichtsgläubig und ein frommes Weltkind, dass ich mit Hilbert finde, der Erfolg ist das Entscheidende.[407]

---

[403][Zariski 1925, p. 70]

[404]'Now, from the point of view of a nominalistic [intuitionistic, DH] mathematician, to give a correspondence (or to say that a correspondence exists) means *to construct it* effectively by virtue of a precise *law*. (...) We see that the dissent [between intuitionism and formalism, DH] on this subject arises from the meaning attributed to the property of existence in mathematics.'

[405][Weyl 1925, p. 529]

[406]Letter from Becker to Mahnke, 22/8/1926; cited from [Mancosu & Ryckman 2002]

[407]'For me, the battle Brouwer–Hilbert has of course a very fundamental significance. I am too much of a mathematician to be able to shut my eyes to the impression that in practice Brouwer's mathematics is not what we need and that it will not prevail. In this respect, I believe in history and I am a pious child of this world, in that, together with Hilbert, I find that success is what is decisive.'

## 4.3. THE DEBATE WIDENED

Weyl's reference to Hilbert probably goes back to Hilbert's 1925 lecture.[408]

In his book *Philosophie der Mathematik und Naturwissenschaft*,[409] which appeared in 1927, Weyl once again expressed his sympathy to constructivist ideas, but he made no positive contribution to it. He devotes a section to Brouwer's 'intuitive mathematics', and he further clarifies his view on existence statements.

**Gonseth in 1926** In his book *Les fondements des mathématiques* ('The foundations of mathematics'), Gonseth discusses intuitionism at length but treats the question of mathematical existence only marginally.[410] Having discussed the value of the limit concept in the infinitesimal calculus, he asks himself if intuitionists do not have the right to pose the following question:[411]

> Admettant que la notion de limite puisse logiquement exister, que même elle soit nécessaire, je vous mets en demeure de m'en donner la preuve. C'est justement ici le point sensible de la question. On ne peut logiquement *prouver* d'aucun concept que finalement il n'entraînera pas contradiction. Si – par un miracle inexpliqué – nous sommes entrés en possession de telle ou telle notion, nous n'avons pas à prouver qu'elle existe, puisque nous la possédons.[412]

Several things go wrong here. In the first place, intuitionists do not ask for a *logical* proof of the existence of a certain concept, but for a proof of its existence as a mental construction. Secondly, one can very well prove logically that a certain concept does not lead to contradictions within a given axiomatic system. Finally, Gonseth's way out is a very easy one and can hardly be said to clarify anything on the issue of mathematical existence.

Gonseth's book drew some attention in the French speaking world,[413] and it was mentioned in the third edition of Fraenkel's *Einleitung in die Mengenlehre*.[414]

**Von Neumann in 1927** In 1927, Von Neumann published a paper on Hilbert's proof theory, in which he gave a consistency proof for a certain part of first-order arithmetic. In the paper, Von Neumann points out that the existence statements he produces in an axiomatic way are not without meaning to intuitionists. However, he is mistaken, for in the axioms he uses Hilbert's $\tau$-operator.[415] Hilbert had

---
[408] See 4.3.1.
[409] [Weyl 1927B]
[410] Gonseth's book, as well his own background, is discussed in more detail in 5.3.1.
[411] [Gonseth 1926, p. 203]
[412] 'Admitting that the notion of a limit can logically exist, that it is even necessary, I urge you to present me a proof of it. This is exactly the sensitive point of the question. One cannot *prove* logically of any concept that in the end it will not lead to a contradiction. If – by some unexplained miracle – we have come to possess such and such a notion, we do not have to prove that it exists, since we possess it.'
[413] [Juvet 1927], [Reymond 1932A], and [Dassen 1933]
[414] [Fraenkel 1928]
[415] [Von Neumann 1927, p. 17]

180     CHAPTER 4. REACTIONS: EXISTENCE AND CONSTRUCTIVITY

introduced the operator in print in 1923.[416] The idea is that $\tau(A)$ picks out a counter-example to a property $A(a)$ with a free variable $a$, if such a counterexample exists. If there is no counterexample, then $\tau(A)$ is chosen arbitrarily. This means that, given $\tau(A) = c$, we can easily check if there exists a counter-example to $A$ or not by checking if $A(\tau(A))$ holds. If this is the case, then we know that there is no counter-example to $A$ and the predicate $A$ applies to all objects involved. If $A(\tau(A))$ does not hold, then $\tau(A)$ is a counter-example. In axiomatic terms, the $\tau$-operator satisfies the transfinite axiom $A(\tau(A)) \to \forall(a)A(a)$. From its definition, it is clear that the operator is a non-constructive device.[417] Von Neumann's remark is all the more strange, since Hilbert himself had explicitly mentioned that the $\tau$-operator belonged to the category of functions that was 'forbidden' by Brouwer and Weyl.[418] Weyl later described Hilbert's $\tau$-operator as a 'divine automaton', and repeatedly called the belief in its existence 'pure nonsense'.[419] (In a 1925 paper,[420] Hilbert replaced the $\tau$-operator by its inverse, the $\epsilon$-operator,[421] which is equally non-constructive.[422])

Von Neumann's remark indicates that even as late as 1927 some of the better informed mathematicians misunderstood the intuitionistic view on mathematical existence.

**Burkamp in 1927**   In 1927, the German philosopher Wilhelm Burkamp (1879–1939) published a book called *Begriff und Beziehung* ('Concept and relation'). Burkamp had a strong interest in biology and psychology and tried to provide empirical foundations for philosophical concepts. He only started as a *Privatdozent* in 1923, at the age of 44, at the university of Rostock. His interest in logic dates back to his *Habilitationsschrift*, published the year before. *Begriff und Beziehung* was his first publication fully devoted to logic.[423]

In the foreword, Burkamp explicitly states that he does not want to go into the battle between formalism and intuitionism. He only maintains that[424]

---

[416] [Hilbert 1923, p. 183]. Hilbert first used the $\tau$-operator in his 1922 lectures, [Sieg 1999, p. 36].

[417] For a further explanation of the $\tau$-operator and its use for the introduction of quantifiers, see [Smoryński 1988, pp. 32–33].

[418] '*verboten*', [Hilbert 1923, p. 185]

[419] '*göttlichen Automaten*', '*der reinste Unsinn*', [Weyl 1927B, p. 46]; [Weyl 1929, p. 164]

[420] and in 1922/23 in his lectures, [Sieg 1999, p. 36]

[421] [Hilbert 1926, p. 178]

[422] The $\epsilon$-operator is a choice function which fulfills the axiom $A(a) \to A(\epsilon A)$. In words: if there exists an element for which $A$ is true, then the $\epsilon$-function picks out such an element, otherwise it chooses an arbitrary element. As Kreisel remarked, the advantage of the $\epsilon$-function is that in any given proof, the $\epsilon$-function is used only finitely many times. Therefore, we do not need all its values, but only the ones used in the proof, [Kreisel 1964, p. 165]. Hilbert pointed out so in [Hilbert 1923, p. 187]. His description of it as 'the transfinite logical choice function', however, seems to imply that he saw it as something transcending our finite knowledge, [Hilbert 1926, p. 178].

[423] [Weller 1994, pp. 133–136].

[424] [Burkamp 1927A, p. VI]

## 4.3. THE DEBATE WIDENED

> Der Existenzbegriff ist ja plötzlich für die Mathematik, für den Streit zwischen Formalisten und Intuitionisten über die Grundlegung der Mathematik hochaktuell geworden. Mir scheint aber nicht bei den Intuitionisten und erst recht nicht bei den Formalisten Klarheit darüber zu herrschen, welche komplizierte Bedeutung dem Existenzbegriff zukommt (...).[425]

Burkamp himself uses a notion of existence which, not surprisingly, differs from those of intuitionism and formalism. In his view, a concept can be considered existing if it satisfies the following two conditions: it should be valid and it should be related to a definite or a 'more or less indefinite'[426] individual position in reality. The requirement of a relationship to an individual position is included in order to exclude laws, such as the law of gravity, which are valid, too, but do not exist. Even though Burkamp adheres to a concept of reality which is independent of human beings,[427] it is his second demand which brings him closer to the intuitionistic view on mathematical existence. For in the formalist opinion, no bond with reality is needed.

Burkamp's book was praised by Fraenkel,[428] but mentioned by very few others who were involved in the foundational crisis.[429] However, it was influential enough in other circles to make the philosophical faculty of the university of Rostock decide to accord Burkamp the title *außerordentlicher außerplanmäßiger Professor*.[430]

In the same year, Burkamp published a paper on 'the crisis of the principle of the excluded middle', in which he again touched upon the issue of mathematical existence.[431] Burkamp refers to Weyl's 1921 paper in which the latter had argued against 'propositional abstracts' such as 'there exists a natural number with a certain property'. Burkamp calls such statements indefinite existence statements, since they do not point out any specific object which the existence statement is about. However, Burkamp maintains, Weyl's expression would too easily 'seduce'[432] one to consider indefinite existence statements not as propositions at all, or only if they are derived from a definite existence statement. That, indeed, is what Weyl meant. But, Burkamp argues, indefinite existence statements can very well have a meaning, as in the case 'there is a senior in the new *Reichstag*[433] (even

---

[425]'For the notion of existence has suddenly become most topical for mathematics, for the conflict between formalists and intuitionists on the foundations of mathematics. But it seems to me that there is no clarity, not among intuitionists and definitely not among formalists, about how complicated the meaning is that belongs to the notion of existence.'
[426]'mehr oder weniger unbestimmten', [Burkamp 1927A, Band I, pp. 146–147]
[427][Ziegenfuss 1949–1950, Vol. 1, p. 160]
[428][Fraenkel 1928, p. 264]
[429]One of the exceptions was Cassirer, [Cassirer 1929, p. 436].
[430][Weller 1994, p. 137]
[431]The part of the paper which deals with the excluded middle is dealt with in 5.3.1.
[432]'verleiten', [Burkamp 1927B, p. 71]
[433]the German parliament

182    CHAPTER 4. REACTIONS: EXISTENCE AND CONSTRUCTIVITY

if nobody has determined who he is).'[434] The example Burkamp uses makes it clear that he has not understood the intuitionistic criticism. Since the number of members of the *Reichstag* is finite, the problem of determining the senior among them is solvable by constructive methods.

**Dresden in 1927**  In December 1927, the American Mathematical Society held a joint meeting with the Mathematical Association of America, in which also intuitionism was discussed. It was presented to the audience in two lectures, one by James Pierpont and the other by Arnold Dresden. The latter had in that year become professor of mathematics at Swarthmore College.[435] Concerning the topic we are dealing with in this chapter, his lecture is the more interesting one. Both lectures attracted little attention; the only one to write about them was Hedrick, who did so more than five years later.[436]

In his lecture, Dresden compares Brouwer's view on logic with the emergence of non-Euclidean geometry.[437] In his view, Brouwer has 'freed the mind from the compulsory use of the Aristotelian base'.[438] He then draws the following conclusion:[439]

> [I]f the entire mathematical structure, not the basis only [i.e., the primitive ideas and postulates, DH], but also the guiding principles for its development, is at the choice of the individual, we have to admit that a mathematic exists only in the minds of the individual, and that without the activity of the human mind there would be no mathematics. This carries with it furthermore, that a mathematic exists only in so far as it has been developed, that is, is invented rather than discovered; (...) that mathematical entities may exist to-morrow which do not exist to-day, that their existence depends upon the construction of a process, which can call them into being.

Like Fraenkel had done before, Dresden presents intuitionism as a coherent philosophy of mathematics instead of a set of more or less arbitrary points of view. In Dresden's presentation, the point of departure for intuitionism is that not all classical logical rules should be considered universally valid. The logical rules which guide the development of mathematics are at the choice of the individual. But if this is the case, Dresden argues, then there can be no mathematical existence independent of the mathematician. In this way, Dresden reasons from Brouwer's rejection of the principle of the excluded middle to the intuitionistic view on mathematical

---

[434][Burkamp 1927B, pp. 70–71]
[435][Poggendorff 1936–1940, p. 600]
[436][Hedrick 1933, p. 336]
[437]Dresden remarks that the controversy which emerged from Brouwer's writings provides us with 'a vivid realization of the reasons which may have led Gauss to withhold his discovery of non-euclidean geometry from publication', [Dresden 1928, p. 441].
[438][Dresden 1928, p. 448]
[439][Dresden 1928, pp. 449–450]; note that Dresden uses the singular 'mathematic'.

## 4.4. LATER REACTIONS

existence. This is exactly the opposite from what Fraenkel had done.[440] The latter departed from the intuitionistic point of view on mathematical existence, and then derived the other intuitionistic conclusions regarding, e.g., logic. Dresden's argumentation departs from Brouwer's own point of view, but still presents a coherent picture of intuitionism.

The fundamental difference, Dresden maintains, between the current which still adheres to an absolute mathematical certainty and the one, inaugurated by Brouwer, which is doubtful of a large part of the mathematical conclusions, lies in the characterisation of mathematical existence. He judges positively the fact that there now are different views on mathematical existence: in diversity, as he puts is, there may lie strength, and we can learn more about mathematics by looking at it in different ways.[441]

Finally, it should be remarked that Dresden was well informed, since he mentioned a large part of the literature published on the foundational controversy in Europe in the references.

## 4.4 Later reactions

From 1928 on, when the foundational crisis was over,[442] some reactions to the intuitionistic view on mathematical existence and constructivity arose which are worth mentioning. I selected these contributions because they either mark an important event in the foundational discussion (the Königsberg conference), or they present the reaction of an important thinker (Wittgenstein), or they show developments which are related to parts of the discussion presented earlier (such as in the cases of Bernays and Weyl). Thus, instead of giving an overview of the debate as I have done so far, I here only highlight some reactions from the later period.

### 4.4.1 The Königsberg conference

The debate about what 'constructivity' should mean in mathematics was discontinued for several years, but reappeared in 1930. In September of that year, the second *Tagung für Erkenntnislehre der exakten Wissenschaften* ('Conference on the Epistemology of the Exact Sciences') was organised in Königsberg.[443] The main theme of the conference was the foundations of mathematics,[444] and nowadays the conference is perhaps mostly known for Gödel's first public announcement of his first incompleteness theorem.

---

[440]See 4.3.1.
[441][Dresden 1928, pp. 450–451]
[442]See 3.4.
[443]Nowadays Kaliningrad; the first Conference was held in Prague in 1929.
[444]The theme of the conference was suggested by Hahn, as mentioned in a letter from Carnap to Schlick, [ASP Carnap, 029-30-10].

**184 CHAPTER 4. REACTIONS: EXISTENCE AND CONSTRUCTIVITY**

***Tagung für Erkenntnislehre*** The conference was organised by the Berlin *Gesellschaft für empirische Philosophie* and the Vienna *Verein 'Ernst Mach'*, and was held in conjunction with the *Deutsche Physiker- und Mathematikertag* and the meeting of the *Gesellschaft deutscher Naturforscher und Ärzte*.[445]

For the first time, the proponents of all *three* currents intuitionism, formalism and logicism were present at one meeting, and a discussion followed afterwards. For the first time, the debate was officially taken out of the journals, books and letters and made the focus of a conference that specifically dealt with foundational issues. It was the younger generation that spoke: formalism was represented by Von Neumann,[446] intuitionism by Heyting, and logicism by Carnap.[447] Hilbert and Brouwer had left the scene (Russell had never been on it), although Hilbert was present at the conference.[448] The organising committee had taken the decision to ask younger researchers to give the lectures without further motivation. But Reidemeister, who was a member of the committee, personally thought that young researchers often presented topics in a more understandable way, and that the possibilities for discussion were greater. Furthermore, he argued, the conference was directed towards young researchers, so it was also an expression of modesty not to ask for the great names.[449]

The lectures and the discussion were published afterwards in the journal *Erkenntnis*, the successor of the *Annalen der Philosophie*, which was published jointly by the same two associations. By organising international conferences and by publishing a journal, the societies tried to strengthen their aim of a scientifically oriented philosophy. Reichenbach and Dubislav, among others, were active in the Berlin group; Schlick, Hahn, Neurath, Carnap and Zilsel were among the Viennese participants. The *Verein 'Ernst Mach'*, founded in 1929, was seen as an externally oriented version of the *Wiener Kreis*.[450]

The published versions of the lectures are rather short. Nevertheless, all three expositors spoke for one hour.[451] Therefore, it seems probable that the published papers were shortened versions of the lectures actually given. From correspondence between Carnap and Von Neumann, it becomes clear that both of them made objections against the publication of the lectures, due to the fact that Gödel had proved the incompleteness theorems.[452] Because of this, Von Neumann even considered the situation as discussed at the Königsberg conference rendered out

---

[445] [Rundschau 1930, p. 80]

[446] In the conference announcement, Von Neumann's contribution went under the title *'Die axiomatische Begründung der Mathematik'*, [Rundschau 1930, p. 80].

[447] Originally, Hahn was going to speak on logicism, as mentioned in a letter from Carnap to Schlick, [ASP Carnap, 029-30-10]. It is not clear why Carnap ended up doing it.

[448] [Van Dalen 2001, p. 358]

[449] Letter from Reidemeister to Heyting, 4/5/1930; [TLI Heyting, B rei3-300504]

[450] [Neurath 1930, p. 312]. More information on the *Wiener Kreis* and the Berlin *Gesellschaft* is given in 3.3.3.

[451] [Rundschau 1930, p. 80]

[452] On the incompleteness theorems, see 5.4.2.

## 4.4. LATER REACTIONS

of date. However, in the end they agreed to publish the lectures as an indication of the state of the art in September 1930.[453] Since the published papers are the only things left to us, I based my interpretation on them.

**Constructivity**  The lectures did not contain many new points of view. However, the way in which the philosophy of each current was presented was striking. All proponents professed to possess the truly constructive version of mathematics. Thus, Von Neumann describes formalism in the following terms:[454]

> Ihr Grundgedanken ist diese: Auch wenn die inhaltlichen Aussagen der klassischen Mathematik unzuverlässig sein sollten, so ist es doch sicher, daß die klassische Mathematik ein in sich geschlossenes, nach feststehenden, allen Mathematikern bekannten Regeln vor sich gehendes Verfahren involviert, dessen Inhalt ist, gewisse, als 'richtig' oder 'bewiesen' bezeichnete, Kombinationen der Grundsymbole sukzessiv aufzubauen. Und zwar ist dieses Aufbauverfahren sicher 'finit' und direkt konstruktiv.[455]

Next Carnap characterises logicism:[456]

> Das Wesentliche an der angedeuteten logizistischen Methode der Einführung der reellen Zahlen ist, daß hier diese Zahlen *nicht 'postuliert'*, sondern *'konstruiert'* werden. (...) Diese *'konstruktivistische'* Auffassung gehört zu den Grundtendenzen des Logizismus.[457]

Heyting only makes mild use of the constructivist image, claiming simply:[458]

> Ein Beweis für ein Aussage ist eine mathematische Konstruktion (...).[459]

These positions require some clarification.

Carnap explains his position by pointing out that logicism starts with certain (undefined) logical concepts, such as disjunction and negation. All mathematical

---

[453] Letter from Von Neumann to Carnap, 7/6/1931; letter from Carnap to Von Neumann, 11/07/1931; [ASP Carnap, 029-08-03], [ASP Carnap, 029-08-02]; the letters are published in [Mancosu 1999A, pp. 39–42].
[454] [Von Neumann 1931, pp. 116–117]
[455] 'Its principal idea is the following: even if the meaningful judgements of classical mathematics would be inadmissable, still it is sure that classical mathematics involves a method which is closed in itself, which proceeds according to fixed rules known to all mathematicians, and which consists in successively building up certain combinations of primitive symbols which are designated as 'correct' or 'proved'. This building method, moreover, is definitely 'finitary' and directly constructive.'
[456] [Carnap 1931, p. 94]
[457] 'The essential point of the indicated logistic method of introducing the real numbers is that, here, the numbers are *not postulated*, but *constructed*. (...) This *constructivistic* conception is one of the fundamental tendencies of logicism.'
[458] [Heyting 1931A, p. 114]
[459] 'A proof of a proposition is a mathematical construction (...).'

notions are given by explicit definitions in terms of logic. Contrary to formalism, implicit or 'creating' definitions do not occur. This way of proceeding, Carnap maintains, is common to both logicism and intuitionism. However, in addition to the construction rules accepted by the intuitionists, logicism also accepts the application of the expression 'for all properties'.[460] What Carnap calls a 'property' is what is nowadays known as a 'predicate'. Carnap's claim thus means that logicists would accept second order predicate logic,[461] whereas intuitionists would not. Whether this is correct depends on the interpretation one gives to the quantification 'for all predicates $P$, $Q(P)$ holds'. If one uses the game theory interpretation ('if you give me a predicate $P$, I will prove that $Q$ holds for $P$') or the proof interpretation ('I have a method which, given a predicate $P$, automatically proves $Q(P)$'), intuitionists could also accept it. This does not mean that one has to know the class of all predicates. However, around 1930 these concepts were not that clearly developed.

Von Neumann clarifies his point of view by giving the example of a meaningful, non-constructive proof. Such a proof may, for example, prove the existence of a real number $r$ with certain complicated properties, without giving a procedure for actually finding such a number. In such a case there is no possibility for checking the contentual statement in a finite way. However, Von Neumann reasons, the formal proof by which the statement was proved can be checked by a finite procedure, and this is what he calls constructive.[462]

Heyting states only that a mathematical construction is a 'method of proof', by which an intention (in the phenomenological sense) can be fulfilled.[463] Throughout his talk, he gives several examples of mathematical constructions, e.g. by means of choice sequences.

It is worth considering the different claims to constructivity. Nowadays, mathematicians would call Heyting's and possibly also Von Neumann's point of view constructive. Carnap's use of Dedekind cuts in the creation of the real numbers is definitely not constructive. More importantly, however, I would argue that to the majority of their contemporaries only Heyting's use of the term 'constructive' was in accordance with its general usage at the time. The discussion on the meaning of the concept 'constructive' had disappeared from the general debate in 1925, and all explanations since then had agreed that intuitionism should be called 'constructive'. This may also explain why both Carnap and Von Neumann put such stress on the 'constructivity' of their theories: they still had to fight for the label, whereas Heyting did not.[464] Possible explanations for why all currents

---

[460][Carnap 1931, pp. 92–94; 104–105]
[461]In second order predicate logic, one can quantify not only over the elements of a given set, but also over subsets and their cartesian products.
[462][Von Neumann 1931, p. 117]
[463]'*Beweismethode*', [Heyting 1931A, p. 114]; as Becker had claimed in 1927, see 5.3.1.
[464]Seen in this light, Heyting's remark thirty years after the Königsberg conference that 'it is undeniable that the appreciation of constructivity has considerably fallen since 1930' seems inappropriate. The concept of constructivity in fact became clearer through the years. Therefore, it was inevitable that the mere term 'constructive' would lose adherents: those who found out that the meaning of 'constructive' was not (or not any more) what they wanted it to stand for.

## 4.4. LATER REACTIONS

wanted to be seen as constructive are the classical use of the term in geometry and the popularity of the constructivity concept in a larger cultural sphere, most notably in art.[465] It should also be noted that the similarity between logicism and intuitionism as pointed out by Carnap is only superficial and somewhat forced. Intuitionism, rather than starting with logic, considers logic to be simply a part of the mathematical language, dependent on mathematics itself.

**Mathematical existence**  The matter of mathematical existence was also discussed, although the claims made in this respect were not as opposed as in the case of constructivity.

Heyting presents the intuitionistic point of view on mathematical existence in a more moderate way than Brouwer used to do. Intuitionists, Heyting maintains, do not ascribe to mathematical objects any existence independent of thought. For, he continues,[466]

> Vielleicht ist es wahr, daß jeder Gedanke auf einen als unabhängig von ihm bestehend gedachten Gegenstand Bezug nimmt; wir können das dahingestellt bleiben lassen. Jedenfalls braucht dieser Gegenstand nicht vom menschlichen Denken überhaupt unabhängig zu sein. Die mathematische Gegenständen, wenn auch vielleicht unabhängig vom einzelnen Denkakt, sind ihrem Wesen nach durch das menschliche Denken bedingt. Ihre Existenz ist nur gesichert, insoweit sie durch Denken bestimmt werden können; ihnen kommen nur Eigenschaften zu, insoweit diese durch Denken an ihnen erkannt werden können. Diese Möglichkeit der Erkenntnis offenbart sich uns aber nur durch das Erkennen selbst. Der Glaube an die transzendente Existenz, der durch die Begriffe nicht gestützt wird, muß als Beweismittel zurückgewiesen werden.[467]

Note that Heyting claims that there is no *necessity* for assuming, as classical mathematics does, that mathematical concepts have an existence independent of our mental activities. Regardless of what kind of metaphysical existence mathematical objects may possess, all we know is conditioned by the epistemological possibilities, and these can only be proved possible by actually being realised. Therefore, only mentally realised judgements can be used in mathematics. In this way, he

---

[465] See 6.5.1.
[466] [Heyting 1931A, pp. 106-107]
[467] 'It might be true that every thought refers to an object conceived to exist independently of this thought; we can let this remain an open question. In any case, such an object need not be completely independent of human thought. Even if they may be independent of individual acts of thought, mathematical objects are by their nature dependent on human thought. Their existence is guaranteed only insofar as they can be determined by thinking; they have properties only insofar as these can be recognized in them by thought. But this possibility of knowledge is revealed to us only by the act of knowing itself. The belief in transcendental existence, not supported by notions, must be rejected as a means of proof.' English translation based on [Heyting 1983, p. 53]. Heyting had already taken such a stand in his dissertation, [Heyting 1925, p. 1].

188    CHAPTER 4. REACTIONS: EXISTENCE AND CONSTRUCTIVITY

places the burden of proof with the classical mathematicians, who believe in a transcendental existence, and use this in their mathematical proofs. They should therefore support this belief with additional arguments.

After all three currents had been presented their contribution, a forum discussion was organised with, besides the speakers, Scholz, Reidemeister and Gödel. The Austrian mathematician Hans Hahn (1879–1934), who chaired the discussion, made an important remark. Having put forward his view that the only possible view of the world is the empirical one, and that logic, interpreted as a system of rules indicating how given linguistic symbol complexes can be changed into others, is consistent with this empiricist view,[468] Hahn turns to the question of existence statements. The meaning of an existence statement, he argues, is not, as the intuitionists insist, a statement of constructivity; but it is not devoid of content either. For assume that the proposition 'there is a continuous function without a derivative' be proved in a non-constructive way. Would anyone then, Hahn asks, try to prove the theorem 'every continuous function has a derivative'? Of course not:[469]

> Und damit hat dieser bloße Existentialsatz eine faktische Bedeutung; nicht die, daß irgendwie eine solche Funktion in der Welt empirisch aufweisbar sei; auch nicht die, daß sie 'konstruierbar' sei; wohl aber die, ich möchte sagen 'wissenschaftstechnische' Bedeutung einer Warnungstafel: Suche nicht den Satz: 'Jede stetige Funktion hat eine Ableitung', zu beweisen, denn es wird dir nicht gelingen.[470]

Like Wavre had done before, Hahn states, in modern terms, that $\exists x \neg \varphi(x)$ should be read as $\neg \forall x \varphi(x)$. Hahn's contribution is a very good example of the degree to which intuitionistic arguments were taken over. The explanation Hahn gives of the meaning of a pure existence statement is completely in line with what intuitionists had claimed, namely that classical mathematicians have only proved that the assumption of the negation of the complementary universal statement gives rise to a contradiction.[471] Hahn only differs from intuitionism in one respect: he continues to call such a statement an existence statement, even though it does not indicate the existence of anything, and refuses to limit this notion to the intuitionistically constructable. This, I think, is also what happened in the mathematical community at large.

---

[468][Hahn et al. 1931, pp. 135–137]

[469][Hahn et al. 1931, p. 140]. Hahn's example is somewhat unfortunate, since in intuitionistic analysis functions behave differently. From the context it is clear that Hahn's example is meant within classical mathematics.

[470]'And thus the mere existence theorem has a factual meaning; not the one that such a function could be empirically pointed out in the world; nor the one that it would be 'constructible'; but the, I would say 'scientific-technical', one of a *caveat*: do not try to prove the theorem 'Every continuous function has a derivative', for you will not succeed.'

[471]Gödel, who wrote his dissertation under Hahn, later (implicitly) used the same idea for his translation between classical and intuitionistic first-order arithmetic. In the Gödel translation, a classical existence statement $\exists(x)P(x)$ is turned into $\neg\forall(x)\neg P(x)$, see 5.4.2.

## 4.4. LATER REACTIONS

**Report** Kurt Hirsch (1906–1986) wrote a report in serial form on the Königsberg conference and the philosophy of mathematics. It is neither clear where the report was published, nor when; however, since Hirsch worked for the liberal newspaper *Vossische Zeitung* since 1928 and since the conference was held in 1930, the articles were presumably published in the *Vossische Zeitung* in 1930. Hirsch had studied mathematics and philosophy in Berlin from 1925, taking lectures from, among other persons, Bieberbach, Von Mises, Schmidt and Schur. His doctoral dissertation was devoted to the foundational debate, which he interpreted in a philosophical way. He passed the oral examination for his dissertation with Bieberbach in 1930. However, he could only receive the degree after the dissertation had been printed, and since Hirsch was married and had a child, he could not afford the costs. The dissertation was finally printed in 1933.[472]

In the report, Hirsch first sketches the background of the Königsberg conference, namely the foundational debate. He stresses the philosophical nature of the controversy and remarks that, if mathematicians are occupying themselves with these questions, that is because philosophers have not found answers to their question. In introducing the different parties in the debate, Hirsch follows a seldomly used distinction between five currents.[473] Besides intuitionism, formalism and logicism, these are conventionalism and empiricism. However, following the organisers of the Königsberg conference, Hirsch left out the two last ones.

Hirsch next describes the lectures delivered by Carnap, Heyting and Von Neumann clearly and accurately. Regarding intuitionism, he remarks that this point of view, by its separation between mathematics and mathematical language, presupposes a certain solution to the problem of the relation between language and thought. He continues:[474]

> Nimmt man aber die Lösung des Problems im gewünschten Sinn vorweg, dann verlieren die Ueberlegungen der Intuitionisten ihre oft empfundene Befremdlichkeit.[475]

He mentions the intuitionistic point of view on existence as construction as a 'pragmatic'[476] component of their thought.

The only mistake Hirsch makes is that he maintains that, from a mathematical point of view, intuitionism, formalism and logicism all amount to the same thing.[477]

---

[472][Gruenberg 1988, pp. 350–351]; the dissertation is [Hirsch 1933].
[473]Only Ackermann had used the same distinction to characterize the foundational debate, [Ackermann 1927]. Hirsch does not refer to Ackermann in the report.
[474][Hirsch 1930]; the clippings I have do not contain page numbers.
[475]'But if one in advance accepts the solution to the problem in the desired way, then the intuitionistic considerations lose their often experienced surprisingness.'
[476]'pragmatistisch', [Hirsch 1930]
[477][Hirsch 1930]

## 4.4.2 Wittgenstein

'The most influential 20th-century philosopher in the English-speaking world',[478] Ludwig Wittgenstein (1889–1951), was born in Vienna in 1889. Wittgenstein studied for six years in Berlin, among other places, to become an engineer. In 1912, he went to Trinity College to study with Russell, on the advice of Frege. In the Great War, he served as a volunteer in the Austrian army, during which time he worked on the Tractatus. The *Logisch-Philosophische Abhandlung*, or Tractatus Logico-Philosophicus, was the only book he published during his lifetime. After its publication in 1921, Wittgenstein worked as an elementary teacher in rural Austria, as a gardener and as an architect. By 1928, he had achieved fame in some circles, notably in the *Wiener Kreis*,[479] as the author of the Tractatus. In 1929 he went back to Cambridge, where he received a Ph.D. for the Tractatus. From then on, Wittgenstein lectured in Cambridge.[480]

**Brouwer's influence** A widely defended thesis is that two phases can be distinguished in Wittgenstein's philosophical work. The first one is characterised by logical analysis and its main representative is the *Tractatus*. The second phase, the most important work of which is the *Philosophische Bemerkungen*, focuses on the concept of meaning.[481] This implies that there must have been a critical period in Wittgenstein's philosophy when he moved from the former position to the latter. It was Brouwer who played a dominant role in this transition.

The general story about how Brouwer influenced Wittgenstein is well-known and has been told by several authors.[482] In 1928, Wittgenstein attended a lecture by Brouwer in Vienna, which was decisive in Wittgenstein's return to philosophy.[483] The value attached to the event, however, differs from person to person. Wright even calls the whole story a 'legend'.[484] Hacker, in his important Wittgenstein study *Insight and illusion*, argues the possible influence of Brouwer on Wittgenstein as follows:[485]

> The fundamental idea that neither language, nor mathematics, nor logic are anything but free creations of the human will imposing an order on reality may well have appeared a deeply liberating conception. (...) Brouwer's schematic outline of his views not only challenged the key doctrines of the tradition of thought in which the *Tractatus* was

---

[478] [Brown, Collinson & Wilkinson 1996, p. 846]
[479] On the *Wiener Kreis*, see 3.3.3.
[480] [Brown, Collinson & Wilkinson 1996, pp. 845–847], [Reeder 1997B, p. 732]
[481] [Mason 1996, pp. 846–848]
[482] Cf, for example, Monk's Wittgenstein-biography [Monk 1990, pp. 249–251].
[483] In fact, Brouwer delivered two lectures in Vienna, see 2.7.3. Menger claims that Wittgenstein only attended the first, [Menger 1994, p. 138]. Schlick, however, maintains that Wittgenstein attended both of them (letter from Schlick to Carnap, 27/3/1928; [ASP Carnap, 029-30-31]). It is not clear to me whose version is correct.
[484] [Wright 1980, p. vii]
[485] [Hacker 1975, pp. 102–104]

## 4.4. LATER REACTIONS

firmly embedded, it pointed the way to a diametrically opposed theory not just of mathematics but of general critical philosophy. (...) The general convergence of ideas between Brouwer's sketch and Wittgenstein's later work, whether causally explicable or not, is crucially important from the point of view of interpreting Wittgenstein's notoriously controversial later philsophy of language.

There seems to be no extensive analysis available as to the more specific question in what way exactly Wittgenstein was influenced by Brouwer.[486] In the following, I present a first version of such an evaluation.

After the publication of the Tractatus in 1921, Wittgenstein had remained silent for years on philosophical issues. In the foreword to the Tractatus, he had claimed that he considered the problems he had treated to be solved now and for ever.[487] He seemed to adhere to this opinion, for he often remarked that all he had to tell was written in the *Tractatus*, and that he had nothing to add.

Apparently, this attitude changed when Brouwer visited Vienna in March 1928 to deliver two lectures. At the suggestion of Menger, Wittgenstein attended the first of these, on 'Mathematics, science and language'.[488] First-hand evidence of the effect Brouwer's lecture had on Wittgenstein comes from Herbert Feigl, who also attended the lecture. In Feigl's words, written down decades afterwards, the following happened:[489]

> Waismann and I, after having overcome considerable resistance on Wittgenstein's part, managed to persuade him to attend this [Brouwer's, DH] lecture. Afterwards W[ittgenstein], Waismann and I spent a few hours in a cafe. It was fascinating to behold the change that had come over W[ittgenstein] that evening. He became extremely voluble and began sketching ideas that were the beginnings of his later writings. (...) As I recall it, W[ittgenstein] agreed essentially with Brouwer's finitism but disagreed in some details which I unfortunately cannot remember. In any case, I believe that evening marked the return of W[ittgenstein] to strong philosophical interests and activities.

To this description one may add Menger's remark that during the talk Wittgenstein's expression changed from one of amazement to one of enjoyment.[490] Feigl's opinion about the change Wittgenstein underwent should be taken seriously, since Feigl belonged to the select group of people who met regularly with Wittgen-

---
[486] Although Richardson pretends to give answers to questions such as how and why Brouwer influenced Wittgenstein in his book *The grammar of justification*, Brouwer's influence in fact plays only a very meagre role in Richardson's a-historical analysis, [Richardson 1976].
[487] [Wittgenstein 1921, p. 9]
[488] [Menger 1994, p. 131], [Monk 1990, p. 249]
[489] Letter from Feigl to Pitcher, 25/09/1962, [ASP Feigl, HF 03-150-02]; a slightly more dramatised version can be found in [Feigl 1969, pp. 638–639].
[490] [Menger 1994, p. 131]

stein in the years 1927 and 1928.[491] In this way, he must have been *au courant* with Wittgenstein's attitude before Wittgenstein heard Brouwer. The impression Brouwer's lecture made on Wittgenstein fits in very well with the later story that, after the lecture, Brouwer and Wittgenstein spent a day together on an island discussing their ideas.[492]

It seems that Brouwer's lecture was Wittgenstein's first real introduction to intuitionistic thought. Although Wittgenstein was interested in questions relating to the foundations of mathematics, the first time he came across Brouwer's intuitionism appears to have been in Ramsey's paper on the foundations of mathematics.[493] The article came to his attention through Schlick, and Wittgenstein reacted to it in July 1927. However, his response concerned the notion of identity, not intuitionism.[494] In the same month, Wittgenstein met with Schlick and Carnap, and they discussed, among other subjects, intuitionism and Esperanto.[495] Whether he at that occasion discussed intuitionism in a more detailed way is unknown.

The new direction of Wittgenstein's thought did not go unnoticed. By 1930, Russell was aware that Wittgenstein was inclined to follow Brouwer's reasonings in the foundations of mathematics. In discussing with Moore the continuation of a Council of Trinity grant to Wittgenstein, Russell noted:[496]

> Then he [Wittgenstein, DH] has a lot of stuff about infinity, which is always in danger of becoming what Brouwer has said, and has to be pulled short whenever this danger becomes apparent. His theories are certainly important and certainly very original. Whether they are true, I do not know; I devoutly hope they are not, as they make mathematics and logic almost incredibly difficult.

It is clear that the broad lines of Brouwer's 1928 lecture had a strong influence on Wittgenstein's thinking. For the following sixteen years, the philosophy of mathematics was one of Wittgenstein's main areas of interest, and large parts of the 1929-1930 and 1932-33 typescripts, nowadays known as the *Philosophische Bemerkungen* and the *Philosophische Grammatik*, deal with this topic.[497] Furthermore, in Wittgenstein's 1937-1944 *Bemerkungen über die Grundlagen der Mathematik*, intuitionistic features such as criticism of the unrestricted use of the principle of the excluded middle and of non-constructive proofs still appear.[498]

These intuitionistic characteristics of Wittgenstein's arguments have made some people place Wittgenstein in Brouwer's camp. Dummett, for instance, la-

---

[491] [Waismann 1993, p. 15]; others who came to these meetings were Schlick, Carnap and Waismann.
[492] [Van Dalen 2001, p. 476]
[493] [Ramsey 1926A]
[494] [Monk 1990, p. 245]
[495] 'Über Wittgenstein', [ASP Carnap, 102-78-07]
[496] [Russell 1968, Vol. II, p. 198]
[497] [Wright 1980, p. vii]
[498] [Fogelin 1968, p. 153]

## 4.4. LATER REACTIONS

belled Wittgenstein a constructivist of a 'much more extreme kind' than intuitionists.[499] However, it should be pointed out that, although Wittgenstein sometimes criticized classical mathematics along intuitionistic lines, he never made any attempt to construct an alternative type of mathematics.[500]

**Wittgenstein's turn – intentionality and grammar**  An extensive work on Wittgenstein's turning-point is Wolfgang Kienzler's dissertation. In this work, Kienzler argues convincingly that, regarding the question of infinity, Wittgenstein moved from an 'intentional' point of view in 1929 to a 'grammatical' position in 1931. In other words, at first Wittgenstein maintained that infinity only existed as an intention, that is, that only potential infinity existed; later, he did not express opinions on existence, but restricted himself to noting the different meanings of the word 'infinity'. Thus, he passed from a normative or prescriptive position to a more descriptive one. Since Wittgenstein's later philosophy is generally characterised by stating that philosophy leaves everything as it is, merely describing the rules of language by using the notion of 'meaning is use',[501] the shift in Wittgenstein's judgement of infinity can be seen as part of Wittgenstein's more general change. As one of the main reasons for this change in position, Kienzler mentions Wittgenstein's discussions with Ramsey in 1929.[502]

In what follows, I move from a general observation of resemblances between Wittgenstein's and Brouwer's views on the foundations of mathematics to more specific points. By means of a historical investigation into Wittgenstein's view on mathematical existence, I argue that Wittgenstein's switch to his later philosophy is likely to have been a two-step event. First, Wittgenstein more or less agreed with Brouwer's normative, intuitionistic line, but later, possibly in reaction to the Brouwer–Hilbert fight, he took up a more descriptive position, simply describing the different uses of mathematical expressions.

In my interpretation of Wittgenstein's turn, I relied on Waismann's notes on conversations which Wittgenstein, Schlick and he had in Vienna from December 1929 to July 1932.[503] These notes were chosen because they cover the whole period which is relevant to this review. Certainly, one has to be prudent in interpreting the views expressed during these conversations in the form in which they are left to us as Wittgenstein's views, because they were not written down by Wittgenstein himself and they were not even published during Waismann's lifetime. But they can, I think, be taken to show some of the directions in which Wittgenstein was thinking, and they reveal some of the themes which influenced him.

With respect to Kienzler's above-mentioned analysis, I would comment as follows. In the first place, Wittgenstein's intentional view on infinity is likely to

---
[499][Dummett 1964, p. 504]
[500][Kienzler 1994, p. 122]
[501][Hacker 1975, p. 124]
[502][Kienzler 1994, pp. 122–152; 206]
[503]Published from Waismann's *Nachlaß* by McGuinness, [Waismann 1993]; the original is in Gabelsberger shorthand.

have been triggered by Brouwer.[504] In the Vienna lecture which Wittgenstein attended, Brouwer made the following statement:[505]

> In dieser Weise wurden auch für die Mathematik der unendlichen Systeme (...) Aussagen 'idealer Wahrheiten' hergeleitet, welche von den Mathematikern für mehr als leere Worte gehalten wurden.[506]

In Brouwer's view, ideal truths are obtained in mathematics by an unjustified application of logical principles, which are valid in finite systems, to infinite systems. This application is carried out without further discussing the meaning of mathematical concepts involved.[507]

Secondly, as far as the discussion notes by Waismann permit any conclusion to be drawn with respect to Wittgenstein's view on mathematical existence, I think Kienzler's thesis can be extended to include this topic. In order to substantiate this idea, I go into the reports of the meetings where Wittgenstein expressed his view on mathematical existence most clearly.

**Existence – normative** The first time Wittgenstein considered the existence question was during the second discussion in the series of conversations, held on December 22, 1929. Wittgenstein here discusses the example where someone has seen two cloths of the same colour. He asks what the statement 'the two cloths had the same colour' should mean.[508]

> Da könnte man glauben, es heißt: 'Es waren beide grün oder beide blau oder ...' Es ist uns allen klar, daß es das nicht heißen kann. Wir können ja eine solche Aufzählung nicht produzieren.[509]

It is clear that Wittgenstein here is also thinking about mathematical existence, since he continues the discussion by applying the same mode of argumentation to the example 'there is a circle in the square'. In this case, too, Wittgenstein maintains that one cannot interpret the statement as '*this* circle is in the square or *this* circle or ...'.

The statement just cited fits in very well with the picture created by Kienzler. Wittgenstein is not just analysing the way in which the existence statement 'there is a colour which is the colour of both cloths' is used, but he is judging whether the statement satisfies his demand that one should be able to actually produce what is stated. That is, he already has a meaning of the words 'they have the

---

[504] Unfortunately, Kienzler does not discuss a possible influence by Brouwer on Wittgenstein.
[505] [Brouwer 1929, p. 424]
[506] 'In this way, propositions of 'ideal truths' were deduced also in the mathematics of infinite systems (...), which were taken by mathematicians as more than empty words.'
[507] [Brouwer 1929, p. 424]
[508] [Waismann 1993, p. 39]
[509] 'Then, one could think this means: 'both were green or both were blue or ...' It is clear to all of us that this cannot be its meaning. For we cannot generate such an enumeration.'

## 4.4. LATER REACTIONS

same colour' in his mind. Since the proposed interpretation involves enumerating an infinite sequence, he considers the statement not to fulfill the demand.

Further on in the same discussion, Wittgenstein criticises existence statements very much along the lines of Weyl. With regard to propositions containing variables, such as 'there is a circle in the square', Wittgenstein remarks:[510]

> Und jetzt ist die Frage: Wie lautet der richtige Ausdruck des Satzes? Ich meine: der Ausdruck lautet nicht: '$\exists(x).\varphi(x)$', sondern '$\varphi(x)$'. (...) '$\varphi(x)$' ist also ein richtiger Satz, nicht erst in Vorbereitung zu einem Satz.[511]

The way in which Wittgenstein arrives at the conclusion, however, differs from Weyl's. In Wittgenstein's argumentation, which I cannot follow completely, the decisive point is that an existence statement can be negated in two ways (namely, $\neg\exists(x)\varphi(x)$ and $\exists(x)\neg\varphi(x)$). Therefore, Wittgenstein concludes, the correct expression is not $\exists(x)\varphi(x)$, but $\varphi(x)$.[512] This point seems to be directed mainly against Russell's formulation of an existence statement.

It should be pointed out that Brouwer nowhere in his Vienna lecture explicitly criticised existence statements. This leaves us with the question how Wittgenstein arrived at his judgement. Two possible explanations come to mind: either he concluded it himself from Brouwer's analysis, as Weyl had done before; or he took it over from Weyl. The latter hypothesis is supported by the fact that we know for certain that Wittgenstein had read Weyl's *Die heutige Erkenntnislage in der Mathematik*[513] on January 2, 1930, and that he was interested in the views on mathematical existence expressed in that work. On the other hand, Wittgenstein's reason for rejecting existence statements differs from Weyl's. Therefore, I cannot make a definite judgement.

At a meeting in January 1930, where Wittgenstein explicitly discusses Weyl's view on mathematical existence, Wittgenstein's normative approach again becomes clear. Weyl's *heutige Erkenntnislage* serves as a basis for the discussion.[514] In the paper, Weyl had repeated his view expressed earlier in the *neue Grundlagenkrise*, namely that existence statements are nothing but propositional abstracts, and that only a construction that has been carried out is worthwhile.[515] Wittgenstein, in reacting, applies his method of verification when distinguishing between several expressions. A universal statement, he maintains, is expressed correctly by induction, and therefore cannot be negated.[516] A fortiori, its nega-

---
[510][Waismann 1993, p. 40]

[511]'And now the question is: what is the correct expression of the theorem? I mean: the theorem does not read: '$\exists(x).\varphi(x)$', but '$\varphi(x)$'. (...) thus, '$\varphi(x)$' is a proper theorem, not just a preparation of a theorem.'

[512][Waismann 1993, p. 40]

[513][Weyl 1925]

[514]discussion on January 2, 1930, [Waismann 1993, pp. 81–82]

[515]See 4.3.2.

[516]This is mentioned without further explanation, which makes it hard to determine what exactly Wittgenstein may have meant.

tion cannot be an existence statement. A statement expressing the existence of a mathematical object at a certain position, for instance at a certain place in the decimal development of $\pi$, can be negated, but its negation simply means that the object does not occur at that particular place. In Wittgenstein's view, the general statement 'the number 7 appears' (e.g. in the decimal expansion of $\pi$) is devoid of meaning, since there is no way of verifying it. What one could state is that the number 7 appears between the positions $p_1$ and $p_2$ in de decimal development of $\pi$; then the verification method is clear. Wittgenstein's judgement here is normative because in his view, a statement that is used in mathematics, like 'the number 7 appears', has no content.[517]

The question of constructivity does not play an important role in Wittgenstein's argumentation. When it appears, however, Wittgenstein does tend to prefer constructive statements. In the summer of 1930, he uses construction in a way similar to verification:[518]

> Die Definition eines Begriffes weist den Weg zur *Verifikation*, die Definition eines Zahlwortes (einer Form) den Weg zur *Konstruktion*.[519]

Presumably, Wittgenstein thought of the 'verification of a notion' as a procedure in which, given a notion $N$, $N$ is compared to the definition of notion $B$, to verify if $N$ is $B$.

It is unclear to me to what extent Wittgenstein's constructions coincide with Brouwer's. On the one hand, he uses 'construction' in the Brouwerian sense, as in the quotation given below on the fundamental theorem of algebra; on the other hand he also seems to include symbolic constructions.

**Existence – descriptive** In September 1931, Wittgenstein returns to the question of mathematical existence. Discussing the fundamental theorem of algebra, he expresses a radically different view on existence statements:[520]

> Wenn ich einmal beweise, daß eine Gleichung $n$-ten Grades $n$ Lösungen haben muß, indem ich z.B. einen der Gauß'schen Beweise gebe, und wenn ich ein zweites Mal die Existenz dadurch beweise, das ich das Verfahren zur Konstruktion der Lösungen angebe, so habe ich nicht etwa zwei verschiedene Beweise für denselben Satz gegeben, sondern ich habe ganz verschiedene Dinge bewiesen.[521]

---

[517]Brouwer used these kind of statements in his counterexamples against the principle of the excluded middle, see 2.6.2.

[518][Waismann 1993, p. 226]

[519]'The definition of a notion shows the way to *verification*, the definition of a number word (a form) the way to *construction*.'

[520][Waismann 1993, pp. 172–173]

[521]'If I once prove that an equation of degree $n$ must have $n$ solutions, for example by giving the Gaußian proof, and if I prove the existence a second time by indicating the construction method of the solutions, then I have not, let us say, given two different proofs for the same theorem, but I have proved completely different things.'

## 4.4. LATER REACTIONS

The point that Wittgenstein is making is that, although the same wording 'there exist $n$ solutions' is used in both cases, there are actually two versions of (in this case) the fundamental theorem of algebra. Since the proofs for them are different, so are the meanings of the theorems. This is completely in line with Wittgenstein's new maxim 'the meaning of a sentence is its mode of verification'.

**Wittgenstein's turn revisited**  I think the foregoing observations may give us a better explanation of Wittgenstein's turning point than has been given so far. The evidence available points in the direction of a two-step shift. At first, when listening to Brouwer's Vienna lecture, Wittgenstein was drawn to intuitionism. This is in accordance with both Feigl's and Russell's report, and with opinions held by Wittgenstein until about 1930 on such themes as the meaning of the infinite and of existence statements.

Later, however, Wittgenstein changed his mind, and moved from a normative to a more descriptive position. Hence, he did not prescribe what 'there is' should mean in mathematics, but he described only the different uses of the phrase and therefore the different meanings.

The accordance noted above between Wittgenstein's judgement of different existence statements and his general 'meaning is use' concept can be interpreted in two ways. First, it can be seen as an application of Wittgenstein's general view to specific problems in the foundations of mathematics. But one could also look at it the other way round. In the light of the evidence that one of the main characteristics of Wittgenstein's later philosophy was its focus on 'meaning is use', that this characteristic applies very well to the foundational debate between Brouwer and Hilbert and that Wittgenstein applied it to the foundational debate, one could interpret Wittgenstein's changing view in the following way. Having considered the matter again, Wittgenstein came to the conclusion that, because Hilbert and Brouwer accepted different proofs in mathematics, in fact the meaning of the mathematics they defended differed. For *the meaning of a statement lies in its use*. Just as the uses of mathematical concepts in intuitionistic and formalistic mathematics differed, so did the concepts themselves, despite the fact that they were described by the same term. Thus, it could well have been the foundational battle between Brouwer and Hilbert which stimulated Wittgenstein to hold his later views.[522]

This explanation, if accepted, would provide us with an answer to two important historical questions: why did Wittgenstein return to philosophy, and where did his new line of thought come from? As to the first question, the conjecture

---

Brouwer had, together with De Loor, given a constructive proof of the fundamental theorem of algebra, see 2.6.2; so had Weyl, independently of them, see 4.3, and Skolem, [Skolem 1924]. The example appears in Brouwer's *Mathematik, Wissenschaft und Sprache*, too; [Brouwer 1929, p. 427].

[522]The interpretation is admittedly quite speculative, since I did not pursue the matter to include all possible materials. Further research by people who are more familiar with Wittgenstein's writings would be needed to put the claim to the test.

# 198  CHAPTER 4. REACTIONS: EXISTENCE AND CONSTRUCTIVITY

put forward here explains better than the traditional account why Wittgenstein again started philosophising. It was not simply because of some kind of inspiration he experienced while listening to Brouwer, but it was because he agreed to some extent with Brouwer's intuitionistic views on such items as the infinite and mathematical existence. Regarding the second question, the origin of Wittgenstein's new line in philosophy can be found in his reaction to the foundational conflict between Brouwer and Hilbert.

It should be pointed out that Wittgenstein's development was not necessarily so straightforward; there is another possibility, namely that the later Wittgenstein was not consistent. Gillies has argued this point, and there is certainly some evidence for a lack of consistency. As examples Gillies mentions that on the one hand Wittgenstein criticised the principle of the excluded middle, which implies a revisionist attitude in the philosophy of mathematics, whereas on the other hand Wittgenstein took the view that philosophy should in no way interfere with the actual use of language, its only aim being to describe it.[523] Since both these remarks stem from the 1940s, it seems likely that from 1928 on Wittgenstein occasionally changed positions, moving between a normative and a descriptive view in the philosophy of mathematics.

### 4.4.3 Others

**Menger from 1928 to 1933** Karl Menger (1902–1985) was the person who gave a prominent place to the protest against the in his view vague meaning of the term 'constructive'. From 1924 to 1927, Menger had worked in Amsterdam on a Rockefeller fellowship as Brouwer's assistant, in the first place as a topologist, but also in order to clarify his thoughts on intuitionism.[524] By 1925, he had become convinced of the intuitionistic point of view regarding existence statements. He wrote to Brouwer that[525]

> mich Ihre Kritik der reinen Existenzialurteile in der Arithmetik nunmehr überzeugt hat. Derartige Sätze sind leere Formen, die nur durch konstruktive Ausfüllung einen sinnvollen Gehalt bekommen können. Daß aber eine solche konstruktive Ausfüllung immer möglich sei,- dafür ist ein Grund bisher nicht angeführt worden und läßt sich, wenn man auf konstruktiven Boden steht, vielleicht überhaupt nicht anführen. Man kann an die Möglichkeit einer solchen Ausfüllung höchstens glauben; aber die Strenge der konstruktiven Schlußweisen hat dann eine Ende erreicht.[526]

---

[523] [Gillies 1982, p. 423]
[524] [Kass 1996, p. 559]
[525] Letter from Menger to Brouwer, 11/2/1925; [MI Brouwer, CB.MEN-1.8]
[526] 'your criticism of pure existence statements in arithmetic has now convinced me. Such propositions are empty forms, which can only obtain meaningful value by a constructive substantiation. But that such a constructive substantiation is always possible, – for that no reason has been given and possibly cannot be given at all, if one stands on constructive ground. One can at most believe

## 4.4. LATER REACTIONS

Although Menger maintained that he had become convinced by reading Brouwer's work, the argumentation he uses resembles more Weyl's 1921 paper than Brouwer's publications.

Apparently, Menger later changed his mind and was not convinced any more by the argumentation he had used himself. Where Fraenkel had spoken about the 'sharp distinction' between constructions and pure existence statements,[527] Menger wrote in a series of papers on foundational questions published in 1928:[528]

> Dabei möchte ich betonen, daß ich das Wort *'Konstruktivität'* für ein wenn überhaupt, so vermutlich *auf verschiedene Arten* und *in verschiedenen Abstufungen* präzisierbares (bisher noch nicht präzisiertes) Wort halte.[529]

Although he denied that constructivity was well-defined, Menger did try to work with the constructivity concept, by using possible definitions of constructivity. For example, he proved that if the definition of a set can be shown to be constructive, then also the set of all its finite subsets is constructively definable.[530]

Two years later, Menger's view had radicalised, and he specifically attacked intuitionism for using vague terms such as 'constructive':[531]

> Was nun die bisherigen intuitionistischen Versuche taten, war, daß jeder von ihnen sich dogmatisch auf einen bestimmten (meist (...) gar nicht klar umschriebenen) Konstruktivitätsbegriff *festlegte* und die zugehörigen Entwicklungen als sinnvoll, die weitergehenden als sinnlos bezeichnete. Nach Ansicht des Verfassers dieses Aufsatzes hat aber eine derartige Aussage *nicht den mindesten Erkenntnisinhalt*. Denn es handelt sich in der Mathematik und Logik nicht darum welche Axiome und Schlußprinzipien man *annimmt*, sondern darum, was man aus ihnen bzw. mit ihrer Hilfe *herleitet*. Ob der Mathematiker *A* das Auswahlsaxiom als 'zulässig' erklärt, an dasselbe 'glaubt' und es anwendet und ob der Mathematiker *B* es als 'unkonstruktiv' oder 'weil er keinen Sinn damit verbinden kann' ablehnt, – diese Tatsachen sind für die Biographie der Mathematiker *A* und *B* von Interesse, eventuell für die Geschichte, keinesfalls aber für die Mathematik und Logik.[532]

---

in the possibility of such a substantiation; but then the rigour of constructive reasoning has come to an end.'

[527] See 4.3.1.
[528] [Menger 1928A, p. 225]
[529] 'Besides, I would like to stress that I take the word *'constructivity'* to be a word that can be specified, if at all, presumably *in various ways* and *in various nuances* (and so far it has not been specified).'
[530] [Menger 1928C, p. 306]
[531] [Menger 1930A, pp. 324]
[532] 'What the intuitionistic attempts did so far was that each of them dogmatically *tied itself down* to a notion of constructivity (mostly (...) not at all clearly described) and designated the according developments as meaningful, those going further as meaningless. In the author's view, however, such an expression contains *no knowledge at all*. For the important thing in

Menger argued the same point in two short papers in the Vienna *Anzeiger* and in his Vien na lecture *Die neue Logik* ('The new logic').[533]

Mannoury was not impressed by Menger's criticism, and countered by observing that such remarks 'apply *mutatis mutandis* to a very large part of the epistemological literature of all ages'.[534]

It is interesting to note that Menger only attacks intuitionism for using the label 'constructive', where in fact Hilbert and Carnap had claimed the same label at different occasions.[535] This may be caused by Menger's wish to attack intuitionism, or because by the time of writing intuitionism was generally considered to be the constructive current.

Menger's criticism of the unspecified use of the term 'constructive' found several echos, for example with Fraenkel and Dubislav.[536]

It is only after 1930 that specifications were put forward to explain what 'constructive' should mean.[537] Until that time, it seemed to fit into the 'cloud of suggestive slogans and fashionable terms'[538] that were widely used in academic lectures. The difference between the 'constructive' label and others, however, is that 'constructive' went *against* the life-philosophical tide, whereas most others were used in order to conform with it.[539]

Menger not only criticised intuitionism, he also came up with a solution. In a 1931 paper on intuitionism, he argued that most discussions to which intuitionism led were terminological ones. Menger's solution to the discussion on mathematical existence and constructivity is, thus, a terminological one. One should simply start using two expressions: one, 'there exists', to indicate classical existence, the other, 'there is', to indicate intuitionistic existence.[540] Menger's solution is similar to the one Dingler had suggested in 1926.[541]

**Heyting in 1930** There are two publications of Heyting in 1930 which are relevant for the discussion on mathematical existence. The first was an important

---

mathematics and logic is not which axioms and rules of inference one *assumes*, but what one can *deduce* from or by means of them. Whether mathematician A declares the axiom of choice to be 'admissable', 'believes' in it and applies it, and whether mathematician B denounces it for being 'not constructive' or 'because he cannot associate any meaning with it' – these facts are interesting for A's and B's biographies, they may be interesting for history, but they definitely are not for mathematics and logic.'

[533][Menger 1930B], [Menger 1931] and [Menger 1933] respectively.
[534]'trifft *mutatis mutandis* für einen sehr großen Teil der erkenntnistheoretischen Literatur aller Zeiten zu', [Mannoury 1934A, p. 293]
[535]See 4.2.2 and 4.4.1.
[536][Fraenkel 1930A, p. 291], [Dubislav 1930, p. 35]
[537]The notion of constructivity was specified by the work of Turing, Church, Kleene and Gödel, see 4.4.3.
[538][Ringer 1969, p. 393]
[539]See 6.4.2.
[540][Menger 1930A, pp. 320–321]
[541]See 4.3.1.

## 4.4. LATER REACTIONS

series of papers in which he put forward a formalisation of intuitionistic mathematics and logic.[542] The second was a paper of a more expository character.

In the formalisation papers, Heyting describes an existence statement as follows:[543]

$(Ex)a$ bedeutet: 'Es kann ein Gegenstand $x$ angegeben werden, für welchen der Satz $a$ gilt.'[544]

He does not elaborate the point, nor does he mention anything about required constructions. What is clear, as Sundholm pointed out to me, is that Heyting gives a more liberal interpretation to an intuitionistic existence statement than Weyl. Where Weyl had argued that the mere possibility of a construction did not suffice,[545] Heyting allowed exactly this.

Heyting next gives three axioms for the existential quantifier, from which he infers a number of theorems. In the discussions that were analysed in the preceding section, a recurring question was whether the negation of a universal statement implies an existence statement, or, in formalised language, does $\neg \forall (x) \varphi(x) \to \exists (x) \neg \varphi(x)$ hold? Classically, this amounts to the same as asking whether $\neg \forall (x) \neg \varphi(x) \to \exists (x) \varphi(x)$? Heyting does not address the question directly, but he proves[546]

**Theorem 2** $\vdash \neg \forall (x) \neg \varphi(x) \to \neg \neg \exists (x) \varphi(x)$.[547]

Furthermore, for propositional logic he had already proved[548]

**Theorem 3** $\vdash p \to \neg \neg p$,

whereas the converse can only be proved if the principle of the excluded middle is assumed to hold for $p$:

**Theorem 4** $\vdash (p \vee \neg p) \to (\neg \neg p \to p)$.

By combining these theorems, one can infer that the classically valid step from $\neg \forall (x) \neg \varphi(x)$ to $\exists (x) \varphi(x)$ can only be made intuitionistically if the principle of the excluded middle is assumed to hold for $\exists x \varphi(x)$ — which intuitionists deny in its general form.

In the expository paper *Sur la logique intuitionniste* ('On intuitionistic logic'), Heyting treats the meaning of logic from the intuitionistic point of view. He reproaches Lévy, in his discussion with Wavre on mathematical existence,[549] of maintaining that everybody understands the word 'existence' in its usual meaning:[550]

---

[542] Heyting's formalisation of intuitionistic logic is treated in more detail in 5.4.1.
[543] [Heyting 1930B, p. 207]
[544] '$(Ex)a$ means: 'an object $x$ can be pointed out, for which the theorem $a$ holds.' '
[545] See 4.2.1.
[546] [Heyting 1930B, p. 213]
[547] All formalised theorems are presented here in the contemporary formulation.
[548] [Heyting 1930A, pp. 198–199]
[549] See 4.3.1.
[550] [Heyting 1930D, p. 233–234]

## 202    CHAPTER 4. REACTIONS: EXISTENCE AND CONSTRUCTIVITY

> Voilà une affirmation bien audacieuse, car dès qu'on sort du domaine de la vie cotidienne, où la signification exacte d'un mot a moins d'importance que son efficacité, pour entrer dans le domaine de la philosophie, le sens du mot 'exister' donne lieu à une controverse des plus profonde; c'est sur ce point que se séparent les grands systèmes.[551]

Heyting's remark is most correct; the different systems he is referring to are the ones we nowadays call realism and anti-realism.[552] He paraphrases Brouwer's demand as follows:[553]

> *L'idée d'une existence hors de notre esprit des entités mathématiques ne doit pas entrer dans les démonstrations.*[554]

Even realists, Heyting continues, should recognize the importance of the question to what extent mathematics can be developed without making use of the assumption of a transcendental existence of mathematical entities.[555]

**Carnap in 1934**  One of the persons most able to grasp both the philosophical and the mathematical implications of the foundational debate was Rudolf Carnap (1891–1970), whose educational background was in philosophy as well as in mathematics and physics. He had studied in Jena and Freiburg; Gottlob Frege was one of his main teachers, whereas also Hugo Dingler had an important influence on his early development. After having finished his dissertation in 1921, Carnap moved closer to Russell's views.[556] Carnap was one of the prominent members of the *Wiener Kreis*,[557] which he joined in 1926, when he became *Privatdozent* in Vienna. His public reactions to intuitionism were scarce until September 1930, when the main fight was over.[558] However, he had been introduced earlier to Brouwer's ideas on the foundations of mathematics. As Carnap reports in his intellectual autobiography, several *Wiener Kreis* members met Brouwer privately when he came to lecture in Vienna in 1928. He recalls that understanding Brouwer, either via written documents or oral presentations, was not always easy. Nevertheless, as Carnap writes, 'the constructivist and finitist tendencies of Brouwer's thinking

---

[551]'That is quite a bold statement, because as soon as one leaves the domain of everyday life, where the exact meaning of a word is of less importance than its efficacy, to enter into the domain of philosophy, the meaning of the word 'to exist' causes one of the most profound controversies; at this point, the great philosophical systems separate.'
[552]See 4.1.1.
[553][Heyting 1930D, p. 234]
[554]'*The idea of an existence of mathematical entities outside our mind should not enter in proofs.*'
[555][Heyting 1930D, p. 234]
[556][Grattan-Guinness 1997, p. 408]
[557]On the *Wiener Kreis*, see 3.3.3.
[558]Carnap's 1930 contribution is treated separately, in the section on the Königsberg conference, see 4.4.1. In an earlier paper, *Eigentliche und uneigentliche Begriffe*, Carnap had made some comments on the intuitionistic point of view, [Carnap 1927, pp. 363-365].

## 4.4. LATER REACTIONS

appealed to us [the *Wiener Kreis* members, DH] greatly'.[559] Already in the following year Carnap acknowledged — at least privately — the validity of much of Brouwer's criticism of universal and existence statements.[560]

In 1931, Carnap became professor of natural philosophy at the German university in Prague. There, a Ph.D. student of his worked on a construction of mathematics with a criticism of the concept of existence which was partially based on Brouwer.[561]

The main part of Brouwer's influence on Carnap's thinking, however, can be found in the Carnap's *Logische Syntax der Sprache* ('Logical syntax of language').[562] This work, published in 1934, is seen as the culmination of the so-called constructional phase in Carnap's career.[563]

The aim of the *Logische Syntax*, according to Carnap, is to provide an exact method for the analysis of propositions about propositions. This was part of the attempt by the *Wiener Kreis* to replace philosophy by the logic of science, the latter to be understood as the logical syntax of scientific language[564] — hence the title. Carnap works out two languages and gives a sketch of a general syntax for arbitrary languages. The first two languages, called Language I and Language II, are of interest to us here.

Carnap describes Language I as a language that only includes elementary arithmetic of the natural numbers to a certain, limited extent, in accordance with views 'that consider themselves constructivistic, finitistic or intuitionistic'.[565] Language II is much broader: it includes classical mathematics (such as real and complex functions and set theory), and it enables one to formulate physical propositions.[566] One of the main differences between Language I and II is the limitation of quantifiers. Carnap introduces limitation of quantifiers in the following way: $\exists(x)3P(x)$ means: there is an $x$ up till and including 3, for which $P(x)$ holds; $\forall(x)3P(x)$ is defined analogously.[567] In Language I, all universal and existence quantifiers are limited; in Language II unlimited quantifiers are allowed, too. The advantage of limiting the scope of the quantifiers, Carnap claims, is that the main demands of Brouwer's intuitionism, such as the rejection of pure existence statements, can be fulfilled without there being any need to give up the principle of the

---

[559][Carnap 1963, pp. 48–49]
[560]Carnap, R., 'Bemerkungen zu Kaufmanns MS 'Das Unendliche in der Mathematik' ', unpublished manuscript, April 1929; [ASP Carnap, 028-26-10]
[561][ASP Carnap, 102-67-01]
[562]Carnap explicitly mentioned Brouwer in the foreword as one of the authors from whom he had learned a lot, even though he did not completely share his views, [Carnap 1934A, p. VII].
[563][Sarkar 1992, pp. 1–2]
[564][Carnap 1934A, pp. III–IV]
[565]'die sich als konstruktivistisch, finitistisch oder intuitionistisch bezeichnen', [Carnap 1934A, p. 10]
[566][Carnap 1934A, p. 74]
[567]Carnap's idea of the limitation of quantifiers is strikingly similar to Wittgenstein's remarks on existence statements regarding the appearance of a certain number in a decimal development, see 4.4.2.

excluded middle.[568] It is true that the principle of the excluded middle holds for limited quantifiers, since the limitation of quantifiers makes all propositions decidable. However, the price which Carnap has to pay is that he can no longer state propositions about infinite collections, something which is allowed in intuitionism.

In his intellectual autobiography, Carnap states that his original intention was to construct only Language I, 'in agreement with the finitist ideas with which we sympathized in the Circle'.[569] But in the book he stresses that Language I was not chosen because it was seen as 'the only possible or the only justified language'.[570] It seems, thus, that Carnap originally was strongly attracted by Brouwer's views on mathematical existence, but later downgraded the philosophical importance of such distinctions. In the *Logische Syntax* Carnap is very clear about his stand with regard to different mathematical languages. Carnap's Principle of Tolerance reads:[571]

> [W]ir wollen nicht Verbote aufstellen, sondern Festsetzungen treffen. (...) In der Logik gibt es keine Moral. Jeder mag seine Logik, d.h. seine Sprachform, aufbauen wie er will.[572]

If accepted, the principle would provide anyone with enough justification to construct any language whatsoever. Therefore, the need to justify a language choice would disappear — one of the explicitly stated main goals of Carnap's book.[573] Nevertheless, Carnap apparently felt obliged to justify his own acceptance of Language II, for he devoted a section to it. As he puts it, the general criticism of an unlimited existence statement is that it is meaningless, since we have no way of looking for the answer; and the meaning of a concept lies only in the method of observing whether the concept applies or not. As one can see, Carnap's formulation is a mixture of Brouwerian and Wittgensteinian terms. Carnap's answer is that we may not have a method of looking for the answer, but we do know what an answer looks like. This gives us the possibility of finding an answer, and therefore there is no ground for rejecting the statement as meaningless.[574]

In a letter to Carnap, Menger claimed priority for the Principle of Tolerance, which Gödel, among other persons, had taken over from him. Menger stressed that, although it admittedly cost him (i.e., Menger) little effort to develop the idea, it was diametrically opposed to both Hilbert's and the intuitionistic view.[575] Judged to Menger's publications on the different interpretations of constructivity,[576] his claim seems to be correct.

---

[568] [Carnap 1934A, p. 43]
[569] [Carnap 1963, p. 55]
[570] [Carnap 1934A, p. 42]
[571] [Carnap 1934A, pp. 44–45]
[572] '[W]e do not want to issue prohibitions, but to make determinations. (...) In logic, there is no moral. Anyone may construct his logic, i.e., his language form, as he wishes.'
[573] [Carnap 1934A, p. V]
[574] [Carnap 1934A, p. 114]
[575] Letter from Menger to Carnap, 15/3/1934; [ASP Carnap, 029-01-09]
[576] See above.

## 4.4. LATER REACTIONS

**Bernays, Behmann and Turing, 1930–1937**   Bernays was one of the mathematicians who had first claimed the label 'constructive' for Proof theory,[577] but later recognised the difference between constructive and axiomatic mathematics.[578] In the aftermath of the debate, Bernays was still occupied with the meaning of the term. He originally disliked what he saw as the vague idea of constructivity, but changed his mind under the influence of Turing and ended up using it again.

Bernays' original negative attitude towards the concept of constructivity can be seen from his reaction to Behmann's work on constructive mathematics. The German logician Heinrich Behmann (1891–1970) had written his dissertation under Hilbert in 1918, with the aim of introducing Russell and Whitehead's *Principia Mathematica* to a wider audience. He next worked as a *Privatdozent* in Göttingen, from 1921 to 1925, and at the university of Halle.[579]

Around 1930, Behmann was involved in investigating which part of mathematics is affected by the demand of constructivity.[580] The philosopher Felix Kaufmann, who frequented meetings of the *Wiener Kreis*, had suggested to him the following problem: is it possible to show that all proofs of existential claims, except those obtained by means of the axiom of choice, in fact contain implicitly an instantiation of the existential statement? Kaufmann's conjecture was that this was the case, and that thus Brouwer's demand for constructivity would to a great extent have been met.[581] Behmann set out to prove the conjecture.[582]

To Behmann, the relevance of constructivism was that it provided a *mathematical* (as opposed to a philosophical or epistemological[583]) way of distinguishing between purely formal mathematics and what he called 'factual' (*sachliche*) mathematics. Thinking of Brouwer's and Weyl's earlier characterisations of intuitionism, it seems reasonable to interpret 'factual' as 'meaningful', which means that for Behmann constructivity was a formal criterion for differentiating between meaningful and purely formal mathematics. Behmann did not consider himself to be a constructivist,[584] and he was not interested in questions about the justification of constructivistic demands.[585] Nevertheless, he thinks that investigations into constructive mathematics are useful[586]

---

[577] See 4.2.2.
[578] See 4.3.1.
[579] [Mancosu 1999B, pp. 305–308], [Poggendorff 1936–1940, Teil 1, p. 161]
[580] Letter from Behmann to Kaufmann, 14/10/1930 [ASP Carnap, 028-06-16]
[581] In his book *Das Unendliche in der Mathematik und seine Ausschaltung*, Kaufmann had made a similar claim, namely that all (non-constructive) classical existential proofs, except those in non-denumerable domains, remain valid intuitionistically, cf. [Kaufmann 1930, p. 67].
[582] The account given here on Behmann's contribution to the debate on constructivity made use of [Mancosu 2002] for filling in the missing parts of the story.
[583] Epistemology, as explained in the glossary, is the study of knowledge and the justification of belief, [Dancy 1989, p. 1].
[584] Behmann even claimed that he took up a position of 'absolute neutrality' towards constructive demands.
[585] Letter from Behmann to Kaufmann, 22/10/1930 [ASP Carnap, 028-06-13]
[586] Letter from Behmann to Bernays, 24/10/1930 [ETH Bernays, 975-275]; published in [Mancosu 2002, p. 23].

## 206   CHAPTER 4. REACTIONS: EXISTENCE AND CONSTRUCTIVITY

> (...) damit die praktische Tragweite der konstruktivistischen Prinzipien nicht überschätzt und infolgedessen unberechtigterweise große Teile der Mathematik in Frage gestellt werden.[587]

Even though Behmann worked on constructivism, he preferred to do so within formal mathematics.

In October 1930, Behmann submitted his paper *Zur Frage der Konstruktivität von Beweisen* ('On the question of the constructivity of proofs') to Bernays, who was then assistant editor of the *Mathematische Annalen*. Behmann points out that two of the main restrictions applied in constructive mathematics are to eliminate non-constructive existential axioms, in particular the axiom of choice, and to eliminate non-constructive existential proofs. Regarding the latter point, he distinguishes between direct existential proofs (by instantiation) and indirect ones (by contradiction). He then presents a demonstration of Kaufmann's conjecture by applying topological and graph-theoretical considerations to proofs.[588] Kaufmann, upon reading the draft paper, proposed to add an appendix claiming that Behmann's proof showed that, although the intuitionistic requirement of constructivity is legitimate, its realisation does not entail a reduction of mathematics. Behmann declined to do so, on the grounds that he did not want to alienate part of the audience for ideological reasons.

Gödel and Bernays, however, were less positive about Behmann's paper. Gödel uses a Brouwerian counterexample based on Goldbach's conjecture to show that Behmann's statement is not correct.[589] Bernays, in turn, points out that Behmann has not realised the proof. What is needed, in order to prove the claim, is a procedure which, given a provable proposition $P$, produces a provable proposition $P'$ in which all existential statements are replaced by instantiations. Moreover, the procedure itself may not be formulated in terms of existential expressions.[590] Note that Bernays is here asking for an explicit procedure, thus for a constructive proof. Behmann acknowledged the value of Bernays' criticism, and the paper was never published.

Bernays was not very pleased to find one of Hilbert's former Ph.D. students involved in constructive mathematics. In a letter to Behmann written in 1930 he complains:[591]

> Warum begnügen Sie sich, der Sie in dem Formalismus der Logik so zu Hause sind, mit solchem unscharfen Begriffen wie 'konstruktiv',

---

[587]'(...) so that the practical import of the constructivistic principles is not overestimated and consequently large parts of mathematics are, unjustifiably, called into question.' Translation based on the English translation in [Mancosu 2002, p. 2].
[588]Behmann's paper is discussed in more detail in [Mancosu 2002].
[589]Carnap, memorandum of 16/10/1930; published in: [Dawson 1997, p. 73]
[590]Letter from Bernays to Behmann, 17/11/1930 [ETH Bernays, 975-5769]
[591]Letter from Bernays to Behmann, 31/10/1930 [ETH Bernays, 975-5768]

## 4.4. LATER REACTIONS

'Aufweisung', an deren Stelle doch die logische Symbolik viel schärfere Unterscheidungen an die Hand gibt?[592]

Behmann, however, was clear about what he meant by constructivity. As he explains to Bernays, he calls a proposition 'constructive' if every existential statement used in its proof can be replaced by an instantiation.[593]

From about 1935, new formulations of constructive mathematics appeared. The British mathematician Alan Turing, known nowadays particularly because of the so-called Turing machine, did much work on constructivity and computability. His concept of constructive mathematics, which was published in 1936 and provided an important step in the development towards the modern computer, was based on the notion of computable numbers. He described computable numbers as those numbers that are calculable by a (finite) computing machine, i.e., numbers for which there exists a computing machine, through which a tape runs with symbols printed on it and which is capable of calculating the number.[594] Computable functions, predicates, etc., are then defined in an analogous way.[595] In Bernays' view, Turing's concept of computability led to a form of constructive analysis which was 'much more winning' than Brouwer's.[596]

In this way, the term 'constructive' in the end obtained a more precise meaning by the work of logicians who had mostly not or scarcely participated in the foundational debate and whose work, so it seems, was hardly influenced by it.[597] Some 15 years after the fight for the term had started, Menger's demand for clarification had finally been fulfilled.[598] However, since by that time the foundational crisis was largely over, the results of Turing, Church, Kleene and Gödel and the like

---

[592]'Why are you, who is so at home in the formalism of logic, content with such vague notions as 'constructive' and 'indication', instead of which logical symbolism puts much sharper distinctions in one's way?'

[593]Letter from Behmann to Bernays, 01/11/1930 [ETH Bernays, 975-276]

[594]A more rigorous description of Turing's definition can be found in the original [Turing 1936, pp. 230–233].

[595]Later, it was proved that Turing's concept of computability coincides with Gödel and Herbrand's notion of general recursiveness and with Church's λ-definability, [Church 1938, p. 227]; the proofs are in [Kleene 1936, Turing 1937]. A history of the origin of the different versions of computability and their interaction can be found in [Davis 1982].

[596]Letter from Bernays to Church, 22/04/1937 [ETH Bernays, 975-781]. After Turing's interpretation, Bernays again used words like 'constructive' and 'construction' without hesitation (cf., e.g., letter from Bernays to Fraenkel, 12/05/1938 [ETH Bernays, 975-1435]). Sometimes, he regarded the term 'constructive' as equivalent to 'intuitive' (*anschaulich*, letter from Bernays to Beth, 18/09/1945 [ETH Bernays, 975-331]) or 'contentual' within the domain of meta-mathematics (letter from Bernays to Fraenkel, 12/05/1938 [ETH Bernays, 975-1435]).

[597]Turing and Kleene did not react to intuitionism at all in the period considered here. Brouwer's influence on Church is unclear. Church had spent some time as a National Research Fellow in Amsterdam between 1927 and 1929, where he visited Brouwer, [Enderton 1995, p. 486]. He did react to intuitionism, but only on the issue of intuitionistic logic, see 5.3.1. Gödel is the exception. Although most of his reactions also focus on intuitionistic logic, Gödel's view on constructivity is clearly influenced by intuitionism, see below. However, Gödel did not stress the point.

[598]See 4.4.3.

did not influence its outcome. Rather, they started a new chapter in the history of mathematics.

**Gödel from 1929 to 1938** Gödel[599] published a number of papers on intuitionism, some of which also deal with the question of existence statements.

In his dissertation, accepted in 1930, Gödel proves the completeness of what we now call first order predicate logic.[600] In the original version, Gödel uses the introduction to discuss the relationship between his work and the views put forward by Brouwer and Hilbert. In the published version, however, the introduction is skipped and there is no reference to the foundational debate.[601] As Gödel maintains in the introduction, the completeness theorem is equivalent to the statement that every consistent axiom system has a model.[602] He remarks that Brouwer has 'emphatically stressed'[603] that one cannot simply conclude that one can construct a model as soon as an axiomatic system is consistent. Now one could, Gödel continues, define the existence of the concepts contained in an axiomatic system by the consistency of the system. However, this presupposes the solvability of every mathematical problem. For, Gödel argues, if one could prove the unsolvability of a certain mathematical problem, for example in the field of the real numbers, then the 'existence as consistency' definition would provide us with two non-isomorphic models of the reals, whereas one can on the other hand prove that every two models are isomorphic.[604]

Gödel's argumentation shows that, concerning the question of mathematical existence, he sided with Brouwer rather than with Hilbert. Furthermore, he was more specific than most participants to the debate by speaking about the construction of a model.[605] Finally, however, his argument about any two models of the reals being isomorphic is not completely correct as it stands, since, as was proved later, there exist non-standard models for the real numbers which are not isomorphic to the standard one. The difference is caused by the question whether one uses first order logic, in which case there exist non-standard models, or second order logic, in which case one can indeed prove that any two models are isomorphic.[606]

In June 1932, Gödel lectured at Menger's colloquium in Vienna on the re-

---

[599] For biographical information and more details on Gödel's work in intuitionistic logic, see 5.4.2.
[600] Nowadays, completeness is defined as $\Gamma \models \varphi \Leftrightarrow \Gamma \vdash \varphi$. Gödel took completeness to mean what we would now describe as $\Gamma \models \varphi \Rightarrow \Gamma \vdash \varphi$.
[601] [Gödel 1930]
[602] The equivalence can be proved (non-constructively) in the following way. Assume the completeness theorem, and assume that the axiomatic system $\Gamma$ has no model. Then $\Gamma \models \bot$, thus, by completeness, $\Gamma \vdash \bot$, which means that $\Gamma$ is inconsistent, contradiction. Conversely, assume that every consistent system has a model, and that $\Gamma \models \varphi$. Suppose $\Gamma \not\vdash \varphi$, then $\Gamma \cup \neg\varphi$ is consistent, and thus has a model. But this is in contradiction with $\Gamma \models \varphi$.
[603] 'mit Nachdruck hingewiesen', [Gödel 1929, p. 60]
[604] [Gödel 1929, pp. 60–62]
[605] Gödel uses both the term *Realisierung* and *Modell*, [Gödel 1929, p. 60].
[606] Cf. [Dreben & Van Heijenoort 1986, p. 170]. Examples of non-standard models of the reals are given in [Van Dalen 1997, pp. 124–126].

## 4.4. LATER REACTIONS

lationship between classical and intuitionistic (first-order) arithmetic.[607] Basing himself on Heyting's formalisation of intuitionistic logic and expanding upon earlier work by Glivenko, Gödel presents a translation from classical arithmetic into its intuitionistic counterpart. This interpretation leads him to the conclusion that intuitionistic arithmetic only *appears* narrower than classical arithmetic. Gödel's explanation is that, even though it is not allowed to use pure existence statements in intuitionism, one can still apply the absurdity predicate to universal statements, which leads formally to the same result as in classical arithmetic.[608]

In December 1933, Gödel lectured to a joint meeting of the American Mathematical Society and the Mathematical Association of America in Cambridge, Massachusetts.[609] Gödel starts by stating that the question of providing foundations for mathematics can be divided into two parts. First, the methods of proof have to be reduced to the axioms and primitive rules of inference, and second, a justification for the axioms has to be given. The first part, Gödel claims, has been solved in a 'perfectly satisfactory' way by the formalisation of mathematics. The second part, however, is in an 'extremely unsatisfactory' situation.[610]

Gödel sees three main difficulties in attaching meaning to the symbols involved in mathematics, one of them being the question of non-constructive existence proofs. He makes it clear that these can only be interpreted as meaningful statements if one presupposes a kind of Platonism, 'which cannot satisfy any critical mind'.[611] His solution is a pragmatic one: maybe we are not able to attach unobjectionable meaning to the symbols we use in our formal systems, but at least we might succeed in giving a consistency proof by using unobjectionable methods.

Gödel remarks that if we restrict ourselves to mathematics we can construct, we get intuitionistic mathematics, but this is not uniquely determined. As Menger had remarked before,[612] Gödel maintains that there are, in fact, different notions of constructivity, each giving rise to different layers of 'intuitionistic or constructive' mathematics. In its strictest form, Gödel maintains, constructive mathematics should satisfy the following two demands. In the first place, universal quantification should only be applied to infinite totalities for which we can give a finite procedure for generating all their elements. And, secondly, existence propositions should only be stated if we have found an example but, for the sake of brevity, do not want to use it explicitly.[613] Note that Gödel's second demand is completely in accordance with what intuitionism claims.

The idea of different layers of constructive mathematics was developed further in Gödel's 1938 Zilsel lecture. In the lecture, Gödel specifies different meanings

---

[607][Troelstra 1986, p. 282]. The lecture was published the following year.
[608][Gödel 1933E, p. 294]; more details are given in 5.4.2.
[609]The analysis that follows was based on the version of Gödel's lecture as published in his Collected Works, [Gödel 1933O].
[610][Gödel 1933O, p. 45]
[611][Gödel 1933O, p. 50]; Feferman calls this statement surprising, in the light of evidence that Gödel had held Platonistic views since his student days, [Gödel 1933O, p. 39].
[612]See 4.4.3.
[613][Gödel 1933O, p. 51]

## 210 CHAPTER 4. REACTIONS: EXISTENCE AND CONSTRUCTIVITY

of constructivity, based among other things on computability and primitive recursiveness, and gives a hierarchy of constructive theories suitable for giving relative consistency proofs for parts of classical mathematics.[614] In this way, he uses the notion of constructivity in order to save a modified version of Hilbert's programme.

Thus, at various occasions Gödel occupied himself with the question of mathematical existence, and took the intuitionistic demand of constructive existence statements seriously. Gödel refrained from going into the more philosophical questions. What he did do, following Heyting and Glivenko, was to further demystify the intuitionistic view on mathematical existence by presenting a translation from classical into intuitionistic arithmetic. Gödel's view on what constructivity should mean comes close to Brouwer's. However, Gödel was able to distinguish between several kinds of constructivity. In this way, he could even use a modified idea of constructivity to present a modified version of Hilbert's programme.

**Weyl's last words, 1930–1953** When Weyl, who had begun as a student of Hilbert, publicly changed camps and supported Brouwer in 1921,[615] the question of mathematical existence was one of his central themes. From 1923 on, he moderated his views. Apparently, he had been looking for a kind of compromise, some way of reconciling the two views, ever since that time.

In October 1930, Weyl delivered a lecture at the occasion of the opening of a conference in Jena. It was published the following year as *Die Stufen des Unendlichen* ('The levels of infinity'). Weyl labels as a 'constructive turn'[616] the recognition that a mathematical sequence not only describes a certain mathematical object the existence of which is secured independently, but that it should first generate the object involved. At the end of the lecture, he sums up his own opinion as follows:[617]

> Nimmt man die Mathematik für sich allein, so beschränke man sich mit Brouwer auf die einsichtigen Wahrheiten, in die das Unendliche nur als ein offenes Feld von Möglichkeiten eingeht; es ist kein Motiv erfindbar, das darüber hinausdrängt. In der Naturwissenschaften aber berühren wir eine Sphäre, die der schauenden Evidenz sowieso undurchdringlich ist; hier wird Erkenntnis notwendig zu symbolischer Gestaltung, und es ist darum, wenn die Mathematik durch die Physik in den Prozeß der theoretischen Weltkonstruktion mit hineingenommen wird, auch nicht mehr nötig, daß das Mathematische sich daraus als ein besonderer Bezirk des anschaulichen Gewissen isolieren lasse: auf dieser höheren Warte, von der aus die ganze Wissenschaft als eine Einheit erscheint, gebe ich Hilbert recht.[618]

---

[614] [Gödel 1938A] and the commentary [Sieg & Parsons 1995]
[615] See 4.2.1.
[616] 'konstruktive Wendung', [Weyl 1931, p. 7]
[617] [Weyl 1931, pp. 17–18]
[618] 'If mathematics is taken by itself, one should restrict oneself with Brouwer to the intuitively

## 4.4. LATER REACTIONS

In the spring of 1931, Weyl delivered the Terry Lectures at Yale University, published the following year as 'The Open World'. In these talks, which were originally written in German, Weyl frequently uses the term 'constructive'. He uses 'construction' in a broader sense, including both intuitionistic mathematics and symbolic constructions.[619] The third and last lecture, on the infinite, is mostly a repetition of his 1930 Jena lecture.

Two years before he died, Weyl stated his opinion in a slightly different form, reducing the extent to which Brouwer was right even more:[620]

> Der im Symbolismus der Quantificatoren sich ausdrückende 'mathematische Existentialismus' ist recht und gut, solange es sich um die Entwicklung *allgemeiner Theorien* handelt. Sobald aber in einem *konkreten Fall* eine bestimmte numerische (...) Voraussage gemacht werden soll, muß man versuchen, die symbolisch sichergestellte Existenz durch eine explizite Auswertung auszufüllen, wie das die Brouwersche Mathematik grundsätzlich verlangt.[621]

And in one of his last lectures, published only recently and probably delivered some time after 1953, he ends with the following words:[622]

> Indeed my own heart draws me to the side of constructivism. Thus it cost me some effort today to follow the opposite direction, putting axiomatics before construction, but justice seemed to require this from me.

---

cognizable truths and consider the infinite only as an open field of possibilities; nothing compels us to go farther. But in the natural sciences we are in contact with a sphere which is impervious to intuitive evidence; here cognition necessarily becomes symbolical construction. Hence we need no longer demand that when mathematics is taken into the process of theoretical construction in physics it should be possible to set apart the mathematical element as a special domain in which all judgements are intuitively certain; from this higher viewpoint which makes the whole of science appear as one unit, I consider Hilbert to be right.' Translation cited from [Weyl 1932A, p. 82].

[619] Actually, constructivity had even a third and more general meaning for Weyl. At the end of the Terry Lectures he states as the first thesis to be drawn from the history of mathematics in connection with the infinite: 'In the spiritual life of man two domains are clearly to be distinguished from one another: on one side the domain of creation (*Gestaltung*), of construction, to which the active artist, the scientist, the technician, the statesman devote themselves; on the other side the domain of reflection (*Besinnung*) which consummates itself in cognitions and which one may consider as the specific realm of the philosopher. The danger of constructive activity unguided by reflection is that it departs from meaning, goes astray, stagnates in mere routine; the danger of passive reflection is that it may lead to incomprehensible 'talking about things' which paralyzes the creative power.' [Weyl 1932A, pp. 82–83]

[620] [Weyl 1953, p. 35]

[621] '*Mathematical existence-ism*' as expressing itself in the symbolism of quantification is nice and good, as long as it is about the development of *general theories*. However, as soon as a specific numerical (...) prediction has to be made in a *concrete case*, one has to try to insert an explicit practice in the symbolically assured existence, as Brouwerian mathematics in principle demands.'

[622] [Weyl 1985, p. 38]

## 4.5 Conclusion

The debate on mathematical existence and constructivity was opened by Weyl's 1921 paper *Über die neue Grundlagenkrise der Mathematik*. Weyl put more stress on an intuitionistic view on mathematical existence than Brouwer had done, rejecting pure existence statements as proposition abstracts. Not the existence statement itself was valuable, in Weyl's view, but the construction carried out in the proof.

Hilbert reacted immediately to Weyl's paper. In lectures delivered in Copenhagen and Hamburg the same year, he opposed Weyl's view by the idea of consistency as a sufficient criterion for existence. Where Weyl had spoken about individual existence statements, Hilbert referred to the existence of a mathematical system as a whole.

From then on, two main lines can be distinguished in the debate on mathematical existence: one on existence itself, the other on the label 'constructive'.

**Existence** From 1924 on, the debate on mathematical existence widened. Dresden brought it to the English-speaking public; Wavre did the same for the French readers.

Wavre presented a clear interpretation of pure existence statements. Put in modern terms, he stated that in the formalistic conception $\exists (x)\neg\varphi(x)$ should be interpreted as $\neg\forall(x)\varphi(x)$. In Wavre's view, intuitionists adhered to the real meaning of mathematical existence. Thus, Wavre reduced the meaning of a pure existence statement to its way of being proved. In the discussion at the 1930 Königsberg conference, Hahn put forward the same interpretation, explicitly stating that such an existence statement did not express anything about constructibility. Gödel later used the same reading in his translation from classical into intuitionistic arithmetic.

In the debate that developed, most participants merely mentioned or explained the intuitionistic point of view on mathematical existence. Most contributions did not treat the matter in detail. The main exception was Becker's voluminous *Mathematische Existenz*, published in 1927. Becker argued that, from a phenomenological point of view, intuitionism was right. However, since his work was hard to read for non-phenomenologists, it played only a marginal role in the debate.

In 1926, Dingler presented the valuable contribution to simply differentiate between 'logical existence', meaning consistency, and 'constructive existence'. However, Dingler hardly played a role of importance in the debate. Menger later made the same suggestion.

Hilbert's position regarding mathematical existence was ambiguous. On the one hand, he took over most of the intuitionistic criticism on pure existence statements already in 1922. Combining Brouwer's and Weyl's view on contentual mathematics, he maintained that in infinite totalities, there is a difference between 'there is' and 'there is available'. On the other hand, Hilbert was the main person putting forward arguments to justify the idea of existence as consistency. In 1921, he still believed that a consistency proof implied the truth of all the mathematical propositions provable in the system under consideration. In 1922, he continued stressing the importance of a consistency proof, but he did not state anything about the character of the mathematical objects in the system proved consistent. By 1925, Hilbert had explicitly moved towards a more formalistic position, admitting that some of the propositions in formalised mathematics did not have any meaning.

Whereas Weyl had already in 1922 dropped the unique claim for intuitionism, he again rose to the defence of intuitionism in 1927. Brouwer's view, Weyl claimed, was the one of holding on to the meaningful character of mathematics. Hilbert only managed to save classical mathematics by radically changing its interpretation. By 1930, Weyl added a judgement to this description. If one takes mathematics by itself, Weyl maintained, Brouwer is right. If, however, one moves on to science, one cannot deny that there is a certain need for a symbolic construction of the world, and from that point of view, he judged Hilbert to be right.

In 1930, Heyting published a formalisation of intuitionistic logic. He described intuitionistic existence as 'one can point out an object which satisfies the required demand'. This runs contrary to Weyl's original interpretation, which had started the whole debate, in which possible constructions were rejected and only constructions actually carried out were accepted.

The question of mathematical existence played an important role in the search for a coherent presentation of the different consequences of the intuitionistic position. In 1924, Wavre put forward the intuitionistic view on mathematical existence as one of its two main principles (the other one being the rejection of the principle of the excluded middle). In 1928, Fraenkel went further in the third edition of his popular *Einleitung in die Mengenlehre*. He derived all intuitionistic statements from one central point of view, namely the rejection of pure existence statements in favour of constructive existence proofs. This was the first time a non-intuitionist gave a completely coherent presentation of intuitionism, rather than to present it as a collection of more or less arbitrary positions.

Wittgenstein returned to philosophy as a result of attending one of Brouwer's Vienna lectures. His second philosophy can be characterised by two periods, as far as his views on mathematics are concerned. At first, he agreed with intuitionism on such items as mathematical existence and the infinite. Later, however, he moved on to his position of 'meaning is use'. He then claimed that if one has given a constructive and a non-constructive proof of an existence theorem, one has not given two different proofs for the same theorem, but one has actually proved two different theorems. Although the idea of 'meaning is use' applies very well to existence theorems and Wittgenstein applied it in that way, it is hard to say whether reflecting on these theorems drove him to take up his new position.

The overview of the debate on mathematical existence just presented leads to the following conclusion. In the first place, the debate was mostly one of clarification. During the debate, the meaning of 'existence' in mathematics underwent the following changes. On the level of meaning, Hilbert basically agreed with Weyl's criticism of pure existence statements. On the level of formal mathematics, existence statements should be interpreted as the negation of a universal statement, as Wavre was the first to point out. Regarding the existence of an axiomatic system, Hilbert at first adhered to the idea that a consistency proof proved the truth of the propositions in the system, i.e., he believed in the existence of a unique model for each consistent system. By 1925, however, he admitted that some of the propositions in a consistent system actually had no meaning. In this way, Hilbert's weakened claim was more in accordance with what Brouwer had already claimed in his dissertation, namely that a consistency proof only provides information about the linguistic system, not about the mathematics that the system is supposed to represent.

It took a long time before a terminological solution was accepted, indicating the problem people had in accepting a plurality of views in mathematics. Secondly, the subject of mathematical existence played an important role in a more unified presentation of intuitionism, whereby the image of a more or less arbitrary collection of positions was countered. The main person who accomplished this was Fraenkel.

**Constructivity** Regarding the constructivity label, the developments for quite some time were about form rather than contents. The label was claimed in 1921 by both Hilbert and Bernays for the proof theoretic point of view. However, when the debate widened, it soon became clear that almost all participants saw intuitionism as the constructive current. Hilbert and Bernays did not repeat their claim. Thus, most contributions mentioned intuitionism as 'constructive' or as demanding 'constructions' for mathematical existence.

It took until 1928 before criticism arose regarding the exact meaning of the word 'constructive'. The change is very noticeable in Menger. In 1925, he had become convinced that the intuitionistic criticism of pure existence statements was correct, and that one should work with constructive existence statements. By 1928, when Fraenkel still spoke about the 'sharp distinction' between pure existence statements and constructions, Menger had become doubtful. He remarked that 'constructivity' could be interpreted in different ways, and that no specifically intuitionistic interpretation had been given yet. In 1930, Menger's position had radicalised further, and he reproached intuitionists of dogmatically holding on to one amongst several possible interpretations of constructivity. Menger's criticism found many echos, amongst others with Gödel.

Nevertheless, 'constructivity' was still used as a label. In the 1930 Königsberg conference on the foundations of mathematics, both Von Neumann, speaking on formalism, and Carnap, speaking on logicism, claimed to be working 'constructively'. Only from 1935 on did constructivity get a more specific meaning, by Turing's interpretation in terms of computability.

## 4.5. CONCLUSION

Thus, regarding the constructivity label, the following can be concluded. Constructivity was a popular concept among mathematicians. It was first claimed by both intuitionists and Hilbert and Bernays, but the former soon won. Apparently, its image as a label was quite strong, for it took until 1928 before the first criticism about the contents of the constructivity concept arose, by Menger. The question why the constructivity label was so popular remains to be answered.[623]

---

[623] See 6.5.1.

# Chapter 5

# Reactions: logic and the excluded middle

> Bei einem Vortrag eines der Führer der Intuitionisten wurde diesem entgegengehalten: Ja wenn es auch die Mathematiker heute noch nicht wissen, so wird es doch z.B. der liebe Gott wissen, wir können also doch annehmen, daß es entweder das eine oder das andere ist. Darauf erwiderte der Intuitionist: Dann müssen Sie den lieben Gott sehr genau kennen, wenn sie wissen, daß er es weiß.[1]
>
> Walter Lietzmann[2]

## 5.1 Introduction

In mathematics one uses various kinds of argumentations. Some are well-known logical rules, such as the syllogism: all men are mortal, Socrates is a man, therefore Socrates is mortal. Others are not formalised in strict rules and rely on 'good understanding'. Mathematicians usually agree on what they consider legitimate arguments in a mathematical proof. But is this sufficient? Does this mean that the argumentations mathematicians have employed for centuries are correct?

Some mathematicians might consider such questions too far-fetched. However, posing these kinds of questions in the past led to important developments in mathematical logic. Furthermore, anyone who regards mathematics as more than

---

[1] 'At a lecture of one of the leading intuitionists, the following was held against him: Even if mathematicians do not know it today, then still for example God will know, so we can assume that it is either the one or the other. Thereupon the intuitionist answered: You must know God very well, if you know that he knows so.'

[2] [Lietzmann 1925, p. 357]. A book Lietzmann wrote later reveals that the anecdote is about Brouwer, [Lietzmann 1949, p. 165].

an arbitrary game should have an answer to the question what makes the inference rules applied in mathematics legitimate.

In the history of mathematics, such questions do not seem to have played an important role. There always was a well-established paradigm, or if it changed, most people agreed that it had to. Not so with Brouwer. Brouwer challenged argumentations that had been employed from the very beginning of mathematics: those of classical logic.

### 5.1.1 A short history of classical logic

**Antiquity** Although Plato was presumably one of the first persons to explicitly formulate some logical principles, classical logic is usually given its full start with the (posthumous) publication of Aristotle's *Organon* in 322 B.C.[3] This work contains a number of Aristotle's treatises, including *De Interpretatione* and *Prior Analytics*. It determined to an important extent what was later to be called logic. In another important work, *Metaphysica*, Aristotle formulated and accepted the principles later known as the law of contradiction and the law of the excluded middle, the latter also designated by its Latin name *tertium non datur*. In modern formulation, the former says that a proposition cannot be true and false at the same time. The second states that for every proposition either the proposition itself or its negation holds; there is no third possibility. Kneale and Kneale argued that Aristotle already questioned the validity of applying the principle of the excluded middle to future events, although he did accept it in the end.[4] Thus, Aristotle may have foreshadowed Brouwer's criticism by almost 25 centuries.

Characteristic of Aristotle's work is his grouping of statements in opposing pairs, one element of the pair being contradictory to the other in the sense that one must be true while the other is false. Thus, Aristotle classified general statements in the following way:

|            | affirmative          | negative              |
|------------|----------------------|-----------------------|
| universal  | Every man is white   | No man is white       |
| particular | Some man is white    | Some man is not white |

The statements in the opposing corners are opposed to each other as contradictories. The distinction is still useful; the question whether certain statements are opposed as contradictories or not played an important role in the foundational crisis.

Finally, an important innovation Aristotle made in his *Prior Analytics* was the introduction of letters as variables. Thus, he could make truly general statements, instead of having to rely on his readers' abilities to abstract general rules from the examples he gave.

---

[3]The account given here was based on [Kneale & Kneale 1964], unless stated otherwise. I highlight those aspects of the history of classical logic which are of importance to the foundational debate.

[4][Kneale & Kneale 1964, pp. 46–54]. They refer to a passage in *De Interpretatione* (the original name is ‾‾‾ ‘‾‾‾‾‾‾‾, but since the Renaissance it has been known by its Latin name).

## 5.1. INTRODUCTION

It is worth mentioning that, on a more philosophical level, Aristotle (like Brouwer) maintains that one should distinguish between words and the mental experiences they represent. Predicates like 'true' and 'false', Aristotle implies, belong primarily to the latter.[5] Thus, Scholz had a point when he later asserted that, in a way, Aristotle was behind Brouwer and Weyl rather than behind Hilbert.[6]

Besides Aristotelian logic, the main current in Antiquity was Stoic logic. This logic, founded by Zeno, put more stress on argumentations used in everyday conversations.

**Middle Ages** Both Aristotelian and Stoic logic was taken up in Boethius' work, which became one of the most important means for the transmission of logic from Antiquity until after the Middle Ages. Boethius, who was active around 600 AD, published Latin translations of Greek works on logic, besides commentaries and treatises of his own.

In the 12th century, when the first universities were founded, Abelard was one of the most influential persons in the field of logic. He based his works on Boethius and Aristotle and stressed the link between logic and oratorical skills.

**Modern times** By the end of the 17th century, Leibniz put forward the idea of logic as a calculus, after the example of algebra. In this way, logic would come to serve as a basis for several forms of mathematics. Even though Leibniz only worked out the idea partially and it took some 200 years for his logical work to become really influential, his idea was to prove to be decisive in the transformation of logic.

Halfway the 19th century, Boole succeeded in realising Leibniz' idea. He explicitly used algebraic formulas and operations in order to express logical relations.[7] The important thing, he stressed, was not the interpretation of the symbols, but the 'laws of thought' to which the symbols obeyed. His book 'An investigation of the laws of thought, on which are founded the mathematical theories of logic and probabilities' opens with the following words:[8]

> The design of the following treatise is to investigate the fundamental laws of those operations of the mind by which reasoning is performed (...) and (...) to collect from the various elements of truth brought to view in the course of these inquiries some probable intimations concerning the nature and constitution of the human mind.

Thus, Boole sees a strong relationship between logic and the reasoning of the human mind.

After Boole, Peirce and Schröder made important contributions to the development of logic. Peirce introduced a theory of logical relationships in 1882,

---
[5] [Kneale & Kneale 1964, p. 45]
[6] [Scholz, H. 1930, p. 42]
[7] [Boole 1854, p. 27]
[8] [Boole 1854, p. 1]

220     CHAPTER 5. REACTIONS: LOGIC AND THE EXCLUDED MIDDLE

which could handle operations such as negation, sum and product. Furthermore, three years later he formulated the universal and existential quantifiers. Schröder took this up to develop classical logical algebra, permitting the use of algebraic derivation rules in logic.[9]

The last big step in logic before Brouwer was made by Frege. In 1879, he published his *Begriffschrift*, in which he used a completely symbolic language to express arithmetic relations. In this way, Frege aimed at freeing logic from everyday associations. Frege even reduced the natural numbers to (contentual) logic. He defined the number 0 as the number belonging to the concept 'not identical with itself' (we would now say: the empty set), the number 1 as the number belonging to the concept 'identical with 0', the number 2 as the number belonging to the concept 'identical with 0 or with 1', etcetera. Frege's work, however, was not widely read at first. In the second part of his *Grundgezsetze der Arithmetik*, published in 1903, in which he further elaborated his ideas, he had to admit that Russell had found a contradiction in his system.

Thus, two important developments in the evolution of logic may be noticed. One is its move away from oratory rhetoric and towards mathematics, by the use of algebraic operations. The other is its formalisation. In this way, logic became more and more important for mathematics, up to the point that people like Frege and Russell put forward logic as a foundation for mathematics. This was a tendency Brouwer protested against. Furthermore, it seems that, since Aristotle, nobody had doubted the validity of the logical principles. Whereas Euclidean geometry had lost its absolute status in the first half of the 19th century due to the emergence of the non-Euclidean variants, similar developments did not take place in the field of logic. Presumably, the main difference was that logic was generally seen as part of philosophy, whereas geometry was part of mathematics. Boole's characterisation of logic as the laws of thought continued to play an important role in the foundational crisis and marks its prominent status.

It is against this background that one should see Brouwer's criticism of the unrestricted use of the principle of the excluded middle.

## 5.2   The beginning of the debate

In the beginning of the debate on the principle of the excluded middle, the same protagonists figure as in that on mathematical existence: Weyl, Hilbert, Bernays, and Fraenkel. The main absentee was Becker, which indicates that, regarding this subject, the discussion entered philosophical circles later. Furthermore, the subject was taken up early in university addresses, as in those by Wolff, Finsler and Baldus.

**Pre-Weyl reactions**   The only publications I found treating Brouwer's objections to the principle of the excluded middle before the publication of Weyl's 1921

---
[9][Beth 1944, pp. 65–67]

## 5.2. THE BEGINNING OF THE DEBATE

paper were two contributions by **De Haan**: his inaugural lecture *Wezen en taak der rechtskundige significa* ('Essence and task of legal significs') and the book *Rechtskundige significa* ('Legal significs').[10] Jacob Israël de Haan (1881–1924) was a poet, lawyer and Zionist who shocked the Dutch society with the publication of his homo-erotic *Pijpelijntjes* in 1904. Like Brouwer, he was a member of the signific movement in the Netherlands,[11] which he had got to know via Van Eeden. De Haan had already touched upon Brouwer's dissertation in his own dissertation. At the defence of De Haan's dissertation in February 1916, where Van Eeden acted as a *paranimf*,[12] Brouwer was one of the persons who opposed.[13] Brouwer published two critiques of De Haan's dissertation, which caused De Haan to study Brouwer's work more closely.[14] The first results thereof can be seen in his inaugural lecture, delivered in October of the same year. However, of the two publications mentioned above, the book is the one in which De Haan treats Brouwer's objections to the principle of the excluded middle in more detail. De Haan found a both special and sensible application of Brouwer's criticism of the status of logical principles, namely in law. Following an argument used by Brouwer to play down the value of logical arguments, De Haan maintains:[15]

> Eveneens overal waar regelmatigheid in de taal, die recht begeleidt, wordt uitgebreid over een taal van juridische woorden, die geen recht begeleidt. Op die manier wordt slechts een taalgebouw verkregen, dat van het recht onherroepelijk gescheiden blijft. De rechts-intuïtie moet het symbolengebouw van het recht op ieder punt van opbouw verifieeren.[16]

Apart from De Haan's monograph, there were two private letters to Brouwer which touched upon the issue before 1921. In 1908, the Dutch physicist **Kohnstamm**, a student of Van der Waals, wrote to Brouwer about his paper *De onbetrouwbaarheid der logische principes*.[17] Kohnstamm argues that he does not consider the fact that there may be unsolvable mathematical problems to be a violation of the principle of the excluded middle. He compares Brouwer's argumentation to the question 'whether a square circle is supposed to be round or angular', that is, he thinks that Brouwer addressed the problem in a way which

---

[10] [De Haan 1916B], [De Haan 1919]. In Vollenhoven's dissertation, both the *tertium non datur* and Brouwer are treated, but not in conjunction, [Vollenhoven 1918].

[11] See 2.6.

[12] A traditional formal helper to the *promovendus*.

[13] [Van Dalen 1999A, pp. 255–256]

[14] [Schmitz 1990, pp. 144–145]. Schmitz has more information on the impact of Brouwer's ideas on De Haan's work, cf. [Schmitz 1990, pp. 144–183].

[15] [De Haan 1919, p. 77]; on Brouwer's argumentation, see 2.3.2.

[16] 'The same applies everywhere where regularities in the language, which accompanies law, is extended to a language of juridical words, which do not accompany law. In such a way only a linguistic building is obtained, which remains irrevocably separated from law. The juridical intuition has to verify the symbolic building of law at every point of its construction.'

[17] See 2.3.2.

## 222  CHAPTER 5. REACTIONS: LOGIC AND THE EXCLUDED MIDDLE

was not fruitful.[18] The other letter came from the philosopher Hugo **Dingler**. Dingler reacts in a pragmatic way, claiming about mathematics without use of the principle of the excluded middle:[19]

> Sicher ist (...) so etwas möglich, ebenso, wie nichteuklidische Geometrieen aufgebaut werden können.[20]

### 5.2.1 Weyl's *Grundlagenkrise*

In 1921, Hermann Weyl published his paper *Über die neue Grundlagenkrise der Mathematik* ('On the new foundational crisis in mathematics').[21] The principle of the excluded middle figures in the context of Weyl's presentation of the intuitionistic continuum and is discussed at length. Weyl describes his disbelief in the *tertium non datur* and the introduction of choice sequences as the two main items he took over from Brouwer.[22]

Already before Weyl was converted to intuitionism, he had recognized the limited practical value of logic. In his well-known work on relativity theory *Raum · Zeit · Materie* ('Space · Time · Matter'), published in 1918, he wrote:[23]

> Was die formale Logik lehrt, gründet gewiß im Wesen der Wahrheit, und keine Wahrheit verletzt ihre Gesetze. Ob aber eine konkrete Behauptung wahr ist oder nicht, darüber lehrt sie schlechterdings nichts, das Inhaltliche der Wahrheit läßt sie gänzlich dahingestellt; der Grund der Wahrheit eines Urteils liegt in der beurteilten Sache und nicht in der Logik.[24]

In his 1921 paper, Weyl went further and followed Brouwer in attacking the principle of the excluded middle. The way he did so was, however, different from Brouwer's. Weyl presented his own thoughts and doubts at length in the article, and I follow him in the presentation below.[25]

Weyl's argumentation starts with the subject of mathematical sequences. He mentions that in his earlier work, i.e., *Das Kontinuum*, he had restricted sequences to ones that he called *umfangsdefinit* ('extent definite').[26] A concept is called

---

[18]'of een vierkante cirkel geacht moet rond te zijn dan wel hoekig', letter from Kohnstamm to Brouwer, 3/1/1908, [MI Brouwer, CB.AKO.1]

[19]Letter from Dingler to Brouwer, 26/7/1920; [MI Brouwer]

[20]'For sure (...) such a thing is possible, just like non-Euclidean geometries can be constructed.'

[21]The background of Weyl's conversion to intuitionism, as well as the contents of his paper as far as mathematical existence is concerned, is treated in 4.2.1. Weyl's use of metaphors is discussed in 6.2.

[22][Weyl 1921, p. 226]

[23][Weyl 1918B, p. 227]

[24]'What formal logic teaches is certainly rooted in the essence of truth, and no truth violates its laws. But it teaches absolutely nothing on whether a concrete assertion is true or not; it leaves the contents of truth undecided. The ground for the truth of a proposition lies in the matter judged and not in logic.'

[25]Weyl's presentation is in [Weyl 1921, pp. 222–225].

[26][Weyl 1921, p. 213]

## 5.2. THE BEGINNING OF THE DEBATE

*umfangsdefinit* if the objects falling under it can be seen as a whole which is closed and determined by itself. Weyl had used the concept in order to avoid impredicative definitions. He elaborates it by demanding that only certain logical construction principles may be used.[27] In such cases, Weyl maintains, the answer to every question is determined and can only be yes or no, which together make up a full disjunction. Weyl does not describe what he means by a full disjunction, but the natural interpretation would be to read it as saying that the union of the two parts of the disjunction make up the whole set of possibilities. Weyl next changes the circumstances about which he reasons. Still restricting himself, as in classical mathematics, to lawlike sequences, he drops the restriction that they have to be *umfangsdefinit*. He then asks whether there is a sequence with a certain property $P$. In other words, since he only considers lawlike sequences, he asks whether there is a law with the property $P$.

In Weyl's view, the answer 'yes' can now only be given if one *has succeeded* in the construction of a law that fulfils $P$. The mere possibility of a construction does not suffice. The answer 'no' has become devoid of content. Weyl does not specify why this is the case, but his argument seems to be that we no longer possess a well-determined set of construction principles, so that we cannot prove the impossibility of the required construction.

Weyl next proposes to interpret the negative answer positively: instead of 'there is no sequence with the property $P$' we read 'all sequences have the property $\neg P$'. In doing so, 'all sequences' should be interpreted as 'all lawless sequences'.[28] In this way, the existence statement comes to designate 'being and law', whereas the universal statement is associated with 'becoming and freedom'.[29] The reason why Weyl switches from lawlike to lawless sequences is that, whereas he thought of 'all sequences' as including lawless sequences, he considered that only a lawlike sequence could be given individually. This, in turn, is justified by Weyl's requirement that individual mathematical objects be codable in natural numbers, which is not the case for lawless sequences.[30] In this respect, Weyl differs from Brouwer, who thought of a choice sequence as an individual object, created by the mathematician.[31] The result is that, regarding sequences, the existence statement and the universal statement of the negated property do not constitute a full disjunction anymore, since they range over different kinds of objects. This, Weyl explains, is

---

[27] The construction principles are specified in 4.2.1.

[28] It seems, as Van Atten pointed out, that Weyl only distinguishes between lawlike and lawless sequences, referring to the latter as 'choice sequences' but leaving out proper Brouwerian choice sequences in which restrictions can be placed on the choices to be made: *'So können wir mit Bezug auf eine Wahlfolge wohl fragen, ob in ihr an vierter Stelle die Zahl 1 auftritt, aber nicht, ob in ihr die Zahl 1 überhaupt nicht auftritt'* ('Thus, regarding a choice sequence we can ask whether the number 1 appears at the fourth position, but not if the number 1 does not appear alltogether'), [Weyl 1921, p. 220]. However, Weyl is not consistent, for he does allow the sum of two choice sequences as a new choice sequence, which is neither lawless nor lawlike, [Weyl 1921, p. 221], [Van Atten 1999, pp. 37–41].

[29] 'dem Sein und dem Gesetz', 'Werden und die Freiheit', [Weyl 1921, p. 223]

[30] [Weyl 1921, p. 228], [Van Atten 1999, p. 38]

[31] [Van Dalen 1995, p. 152]

where Brouwer says the principle of the excluded middle fails. Weyl himself prefers to conclude that the propositions 'there is a sequence with the property $P$' and 'all sequences have the property $\neg P$' are not the negation of one another.

But this is not the whole story. Weyl remarks that Brouwer goes further and even denies the validity of the principle of the excluded middle for statements on natural numbers, which are *umfangsdefinit*. Given a question which is to be answered by yes or no, Brouwer, Weyl explains, is of the opinion that there should be a general method for deciding which of the two alternatives is the right answer. Weyl had always stuck to the view that it is not our ability to decide that matters, but how the situation *an sich* is.

Weyl proceeds with an objection to his own thought. Let $P$ be a decidable property of the natural numbers, i.e., for every natural number $n$ it can be decided whether $P(n)$ holds or not. The idea, Weyl explains, that it would be determined *an sich* if there is a number $n$ fulfilling $P$ rests on the following procedure. We can go through the natural numbers $1, 2, 3, \ldots$ and each time check whether $P(n)$ holds or not. Thus, we can answer the question positively if we have found such a number, and negatively if we have not found it. But giving the negative answer requires going through an infinite sequence, a concept which is meaningless. Thus, the negative answer could only be produced by proving that it lies in the essence of the concept 'natural number' that $P$ does not apply. But then we have two propositions which are not opposed as contradictories, and thus the *tertium non datur* does not hold.

However, Weyl is still thrown back into his old belief, maintaining that in going through the sequence of natural numbers either the process will stop or it will not stop, 'definitely and without a third possibility'.[32] Finally he finds the solution: the existence statement 'there is a natural number $n$ such that $P(n)$' is not a proper proposition at all, but a proposition abstract, and thus cannot be negated.[33] Therefore, his former idea that it has to be so or not, even if one may be unable to decide which of the alternatives is the right one, does not apply.

Thus, Weyl's argumentation differs markedly from Brouwer's. The process Weyl goes through includes Brouwer's argument about decidability being crucial,[34] which Weyl rejects. This marks the philosophical difference between Brouwer, who was a metaphysical idealist, and Weyl, who was not. Weyl's own solution, as Majer rightly pointed out, is to exclude statements which are not proper propositions from the scope of the principle of the excluded middle, namely existence and universal statements.[35] Presumably, Weyl only accepted $\forall x \varphi(x) \to \varphi(a)$ (a universal statement as '*Anweisung auf Urteile*', 'rules for propositions') and $\varphi(a) \to \exists x \varphi(x)$ (an existence statement as '*Urteilsabstrakt*', 'proposition abstract') as axioms of quantification theory.[36]

---

[32] 'ohne Wandel und Wank und ohne eine dritte Möglichkeit', [Weyl 1921, p. 224]
[33] This point is worked out in 4.2.1.
[34] As in the Brouwerian counter-examples; see 2.6.2.
[35] [Majer 1988, p. 548]
[36] [Majer 1989, pp. 245–246]

## 5.2.2 Hilbert's first reactions

David Hilbert[37] had been working on logic before the foundational crisis erupted. In his 1904 Heidelberg lecture, he had pleaded to develop the laws of logic and those of arithmetic to some extent simultaneously.[38] In the same lecture, he for the first time used formal language to investigate logic. The logic he employed was part of what we now call first-order logic, with, however, infinitary formulas and quantifiers restricted to a fixed domain.[39] In his Göttingen lectures of 1905 on the logical principles of mathematical thougt, Hilbert developed propositional logic algebraically.[40]

From 1905 to 1917, Hilbert's Göttingen lectures on the foundations of mathematics treated 'elementary mathematics from a higher standpoint',[41] leaving logic to colleagues such as Behmann, Bernstein and Grelling.[42] From 1917 on, however, with the return of Bernays to Göttingen, logic again became a more prominent subject in Hilbert's foundational lectures.[43]

Hilbert's 1917 Zürich lecture *Axiomatisches Denken* ('Axiomatic thinking') marks his (published) return to foundational issues. Explicitly following Russell, Hilbert mentions the axiomatisation of logic as the crowning of the axiomatic method.[44] A lecture course given in Göttingen in 1917–1918 on the principles of mathematics witnesses Hilbert's more mature and more elaborate idea of logic. There, he applied the axiomatic method to the 'functional calculus' (first-order logic) and proved its consistency. A division between syntax and semantics appears, although not a complete one.[45] Also, Hilbert for the first time called for a proof of the completeness of the axiom system (in the sense of: if we add to the system a formula which is not derivable within the system, the system becomes inconsistent), rather than simply postulating it.[46] He argued for a more extended logic, namely a logic at least as strong as second-order logic, in order to investigate the foundations of mathematics themselves.[47] Thus, to Hilbert the importance of a strong logic for the foundations of mathematics was beyond doubt.[48]

Hilbert also pointed out the limitations of logic. In lectures delivered in 1919, he renounced the view that mathematics was an arbitrary accumulation of con-

---

[37] Biographical information on Hilbert is given in 4.2.2.
[38] [Hilbert 1905A, p. 266]
[39] [Moore 1987, p. 112]
[40] [Zach 1999, p. 333]
[41] [Zach 1999, p. 358]
[42] [Mancosu 1999B, p. 304]; Mancosu also has some interesting passages from Behmann's early work which show a remarkable similarity to argumentations Hilbert later put forward.
[43] [Sieg 1999, pp. 8–12]
[44] [Hilbert 1918, p. 153]
[45] Zach argued the case that Bernays was the first to completely distinguish between syntax and semantics, in his *Habilitationsschrift* of 1918, [Zach 1999, p. 342].
[46] [Zach 1999, pp. 339–340]
[47] [Moore 1987, pp. 117–118]
[48] From 1917 to about 1928 Hilbert worked in a version of the so-called ramified theory of types, [Moore 1987, p. 121].

clusions, only driven by logic:[49]

> Von einer solchen Willkür ist aber tatsächlich keine Rede; vielmehr zeigt sich, daß die Begriffsbildungen in der Mathematik beständig durch Anschauung und Erfahrung geleitet werden (...).[50]

During the Summer semester of 1920, Hilbert lectured on *Probleme der mathematischen Logik* ('Problems of mathematical logic'). Extracts of these lectures have been published in Ewald's most informative source book in English translation. In the lectures, Hilbert argued against the 'dictatorial' tendencies of Kronecker and Poincaré, accusing them of throwing out the baby with the bathwater. He specifically mentioned items such as the *tertium non datur* and propositions involving the infinite as falling prey to their restrictve behaviour.[51] These lectures are interesting because they show that, even before the foundational controversy with Brouwer and Weyl had started, Hilbert was using argumentations which would re-appear in the foundational crisis.

**Hilbert's 1921 lectures** As was the case with the discussion on mathematical existence, Hilbert was the first person to react to Weyl's paper. He did so in his 1921 Copenhagen and Hamburg lectures, which were published the following year.[52] The question of the principle of the excluded middle does not figure prominently. At first, Hilbert only mentions it as one of the 'forbidden propositions' of Brouwer and Weyl.[53] Towards the end of the paper, however, he recognizes that there is a problem. Hilbert presents a set of axioms for arithmetic as an example of how his foundational theory should work. He remarks that one can prove the consistency of the axioms, thereby securing them and thus the theory represented by them. But the 'most fundamental'[54] step to be done is to prove that it is allowed to use the principle of the excluded middle also for an infinite number of objects. Only in this way, Hilbert maintains, one can bridge the gap to analysis and set theory and give a foundation for the theory of the real numbers. Hilbert claims that this succeeded by adding certain functions to the axiomatic system and proving that the enlarged system is consistent.[55]

---

[49][Hilbert 1992, p. 5]
[50]'But such an arbitrariness is indeed not the case; rather, it turns out that the making of concepts in mathematics is constantly guided by intuition and experience.'
[51][Ewald 1996, vol. II, pp. 943–946]
[52]The background of these lectures, as well as their contents regarding mathematical existence, is treated in 4.2.2.
[53]'verbotener Sätze', [Hilbert 1922, p. 160]
[54]'wesentlichste', [Hilbert 1922, p. 176]
[55][Hilbert 1922, p. 176]. In his report on Hilbert's Hamburg lectures, Reidemeister remarks (without further explanation) that Hilbert 'gave a positive turn' to the principle of the excluded middle, (*'positiv gewendet'*, [Reidemeister 1921A, p. 107]). Reidemeister repeated the remark in a newspaper article, where he labelled the step 'curious' (*merkwürdig*), [Reidemeister 1921B]. I do not know what positive turn Reidemeister had in mind, nor can I find a place in the paper where it could refer to. Probably it referred to something that was in one of the lectures but not in the published version.

## 5.2. THE BEGINNING OF THE DEBATE

**Bernays in 1921** In general, it is hard to say what role exactly Bernays played in the reception of intuitionism, since a lot of his work was done behind the scenes, as Hilbert's assistant. It may be pointed out again that Bernays made not a single direct public reference to intuitionism between 1922 and 1929.[56] Zach argued the case that Bernays' role in the development of propositional logic was in fact much greater than generally perceived, especially regarding concepts such as the division between syntax and semantics, and on questions of completeness and decidability,[57] all subjects which were closely related to the discussion on the intuitionistic view on logic. I will leave that as it may and focus mostly on Bernays' published contributions to the debate.

In his 1921 lecture at the *Mathematikertagung* in Jena on Hilbert's thoughts on the foundations of mathematics, Bernays also gives the principle of the excluded middle only a marginal place. But he shows himself willing to think from Brouwer's perspective by admitting that, from the intuitionistic point of view, the application of the principle of the excluded middle to infinite totalities is dubious, to say the least.[58]

**Hilbert's 1922 lecture** In a lecture given to the *Deutsche Naturforscher-Gesellschaft* ('German Society of Natural Scientists') in September 1922, Hilbert treats the question of the principle of the excluded middle in more detail. He starts by analysing universal and existence statements for finite totalities. In this case, he argues, the universal statement that all objects of a finite totality have a certain property is logically equivalent to a conjunction: this object has the property *and* this object *and* ... *and* that object has the property. Similarly, an existence statement is equivalent to a disjunction: this object has the property *or* this object *or* ... *or* this object has the property. On this basis, Hilbert continues, we conclude that the principle of the excluded middle holds for finite totalities: either all objects have a certain property, or there exists an object which does not have the property. At the same time, we obtain the rigorous validity of the equivalences (in modern notation)

$\neg \forall a A(a)$ is equivalent to $\exists a \neg A(a)$ and
$\neg \exists a A(a)$ is equivalent to $\forall a \neg A(a)$, where $A$ is a (unary) predicate.[59]

But Hilbert is not satisfied with only statements about finite totalities; he also wants to obtain 'such provable formulas (...), which are the representations of transfinite theorems of ordinary mathematics'.[60] Note that Hilbert is here dis-

---

[56] See 3.3.3.
[57] [Zach 1999, pp. 344–348]
[58] [Bernays 1922A, p. 14]
[59] [Hilbert 1923, pp. 181–182]
[60] 'solche beweisbaren Formeln (...), die die Abbilder transfiniter Sätze der gewöhnlichen Mathematik sind.', [Hilbert 1923, p. 181]

## 228 CHAPTER 5. REACTIONS: LOGIC AND THE EXCLUDED MIDDLE

cussing the level of formalised mathematics, not that of meta-mathematics. He then continues:[61]

> Diese Äquivalenzen werden aber gewöhnlich in der Mathematik auch bei unendlich vielen Individuen ohne weiteres als gültig vorausgesetzt; damit aber verlassen wir den Boden des Finiten und betreten das Gebiet der transfiniten Schlußweise. Wenn wir ein Verfahren, das im Finiten zulässig ist, ohne Bedenken stets auf unendliche Gesamtheiten anwenden würden, so öffneten wir damit Irrtümer Tor und Tür.
> (...) Bei unendlich vielen Dingen hat die negation des allgemeinen Urteils $(a)Aa$ [$\forall a A(a)$, DH] zunächst gar keinen präzisen Inhalt, ebensowenig wie die Negation des Existenzialurteils $(Ea)Aa$ [$\exists a A(a)$, DH]. Allerdings können gelegentlich diese Negationen einen Sinn erhalten, nämlich, wenn die Behauptung $(a)Aa$ durch ein Gegenbeispiel widerlegt wird oder wenn aus der Annahme $(a)Aa$ bzw. $(Ea)Aa$ ein Widerspruch abgeleitet wird. Diese Fälle sind aber nicht kontradiktorisch entgegengesetzt; denn wenn $A(a)$ nicht für alle $a$ gilt, wissen wir noch nicht, daß ein Gegenstand mit der Eigenschaft Nicht-$A$ wirklich vorliegt; ebensowenig dürfen wir ohne weiteres sagen: entweder gilt $(a)Aa$ bzw. $(Ea)Aa$ oder diese Behauptungen weisen einen Widerspruch wirklich auf.[62]

Hilbert's solution is to add transfinite axioms, which express the transfinite reasonings used in classical mathematics, and to prove that the resulting system is consistent.

Thus, once again Hilbert acknowledges that there is a problem with the principle of the excluded middle. I have quoted the paper at length for two reasons. In the first place, it makes it clear that Hilbert applies a contentual argumentation to the level of 'ordinary' mathematics, which means that he at the time did not simply put forward formalised mathematics as a formal system. In the second place, the quote shows that Hilbert's argumentation is some mixture of views which Brouwer and Weyl also held. As Brouwer had written in a paper in 1920,[63]

---

[61] [Hilbert 1923, p. 182]

[62] 'But in mathematics these equivalences are usually, without due consideration, assumed to be valid for infinitely many individuals as well. But in doing so we are leaving the ground of the finite and entering the domain of the transfinite modes of inference. If we were without further consideration always to apply to infinite totalities a method which is admissible in the finite, then we would open the floodgates of error.
(...) For an infinite number of things the negation of the universal proposition $(a)Aa$ [$\forall a A(a)$, DH] initially does not have a precise meaning at all, as little as the negation of the existential proposition $(Ea)Aa$ [$\exists a A(a)$, DH] does. To be sure, these negations can occasionally obtain a meaning, namely, if the statement $(a)Aa$ is refuted by a counterexample or if a contradiction is derived from the assumption $(a)Aa$ or $(Ea)Aa$ respectively. But these cases are not opposed as contradictories; for if $A(a)$ does not hold for all $a$, we do not know yet that an object with the property Not-$A$ really is available, as little as we can without due consideration say: either $(a)Aa$ or $(Ea)Aa$ holds, respectively, or these statements really present a contradiction.' English translation based on the translation in [Ewald 1996, vol. II, pp. 1139–1140].

[63] See 2.6.2.

## 5.2. THE BEGINNING OF THE DEBATE

Hilbert recognizes that one cannot simply use all reasonings that are valid in finite sets in infinite ones, too. This is not to say that Hilbert took the idea from Brouwer, since it can already be found in Hilbert's lecture notes for Easter 1896.[64] It does show that Hilbert and Brouwer agreed on this particular argumentation. As Weyl, furthermore, Hilbert maintains that in infinite sets the negation of a universal or existence statement does not have a precise meaning. It is interesting to note that, where Weyl had been thrown back to his old idea that, even in infinite sets, a proposition has to be either true or false, Hilbert seems to agree with Brouwer that this is not the case. Where Weyl had introduced proposition abstracts for existence and universal statements, Hilbert still seems to regard them as proper propositions.[65]

### 5.2.3 Addresses: Wolff, Finsler and Baldus

The principle of the excluded middle soon spread as a theme. It figured in inaugural lectures in the Netherlands and Germany as early as 1922 and 1923. Already in 1922, its fame had spread as far as the Ukraine, where professor **Kagan**[66] tried to decipher Brouwer's Dutch papers in order to understand his thoughts on the excluded middle.[67] Brouwer had sent Kagan some of his papers at the request of Ehrenfest's wife.[68]

**Wolff** The first to use the principle of the excluded middle in an inaugural lecture was the Dutch mathematician Julius Wolff (1882–1945), who spoke on it when he became professor of integral and differential calculus at the *Rijks-universiteit Utrecht* in 1922. Wolff had studied mathematics in Amsterdam and finished his dissertation under Korteweg in 1907. In this way, he must have known Brouwer personally. After his studies, Wolff had worked as a teacher and, from 1917 on, as a professor at the university of Groningen.[69]

In his inaugural lecture, Wolff wants to show that, even in mathematics, subjective influences play a role.[70] He chose the title accordingly: 'On the subjective in mathematics' (*'Over het subjectieve in de wiskunde'*). Wolff presents the fight between what he calls axiomaticians, led by Hilbert, and syntheticians, led by Brouwer. One of the arguments the latter use is that it is meaningless to pronounce the *tertium non datur* in cases where it is not sure that one can ever decide

---

[64] [NSUB Hilbert, 597]
[65] I translated Hilbert's *'Urteil'* as 'proposition', since I think he took the term from Weyl. For Weyl's use of the term, see 4.2.1.
[66] Presumable, this was Veniamin Fedorovich Kagan (1869-1953), who had studied mathematics in Odessa and Kiev. He specialised in non-Euclidean geometry and had worked at Odessa university from 1897 on; [Lopshitz & Rashevskii 1969].
[67] Letter from Kagan to Brouwer, 25/6/1922; [MI Brouwer]
[68] [Van Dalen 2001, p. 242]
[69] [Poggendorff 1936–1940, vol. VI, p. 2922]
[70] [Wolff 1922, p. 3]

between the two alternatives.[71] Wolff does not explicitly state his own view, but he seems to have sympathy for intuitionism.

**Finsler** The following year, the German mathematician Paul Finsler (1894–1970) touched upon the same subject in his inaugural lecture at the *Universität Köln*. Finsler had written his dissertation in Göttingen under Carathéodory. It was finished in 1918 and treated the foundations of what became known as the theory of Finsler manifolds. Besides the foundations of mathematics, Finsler's interests lay in algebraic geometry and in astronomy.[72]

Finsler's inaugural lecture is devoted to the question whether there are contradictions in mathematics. He mentions Brouwer and Weyl as the two mathematicians who reject the principle of the excluded middle, because there could be a third possibility. For example, two numbers need not be identical or different, they may also be indistinguishable. Finsler quickly rejects the intuitionistic point of view, saying that such assumptions may lead to interesting research, but one cannot base a rigorous science on them.[73]

**Baldus** On December 1, 1923, Richard Baldus (1885–1945) delivered his rector's address at the *Technische Hochschule Karlsruhe*. It was published the next year as *Formalismus und Intuitionismus in der Mathematik* ('Formalism and intuitionism in mathematics') in the series *Wissen und Wirken*,[74] and it was one of the more influential publications in the foundational debate. Baldus lectured in Karlsruhe as a full professor of geometry.[75]

Baldus presents the principle of the excluded middle as one of the subjects on which the controversy between intuitionism and formalism focuses. Whereas the formalist only recognizes the either – or, the intuitionist, Baldus maintains, takes into account the case of undecidability, too, next to that of 'as yet undecided'. Baldus states that the question of the *tertium non datur* plays an important role in the publications for or against intuitionism, a statement that is even more applicable to the period after Baldus' address appeared. Baldus finishes his address with the remark that there is no logical way to decide between intuitionism and formalism; it is a matter of feeling, of inner conviction.[76]

Thus, both Wolff and Finsler follow Brouwer in their explanation of the intuitionistic criticism of the principle of the excluded middle, whereas Baldus merely mentions the intuitionistic rejection.

---

[71] [Wolff 1922, p. 16]
[72] [Gottwald, Ilgauds & Schlote 1990, p. 150]
[73] [Finsler 1926A, pp. 147–148]
[74] See also 4.2.5.
[75] [Poggendorff 1936–1940, vol. VI, p. 116]
[76] [Baldus 1924, p. 28; 35]

## 5.2.4 Fraenkel's early commentaries

In 1923, Fraenkel[77] entered the foundational debate with some contributions in which he treated the principle of the excluded middle. In the second edition of his well-read *Einleitung in der Mengenlehre* ('Introduction to set theory'), Fraenkel follows Weyl's explanation of the intuitionistic rejection of the principle of the excluded middle. Fraenkel's general conclusion regarding the intuitionistic attacks on classical mathematics is that he does not think that the far-reaching intuitionistic amputations are necessary, but he does agree that mathematics is in need of better foundations.[78]

In September of the same year, when he had finished writing the *Einleitung*, Fraenkel delivered a lecture before the *Deutsche Mathematikertagung* in Marburg. Here, he was more pronounced on the question of the *tertium non datur*. He puts forward the principle of the excluded middle as the main point dividing, as he calls them, classical and intuitionistic mathematicians. Fraenkel rightly points out that, where Hilbert aims at proving that the application of the principle of the excluded middle is without danger, this is not what the intuitionists contest. Their question is whether the principle is justified.[79]

## 5.3 The debate widened

From 1924 onwards, the debate (both in its general form and specifically regarding the principle of the excluded middle) extended beyond the initial group of the directly involved Brouwer, Weyl, Hilbert, and Bernays, the commentator Fraenkel and the relative outsiders Wolff, Finsler and Baldus. Not only the number of people involved increased, but also the languages used, a fact that considerably widened the group of people that could become involved in the discussion. For the first time after the publication of Weyl's paper, the debate was brought to the English and French reading public, by Dresden and Wavre respectively.[80] By 1924, the principle of the excluded middle had been recognised as one of the central themes of the debate: almost all contributions of that year treated the subject.[81]

The attention paid to the subject of the principle of the excluded middle varied substantially per paper. In some of the contributions, the subject is only mentioned or explained briefly. For example, **Von Neumann** in his paper on an axiomatisation of set theory simply mentions Weyl's and Brouwer's criticism of the *tertium non datur*.[82] Similarly, **Grelling** discusses Weyl's and Brouwer's criticism briefly, where he presents the third possibility as the case where both a proposition and its

---

[77]Biographical information on Fraenkel is given in 4.2.4.
[78][Fraenkel 1923A, pp. 166–168; 173]
[79][Fraenkel 1924A, pp. 98–99]
[80][Dresden 1924] and [Wavre 1924]
[81]The exception is Hölder's *Die mathematische Methode*, [Hölder 1924].
[82][Von Neumann 1924, p. 220]

negation are meaningless.[83] In his lectures before the *Kant-Gesellschaft* in Halle and Magdeburg in January 1924, **Doetsch** mentions Brouwer's and Weyl's criticism of the principle of the excluded middle as a consequence of their constructivistic method.[84] And **Mannoury**, in his 'signific-communist' *Mathesis en mystiek* ('Mathematics and mysticism'), gives a more popular presentation of the doubts concerning the principle of the excluded middle, loosely based on Brouwer.[85]

The other contributions from 1924 are more substantial, and I deal with them in more detail.

**Weyl in 1924** In 1924, Weyl published a paper in the *Mathematische Zeitschrift*, in which he reacted to Hilbert's reactions. He puts the principle of the excluded middle in the centre of the discussion:[86]

> In seiner ersten Mitteilung zur 'Neubegründung der Mathematik' hat sich Hilbert in heftiger Polemik gegen die von Brouwer und mir vertretene Auffassung gewendet. Mir scheint, selbst von seinem Standpunt mit geringem Recht; denn soviel ich sehe, stimmen wir in dem entscheidensten Punkte miteinander überein. Auch für Hilbert reicht die Kraft des inhaltlichen Denkens nicht weiter als für Brouwer; es ist für ihn ganz selbstverständlich, daß sie die 'transfiniten' Schlußweisen der Mathematik nicht trägt (...). Er wird nicht leugnen wollen, daß Brouwer hier im 'Axiom des ausgeschlossenen Dritten' den wesentlichen Punkt getroffen hat.[87]

Weyl rightly concludes that, in fact, there is no difference between Brouwer and Hilbert regarding the reach of meaningful reasoning. Earlier, Von Neumann had drawn the same conclusion.[88]

Towards the end of the paper, however, Weyl significantly modifies his earlier position. Now, he maintains that Brouwer and Hilbert together demarcate a new period in modern foundational research, and that one should not only do mathematics in Brouwer's way, but also in Hilbert's symbolic way. Thus, he drops the exclusive claim for intuitionism which he had defended before. The reason he

---

[83] [Grelling 1924, p. 47]
[84] [Doetsch 1924, p. 449]
[85] [Mannoury 1924, pp. 16–17; p. 31; pp. 39–40]
[86] [Weyl 1924, p. 146]
[87] 'In his first communication on the 'new foundation of mathematics', Hilbert turned in heated polemics against the stand taken by Brouwer and me. I think with little justification, even from his own point of view; for as far as I see we agree on the decisive points. For Hilbert, too, the power of meaningful thinking does not extend further than for Brouwer; it is totally self-evident to him, that it does not support the 'transfinite' reasonings in mathematics (...). He will not want to deny that Brouwer has hit the essential point here in the 'axiom of the excluded middle'.'
[88] In a letter to Fraenkel, Von Neumann wrote in 1923: *'um die Mengenlehre zu reconstruiren, muß man sich, wie Hilbert, rückhaltlos auf den Boden des intransigentesten Brouwerschen Intuitionismus stellen. (Solange man 'inhaltlich schließt'.)'* ('to reconstruct set theory, one has to take the most intransigent Brouwerian intuitionistic position, as Hilbert does. (As long as one 'derives contentually'.)' Letter from Von Neumann to Fraenkel, 26/10/1923, [BF Fraenkel]

## 5.3. THE DEBATE WIDENED

gives is that there is a theoretical need in us to create a symbolical image of the transcendental.[89]

Thus, using Mehrtens' terms,[90] we can say that Weyl by 1924 had become a modernist mathematician, who accepted both intuitionistic (counter-modernist) and more formalistic (modernist) mathematics, but rejected classical mathematics.

**Fraenkel in 1924** In June 1924, Fraenkel delivered a lecture before the *Gesellschaft zur Beförderung der gesamten Naturwissenschaften* ('Society for the advancement of all natural sciences') in Marburg. It was published in two slightly different versions. In these texts, Fraenkel puts forward the criticism of the principle of the excluded middle as the newest and most far-reaching action of the intuitionists. His explanation differs from the one he had given before.[91] This time, he uses a counter-example based on the decimal expansion of $\pi$. Fraenkel asks whether the digit 7 appears seven consecutive times in the decimal expansion. He maintains that one could answer the question positively by indicating a place where such a sequence of digits appears, and negatively by proving that such a property is incompatible with the mathematical properties of the number $\pi$. But these possibilities, Fraenkel points out, do not make up a full disjunction. There is a third possibility, namely that one has proved neither the positive nor the negative answer. As long as this is the case, Fraenkel maintains, the intuitionists reject the use of a full disjunction to which use of the principle of the excluded middle leads.[92]

**Dresden in 1924** Also in the United States, some reaction came up. The Dutch-born Arnold Dresden, who had studied and written his dissertation at the university of Chicago,[93] was the first in the region to do so. I do not know whether his Dutch background played a role in his interest for intuitionism. Dresden had translated Brouwer's inaugural lecture into English,[94] so it is reasonable to assume that they had been in contact before. It is known that Dresden and Brouwer corresponded in 1922 and 1923 on the contents of Dresden's paper.[95] In a paper presented to the American Mathematical Society in December 1923 and published the following year, Dresden treated Brouwer's contributions to the foundations of mathematics.

Dresden presents Brouwer's rejection of the principle of the excluded middle as a consequence of his constructivistic view on mathematics. He explains that, for

---

[89] [Weyl 1924, pp. 147–150]
[90] See 6.7.
[91] See 5.2.4.
[92] [Fraenkel 1924B, pp. 123–124], [Fraenkel 1925A, p. 253]
[93] [Poggendorff 1936–1940, p. 600]
[94] The translation was published in the Bulletin of the American Mathematical Society, [Brouwer 1913].
[95] Dresden's letters have survived: letter from Dresden to Brouwer, 10/10/1922 and 17/9/1923; [MI Brouwer]. In the first of these letters, Dresden does not introduce himself, which indicates that there had indeed been some contact before.

234    CHAPTER 5. REACTIONS: LOGIC AND THE EXCLUDED MIDDLE

Brouwer, accepting the *tertium non datur* amounts to believing in the solvability of every mathematical problem, which is problematic in infinite systems.[96] Since the search for a mathematical construction, which is needed for establishing the truth of a proposition, cannot be carried out systematically, Dresden argues,[97]

> it is uncertain whether for an arbitrary proposition concerning an infinite system either the construction or the obstruction can be established, and hence it is equally uncertain whether the L[aw of the] E[xcluded] M[iddle] is valid in such a case.

Although Dresden does not state his own opinion explicitly, he seems to have sympathy for Brouwer's view. In any case he provided a comprehensible and adequate exposition of the intuitionistic point of view.

**Wavre in 1924**   In a paper published in 1924, the Genève professor Rolin Wavre presented a survey of the foundational debate as it had developed until then. Wavre's own field was differential and integral calculus, and most of his publications were on mathematics applied to physics or astronomy. However, he also wrote three papers on foundational questions in the *Revue de Métaphysique et de Morale*, of which this was the first. Wavre based his presentation on publications by Brouwer, Weyl, Hilbert and Bernays. The purpose of the paper was to[98]

> (...) faire entrevoir que l'opposition [between formalism and intuitionism, DH] devient tout à fait nette 'à propos de la notion d'existence et d'une application suspecte du principe du tiers exclu'.[99]

Note that Wavre presents the principle of the excluded middle as one of the two issues where the opposition between intuitionism and formalism becomes clear, the other one being the subject of mathematics existence. Thus, he correctly identified the main themes of the debate that was developing.[100]

In order to discuss the question of the *tertium non datur*, Wavre introduces, following Du Bois-Reymond's terminology, a fictional 'idealist' (formalist) and 'empiricist' (intuitionist).[101] Their discussion starts as follows:[102]

---

[96][Dresden 1924, p. 39]
[97][Dresden 1924, p. 40]
[98][Wavre 1924, p. 436]
[99]'(...) show that the opposition [between formalism and intuitionism, DH] becomes clear 'regarding the notion of existence and a dubious application of the principle of the excluded middle'.' The quotation marks indicate the subtitle of Wavre's paper.
[100]It must be taken into account that Wavre's characterisation may also have influenced the development itself and thus obtained somewhat the character of a self-fulfilling prophecy, since his paper was one of the better read ones.
[101]Wavre uses 'empiricist' and 'intuitionist' on the one hand, 'idealist' and 'formalist' on the other more or less as synonyms, [Wavre 1924, p. 435]. For the sometimes confusing terminology, see 4.1.1.
[102][Wavre 1924, pp. 443–444]

## 5.3. THE DEBATE WIDENED

L'Idéaliste. – Existe-t-il, oui ou non, un nombre $m_n$ de la forme $m_n = 2^{2^{n+8}} + 1$ qui soit décomposable;[103] on donne à $n$ les valeurs $1, 2, 3$, etc.
L'Empiriste. – Je ne puis répondre immédiatement à votre question avec mes connaissances actuelles; mais je tiens à vous dire que je trouve la question mal posé, voire absurde.
L'Idéaliste. – Vous me surprenez!
L'Empiriste. – Je distingue, en effet, deux parties dans la question; la première: Existe-t-il un nombre $m_n$ décomposable? c'est la question proprement dite; mais vous ajoutez l'alternative oui ou non, par laquelle vous préjugez de ma réponse. Je ne puis répondre par oui, il en existe un, qu'en vous présentant un tel nombre, supposons $m_{1000}$, qui soit décomposable; et ne puis répondre par non qu'en déduisant de la définition des nombres $m_n$ qu'ils sont tous premiers.
Mais je ne vois pas que le rejet d'une des parties de l'alternative me contraigne à affirmer l'autre; comme je ne puis exclure *a priori* tout *tertium*, je me refuse à me laisser réduire à votre alternative.[104]

After a discussion on what mathematical existence should mean,[105] they continue:[106]

L'Idéaliste. – Enfin vous ne me refuserez pas que, si la réponse à la question que je vous pose est donnée un jour, ce ne sera que par oui ou par non.
L'Empiriste. – Qu'en savez-vous? Ce n'est pas tout à fait évident, de cette évidence que nous avons coutume d'exiger en mathématique.[107]

---

[103]The numbers Wavre uses are the Fermat-numbers $F_n = 2^{2^n} + 1$ from $n = 8$ on. Fermat had conjectured these numbers to be prime on the basis of this being so for $n = 1, \ldots, 4$. Later, Euler proved that $F_5$ is composite. Until now, all the other Fermat numbers that have been checked turned out to be composite. There is, however, no proof that $F_1$ to $F_4$ are the only primes.

[104]'The Idealist. – Does there exist, yes or no, a factorable number $m_n$ of the form $m_n = 2^{2^{n+8}} + 1$; on assigns to $n$ the values $1, 2, 3$, etc. . The Empiricist. – I cannot answer your question immediately with my present knowledge; but I insist on telling you that I find the question badly posed, even absurd. The Idealist. – You surprise me! The Empiricist. – In fact, I distinguish between two parts of the question; the first: Does there exist a factorable number $m_n$? that is the question properly speaking. But you add the alternative yes or no, by which you prejudice my answer. I can only answer yes, there exists one, by exhibiting such a number to you, say $m_{1000}$, which is factorable; and I can only answer no by deducing from the definition of the numbers $m_n$ that they are all prime.
But I do not see that the rejection of one part of the alternative compels me to affirm the other; since I cannot exclude *a priori* every *tertium*, I refuse to be reduced to your alternative.'

[105]See 4.3.1.

[106][Wavre 1924, p. 445]

[107]'The Idealist: In short, you will not deny that, if the answer to the question that I put to you will be given one day, it will only be by yes or by no. The Empiricist. – What do you know about it? That is not evident at all, of the sort of evidence we are used to demanding in mathematics.'

Indeed, from the intuitionistic point of view, 'yes' or 'no' are not the only answers that can be given to the question, since 'not no' presents a further possibility.[108]

Thus, Wavre uses the fictional dialogue to explain the intuitionistic stand against the one normally held in mathematics. He even makes it clear that the intuitionist is the one who is more rigorous, by not making use of assumptions about future events of which we cannot be sure.

Wavre moves on to explain that, for an intuitionist, the principle of the excluded middle only holds for a finite number of elements, not for an infinite one. In the latter case, the intuitionist denies that the alternative should only be between the positive universal statement and its existential negation.[109]

In his conclusion, Wavre shows an early understanding of the status of the rejection of the principle of the excluded middle in intuitionistic mathematics:[110]

> M. Brouwer, en rejetant l'application du principe du tiers exclu, ne fait que tirer de la thèse empiriste [intuitionistic, DH] une conséquence peut-être paradoxale, mais inévitable, croyons-nous.[111]

Wavre claims that the intuitionists occupy the stronger position, since their demands in terms of rigour are stricter than those of the formalists. His conclusion, however, is quite paradoxical: he maintains that one should neither give up the formalistic language nor even its reasonings, but one should insist on intuitionistic verifications. It is hard to see how these two could be reconciled, since working with intuitionistic verifications in fact implies giving up formalistic reasonings. The very end of the paper shows Wavre's way out: there, he expresses the hope that one could replace formalistic proofs using the principle of the excluded middle by intuitionistic ones.[112] Thus, it seems that Wavre hoped that one could turn to intuitionistic proofs without having to give up any mathematical theorems – something which Brouwer and Weyl had explicitly announced as an inevitable consequence of the intuitionistic point of view.

### 5.3.1 The excluded middle in a central position

In 1925, the first substantial contribution to intuitionistic logic appeared, by Kolmogorov. However, the paper remained almost completely unnoticed and thus did not influence the course of the debate. In the discussion that went on, others

---

[108] In Heyting's formalisation: one can prove $p$, $\neg p$, or $\neg\neg p$, where the latter possibility may later be replaced by a proof of $p$; on Heyting's formalisation, see 5.4.1.
[109] [Wavre 1924, pp. 446–447]
[110] [Wavre 1924, p. 467]
[111] 'By rejecting the application of the principle of the excluded middle, Mr. Brouwer only draws, we think, a maybe paradoxical, but inevitable consequence of the empiristic [intuitionistic, DH] thesis.'
[112] [Wavre 1924, pp. 468–469]. Wavre does not indicate whether he thought that all such proofs could be replaced by intuitionistic ones, or that he merely hoped that one could do so for as many as possible.

## 5.3. THE DEBATE WIDENED

tried to formalise intuitionistic logic, unaware of Kolmogorov's paper. Furthermore, Hilbert changed his view on formalised mathematics in order to save the rules of classical logic.

**Kolmogorov in 1925** The first technical contribution to the debate on the principle of the excluded middle came from the Soviet Union. In 1925, the young mathematician Andrei Kolmogorov (1903–1987) published a paper under the title 'О принципе tertium non datur' ('On the principle of the excluded middle'). The paper presents the first published formalisation of a fragment of intuitionistic logic, and anticipated to a large extent later works by Heyting and Gödel. Kolmogorov was well aware of the symbolism used in Russell and Whitehead's *Principia Mathematica*.

When his paper was published, Kolmogorov was still studying mathematics at the university of Moscow, where Luzin was one of his main lecturers. Luzin was later characterised as a semi-intuitionist, and is said to have taken over many of Borel's points of view.[113] Luzin's semi-intuitionistic attitude may have fostered Kolmogorov's interest in Brouwer's work. At the same time, Kolmogorov taught at the experimental model school of the People's Commissariat for Education. Kolmogorov had a broad interest and had, among other things, put forward a hypothesis on the history of Novgorod (which was later confirmed) and stood as a candidate at the age of fourteen at the 1917 Constituent Assembly elections. Also inside mathematics Kolmogorov had a broad interest, and during the rest of his life he was to contribute to almost all fields of mathematics. When he had started his studies in Moscow in 1920, he had already gathered some mathematical knowledge as an autodidact. In 1922, at the age of 19, Kolmogorov published a paper in which he presented a Fourier series which was divergent almost everywhere. It suddenly brought him an international reputation.[114]

It is not clear how Kolmogorov became acquainted with intuitionism. Possibly, as Van Dalen suggested, Alexandroff or Urysohn, who may have met Brouwer in 1923 at the meeting of the German Mathematical Society in Marburg where all of them were present, told him something about the foundational debate. In the paper, Kolmogorov mentions several works by Brouwer, including some which had only been published in the Dutch *KNAW Verhandelingen*, and one of Hilbert's lectures. Notably absent are Weyl's 1921 paper and Fraenkel's expository works, through which most people got to know intuitionism.

Kolmogorov opens the paper[115] by stating that Brouwer has shown that it is illegitimate to use the principle of the excluded middle when dealing with transfinite arguments. Thus, Kolmogorov clearly supports Brouwer's argument. At other places in the paper, he explicitly agrees with Brouwer's view of the time- and knowledge-dependency of mathematics, and he values the Brouwerian

---
[113] [Bockstaele 1949, pp. 40–41]
[114] [Tikhomirov 1993, pp. 103–104], [Vitányi 1988, pp. 5–14]
[115] The exposition given here of Kolmogorov's paper was based on the English translation in [Van Heijenoort 1967, pp. 416–437], since my Russian is non-existent.

counterexamples as Brouwer had meant them: as counterexamples against the principle of the excluded middle.[116]

Nowadays, one mostly remembers the paper because of its formalisation of parts of intuitionistic logic and its translation of classical into intuitionistic logic.[117] Kolmogorov, however, describes the goal of the paper as showing why an illegitimate use of the principle of the excluded middle does not lead to contradictions, and why the illegitimacy has hardly been noticed before.[118]

Kolmogorov characterises intuitionism by its 'recognition of the real meaning of mathematical propositions'.[119] Formalism does not do so. He thus feels the need to justify the use of a formalised language, and does so by maintaining that intuitionism tolerates the formalistic method as one among various possible ones. Kolmogorov does not refrain from pointing out that also for Hilbert, the use of the principle of the excluded middle in transfinite areas is not intuitively obvious. He explicates Brouwer's negation as absurdity as one of the possible forms of negation.[120]

Kolmogorov presents the axioms for part of what we now call the intuitionistic propositional calculus, restricting himself to the implication and negation connectives. Following the axioms put forward in Hilbert's 1921 lecture,[121] Kolmogorov presents an axiom system called $\mathfrak{B}$, presumably after Brouwer, consisting of the following axioms:[122]

(1) $\varphi \to (\psi \to \varphi)$
(2) $(\varphi \to (\varphi \to \psi)) \to (\varphi \to \psi)$
(3) $(\varphi \to (\psi \to \sigma)) \to (\psi \to (\varphi \to \sigma))$
(4) $(\psi \to \sigma) \to ((\varphi \to \psi) \to (\varphi \to \sigma))$
(5) $(\varphi \to \psi) \to ((\varphi \to \neg\psi) \to \neg\varphi)$

One of the differences with Heyting's later formalisation is that the axiom $\varphi \to (\neg\varphi \to \psi)$ is lacking, the status of which was disputed among intuitionists. Kolmogorov mentions that he does not know whether the system is complete.[123]

A different axiom system, called $\mathfrak{H}$ (presumably after Hilbert), is obtained from $\mathfrak{B}$ by adding a sixth axiom, the principle of double negation. It is equivalent to the principle of the excluded middle and is formulated as:

(6) $\neg\neg\varphi \to \varphi$.

---

[116][Kolmogorov 1925, pp. 416–421]
[117]Cf. [Wang 1967].
[118][Kolmogorov 1925, p. 416]
[119][Kolmogorov 1925, p. 417]
[120][Kolmogorov 1925, pp. 417–421]
[121][Hilbert 1922, p. 175]. The axioms (5) and (6) are formulated somewhat differently.
[122]In what follows, I have modernised Kolmogorov's formalism, for the sake of convenience. I do not, however, use formal notations where Kolmogorov does not.
[123][Kolmogorov 1925, p. 422]. In fact, it is complete, cf. [Wang 1967, p. 414].

## 5.3. THE DEBATE WIDENED

Kolmogorov proves that $\mathfrak{H}$ is equivalent to the formalisation of the propositional calculus used by Hilbert.[124]

Next, Kolmogorov introduces a class of propositions for which the principle of double negation holds. These propositions are denoted by $\dot\varphi, \dot\psi$, etc. He does not demarcate which propositions fall in this class, but simply notes that all finitary propositions belong to it. Kolmogorov further remarks that Brouwer has shown that the same goes for all negative propositions, and he proves the statement on the basis of the axiom system $\mathfrak{B}$.[125] By defining the class of propositions $\dot\varphi$ in this way, Kolmogorov manages to differentiate between domains in which the principle of the excluded middle holds and domains in which it does not, without having to go into a more philosophical discussion.

The difference between $\mathfrak{B}$ and $\mathfrak{H}$, Kolmogorov maintains, is that the former is universally applicable, the latter not. The reason for this is that Hilbert's axiom system makes general use of axiom (6), whereas this is only valid for a certain class of propositions. Kolmogorov shows that, as we would now call it, the class of propositions $\dot\varphi$ is closed under composition with negation and implication. In Kolmogorov's view, one can apply $\mathfrak{H}$ only to this class.[126]

The question Kolmogorov then asks is whether we can still give meaning to the formulas obtained by using $\mathfrak{H}$ outside its proper domain of application. He thinks that this is possible by constructing, alongside ordinary mathematics, a new field called pseudo-mathematics. This is done is such a way that to every formula of the former corresponds a formula of the latter and that, moreover, every formula of pseudo-mathematics is of the type $\dot\varphi$.[127]

Kolmogorov explicates the idea by giving what we would now call a translation of mathematics into pseudo-mathematics. The translation is denoted by * and is defined inductively as follows:[128]

$$\varphi^* = \neg\neg\varphi \qquad \text{for atomic } \varphi;[129]$$
$$F(\varphi_1, \varphi_2, \ldots, \varphi_k)^* = \neg\neg F(\varphi_1^*, \varphi_2^*, \ldots, \varphi_k^*) \qquad \text{for composed formulas.}$$

This translation anticipated Gödel's translation to a large extent.[130] However, Kolmogorov uses the translation differently. In our modern interpretation, following Gödel, we use such a translation as a translation from classical into intuitionistic mathematics. Kolmogorov considers it as a translation into a new domain called pseudo-mathematics. After he has used the translation, he does conclude that the formulas of pseudo-mathematics are true in the usual sense, but he does not identify the domain of pseudo-mathematics as a sub-domain of intuitionistic mathematics.[131]

---

[124][Kolmogorov 1925, pp. 422–424]
[125][Kolmogorov 1925, pp. 425–426]
[126][Kolmogorov 1925, pp. 426–427]
[127][Kolmogorov 1925, pp. 427–428]
[128][Kolmogorov 1925, p. 428]
[130]For Gödel's translation, see 5.4.2.
[131]Thus, what Wang presented in the introduction to Kolmogorov's paper was more a modern interpretation of the paper than a representation of what Kolmogorov had proved; [Wang 1967].

Next, Kolmogorov shows that the translation leaves substitution and the modus ponens rule intact. Furthermore, if we apply the translation to the Brouwer-axioms, we get a set of axioms which are provable in $\mathfrak{H}$. In this case, it is admissible to use $\mathfrak{H}$ since we are in a domain where the principle of double negation holds. Therefore, the formulas of pseudo-mathematics corresponding to the Brouwer-axioms are true. Kolmogorov maintains that the same goes for every set of mathematical axioms we know.[132] This is a rather bold statement, which we are far from proving even today.[133]

Finally, Kolmogorov states the result of using the translation in the most general terms. Let a set of axioms be given the *-translation of which are true in pseudo-mathematics. In Kolmogorov's view, we know of no other axioms. Suppose that we have proved from this set of axioms some formulas by illegitimately using the principle of double negation. Then, the formulas of pseudo-mathematics corresponding to the formulas proved can be derived from the *-translation of the axioms.[134] In current terminology, we would say that if $A \vdash_{\mathfrak{H}} \varphi$, then $A^* \vdash_{\mathfrak{B}} \varphi^*$.[135]

Kolmogorov concludes that formulas based on the use of the principle of double negation outside the domain of the finitary cannot be seen as firmly established. However, if the *-translation of the axioms from which the formulas were derived are true in pseudo-mathematics, then the corresponding formulas are true in pseudo-mathematics.[136]

In the Addenda, Kolmogorov draws two important conclusions. First, he disclaims a statement made by Brouwer. Brouwer considered finitary statements which were proved by using the principle of the excluded middle in the domain of the transfinite unreliable.[137] Kolmogorov correctly points out, however, that in finitary domains the principle of double negation holds. Therefore, truth and pseudo-truth coincide, and thus the formulas can also be proved without using the principle of the excluded middle. Kolmogorov's remark amounts to stating that, if we have proved a finitary statement $\varphi$, then, by using the *-translation, we can turn it into a statement $\varphi^*$ which is provable in pseudo-mathematics. However, since $\varphi$ is finitary, $\neg\neg\varphi \to \varphi$ holds, thus $\varphi^*$ is equivalent to $\varphi$, and therefore $\varphi$ itself is provable in pseudo-mathematics. Our modern interpretation would be that classical mathematics is conservative over intuitionistic mathematics regarding finitary propositions.

---

[132] [Kolmogorov 1925, p. 428–429]
[133] [Wang 1967, p. 415]
[134] [Kolmogorov 1925, p. 430]
[135] Since the translation of axiom (6) holds in $^-$, $^-$ and $^-$ coincide in pseudo-mathematics. In fact, one can prove the stronger result that the translation also works for predicate logic (if it is extended by defining $(Qx\varphi)* = \neg\neg Qx(\varphi*)$ for $Q = \forall, \exists$), and that it works both ways, i.e., $A \vdash_c \varphi \Leftrightarrow A* \vdash_m \varphi*$, where $\vdash_c$ indicates provability in classical predicate logic, $\vdash_m$ provability in the so-called minimal predicate logic, i.e., intuitionistic predicate logic without the rule $ex$ falso sequitur quodlibet; cf. [Troelstra & Van Dalen 1988, vol. 1, pp. 56–59].
[136] [Kolmogorov 1925, p. 430]
[137] [Brouwer 1925E, p. 252]; the footnote Kolmogorov refers to starts on page 251.

## 5.3. THE DEBATE WIDENED

Second, Kolmogorov comes to what he himself stated as the main goal of his paper. Using the principle of the excluded middle illegitimately, he concludes, will never lead to a contradiction. For if one in this way could derive a contradiction, the *-translation would transfer the contradiction to pseudo-mathematics, which Kolmogorov considers impossible.[138] Thus, formulated in current terminology, Kolmogorov presented a relative consistency proof[139] for formalistic mathematics, since a supposed contradiction in formalistic mathematics would be transferred to intuitionistic mathematics. Note that Kolmogorov thus proved Brouwer right when he claimed in 1908 that one will never get caught in a contradiction by illegitimately using the principle of the excluded middle in classical mathematics.[140]

Finally, Kolmogorov sketches how his approach can be applied to predicate calculus. He finishes the paper by stating that, besides mathematics in Brouwer's way without an illegitimate use of the excluded middle, one should also continue with ordinary mathematics, if only as the mathematics of pseudo-truth.[141]

Kolmogorov's paper stands head and shoulders above other contributions to the debate at the time. Kolmogorov formalised a part of intuitionistic logic, explicitating the intuitionistic negation as absurdity, clarified the interpretation of formalistic logic, provided a relative consistency proof of formalistic logic, and corrected Brouwer on the issue of finitary statements proved by using the principle of the excluded middle in the transfinite. However, since the paper was written in Russian, it remained almost completely unnoticed. The only reference I found to it in the whole foundational debate was a letter from Glivenko to Heyting in October 1928.[142] There is not a single reference to it in any of the public contributions until 1933.

**Hilbert from 1925 to 1927** In June 1925, Hilbert lectured on the infinite before the *Westfälische Mathematische Gesellschaft* in Münster at the commemoration of Weierstraß.[143] His talk was published the following year in the *Mathematische Annalen*.

As in his 1922 lecture,[144] Hilbert argues that certain statements about an infinite number of elements cannot be negated. Thus, we cannot apply the principle of the excluded middle to such statements, since such an application assumes that

---

[138] [Kolmogorov 1925, p. 431]

[139] A relative consistency proof for a certain system $S_1$ is a proof of its consistency, assuming that another system $S_2$, usually considered to be more firmly established than $S_1$, is consistent.

[140] See 2.3.2. This reasoning does not work for intuitionistic mathematics, which includes choice sequences; cf. 2.7.1

[141] [Kolmogorov 1925, pp. 432–437]

[142] Letter from Glivenko to Heyting, 13/10/1928; [TLI Heyting, B gli-281013]

[143] Weierstraß was born in Westphalia on October 31, 1815; he had studied some months in Münster and taught there for one year. He died in 1897, [Biermann 1976, pp. 219–220]. The reason why such a commemoration should be held in June 1925 is not clear to me. Perhaps it had been planned in 1915, but was postponed because of the war.

[144] See 5.2.2.

# 242  CHAPTER 5. REACTIONS: LOGIC AND THE EXCLUDED MIDDLE

the statements involved can be negated. This time, however, Hilbert explicitly restricts this argumentation to what he calls the 'finitary'[145] point of view. He does not specify what should count as 'finitary', but describes it by words as 'meaningful' and 'intuitive'.[146] Hilbert concludes that, in the domain of finitary statements, the Aristotelian laws of logic do not hold.[147]

Then, Hilbert makes the decisive step:[148]

> Nun könnte man darauf ausgehen, die für den Bereich der finiten Aussagen gültigen logischen Gesetze aufzustellen; aber damit wäre uns nicht gedient, da wir eben auf den Gebrauch der einfachen Gesetze der Aristotelischen Logik nicht verzichten wollen, und niemand, auch wenn er mit Engelszungen redete, wird die Menschen davon abhalten, beliebige Behauptungen zu negieren, Partialurteile zu bilden und das Tertium non datur anzuwenden. Wie werden wir uns nun verhalten?
>
> Erinnern wir uns, *daß wir Mathematiker sind* und als solche uns schon oftmals in einer ähnlichen mißlichen Lage befunden haben und wie uns dann die geniale Methode der idealen Elemente daraus befreit hat.[149]

Hilbert's solution consists of[150]

> *zu den finiten Aussagen die idealen Aussagen [zu] adjungieren*, um die formal einfachen Regeln der üblichen Aristotelischen Logik zu erhalten.[151]

The only demand that has to be fulfilled when adding ideal elements is the proof of their consistency.[152]

Thus, Hilbert now introduces the idea of using meaningless elements in mathematics.[153] The reason for seeing mathematics in this way is the preservation of the rules of classical logic. And the reason for this preservation is simply that we do not want to lose them, because they are so simple. This is in line with Wavre's claim that the only reason why formalists wanted to add ideal elements was to

---

[145]'finit', [Hilbert 1926, p. 171]
[146]'inhaltlich', 'anschaulich', [Hilbert 1926, pp. 171–172]
[147][Hilbert 1926, p. 174]
[148][Hilbert 1926, p. 174]
[149]'Now one could try to develop the logical laws which hold for the domain of finitary statements. But it would do us no good, for we do not want to give up the use of the simple laws of Aristotelian logic. No one, though he may speak with the tongues of angels, could keep people from negating arbitrary statements, or from forming partial propositions, or from using the tertium non datur. How, then, are we to behave?
Let us remember *that we are mathematicians* and that as such we have often been in a similarly precarious situation from which we have been rescued by the ingenious method of ideal elements.'
[150][Hilbert 1926, p. 174]
[151]'*supplementing the finitary statements with ideal statements*, to preserve the simple formal rules of ordinary Aristotelian logic.'
[152][Hilbert 1926, p. 179]
[153]See also 4.3.1.

## 5.3. THE DEBATE WIDENED

extend the kingdom of classical logic.[154] Carnap later identified the acknowledgement that not all propositions of (classical) mathematics are meaningful as one of the things Hilbert had learned from Brouwer.[155]

The reactions to Hilbert's paper seem to have been mixed. Lévy reports that Zermelo told him in 1928 that, even in Germany, nobody understood what Hilbert wanted with the paper.[156] On the other hand, Skolem later reported that Hilbert's paper was discussed at the mathematical conference in Copenhagen in 1925 as something epoch-making. He himself did not agree, and attributed the exaggerated appreciation mathematicians showed for Hilbert's paper to his general standing as a mathematician:[157]

> Det forundret mig i høi grad, at matematikerne, selv om de ikke er spesialister i grunnlagforskning, ikke fra første stund forstod, hvor problematisk – for å bruke et mildt uttrykk – det vesentlige innhold av denne avhandling var. Nu er visst alle omsider blitt klar over, at den ikke er bra. Da jeg var i Vienna ifjor høst, hørte jeg en av matematikerne der bruke uttrykket 'kompromitterende' om den. Jeg nevner dette, fordi det er en utbredt overtro blandt matematikerne, at først og fremst eldre folk med betydelig navn som matematikere i ordinær betydning kan utrette nogen ting i grunnlagforskningen. I virkeligheten har det vist sig, at omtrent alle fremskritt i grunnlagforskningen i den senere tid skyldes yngre folk og vesentlig spesialister.[158]

Skolem's judgement that it were mostly younger people that made a positive contribution to foundational research is essentially correct, as becomes clear from the remainder of this chapter.[159]

In July 1927, Hilbert again lectured at the mathematical seminar in Hamburg. He basically repeats the same view on the excluded middle. He states that one cannot renounce the principle of the excluded middle, since without it the construction of analysis is impossible, and the solution is again provided by adding

---

[154] See 4.3.1.
[155] Notes on a lecture before the mathematical circle in Prague, 22/1/1932; [ASP Carnap, 110-07-16]. The other thing Carnap mentioned which Hilbert learned from Brouwer (though Carnap was not sure about this one) was the separation between mathematics and meta-mathematics.
[156] [Lévy 1964, p. 89]
[157] [Skolem 1934, p. 91]
[158] 'I was highly surprised that the mathematicians, even if they are no specialists in foundational research, did not understand from the first moment how problematic – to use a mild expression – the essential contents of that treatise was. Now, I think, it has finally become clear to all, that it is not good. When I was in Vienna last autumn, I heard one of the mathematicians use the expression 'embarrassing' about it. I mention this because there is a widespread superstition among mathematicians that in the first place elderly people with an important name as mathematicians in the ordinary sense can achieve something in foundational research. In fact, it has turned out that almost all progress in foundational research in recent times was owing to younger people and essentially specialists.'
[159] See also 3.3.4.

## 244 CHAPTER 5. REACTIONS: LOGIC AND THE EXCLUDED MIDDLE

ideal elements.[160] However, this time he is more explicit about the significance of the formal system of mathematics. Apart from its mathematical value, Hilbert maintains, it also has an important philosophical meaning:[161]

> Dieses Formelspiel vollzieht sich nämlich nach gewissen bestimmten Regeln, in denen die *Technik unseres Denkens* zum Ausdruck kommt. (...) Die Grundidee meiner Beweistheorie ist nichts anderes, als die Tätigkeit unseres Verstandes zu beschreiben, ein Protokoll über die Regeln aufzunehmen, nach denen unser Denken tatsächlich verfährt.[162]

Thus, Hilbert still adhered to a contentual interpretation of logic, namely the idea of logic as representing the laws of thought.

Towards the end of the lecture, Hilbert turns to Brouwer's intuitionism. He claims that he does not do so in order to polemicise,[163] only to enter into one of the most polemical parts of the foundational debate:[164]

> Die schärfste und leidenschaftlichste Kampfansage des Intuitionismus ist diejenige, die er gegen die Gültigkeit des Tertium non datur (...) richtet (...). Dieses Tertium non datur ist eine Folgerung aus dem logischen $\epsilon$-Axiom[165] und hat noch niemals den geringsten Fehler hervorgerufen. Es ist zudem so klar und faßlich, daß eine mißbräuchliche Anwendung ausgeschlossen ist. Insbesondere trägt das Tertium non datur an dem Zustandekommen der bekannten Paradoxien der Mengenlehre nicht die geringste Schuld.[166]

Hilbert's defence of the principle of the excluded middle here becomes not merely polemical, but even seems exaggerated. It is not even clear which reproach he is defending himself against. What comes closest to it is a remark Brouwer had made in his dissertation, when he claimed that erecting logical buildings without having recourse to the mathematical intuition lead to the paradoxes.[167] However, the following year already Brouwer explicitly stated that using the principle of the

---

[160][Hilbert 1927, p. 73]
[161][Hilbert 1927, p. 79]
[162]'For this formula game proceeds according to certain rules, in which the *technique of our thinking* is expressed. (...) The main idea of my proof theory is nothing else but to describe the activity of our mind, to include a protocol about the rules according to which our thinking in fact proceeds.'
[163][Hilbert 1927, p. 77]; Weyl had reproached Hilbert of being polemic, see 5.3.
[164][Hilbert 1927, p. 80]
[165]The $\epsilon$ axiom is explained in 4.3.2.
[166]'The strongest and most passionate declaration of war of intuitionism is the one which is directed against the validity of the *tertium non datur* (...). The *tertium non datur* is a consequence of the logical $\epsilon$-axiom and has never led to the slightest mistake. It is so clear and comprehensible, that an improper application is excluded. In particular, the *tertium non datur* does not bear the slightest responsibility for the coming into being of the well-known paradoxes of set theory.'
[167][Brouwer 1907, pp. 186-191]; see 2.3.1.

## 5.3. THE DEBATE WIDENED

excluded middle illegitimately does *not* lead to a contradiction.[168] Furthermore, the remark just referred to played hardly any role in the debate.

Hilbert continues his passionate defence:[169]

> Dieses Tertium non datur dem Mathematiker zu nehmen, wäre etwa, wie wenn man dem Astronomen das Fernrohr oder dem Boxer den Gebrauch der Fäuste untersagen wollte. Das Verbot der Existenzsätze und des Tertium non datur kommt ungefähr dem Verzicht auf die mathematische Wissenschaft überhaupt gleich. Denn was wollen die kümmerlichen Reste, die wenigen unvollständigen und unzusammenhängenden Einzelresultate, die von den Intuitionisten ohne den Gebrauch des logischen $\epsilon$-Axioms erarbeitet worden sind, gegenüber der gewaltigen Ausdehnung der modernen Mathematik bedeuten![170]

Finally, Hilbert gives his personal view on the cause of the popularity of intuitionism:[171]

> Ich staune unter diesen Umständen darüber, daß ein Mathematiker an der strengen Gültigkeit der Schlußweise des Tertium non datur zweifelt. Ich staune noch mehr darüber, daß, wie es scheint, eine ganze Gemeinde von Mathematikern sich heute zusammengefunden hat, die das gleiche tut. Ich staune am meisten über die Tatsache, daß überhaupt auch im Kreise der Mathematiker die Suggestivkraft eines einzelnen temperamentvollen und geistreichen Mannes die unwahrscheinlichsten und exzentrischsten Wirkungen auszuüben vermag.[172]

Hilbert's reference to a group of mathematicians that joined Brouwer in doubting the principle of the excluded middle seems too strong an interpretation of the commotion caused by the intuitionistic criticism. The one main convert that Brouwer had made, Weyl, had by that time already taken a kind of intermediate position, recognising both intuitionistic and more formalistic mathematics. Mathematicians who, like Brouwer, rejected the universal validity of the principle of the excluded middle were rare: Kolmogorov did so in 1925, but Hilbert certainly

---

[168][Brouwer 1908, p. 258]; see 2.3.2.
[169][Hilbert 1927, p. 80]
[170]'Taking away the *tertium non datur* from the mathematician would be about the same as if one would forbid the telescope to the astronomer or the use of his fists to the boxer. Prohibiting existence statements and the *tertium non datur* is tantamount to relinquishing the mathematical science altogether. For what are the poor leftovers, the few incomplete and incoherent single results which the intuitionists have worked out without the use of the logical $\epsilon$ axiom, to signify against the enormous size of modern mathematics!'
[171][Hilbert 1927, pp. 80–81]
[172]'Under these circumstances, I am astonished that a mathematician doubts the rigorous validity of the derivation rule of the *tertium non datur*. I am even more astonished to see that, as it seems, a whole congregation of mathematicians has now come together, which does the same. I am astonished most by the fact that even in the circle of mathematicians the suggestive force of a single temperamental and penetrating man can exercise the most unlikely and eccentric influences.'

had not read his paper; Bieberbach did so in 1926; and one could also consider Wavre's papers as defending the intuitionistic point of view.[173] One could easily find an equal number of mathematicians who still adhered to the excluded middle. The majority of those who reacted to intuitionism explained the different points of view rather than expressing their opinion on this issue clearly.

Hilbert's final argument is very much *ad hominem*, revealing once more how deeply he felt about the matter.

In Hilbert and Ackermann's *Grundzüge der theoretischen Logik*, which appeared in 1928, remarks on the intuitionistic view on logic are notably absent.[174]

In the same year, Hilbert presumably lectured before the *Gesellschaft der Wissenschaften* ('Society of the sciences'). His notes for that lecture, which are left in the Hilbert archive, show that he had by then moved more towards a purely formalistic position.[175] Under the title *Formalismus*, we find the following passage:[176]

> (...) z.B. die üblichen logischen Regeln, das Tertium non datur, den Kettenschluss sind willkürlich aufgelesen, uns von unseren Kinderfrauen beigebracht, sie könnten sehr gut unvollständig sein oder, was viel schlimmer [ist], zu einander in Widerspruch treten.[177]

**Skolem in 1925** In September 1925, the Norwegian mathematician Thoralf Skolem (1887–1963) lectured for the *Norsk Matematisk Forening* ('Norwegian Mathematical Society') on the most important recent discussions on the foundations of mathematics. Skolem had studied mathematics and science at the university of Oslo, which he had finished with the best possible marks. During the First World War he had spent a semester in Göttingen, discussing set theory with Bernstein. From 1918 on, he lectured at the university of Oslo, even though he had not written a dissertation.[178] In the beginning of the 1920s Skolem published a series of important papers in the field of logic, in which he among other things proved what is now known as the Skolem-Löwenheim theorem. Also, the technique

---

[173] I only included contributions made before Hilbert's lecture.
[174] [Hilbert & Ackermann 1928]
[175] The notes state that the lecture was to take place on June 8, 1928; I do not know if the lecture was indeed delivered. The notes bear two dates, the other one saying that they were used for lectures in the winter semester 1931/32. I take the earlier date to indicate the time when the notes were first written down.
[176] [NSUB Hilbert, 607].
[177] '(...) for example the usual logical rules, the *tertium non datur*, the *Kettenschluss*, are arbitrarily picked up, instilled into us by our nannies; they could very well be incomplete or, what [is] much worse, be in contradiction with one another.' The *Kettenschluss* is the rule that from $\varphi \to \psi$ and $\psi \to \chi$, one can conclude $\varphi \to \chi$, cf. [Hilbert & Bernays 1934–1939, vol. 1, p. 85].
[178] Skolem made up for this in 1926.

## 5.3. THE DEBATE WIDENED

of Skolem functions for the elimination of quantifiers was introduced in that period.[179] Skolem worked in constructive arithmetic and published, independently of Weyl and Brouwer & De Loor, a constructive proof of the fundamental theorem of algebra in 1924.[180]

Skolem uses the set theoretical paradoxes as an introduction to different currents in the foundations of mathematics. He states that Russell and Whitehead's theory has not found any recognition, and then turns to intuitionism. Intuitionism, Skolem maintains, is characterised by its rejection of the principle of the excluded middle as a mathematical proof method when dealing with an infinite number of things. Skolem follows Weyl in using a decidable property for natural numbers in explaining what the problem with the *tertium non datur* is. A mathematical theorem, Skolem continues, is not automatically either correct or incorrect. Until a proof for the theorem has been found, the theorem is nothing but a way of speaking, devoid of contents. If one uses the alternative either $P$ or not $P$ in a proof, the intuitionist will demand a method for deciding between the two.[181]

Next, Skolem considers the consequences of the intuitionistic point of view. As a rule, giving up the principle of the excluded middle has, according to him, no consequences in elementary arithmetic and algebra. In some cases, one will have to improve the proofs used until now. In analysis, however, the consequences are substantial. As an illustration, Skolem uses a Brouwerian counterexample against the ordering of the real numbers.[182] It seems that Skolem thinks that every classical proof in elementary arithmetic and algebra can be replaced by an intuitionistic version proving the same theorem – which, however, is not the case.

Skolem is one of the few people who includes some intuitionistic analysis in his contribution. After the explanation, he continues with a short exposition of choice sequences, the intuitionistic theory of the real numbers, and intuitionistic function theory.[183]

Skolem finishes his lecture with a short description of Hilbert's axiomatics, remarking that, if Hilbert's axioms are taken as mere formal rules, this is not mathematics in the usual sense any more. In concluding, he criticises Hilbert's attitude:[184]

> Nogen gjendrivelse av intuitionismen kan der ikke være tale om ved at gaa frem paa Hilberts maate, det jeg kan skjønne. Det er i det hele ikke godt at forstaa, hvordan intuitionismen skulde kunne gjendrives, da den er en viljessak. Den beror jo paa den *beslutning*, at hver sats, vi opstiller, skal være uttryk for en evne, vi virkelig sitter inde med. Man kan vel ikke gjendrive en teori, som er basert paa visse utgangspunkter,

---

[179] [Fenstad 1970, pp. 9–12], [Moore 1987, p. 125]
[180] [Skolem 1924]
[181] [Skolem 1926, pp. 4–8]
[182] [Skolem 1926, pp. 9–10]
[183] [Skolem 1926, pp. 10–12]
[184] [Skolem 1926, p. 13]

ved at opstille teorier basert paa andre utgangspunkter. I valget av utgangspunkter ligger noget subjektivt.[185]

Skolem was one of the few persons who not only paid attention to the intuitionistic criticism of the excluded middle, but also to the positive contributions of intuitionism. His conclusion regarding the impossibility of Hilbert's attempt is correct. However, the lecture, which was published only in Norwegian, had no influence on the course of the debate. People like Von Neumann, Ramsey, Study, and especially Fraenkel did refer to Skolem's work, but only to his technical papers.[186]

**Fraenkel from 1925 to 1927**  In two papers published in 1925, Fraenkel touched upon the intuitionistic criticism of the principle of the excluded middle. In the one he merely mentions it; the other is a repetition of the explanation given in 1924 using the decimal expansion of $\pi$.[187]

In June of the same year, Fraenkel gave a series of lectures for the *Kant-Gesellschaft* in Kiel on the foundations of set theory. This happened on the invitation of Heinrich Scholz, a philosopher at the university of Kiel who later was to specialise in mathematical logic. Scholz had contacted Fraenkel after he had read the second edition of his *Einleitung in die Mengenlehre*, and a friendship for life developed.[188] The Kant Society was an important one, since virtually all German philosophers were a member of it. Apparently, Fraenkel delivered as many as ten lectures on the foundations of set theory to philosophers, since the lectures were published as *Zehn Vorlesungen über die Grundlegung der Mengenlehre* in 1927.[189]

One of the lectures was about intuitionism. In this lecture, Fraenkel presents the rejection of the tertium non datur as one of the main consequences of the intuitionistic view that mathematical existence should coincide with constructivity.[190] Fraenkel again uses the example of the decimal expansion of $\pi$. He then moves on to argue that many will have the feeling that, even if we cannot decide the answer to a definite mathematical question, *an sich* it has to be either yes or no. But this, Fraenkel explains, is seen as a prejudice by the intuitionists. Next, he mentions

---

[185]'As I see it, it is out of the question to refute intuitionism by pursuing in Hilbert's way. It is actually not clear at all how intuitionism could be refuted, since it is a matter of will. For it is based upon the *decision* that every proposition we form has to be an expression of an ability we actually have. One cannot refute a theory that is based on certain starting points by making theories which are based on other starting points. There is something subjective in the choice of the starting points.'

[186]Skolem's works most referred to in the debate are mentioned in the bibliography in [Fraenkel 1928].

[187][Fraenkel 1925C, p. 252] and [Fraenkel 1925B, pp. 210–211] respectively

[188][Fraenkel 1967, pp. 179–181]. Scholz later lobbied to have Fraenkel appointed as a professor at the university of Kiel, which indeed happened in 1928. (Until that time, Fraenkel worked in Marburg.)

[189][Fraenkel 1927A]; the published text is the one on which the analysis was based. It is likely to differ from the lectured one, since it took Fraenkel until December 1926 to finish completing the manuscript.

[190]Fraenkel's exposition of the intuitionistic view on mathematical existence is presented in 4.3.1.

## 5.3. THE DEBATE WIDENED

that with the principle of the excluded middle also the belief in the solvability of every mathematical problem loses ground.[191]

Discussing Hilbert's view on the foundations of mathematics, Fraenkel remarks:[192]

> Die intuitionistischen Bedenken gegen den Gebrauch des 'tertium non datur' und der Begriffe 'alle' und 'es gibt' innerhalb der transfiniten Mathematik werden von der Schule Hilberts methodisch anerkannt und übernommen, ja sogar erweitert.[193]

As is clear from Hilbert's lectures, Fraenkel described the situation correctly, even though Hilbert would probably not have admitted so. Fraenkel adds that Hilbert uses the intuitionistic insights in order to argue against intuitionism.[194]

**Bieberbach in 1926** In February 1926, Ludwig Bieberbach (1886–1982) lectured before the *Deutsche Verein für Förderung des mathematischen und naturwissenschaftlichen Unterrichts* ('Association for the Advancement of Education in Mathematics and Natural Sciences'). Taking Boutroux' well-read book *L'idéal scientifique* ('The scientific ideal') as his source of inspiration, he devoted the lecture to the scientific ideal of mathematicians. Bieberbach had written his dissertation in Göttingen. He had lectured in Zürich, Königsberg, Basel and Frankfurt, and was now a professor in Berlin, where he had succeeded Carathéodory. He worked mostly in function theory, but also contributed to such fields as geometry, group theory and topology. Bieberbach had been a member of the editorial board of the *Mathematische Annalen* since 1920. All in all, Bieberbach was an esteemed mathematician.

In the lecture, which was not published, Bieberbach presents intuitionism as a fusion of Klein's view, in which logical truth does not equal mathematical truth, and Hilbert's formal view, which looks systematically at the intuitive objects of mathematics. Regarding the principle of the excluded middle, Bieberbach explains that intuitionists consider it not applicable to infinite totalities, since these are outside our control. Furthermore, there are counterexamples to the excluded middle in Brouwer's continuum theory, so that formalists have to renounce the intuitionistic theory of the continuum, which is the most intuitive one, if they want to stick to the principle of the excluded middle. Instead of the usual Brouwerian counterexamples, Bieberbach takes statements about choice sequences, claiming (correctly) that the proposition 'either two choice sequences are equal or they are different' is not true.[195] Reacting to a popular argument against intuitionism, Bieberbach

---
[191][Fraenkel 1927A, pp. 38–42]
[192][Fraenkel 1927A, p. 53]
[193]'The intuitionistic objections against the use of the 'tertium non datur' and the concepts 'for all' and 'there is' in transfinite mathematics are recognised methodologically and taken over by Hilbert's school, even extended.'
[194][Fraenkel 1927A, p. 154]
[195][Bieberbach 1926, pp. 20–24a]

states:[196]

> Will man diesen Sachen gleichwohl in die Worte kleiden, die Brouwersche Theorie führe zu einer Verarmung der Mathematik, so steht dem doch der Gewinn gegenüber, dass sie uns vor logischen Fehlern bewahrt.[197]

Thus, Bieberbach presents intuitionism as a 'poor but honest' theory. Furthermore, Bieberbach claims, Brouwer's theory is much closer to common sense than the classical one.

Bieberbach brings in a new argument by pointing at the geometrical work of the Danish mathematician Hjelmslev. Hjelmslev, Bieberbach explains, aims at providing a geometry of reality, as opposed to abstract geometry. Hjelmslev also found examples where the principle of the excluded middle does not hold, and it is remarkable that he came to such an idea independently of Brouwer.[198]

Bieberbach concludes that the scientific ideal one should present at schools is that of intuitionism, which stands close to life, and not that of formalism.[199] This is quite different from the position he had held before the war, when he explicitly supported the formalistic point of view.[200] Mehrtens has argued convincingly that Bieberbach's change from a formalistic to an anti-formalistic position had little to do with the foundations of mathematics proper, but rather indicated a substantial shift in general values after World War I.[201]

**Ramsey in 1926** In August 1926, the young British logician Frank Plumpton Ramsey (1903-1930) read a paper before the British Association in which he for the first time seriously considered Brouwer's and Weyl's criticism of the excluded middle.[202] Ramsey had studied mathematics and logic in Cambridge and was a member of King's College. He worked in the tradition of Russell and Whitehead's *Principia Mathematica*, for which he tried to provide a new foundation.

In the paper on mathematical logic, Ramsey explains Brouwer's denial of the principle of the excluded middle by referring to our incomplete knowledge:[203]

> Brouwer would refuse to agree that it was raining or it was not raining, unless he had looked to see. Although it is certainly difficult to give

---

[196][Bieberbach 1926, p. 24b]

[197]'If one wants to put these things into words by saying that Brouwer's theory leads to an impoverishment of mathematics, then the positive thing to say is that it keeps us from making logical errors.'

[198][Bieberbach 1926, pp. 25-27]. Bieberbach refers to a textbook Hjelmslev had written, which was used at several schools in Denmark. Presumably, this was [Hjelmslev 1916] (which I have not seen).

[199][Bieberbach 1926, p. 28]

[200][Bieberbach 1914, p. 901]

[201][Mehrtens 1987, pp. 203-204]

[202]In 1925, Ramsey had read a paper on the foundations of mathematics to the London Mathematical Society in which he merely mentioned the 'prejudices' of the intuitionists and the 'Bolshevik menace of Brouwer and Weyl', [Ramsey 1926A, p. 339; 380].

[203][Ramsey 1926B, pp. 216-217]

## 5.3. THE DEBATE WIDENED

a philosophical explanation of our knowledge of the laws of logic, I cannot persuade myself that I do not know for certain that the Law of the Excluded Middle is true; of course, it cannot be proved (...).

He then continues:

(...) I do not see how any common basis can be found from which to discuss the matter. The cases in which Brouwer thinks the Law of the Excluded Middle false are ones in which, as I should say, we could not tell whether the proposition was true or false (...).

Here, Ramsey shows himself to be one of the more open participants to the debate. He understands Brouwer's argument, and he frankly states that he still cannot refrain from the view that he knows the principle of the excluded middle to be true, even though he has no proof. In other words, he believes in it. Therefore, his conclusion that there is no common ground for discussion is most justified.

**Gonseth in 1926** In 1926, Ferdinand Gonseth (1890–1975) published the most extensive book in French on the foundational crisis, *Les fondements des mathématiques* ('The foundations of mathematics'). The book grew out of a series of lectures Gonseth had given in 1924. Gonseth was a Swiss mathematician who lectured at the universities of Zürich and Bern. From 1910 to 1914, Gonseth had studied mathematics at the *Eidgenössische Technische Hochschule* in Zürich. He had written his dissertation in projective geometry in 1915, and he had worked at the ETH as an assistant from 1915 to 1920. Therefore, it is more than probable that he knew Weyl personally, even though they had different mother tongues. Gonseth attended Weyl's 1920 lectures, which formed the basis for the latter's *Grundlagenkrise* paper.[204] In 1919, Gonseth became professor at the universities of Zürich and Bern.[205]

Hadamard wrote the foreword to Gonseth's book. He is very negative about the whole discussion on first the axiom of choice and now the principle of the excluded middle. Referring to Comte's theory of the different stages through which theories develop, Hadamard maintains that mathematics is returning from the positive stage to the metaphysical stage, something unprecedented in the history of science. He sees the whole controversy between Brouwer, Weyl and Hilbert as useless. Neither does he appreciate the intuitionistic criticism of the excluded middle; to him, this principle is simply something one cannot 'forget'.[206]

Gonseth himself devotes a separate section to 'Brouwer and the principle of the excluded middle', basing himself on Brouwer, Weyl and Wavre. Since Gonseth's

---
[204] A comparison between Gonseth's lecture notes and the published version of Weyl's lectures is made in 4.2.1.
[205] 'Zum 70. Geburtstag von Ferdinand Gonseth', Bernays, P., [ETH Bernays, HS 973-25]
[206] 'oublier', [Hadamard 1926, p. XI]

reasoning is quite lengthy, I first present a paraphrase of his own reasoning below, and comment only afterwards.

Gonseth cites Brouwer to make it clear that intuitionism only allows for the excluded middle to be used in finite domains. Now Gonseth has no problem in admitting that there are domains in which the excluded middle does not hold. As an example, he uses the set of French words. This set is finite, therefore, in Brouwer's view, the principle of the excluded middle should be applicable. However, the word 'impredicable' provides us with a counter-example. An adjective is called 'predicable' if it applies to itself; if not, it is called 'impredicable'. Now, we cannot answer the question 'Is the adjective 'impredicable' itself predicable of impredicable? Therefore, Gonseth argues, it seems that the principle of the excluded middle does not hold, even though the domain is finite.[207] In fact, Gonseth is here arguing about the wrong domain, since the domain in question is not that of words, but that of sentences, which, at least in principle, is infinite.

Gonseth's conclusion, however, is different. He argues that the opposition between 'predicable' and 'impredicable' is purely formal and arbitrary. We could just as well not renounce the principle of the excluded middle for finite categories, but conclude that our definitions do not divide the set of words into two categories of which one possesses the attribute $A$ and the other not-$A$, but that there is a third category, namely the words to which neither of these applies.[208]

Regarding infinite sets, Gonseth simply argues that a statement such as 'every natural number is even or odd' is true because of the direct and intuitive knowledge we have of the set of natural numbers.[209]

Having next described the intuitionistic theory of the continuum, following Weyl but without, so it seems, fully understanding the concept of a choice sequence, Gonseth returns to the question of the excluded middle. In his view, logic is not the expression of absolute abstractions, but a description of certain relations, physical laws or laws of thought. He continues by stating that it makes no sense to claim that everything which is not true is false (or the other way round), since there may be things which could be true or false but which are neither. Nevertheless, Gonseth maintains, we imagine a scheme of relations between two statements $a$ and $\neg a$, for which the principle of contradiction and the principle of the excluded middle hold. And these principles, Gonseth claims, in fact are one.[210]

Gonseth tries to give a formalisation of intuitionistic logic. In doing so, he introduces a third truth-value, called 'indifferent'.[211] He concludes that classical logic proves more than intuitionistic logic, and formulates the intuitionistic claim as saying that, if we have no proof that we could use the stronger logic, we should use the weaker.[212]

---

[207][Gonseth 1926, pp. 191–192]
[208][Gonseth 1926, p. 193]
[209][Gonseth 1926, pp. 194–195]
[210][Gonseth 1926, pp. 213–214]
[211]'indifférent', [Gonseth 1926, p. 225]
[212][Gonseth 1926, pp. 225–230]

## 5.3. THE DEBATE WIDENED

Finally, Gonseth takes position. He finds the intuitionistic thesis not founded, neither in its attitude towards the infinite, nor regarding the truth value of the scheme of classical logic. He concludes:[213]

> Les intuitionistes (...) font preuve d'une singulière timidité en face de l'infini, qu'ils considèrent simplement comme inconcevable. Ils acceptent comme un dogma l'assertion suivante: 'L'esprit humain n'est capable que d'un nombre fini d'actes de pensée!' Le plus étonnant nous paraît être qu'on s'accorde à croire que cette phrase signifie quelque chose. (...)
>
> La formation des concepts est justement un acte de la pensée absolument irrationel; vouloir qu'il soit fini, c'est vouloir qu'il ne soit pas.[214]

In Gonseth's view, intuitionism means no threat to traditional mathematics, and there is no crisis in mathematics.[215]

As a commentary, I must say that I find Gonseth's reasonings markedly chaotic and badly supported. First, Gonseth seems to use an impredicative definition ('impredicable') as a counter-example against Brouwer's claim that the excluded middle *does* hold in finite sets. But intuitionists reject impredicative definitions anyway, since they are not constructive, therefore such an example could never be decisive. Second, Gonseth gives a very weak argumentation for the validity of the principle of the excluded middle in infinite sets, namely by simply stating that it is intuitively valid. Third, Gonseth claims that the principle of contradiction and of the excluded middle are one, but that is exactly what intuitionism disclaims. Fourth, Gonseth introduces a third truth value in formalising intuitionistic logic – in itself an accomplishment at the time, but it is not a formalisation of intuitionistic logic.[216] And fifth, Gonseth confuses the process of thinking with that of proving or verifying a proposition. It is the latter that is always finite, as Brouwer had claimed.

I do not know many works in which the confusion surrounding intuitionism was bigger than in Gonseth's work. It is all the more remarkable since his book appeared quite late in the debate, which means that there was a lot of much better literature available. Gonseth's book drew some attention in the French speaking world,[217] and it was mentioned in the third edition of Fraenkel's *Einleitung in die Mengenlehre*.[218]

---

[213] [Gonseth 1926, pp. 230–231]

[214] 'The intuitionists (...) show a strange timidity in face of the infinite, which they consider simply inconceivable. They accept as a dogma the following assertion: 'The human spirit is only capable of a finite number of acts of thinking!' The most surprising, so it seems to us, is that one agrees to believe that that phrase means something. (...)
The forming of concepts is on the contrary an absolutely irrational act of thinking; to wish that it is finite means to wish that it is not.'

[215] [Gonseth 1926, p. 232]

[216] Gödel later proved that intuitionistic logic is not an $n$-valued logic; see 5.4.2.

[217] [Juvet 1927], [Reymond 1932A], and [Dassen 1933]

[218] [Fraenkel 1928]

**Wavre vs. Lévy, 1926** In 1926, a discussion on intuitionistic logic took place between Rolin Wavre and Paul Lévy in the *Revue de Métaphysique et de Morale*.[219] Wavre opened the discussion by publishing a paper that continued the exposition he had given earlier on intuitionism. He again states that there are propositions of which we can neither prove the positive expression, nor their absurdity. Therefore, it *could* be the case that such a proposition is neither true nor absurd. However, Wavre rightly remarks, it makes no sense trying to establish a proposition that *is* neither true nor absurd, for in order to do so we would have to prove that the proposition is not true, which we can only prove by proving its absurdity, in which case it is absurd.[220]

Unaware of Kolmogorov's paper, Wavre proceeds with a formalisation of formalistic and, as he calls it, empiricist (intuitionistic) logic. He correctly takes the principle of the excluded middle as one of the axioms in which these logics differ. Wavre remarks that Brouwer only mentions the principles he rejects, not the ones he admits, but he adds that this follows logically from Brouwer's view of logic as a property of language, not as part of mathematics.[221] Wavre thinks the extra principles admitted in formalistic logic do not lead to much:[222]

> Les démonstrations par l'absurde ne nous apprenaient, en somme, ni le pourquoi ni le comment; elles répondaient comme le sphinx: oui, ou bien: non; qu'elles ne soient plus valables en logique empiriste, où l'on exige le comment, il ne faut pas s'en étonner.[223]

Wavre finishes his paper with a Brouwerian counter-example. He rightly concludes that, if one has succeeded in proving for instance that a number is rational, Brouwer would say that it *has become* rational. But these kind of questions, Wavre maintains, should be sent back to philosophers.[224]

Lévy reacts by a paper called *Sur le principe du tiers exclu et sur les théorèmes non susceptibles de démonstration* ('On the principle of the excluded middle and on theorems which are not capable of demonstration'). He argues that there can be theorems which are true, but not provable.[225] This applies in particular to theorems about an infinite number of particular cases. For such theorems, verification may indeed be impossible. By means of the example of Fermat's last theorem,

---

[219] The part of the discussion that focuses on mathematical existence is treated in 4.3.1.
[220] [Wavre 1926A, p. 66]
[221] [Wavre 1926A, pp. 69–71]
[222] [Wavre 1926A, p. 73]
[223] 'After all, proofs by contradiction did not teach us why or how; they answered like the sphinx: yes, or: no; one should not be surprised that they are no longer valid in empiristic logic, where on demands the how.'
[224] [Wavre 1926A, p. 74]
[225] It is rather unclear what it means, in Lévy's view, for a theorem to be true without us being able to prove so. It seems that he is thinking of some kind of truth in a Platonic sense, and not in a semantical sense which Gödel later used for his incompleteness theorem (on the incompleteness theorem, see 5.4.2). Lévy sticked to the conviction he presented here for the rest of his life, cf. [Lévy 1970, p. 219].

## 5.3. THE DEBATE WIDENED

Lévy argues that for some cases, where the negation of a theorem can be proved by a definite counter-example but where the positive proof would require an infinite verification, there are only three logical possibilities: either the theorem is provably false, or it is provably true, or it is unprovably true. In other cases, where both the positive and the negative proof require an infinite number of operations, four possibilities occur.[226] Thus, Lévy acknowledges the value of verifications, but does not draw the same conclusion as Brouwer.

In his reaction, Wavre once more stresses that doing mathematics in the formalistic way will never lead to contradictions, since a *tertium* cannot be pointed out; it should even be continued, since it leads to beautiful proofs. But the centre of the debate, he maintains, is whether one should proclaim a proposition true or false *a priori*, even if there is no means of deciding which of the two applies. Wavre gives no definite answer, but expresses his sympathy for Brouwer who tries to pose the question.[227]

In the last paper in the series, Lévy states that he now understands Wavre's attitude, but he does not agree with it. As he sees it, it amounts to arbitrarily forbidding the proclamation of certain results which are obvious. Furthermore, he does not want philosophy to stop him from doing science. In Lévy's view, the principle of the excluded middle applies to predicates such as 'rational' and 'irrational', against which Brouwer and Wavre had constructed counter-examples, since these *by definition* divide all numbers into two disjoint classes. Lévy concludes that the difference between Wavre and him lies in whether one wants to describe our actual knowledge or what he calls the objective state of affairs, and he adds:[228]

> (...) je n'aime pas un language qui, projetant en quelque sort notre ignorance sur les faits eux-mêmes, arrive à masquer ce que nous savons.[229]

This is the crucial point. Wavre did not react again, but the difference has now become crystal clear. Lévy does not want mathematics to be an expression of our knowledge, as Brouwer wants, but sees it as a description of something which in true already, perhaps in a Platonic world of ideas.[230] Thus, despite Lévy's dislike of philosophy entering into mathematics, it is precisely at a philosophical point where his view diverges from the intuitionistic one.

---

[226][Lévy 1926A, pp. 255–258]
[227][Wavre 1926B, pp. 427–429]
[228][Lévy 1926B, p. 548]
[229]'I do not like a language which, by in one way or the other projecting our ignorance on the facts themselves, succeeds in hiding what we know.'
[230]The citation is somewhat ambiguous, since Lévy at the same time speaks about 'our ignorance' and about 'what we know'. I take the first to mean the incompleteness of our present (mathematical) knowledge, the second Lévy's idea of an objective reality, which he thinks we know.

## 256   CHAPTER 5. REACTIONS: LOGIC AND THE EXCLUDED MIDDLE

**Reymond, Lévy, Brunschvicg, and Lenoir, 1927**   On January 29, 1927, the *Société française de Philosophie* ('French society of philosophy') devoted a session to axiomatic logic and the principle of the excluded middle. Arnold Reymond, professor at the university of Lausanne, delivered the introductory lecture. He describes logic as the normative science of the formal rules of correct thought. Reymond correctly argues that the question of the general validity of the principle of the excluded middle amounts to the question whether there are mathematical facts different from the intellectual activity carried out in doing mathematics. He concludes in favour of classical logic, which in his view more than suffices for the problem of truth.[231]

Reymond's lecture was followed by a general discussion. There, both Lévy and Brunschvicg state their agreement with Reymond's point of view. Only Lenoir defends a more intuitionistic point of view. Reymond concludes by maintaining that the principle of the excluded middle is one of those principles without which thought is impossible.[232]

**Petzoldt in 1927**   In May 1927, Joseph Petzoldt (1862–1929) lectured before the Berlin section of the *Internationale Gesellschaft für empirische Philosophie*[233] on 'rational and empirical thinking'. Petzoldt maintains that the whole of science is threatened, since one questions the principle of causality and that of the excluded middle.[234] Regarding the latter, Petzoldt describes the doubt, expressed by eminent mathematicians, as a sign of the 'too bold doubting of our time'.[235] He briefly presents a Brouwerian counter-example based on the decimal development of $\pi$, and then asks:[236]

> (...) wie ist es überhaupt denkbar, daß der Satz vom ausgeschlossenen Dritten nicht gilt? Und es ist im Grunde recht leicht zu zeigen, daß es nicht denkbar ist, falls man nicht das Denken überhaupt zerstören will.[237]

After this rather unconvincing argument he arrives at the conclusion that, in cases such as the Brouwerian counter-examples, what is at stake is not the principle of the excluded middle, but the matter of decidability. Petzoldt rejects Brouwer's view that logical principles should be verifiable[238] – even though one might think that such a stand should appeal to someone from a society for empirical philosophy.

---
[231][Leclerc 1927, pp. 3–4; 18]
[232][Leclerc 1927, pp. 18–23]
[233]The Society is described in 3.3.3.
[234]On possible links between questions of causality and logic, see 6.4.2.
[235]'allzukühnen Zweifelns unserer Zeit', [Petzoldt 1927, p. 157]
[236][Petzoldt 1927, p. 157]
[237]'(...) how is it at all conceivable that the law of the excluded middle does not hold? And it is indeed quite easy to show that this is not thinkable, if one does not want to annihilate thinking altogether.'
[238][Petzoldt 1927, pp. 154–158]

## 5.3. THE DEBATE WIDENED

**Härlen in 1927** In September 1927, Härlen lectured before the fourth *Deutsche Mathematikertagung* in Bad Kissingen on completeness and decidability. He starts by stating that what Brouwer ascertained about the principle of the excluded middle is still fiercely rejected, partly because of the subjective way in which Brouwer takes position. Härlen refers to the transcendental character of the principle of the excluded middle, as in the often-heard argument against Brouwer: an object has to have a property or not, regardless of whether the property is provable.[239] What Härlen wants to investigate is not decidability in the usual sense of provability, but whether a proposition is true in all the interpretations of the axiomatic system it appears in, or false in all these interpretations.[240] As we would nowadays say, Härlen wants to investigate the semantical side, not the syntactical one. Unfortunately, the report on Härlen's lecture in the *Jahresbericht der deutschen Mathematikervereinigung* is rather short, whereby it is unclear what the results of Härlen's idea are.

**Burkamp in 1927** In 1927, the philosopher Wilhelm Burkamp (1879–1939) published a paper in the *Beiträge zur Philosophie des Deutschen Idealismus* ('Contributions to the philosophy of German Idealism') on the 'crisis of the theorem of the excluded middle'. Burkamp had a strong interest in biology and psychology and tried to provide empirical foundations for philosophical concepts. He was a late starter; he became *Privatdozent* in 1923, at the age of 44, at the university of Rostock. His interest in logic dates back to his *Habilitationsschrift*, published the year before. In 1927 his first publication fully devoted to logic, the book *Begriff und Beziehung*, was published.[241]

Burkamp based his paper on works by Brouwer, Weyl, Hilbert and Baldus. He correctly relates Brouwer's rejection of the excluded middle to matters of undecidability in infinite sets. In Burkamp's view, the only reason why one could reject the excluded middle would be that one thinks it does not reflect our way of thinking. It is clear that Burkamp adheres to the idea that logic and axiomatic systems should describe something in reality. His solution to the problem that one does not always obtain propositions which are opposed as contradictories is to give the negation a wider meaning than normally, so as to obtain opposing contradictories in all cases. Burkamp describes his interpretation of negation as differing only marginally from absolute neutrality. In his proposal, Fermat's theorem, at the time still unproved, becomes false.[242]

Still, Burkamp seems to be somewhat confused about his own idea of negation. Formalising what he stated before, he claims that (in modernised symbolism) $\forall x(\varphi(x) \to \psi(x))$ should be negated as $\forall x \neg(\varphi(x) \to \psi(x))$.[243] But in this sense

---
[239] The argument was used, for example, by the idealist in Wavre's fictional dialogue (see 5.3) and by Ramsey (whose 1926 paper is treated above).
[240] [Härlen 1928, pp. 226–228]
[241] See 4.3.2.
[242] [Burkamp 1927B, pp. 59–64]
[243] [Burkamp 1927B, p. 65]

proposition and negation are definitely not opposed as contradictories.

Next, Burkamp turns to the real world. He takes up the example of Goethe and asks whether *'Goethe ist ein Sturm- und Drangdichter'* ('Goethe is a *Sturm und Drang* poet')[244] or 'Goethe ist kein Sturm und Drangdichter' (its negation) is true.[245]

> 'Intuitionistische Mathematiker' wie Brouwer werden sagen, hier gelte auch der Satz vom ausgeschlossenen Dritten nicht. Aber denken wir an den ursprünglichen Sinn dieses Satzes (...). Der Satz vom ausgeschlossenen Dritten kann nicht falsch sein. Aber die *individuelle Disjunktion* gilt nicht.[246]

The reason why the individual disjunction does not hold, Burkamp maintains, is that the two statements are not opposed as contradictories.

In applying Brouwer's criticism to natural language, Burkamp drifts far away from what Brouwer held. Furthermore, Burkamp does not differentiate between a proposition that does not hold and one that is false, thus using the principle of the excluded middle in defence of this same principle.

Paradoxically, Burkamp concludes that in the foundational crisis, Brouwer and Weyl are right. However, Brouwer's criticism, which Burkamp finds 'factually justified',[247] in his view does not hit the principle of the excluded middle. Logic is not, as Brouwer thinks, something abstracted from the mathematics of finite sets, but is about the validity of relations between concepts. Therefore, Brouwer's argumentation which leads to the rejection of the principle of the excluded middle does not hold. Mathematics, Burkamp maintains, is a science about forms, for which logic is *a priori* justified.[248] Even though Burkamp claims to be supporting Brouwer, their views differ substantially, and on the latter point they are even diametrically opposed.

**Becker in 1927** In 1927, Becker's voluminous *Habilitationsschrift 'Mathematische Existenz'* appeared.[249] This time, Becker also treats the question of the principle of the excluded middle.[250] He explains the intuitionistic criticism of the excluded middle, claiming that the negation of the existence statement 'there is a number in the sequence $S$ with the property $P$' is devoid of meaning. For, Becker repeats after Weyl, the negative sentence 'there is no number in the sequence $S$

---

[244] *'Sturm und Drang'* was a German literary style in the end of the 18th century, when young poets wanted to break with the established poetical forms.

[245] [Burkamp 1927B, p. 66]

[246] 'Intuitionistic mathematicians' like Brouwer will say that the law of the excluded middle does not hold here. But let us think about the original meaning of this law (...). The law of the excluded middle cannot be false. But the *individual disjunction* does not hold.'

[247] 'sachlich berechtigt', [Burkamp 1927B, p. 79]

[248] [Burkamp 1927B, pp. 76–80]

[249] The work was mainly concerned with mathematical existence, as the title indicates, which is treated in 4.3.1.

[250] The subject was not dealt with in [Becker, O. 1923].

## 5.3. THE DEBATE WIDENED

with the property $P$' does not make sense for choice sequences. These questions, Becker continues, are related to the question of decidability. In the case of an undecided question, intuitionists find that one should not state anything.[251]

By only explaining the criticism of the excluded middle be means of choice sequences, Becker provides a narrower interpretation of the intuitionistic view than what Brouwer and Weyl had presented.

Becker next treats Hilbert's 'formalism'.[252] He does not forget to mention that Hilbert, too, thinks that one cannot simply transfer reasonings from finite to infinite domains. After he has given the relevant citation from Hilbert, he remarks:[253]

Diese Darlegungen könnte wörtlich ein Intuitionist geschrieben haben.[254]

Furthermore, Becker also discusses the matter of the negation of a universal statement. He correctly maintains that Hilbert's distinction between 'there is no exception' and 'there is an exception but I cannot point it out' is not verifiable.[255]

In his analysis of the intuitionistic theses, Becker introduces the idea that one can negate a sentence in different ways. If we express 'p holds'[256] by $+p$, then there are two possible negations: 'not-$p$ holds' $(+-p)$ or '$p$ does not hold' $(-+p)$. In the cases in which the different negations do not coincide, we have three different possibilities; quartum non datur. Thus, Becker interprets intuitionistic logic as a three-valued logic. He further supports the distinction by means of Husserl's theory of judgement.[257] Interesting as the idea may be, it is not an interpretation of Brouwer's intuitionistic logic. Becker's first negation is the strong, intuitionistic one, the second one is weaker. Therefore, there is no *quartum non datur*.[258]

Becker next returns to Brouwer's criticism based on choice sequences. He introduces statements of the second order, which are statements about ('ordinary') statements. Only when including statements of the second order Becker modifies his scheme and allows as the third possibility 'neither $p$ nor $\neg p$ has been proved'. However, if this is the interpretation he had in mind, something goes wrong in the symbolism. He presents the three possibilities when including statements of the second order as $+p$, $+-p$ and $-(p \vee \neg p)$. In the interpretation just given, the last one should have been $-(+p \vee +\neg p)$. Becker maintains that, after all, statements of the first order are the essential thing.[259]

---

[251] [Becker, O. 1927, pp. 449–452]
[252] The quotation marks are Becker's.
[253] [Becker, O. 1927, p. 466]
[254] 'These expositions could literally have been written by an intuitionist.'
[255] [Becker, O. 1927, p. 496]
[256] Becker does not specify whether 'p holds' should be interpreted as 'there is a proof for $p$' or as 'we have a proof for $p$'.
[257] [Becker, O. 1927, pp. 497–503]
[258] Later, Gödel proved that there is no *n-tum non datur* in intuitionistic logic by proving that intuitionistic logic is not a many-valued logic; see 5.4.2.
[259] [Becker, O. 1927, pp. 504–505]

In the mathematical appendix, Becker returns to the matter of Brouwerian logic. Relying on publications by Brouwer and Wavre,[260] Becker remarks that in intuitionistic logic one does not speak of true and false, but of true and absurd, where the latter means that a contradiction can actually be derived. In phenomenological terms, true means the fulfillment of an intention, absurd means its deception. Between these possibilities, of course, there is no full disjunction.

Returning to his scheme of $p$, $\neg p$ and the third possibility, Becker remarks that both Brouwer and Wavre clearly reject such a principle of the excluded fourth. Brouwer does introduce another possibility, namely the absurdity of the absurdity of a propostion, but, Becker correctly remarks, this does not exhaust all possibilities. For it is also possible that we neither have a proof for $p$, nor for $\neg p$, nor for $\neg\neg p$. And even if we add as a new possibility 'we have a proof that neither of these three possibilities is the case', we have still not exhausted all possibilities. For we could then add the proposition that neither of these four is the case, etc. Thus, Becker concludes, it seems that in intuitionistic logic there is no '$n$-$tum$ $non$ $datur$'.[261]

Becker next gives Brouwer's proof that absurdity of absurdity of absurdity equals absurdity. He finishes this part with the remark that, whereas he has presented a phenomenological interpretation of the most important principles which holds specifically for Brouwerian logic, the problem of a calculus for the intuitionistic logic remains to be solved.[262] Becker, too, had not read Kolmogorov's paper, and he thus had to wait until Heyting published his formalisation. Nevertheless, Becker's analysis is a good example of how someone who was used to classical logic struggled to interpret the new, intuitionistic logic.

**Barzin & Errera, Lévy, and Church, 1927–1928** In a paper published in the *Académie Royale de Belgique, Bulletins de la Classe des Sciences* in 1927, Barzin and Errera reacted for the first time to Brouwer's intuitionism. Alfred Errera (1886–1960) had studied mathematics in Brussels before heading for Göttingen in 1909. For 3 years, he had followed lectures from, among others, Klein (projective geometry), Hilbert (partial differential equations), Zermelo (logical foundations of mathematics), Toeplitz (set theory), Weyl (function theory), and especially a number of algebraic lectures and seminars from Landau. He had written his dissertation in 1920 on the map-colouring problem. From 1921 onwards, Errera lectured at the *Université Libre de Bruxelles* and the *Ecole Militaire*. Most of Errera's work was in topology, number theory and set theory. Barzin was a friend and colleague of Errera, as a professor of logic at the *Université Libre de Bruxelles*. The cooperation between Barzin and Errera dates back to at least March 1927, of

---

[260] For some reason, Becker spells Wavre's name as 'Wawre'.

[261] [Becker, O. 1927, pp. 775–777]. Gödel later proved the statement, see 5.4.2. Becker's argumentation resembles the one used to set up the so-called Rieger-Nishimura lattice; cf. [Van Dalen 1997, pp. 189–190].

[262] [Becker, O. 1927, pp. 779–780]

## 5.3. THE DEBATE WIDENED 261

which time notes from Errera on a lecture of Barzin on logic are extant.[263]

Barzin and Errera present Brouwer as the one drawing the logical consequences of Kronecker's position of arithmetising mathematics, by rejecting the principle of the excluded middle. The argument they give is that the *tertium non datur* leads to non-constructive existence proofs, which intuitionists reject. If one interprets existence in a constructive way, Barzin and Errera argue, it is clear that 'for all' and 'there exist' do not make up the only possibilities. Therefore, they conclude, there has to be a *third*.[264]

Next, Barzin and Errera try to interpret what, in Brouwer's view, the third possibility should mean. First, they reject what we know to be the correct interpretation, namely that the third possibility is simply that which is not proved yet (and possibly improvable), with the following argument:[265]

> Si c'était vraiment là le sens du mot tiers, la réforme de M. Brouwer se réduirait à peu de chose. Nous savions depuis longtemps qu'une proposition incertaine ne pouvait être ni affirmée vraie, ni affirmée fausse. Mais nous avions l'habitude d'ajouter qu'assurément elle était l'un des deux. Si toute proposition devait devenir vraie ou fausse dans un avenir quelque éloigné qu'il fût, n'aurions-nous pas le droit d'affirmer le principe du tiers exclu?[266]

The argument resembles very much the one Wavre presented in his fictional dialogue.[267] It differs in that Barzin and Errera silently make the step from 'we have taken the habit of saying that a proposition is either true or false' to 'a proposition has to become true or false one day'. It is exactly the latter point that we do not know, and against which Brouwer protested.

Thus, Barzin and Errera conclude, the third is a value like true and false. It does not depend on our subjective knowledge, but it is an objective logical fact.[268] It is not clear in what sense the negation should be taken; Barzin and Errera do not mention Brouwer's 'negation as absurdity'.

Barzin and Errera next remark that, if Brouwer would obey his own criteria of constructive existence, he should have put forward a construction of a proposition which is third. However, they agree with Wavre that the Brouwerian counter-examples only show the possibility of a proposition which is third.[269]

---

[263] [Godeaux 1960]; '*Anmeldungs-Buch des stud. Math. Herrn Alfred Errera aus Brüssel*'; notes 'Barzin: log. math. 23.3.27', [ULB Errera]
[264] 'tiers', [Barzin & Errera 1927, pp. 56–58]
[265] [Barzin & Errera 1927, p. 59]
[266] 'If that really was the meaning of the word third, Mr. Brouwer's reform would be reduced to very little. We have known since a long time that an incertain proposition could neither be affirmed true nor be affirmed false. But we had the habit of adding that surely it was one of the two. If every proposition had to become true or false in a future how far away it might be, would we not have the right to affirm the principle of the excluded middle?'
[267] See 5.3.
[268] [Barzin & Errera 1927, p. 59]
[269] [Barzin & Errera 1927, pp. 59–60]

## 262  CHAPTER 5. REACTIONS: LOGIC AND THE EXCLUDED MIDDLE

Barzin and Errera easily step over this point, since they believe to have found a different way of refuting Brouwerian logic: by pointing out a contradiction. In order to do so, they formalise intuitionistic logic in the way which seems most logical to them. They then arrive at the principle of the excluded fourth: a proposition has to be either true, or false, or else third. They note that Brouwer has never stated this principle, but to them it seems impossible that he could refuse it. Thus, Barzin and Errera draw a different conclusion than Becker did, who was confronted with the same problem.[270] Finally, they prove the theorem that the notion of a third truth value implies contradiction.[271]

The conclusion, Barzin and Errera maintain, is that the contradiction was already present in Brouwer's postulates. They trace back what they see as the germ of the problem as follows. Brouwer admits simultaneously a third truth value (which they indicate by $p'$, meaning '$p$ is third'), the principle of double negation ($p \rightarrow \neg\neg p$), and the principle of transposition (($p \rightarrow q$) $\Rightarrow$ ($\neg q \rightarrow \neg p$)). But, Barzin and Errera argue, if one knows, in Brouwerian logic, that $p$ is true, one can only conclude about $\neg p$ that it is either false or third. A similar argument goes for the principle of transposition. Thus, one should adapt these two principles to read $p \rightarrow (\neg(\neg p) \vee (\neg p)')$ and ($p \rightarrow q$) $\Rightarrow$ ($\neg q \rightarrow (\neg p \vee p')$) respectively. Now the classical relationship between true and false has disappeared, only to be replaced by a similar one between true and not-true propositions, the latter comprising both false and third propostions. Barzin and Errera conclude that Brouwer's distinction either leads to a contradiction, if one sticks to the original formulation of the principles of double negation and transposition, or, if one adapts these principles, it loses all value. In this way, they consider it shown that arithmetisation is not the universal method of demonstration. The only universal criterion of rigour are the immovable laws of logic.[272]

Barzin and Errera's paper led to a surprising number of reactions. The ones by Lévy and Church are treated here; the ones by Glivenko and Heyting are dealt with in separate sections.[273]

Lévy once more entered the discussion, by reacting to Barzin and Errera's paper. He starts by remarking that the Brouwerian logic cannot lead to a contradiction, for the simple reason that it rests upon a choice among the propositions used in classical logic.[274]

Lévy proceeds in the same way as Barzin and Errera, namely by formalising intuitionistic logic. His formalisation, however, is more subtle. Since Lévy had a

---

[270] See 5.3.1.
[271] [Barzin & Errera 1927, pp. 60–68]
[272] [Barzin & Errera 1927, pp. 68–71]. In a letter to Church, Errera added that changes to logic would be 'detrimental to human thought', letter from Errera to Church, 30/9/1927, [AC Church]. I return to this point in 5.4.2.
[273] See 5.4.1 and 5.4.3 respectively.
[274] [Lévy 1927, p. 256]

## 5.3. THE DEBATE WIDENED

rather unfortunate choice of symbols, I here employ a symbolism which only in its basic elements is similar to his.[275]

Lévy starts with a proposition $\alpha$ and its opposite $\neg\alpha$. Classically, one of these has to be true: *tertium non datur*. He next introduces what he calls 'true in the sense of Brouwer', which, he adds, could also be called 'provable', designated by $+\alpha$. He remarks that the predicate 'provable' does not add anything to a proposition that is provable already, whereby two plusses reduce to a single one.[276] There are two problems with Lévy's suggestion. The first is that he does not differentiate between the semantical and the syntactical level. At the time, these distinctions were not made that clearly. The second is that, as Heyting later pointed out, stating that 'a proposition is provable' does not satisfy the intuitionistic demands. For it is equivalent to maintaining that there exists a proof of the proposition, which again contains the idea of transcendental existence which intuitionists reject.[277]

In Brouwerian terms, Lévy continues, there are three possibilities: either we have a proof for $\alpha$, or we have a proof for $\neg\alpha$, or 'it is not a solvable problem'.[278] Lévy calls the latter possibility 'third' and designates it by $\alpha'$. These are the only possibilities, so that the *quartum non datur* holds.[279] The description of the three possibilities is not correct, since the third possibility should include unsolved (and possibly unsolvable) problems. Furthermore, this is not a truth value in the ordinary sense, hence there is no *quartum non datur* in intuitionistic logic.[280]

But, Lévy maintains, one can go further and divide the case $\alpha'$ into two: either we can prove that we are in this case, indicated by $+\alpha'$, or we cannot, in which case we have $\alpha''$. Then the process of dividing stops, since if we would make $+\alpha''$, we would have a proof that we are in the case $\alpha'$, because $\alpha''$ is a subcase of $\alpha'$, which contradicts the fact that $\alpha''$ indicates that we are in $\alpha'$ but we cannot prove so. Lévy does not interpret these cases, but it is clear that $+\alpha'$ indicates a proved unsolvable problem, whereas $\alpha''$ indicates an unsolvable problem of which we cannot prove that it is unsolvable. Lévy doubts if $+\alpha'$ exists, since in that case one would have to prove both $\alpha$ and $\neg\alpha$ not to be true. But the only way one could prove that $\alpha$ is not true is by proving it false. This was exactly Brouwer's argument for rejecting this case.[281] However, Lévy still keeps open the possibility and thus leaves room for $+\alpha'$.[282]

Thus, Lévy continues, there are in fact four possibilities: $+\alpha$, $+\neg\alpha$, $+\alpha'$ and $\alpha''$. If one adds to this the classical possibility of distinguishing between true and false propositions even if there is no means of deciding, then, Lévy concludes, we

---
[275] An indication of Lévy's symbolism is given in the footnote below attached to the enumeration of the six cases.
[276] [Lévy 1927, p. 257]
[277] [Heyting 1930D, p. 959]; see 5.4.3.
[278] 'n'est pas un problème résoluble', [Lévy 1927, p. 257]
[279] [Lévy 1927, pp. 257–258]
[280] Later, Gödel proved that intuitionistic logic cannot be seen as a system of many-valued logic; see 5.4.2.
[281] See 2.3.2.
[282] [Lévy 1927, pp. 258–259]

end up with six cases: besides $+\alpha$, $+\neg\alpha$, these are $+\alpha'_1$, $+\alpha'_0$, $\alpha''_1$ and $\alpha''_0$, where the subscript indicates whether $\alpha$ is, classically seen, true (1) or false (0).[283]

Let us now look at an intuitionistic proposition $A = +\alpha$. In intuitionism, Lévy maintains, the leading idea is to treat only provable propositions. Let us now ask when $A$ is provable, when $\neg A$ is provable, and when $A$ is third, in terms of $\alpha$. Since $A = +\alpha$, $+A$ amounts to $+ + \alpha$ which is equivalent to $+\alpha$. So $A$ is provable in case $+\alpha$ holds. When is $\neg A$ provable? $\neg A$ is equivalent to $\neg + \alpha$, thus $\neg A$ is provable means that we have a proof, and the proof shows that $\alpha$ is not provable. Thus, $+\alpha$ and $\alpha''$ are excluded, and $+\neg\alpha$ and $+\alpha'$ remain. Finally, $A$ is third if $\alpha''$. But, Lévy argues, this means that, if we only consider provable propositions, we have no means of distinguishing between $+\neg\alpha$ and $+\alpha'$. So we can just as well introduce a new negation, the Brouwerian negation, designated by $\sim$, which comprises both these cases. In this way, Brouwerian logic is reduced to three cases: $A$, $\sim A$ and $A'$, the latter of which is not provable.[284] Note that, if Lévy had concluded with Brouwer that $+\alpha'$ does not occur, the whole problem would not have existed.

Lévy comes to the conclusion that there are three possible logics:

- classical logic, which accepts $+\alpha$, $+\neg\alpha$, $+\alpha'_1$, $+\alpha'_0$, $\alpha''_1$ and $\alpha''_0$;

- mixed logic, which accepts $+\alpha$, $+\neg\alpha$, $+\alpha'$ and $\alpha''$;

- Brouwerian logic, which accepts $+\alpha$, $\sim \alpha$ and $\alpha''$.

He next investigates what happens to the principle of double negation in the different logics. He argues that in Brouwerian logic the 'Brouwerian affirmation'[285] is equivalent to the double Brouwerian negation. His argumentation, however, goes wrong. He argues that, if we have a proposition $\alpha$, its Brouwerian negation is $+\neg\alpha$, and therefore its double Brouwerian negation indicates the cases $+\alpha$ and $+\alpha'$.[286] This, however, is not the case. If we have a proof that $+\neg\alpha$ is not the case, then we can also be in case $\alpha''$.

Lévy concludes that, first, in Brouwerian logic one does not remain faithful to the ideal of dealing only with provable propositions. Second,[287]

> (...) le langage nouveau me paraît plus nuisible qu'utile, et en tout cas il n'apporte rien de nouveau: tous les résultats de la Logique brouwerienne peuvent s'énoncer avec le langage usuel, en se servant des mots *démontrable* et *indémontrable*, et même, à mon avis, deviennent ainsi singulièrement plus compréhensibles.[288]

---

[283][Lévy 1927, p. 258]. As an indication of Lévy's symbolism: Lévy designated the six cases as $A$, $B$, $\gamma_1$, $\gamma_2$, $\alpha''_1$ and $\alpha''_2$.
[284][Lévy 1927, pp. 258–261]
[285]'affirmation brouwerienne', [Lévy 1927, p. 263]
[286][Lévy 1927, pp. 261–263]
[287][Lévy 1927, p. 266]
[288]'(...) the new language seems to me more harmful than useful, and in any case it does

## 5.3. THE DEBATE WIDENED

Lévy's first conclusion rests on the incorrect idea that the third possibility is something like a truth value that can actually be posited. His second conclusion is supported by an incorrect proof that in Brouwerian logic the same rule regarding the principle of double negation holds as within classical logic. Nevertheless, Lévy's paper is a step towards a formalisation of intuitionistic logic. It shows that people took the problem seriously, and it once more shows the problems they had in interpreting it.

The following year, Alonzo Church (1903–1995) reacted to Barzin and Errera's paper in the *Bulletin of the American Mathematical Society*. Church had obtained his doctorate in Princeton in 1927. At the time he wrote the paper, he was staying at Harvard as a National Research Fellow. In the same capacity, he was to spend some time in Göttingen, and in Amsterdam in 1929, where he visited Brouwer.[289]

Church states as the object of his paper to discuss the possibility of a system of logic in which the principle of the excluded middle is not assumed, and to point out some errors in Barzin and Errera's paper. Church correctly describes Brouwer's position as not regarding the principle of the excluded middle as an admissible logical principle. He notes that Brouwer's identification of this principle with the solvability of all problems depends on the identification of the truth of a proposition with our possibility to prove so. But, Church remarks pragmatically,[290]

> it seems more in accord with our usual ideas to think of truth as a property of a proposition independent of our ability to prove it.

Like Dresden, Church takes up a pragmatic position. Referring to the different geometries, he argues that also in logic, it is meaningless to ask about the absolute truth of a proposition. The main criteria for choosing between different systems are, in Church's view, simplicity and serviceability.[291]

Turning to Barzin and Errera's paper, Church convincingly argues that their reasoning that Brouwer has to admit a third truth value is incorrect. For the introduction of a third truth value is motivated by the perceived need to explicitly deny the principle of the excluded middle, and[292]

> (...) the insistence that one who refuses to accept a proposition must deny it can be justified only by an appeal to the law of the excluded middle, the very principle in doubt.

The proof method Barzin and Errera use, as Church points out, is the *reductio ad absurdum*, which can only be justified by accepting the principle of the excluded middle.

---
not bring anything new: all the results of Brouwerian logic can be expressed in the usual language, using the words *provable* and *not provable*, and even, in my view, become much more comprehensible in that way.'

[289] [Enderton 1995]; Letter from Brouwer to Mannoury, 10/4/1929, [MI Brouwer, CB.GMA.15]
[290] [Church 1928, p. 76]
[291] [Church 1928, pp. 75–77]
[292] [Church 1928, p. 77]

266    CHAPTER 5. REACTIONS: LOGIC AND THE EXCLUDED MIDDLE

Furthermore, Church argues, one cannot assume that Barzin and Errera's *tiers* covers all possibilities other than true and false. For, designating '$p$ is tiers' by $p'$, a different possibility would be $(p')'$. And even if we would group all (constructed) possibilities under a new name, say $p*$, we could again apply the $'$ to obtain the not-included case $p*'$.[293]

Before the publication of his paper, Church corresponded with Errera on the issue of intuitionistic logic and the possibility of a third truth value. The correspondence shows that Errera was not impressed by Church's second objection, claiming that 'everybody agrees about a classification being exhaustive, if the last compartment is defined in a purely negative manner, as comprising any object not belonging to any other compartment.'[294] The first objection, formulated so clearly in Church's paper, is not similarly present in the correspondence. Church does point out to Errera that[295]

> if your argument is to be conclusive not only against those who assert the existence of tiers propositions, but also against those who merely omit the law of excluded middle from among their logical principles, then you must have made an actual proof of the law of excluded middle, and this law would therefore be a theorem, and not a postulate (...).

Indeed: only if the principle of the excluded middle is proved to be a theorem, it cannot simply be omitted by those who object against its use.

Interestingly, the correspondence further shows that Church was, at the time, inclined to agree with Barzin and Errera's interpretation of Brouwerian logic:[296]

> I think it quite probable that you are correct in saying that Prof. Brouwer actually does assert the existence of tiers propositions. It was partly my uncertainty as to what position he did take that caused me to write to him at the time that I wrote to you. I have not, however, heard from him in reply.

It is not clear what convinced Church that Barzin and Errera's interpretation was wrong, as he claimed outspokenly in his paper.

Barzin and Errera had also sent their paper to Brouwer for a reaction, but received no answer either. The 'official' intuitionistic reaction only came three years later, when Heyting entered the discussion.[297]

**Dresden and Pierpont in 1927** In a lecture delivered before the American Mathematical Society, the Mathematical Association of America and the American Association for the Advancement of Science at Nashville in December 1927,

---
[293][Church 1928, p. 78]
[294]Letter from Errera to Church, 30/9/1927, [AC Church]
[295]Letter from Church to Errera, 15/8/1927, [AC Church]
[296]Letter from Church to Errera, 15/8/1927, [AC Church]
[297]See 5.4.3.

## 5.3. THE DEBATE WIDENED

Dresden[298] reacted to Barzin and Errera's interpretation of Brouwerian logic. He correctly points out that their interpretation is not right, whence the contradiction they found does not affect intuitionistic logic. Like Lévy, Dresden argues that if intuitionistic logic would lead to a contradiction, this would leave little hope for classical logic.[299]

In Dresden's view, the significance of Brouwer's investigations lies in the fact that he has 'freed the mind from the compulsory use of the Aristotelian base'.[300] Dresden pleads the pragmatic case of various types of logic which may be used at various occasions, just as one can choose to work in Euclidean geometry one day and in non-Euclidean geometry the next. He doubts if he will personally be able to work without the principle of the excluded middle, but he is ready to admit that 'any one who can, may do important work in that way'.[301] Dresden met Errera in the Summer of 1928 at a meeting in Amherst, in which they discussed their points of view regarding Brouwer's intuitionistic logic.[302]

At the same occasion, Pierpont lectured on 'mathematical rigor, past and present'. James Pierpont (1866–1938) was a professor at Yale University, a position he had held since 1898. He had studied both in the United States and in Europe, mostly in Berlin and Vienna. In Berlin, he got to know Kronecker's work, which influenced him greatly. It is reasonable to assume that Pierpont was influenced by Kronecker personally, since at the time he presumably was in Berlin Kronecker was still alive and held a position at the Berlin university.[303] Pierpont's own mathematical work was mostly in algebra, the theory of functions and the theory of relativity.[304]

At the very end of his lecture, Pierpont arrives at 'intuitionalism'. He characterises Brouwer as 'very much of a philosopher',[305] just like Kronecker and Weyl. Brouwer, Pierpont argues, carries Kronecker's idea of constructive mathematics to its logical end, which is that the principle of the excluded middle is only legitimate in finite sets. Pierpont motivates Brouwer's decision by the explanation that the logical principles originated as abstractions from procedures applied to finite sets. As long as this is posited as a different way of doing mathematics, like swimming the Channel with one's hands tied, nobody would object. However, Pierpont continues, Brouwer does not hesitate to tell other mathematicians that they are wrong. In opposition to many of his contemporaries, Pierpont views the complications that arise from giving up the excluded middle as 'an element of strength rather than of weakness' in Brouwer's theory. Unfortunately, he does not specify why this would be the case. Pierpont ends his paper with the question

---
[298] For biographical information on Dresden, see 5.3.
[299] [Dresden 1928, p. 441]
[300] [Dresden 1928, p. 448]
[301] [Dresden 1928, p. 448]
[302] Letter from Dresden to Errera, 7/10/1928; [ULB Errera]
[303] Pierpont left for Europe shortly after his graduation in 1886, and spent the last years in Europe in Vienna where he obtained his doctor's degree in 1894. Kronecker died in 1891.
[304] [Ore 1939]
[305] [Pierpont 1928, p. 51]

### 5.3.2 The excluded middle as a minor subject

Between 1925 and 1928, there were a number of contributions in which the principle of the excluded middle was touched upon briefly. For the sake of completeness, these are listed below.

**1925** In **Heyting**'s dissertation, published in 1925, the intuitionistic criticism is linked to its view on mathematical existence.[307] The philosopher and physician Hans **Lipps** mentioned Brouwer's and Weyl's criticism of the excluded middle in a footnote, in which he interpreted it in terms of decidability.[308] At the Scandinavian Mathematical Conference in Copenhagen in the fall of 1925, the Swedish mathematician Torsten **Brodén** did not deny that the rejection of the principle of the excluded middle was a consequence of the intuitionistic point of view, but he reproached the intuitionists of giving 'a too dominating role to the human capacity of thought'.[309] **Oikonomou** should be mentioned as the first (and last) person (in the period covered here) to bring the debate to the Greek-speaking world, by mentioning Brouwer and Weyl's criticism of the excluded middle in a lecture to the Greek Mathematical Society in November 1925.[310] The Polish mathematician Oscar **Zariski**, at the time a student of Castelnuovo in Rome, did the same for the Italian readers, touching on Brouwer's and Weyl's disputing the absolute validity of the principle of the excluded middle.[311] The mathematics gymnasium teacher Walter **Lietzmann**, who also lectured at the university of Göttingen and who was president of the *Deutsche Verein für Förderung des mathematischen und naturwissenschaftlichen Unterrichts* ('German Association for the Advancement of Education in Mathematics and Natural Sciences'),[312] presented the intuitionistic view on the principle of the excluded middle in an article for a German didactical journal. Lietzmann explains it by pointing out that there may be cases in which we cannot decide whether a proposition or its negation is the case.[313] In a theological-philosophical journal, **Rivier** explains Brouwer's criticism of the principle of the excluded middle by pointing out that for Brouwer Aristotelian logic only holds for finite totalities. Rivier concludes in favour of Hilbert's 'idealistic'

---

[306][Pierpont 1928, pp. 50–53]

[307][Heyting 1925, p. 1]. Heyting's view on mathematical existence is treated in 4.3.2.

[308][Lipps 1925, p. 71]

[309]'der menschliche Gedankenkapazität eine allzu dominierende Rolle erteilt', [Brodén 1925, p. 236]

[310][Oikonomou 1926, p. 80]

[311][Zariski 1925, p. 73]. Zariski devoted more attention to the subject of mathematical existence; see 4.3.2.

[312][Lietzmann 1960, p. 5]

[313][Lietzmann 1925, p. 357]. The quote which is used as a motto for this chapter comes from this paper.

## 5.3. THE DEBATE WIDENED

position rather than of Brouwer's 'empiricist' one. However, he seems to have misunderstood Brouwer, since he characterises the empiricist point of view by stating that the progress of the mind cannot add or deduce anything from the universe.[314]

**1926** In 1926 **Finsler**, who had treated the subject in his inaugural lecture in 1923,[315] returns to the intuitionistic criticism of the excluded middle, only to mention that he will not go into the matter.[316] The same applies to the German mathematician Otto **Hölder** (1859–1937), who had written his dissertation in Göttingen and who at the time lectured in Leipzig. He dryly remarks that the future will learn whether Brouwer's investigations into the development of mathematics without using the principle of the excluded middle in infinite totalities is fruitful.[317] The Viennese physicist Heinrich **Löwy** gives a popular presentation of Brouwer's criticism of the use of the principle of the excluded middle for infinite totalities, and rightly concludes that to Brouwer a logical theorem is only true in as far as it is verified.[318] The German philosopher Walter **Dubislav** (1895–1937) places Brouwer's rejection of the principle of the excluded middle in the framework of Brouwer's view that logic depends on mathematics and not the other way round. Dubislav repudiates the idea on the grounds that, if one supposed that logic could be derived from mathematics, as Brouwer indeed maintains, then one would need arguments in the derivation which were definitely logical.[319] Dubislav's argumentation is rather circular and in any case more a matter of definition.[320] In the Netherlands, Johan **Barrau** (1873–1946), one of the public opponents at Brouwer's promotion,[321] touched on the intuitionistic criticism of the excluded middle in his rector's address at the university of Groningen.[322] **Betsch**, in his *Fiktionen in der Mathematik* ('Fictions in mathematics'), mentions the intuitionistic rejection of the principle of the excluded middle in infinite totalities, explaining that the principle is seen as illegitimately transposed from finite to infinite totalities.[323] **Dingler** explains Brouwer's rejection of the principle of the excluded middle by means of the possibility of unsolvable problems. He reproaches Brouwer to have started making a new mathematics based on the unproved possibility of unsolvable problems, instead of having devoted his attention to the clarification of the mathematical methodology.[324] **Kohnstamm** considers Brouwer's counter-examples not

---

[314] [Rivier 1925]
[315] See 5.2.3.
[316] [Finsler 1926B, p. 683]
[317] [Hölder 1926, p. 245; p. 250]
[318] [Löwy 1926, pp. 707–708]
[319] [Dubislav 1926, p. 196]
[320] Dubislav seems to have been rather confused on the matter. In 1929, he wrote a letter to Brouwer claiming that it was fairly easy to establish a definite counter-example to the principle of the excluded middle, only to mix up questions of truth and of provability (letter from Dubislav to Brouwer, 16/1/1929, [MI Brouwer]).
[321] See 2.3.1.
[322] [Barrau 1926, p. 15]
[323] [Betsch 1926, p. 339]
[324] [Dingler 1926, pp. 92–93]

270    CHAPTER 5. REACTIONS: LOGIC AND THE EXCLUDED MIDDLE

to be counter-examples against the excluded middle, since the definition Brouwer uses does not, in Kohnstamm's view, contain enough information to answer the question. Kohnstamm argues that the excluded middle does not say anything about objects of thought, but only about propositions.[325] **Larguier des Bancels** reacts to a paper by Rivier the year before, also in the *Revue de théologie et de philosophie*, on the principle of the excluded middle. He concludes that there are no new questions in philosophy anymore, since the same subject that is discussed by mathematicians was debated already in Antiquity. Larguier des Bancels seems to be in favour of intuitionism, since he suggests to leave the indeterminable undetermined.[326]

**1927**  By 1927, less than half of the contributions to the foundational debate treated the principle of the excluded middle.

The young German philosopher Robert **Heiss**, who had taken his doctor's degree in Göttingen, presents Brouwer's intuitionism as a reaction to the set theoretical paradoxes. He describes his point of view as 'simply'[327] giving up the principle of the excluded middle, and argues:[328]

> Aber wenn sich der Mathematiker über die Gefahr dieses radikalsten Vorgehens vielleicht auch täuschen kann, so weiß der Philosoph um so genauer, daß das Kind hier mit dem Bade ausgeschüttet wird. Denn wenn man den Satz vom ausgeschlossenen Dritten *überhaupt* aufgibt, so gibt man viel von der Logik auf.[329]

Just like some mathematicians argued that philosophers should not interfere with their business, so Heiss claims a superior position for philosophers, without, however, in any way supporting his position.

In his *Philosophie der Mathematik und Naturwissenschaft* ('Philosophy of mathematics and natural science'), **Weyl** devotes little attention to the excluded middle. Basing himself on Brouwer, Becker and his own writings, he merely mentions the rejection of the *tertium non datur* as one of the consequences of 'intuitive mathematics'.[330]

In his text-book for secondary schools, **Lietzmann** mentions the intuitionistic rejection of the principle of the excluded middle for infinite totalities.[331]

**Hartmann** published a short note in the *Philosophisches Jahrbuch der Görres-Gesellschaft*. He claims that Brouwer's view on the principle of the exclude middle

---
[325] [Kohnstamm 1926, p. 59]
[326] [Larguier des Bancels 1926]. Rivier, in turn, reacted to Larguier des Bancels in [Rivier 1930].
[327] 'einfach', [Heiss 1928, p. 405]
[328] [Heiss 1928, p. 405]
[329] 'But although the mathematician may perhaps be mistaken about the danger of this most radical proceeding, the philosopher knows all the more precisely that here the baby is thrown out with the bath-water. For if one gives up the law of the excluded middle at all, then one gives up a good deal of logic.'
[330] [Weyl 1927A, p. 42]
[331] [Lietzmann 1927, p. 13]

## 5.4. LATER REACTIONS

has to be rejected, since this principle possesses the highest evidence and thus cannot be doubted.[332]

Finally, Harald **Landry** used Brouwer's Berlin lectures and an accompanying lecture by Scholz for the *Kant-Gesellschaft* to write a very positive newspaper article about intuitionism. He explains the rejection of the principle of the excluded middle by means of the intuitionistic continuum, and maintains that the most positive consequences of Brouwer's ideas are to be found not in philosophy, but in mathematics.[333]

## 5.4 Later reactions

Between 1928 and 1933, the reactions to the intuitionistic view on the principle of the excluded middle that draw most attention are the following. In the first place, the formalisation of intuitionistic logic and its interpretation was developed substantially by Glivenko, Heyting and Kolmogorov. Secondly, building on this work, Gödel proved a most clarifying result on the relationship between intuitionistic and classical mathematics. Finally, Barzin and Errera's contributions should be mentioned, not because there are so valuable in themselves, but because they have become notorious in foundational circles.

### 5.4.1 Glivenko, Heyting and Kolmogorov

**Glivenko** The second person from the Soviet-Union who contributed to the debate and who, like Kolmogorov,[334] did so in a positive way, was Glivenko. Unfortunately, I know very little about his personal background. More specifically, I do not know how he became interested in intuitionistic logic. What is clear is that Glivenko was the first person involved in the foundational debate who knew of Kolmogorov's work, by the end of 1928.[335]

In reaction to Barzin and Errera's paper,[336] Glivenko published a short note in the *Académie Royale de Belgique, Bulletin de la Classe des Sciences* in 1928. As he wrote in a letter to Heyting, he does not go deeply into the matter:[337]

> Dans ma note polémique de Bruxelles, je n'avais pas pu de [sic] poser ce problème assez nettement, parce que là, conformément au but spécial de cette note, j'ai eu besoin d'un language qui pourrait être comprise sans peine par des savants qui n'ont pas pénétrés, à mon avis, assez profondement aux idées intuitionnistes.[338]

---
[332][Hartmann 1927]
[333][Landry 1927]
[334]See 5.3.1.
[335]Letter from Glivenko to Heyting, 13/10/1928; [TLI Heyting, B gli-281013]
[336]See 5.3.1.
[337]Letter from Glivenko to Heyting, 4/7/1928; [TLI Heyting, B gli-280704]
[338]'In my polemic Brussels notice, I could not pose the problem sufficiently clearly, since there,

## 272   CHAPTER 5. REACTIONS: LOGIC AND THE EXCLUDED MIDDLE

Despite Glivenko's description of the paper as polemic, it actually contains hardly anything more than a very to-the-point proof that Barzin and Errera's assumption that Brouwer uses a three-valued logic is false.

Glivenko proceeds in the following way. First, he gives an (incomplete) axiomatic system for classical propositional logic,[339] noting that Brouwerian logic differs from classical logic by its rejection of the principle $\neg p \vee p$.[340] Then, he proves two propositions Brouwer had stated or proved before without the use of an axiomatic system, namely $\neg\neg(\neg p \vee p)$ and $\neg\neg\neg q \rightarrow \neg q$. Note that with the proof of the double negation of the principle of the excluded middle, Glivenko, like Kolmogorov before but in a more direct way, proved Brouwer right when he claimed in 1908 that one will never get caught in a contradiction by illegitimately using the principle of the excluded middle.[341] Next, Glivenko puts forward a new theorem, namely

**Theorem 5** *In Brouwerian logic, the proposition 'the proposition $\neg p \vee p$ implies the falsity of a proposition $q$' implies the falsity of the proposition $q$.*

The theorem is proved in a straightforward way. Finally, Glivenko can refute Barzin and Errera's claim. Suppose that there is a third truth value in Brouwerian logic, expressed by ' (i.e., $p'$ indicates that $p$ has the truth value third). Then the falsity of a proposition implies that its thirdness is false: $\neg p \rightarrow \neg p'$. Similarly, the truth of a proposition implies the same: $p \rightarrow \neg p'$. Thus, we have $(\neg p \vee p) \rightarrow \neg p'$, which, by the foregoing theorem, means that $\neg p'$.[342]

In a second paper in the same journal, published the following year, Glivenko goes somewhat deeper into intuitionistic logic. He now gives a full axiomatisation of intuitionistic propositional logic, and proves the following results:[343]

**Theorem 6** *If a certain expression in propositional logic is provable in classical logic, then the falsity of its falsity is provable in Brouwerian logic.*

This theorem is also easily concluded from Kolmogorov's 1925 paper, although Kolmogorov himself did not state it in these terms.[344]

**Theorem 7** *If the falsity of a certain expression in propositional logic is provable in classical logic, then the same falsity is provable in Brouwerian logic.*

---

in accordance with the special goal of that notice, I needed a language which could be understood without difficulty by scholars who have not, in my view, sufficiently deeply penetrated the intuitionistic ideas.'

[339] Glivenko later noted that the system is not complete. (Letter from Glivenko to Heyting, 4/7/1928; [TLI Heyting, B gli-280704])

[340] When I use symbolic expressions, Glivenko does so too, unless indicated otherwise. The symbols were adapted to modern standard notation.

[341] See 2.3.2.

[342] [Glivenko 1928]

[343] [Glivenko 1929, p. 183]

[344] See 5.3.1.

## 5.4. LATER REACTIONS

At first, Glivenko had planned not to publish the note if Heyting would include it in his paper on the formalisation of intuitionistic logic, but later he decided to publish his work independently of Heyting's, since the proof for the statements is rather long.[345]

If we look at the paper in terms of providing a translation from classical into intuitionistic logic, the result is the same as what one can derive from Kolmogorov's 1925 paper,[346] which was apparently unknown even to Glivenko at the time he wrote the paper. The translation used, however, differs.

Glivenko proves the theorems in the following way. First, he remarks that four axioms have to be added to the ones he presented in his paper the year before in order to make the system complete. The two most disputed ones, in his view, are $p \to (q \to p)$ and $\neg q \to (q \to p)$. On the basis of the interpretation of the axioms, he argues that they are admissible in intuitionistic mathematics.[347] He then proves the first theorem by showing that the translation of all the classical axioms are true in intuitionistic propositional logic and that the translation preserves the derivation rules. The translations of the classical axioms are true for those axioms which also appear in intuitionistic mathematics, Glivenko argues, since in intuitionistic propositional logic $p \to \neg\neg p$ holds, as he had proved in his 1928 paper. For the remaining axiom, $\neg p \vee p$, the translation also holds, since he had proved $\neg\neg(\neg p \vee p)$ in the same paper. Finally, Glivenko proves that the translation preserves the derivation rules for substitution and *modus ponens*. The second theorem follows directly from the first by remarking, as Glivenko does, that in intuitionistic logic $\neg\neg\neg p \to \neg p$.[348]

**Heyting** In 1930, Arend Heyting (1898–1980) published three papers in which he presented a formalisation of intuitionistic logic and mathematics.[349] The fact that a number of people before him had tried to formalise intuitionistic logic makes it clear that there was a need for such a work. Heyting's papers soon came to represent the standard formalisation of intuitionistic logic.

Heyting was born in 1898 in Amsterdam, where he also grew up. He finished the HBS[350] in 1915 with excellent grades, and passed the state examination which gave him access to university studies the following year. He chose to study mathematics at the university of Amsterdam. At the time, the mathematicians who lectured there were Brouwer, Mannoury, Korteweg and De Vries. Heyting took about all Brouwer's lectures he could, none of which was about intuitionism.[351] In 1922, Heyting did his master's, *cum laude*. He took a job as a mathematics teacher

---

[345]Letters from Glivenko to Heyting, 30/10/1928 and 13/11/1928; [TLI Heyting, B gli-281030; B gli-281113]. Heyting's paper is treated below.
[346]See 5.3.1.
[347]In fact, he mentions that Heyting pointed out the admissibility of these axioms to him.
[348][Glivenko 1929], which also contains details on the proof for the modus ponens rule.
[349][Heyting 1930A], [Heyting 1930B] and [Heyting 1930C]
[350]The *Hogere Burgerschool* ('Higher Citizen's School'), a secondary school type created by the Dutch 1863 educational reform, focused on middle class youngsters.
[351][Van Dalen 2001, p. 287]

in Enschede, in the far east of the Netherlands.³⁵² In his spare time, he worked on his dissertation on an axiomatic system for intuitionistic projective geometry under Brouwer. He finished it in 1925, again obtaining the degree 'cum laude'. Since university positions in mathematics were very scarce in the Netherlands at the time, Heyting became *privaat-docent*, at the university of Amsterdam, only in 1936.³⁵³ A prize competition in 1927 placed Heyting at the center of attention in foundational circles.

Every year, the *Wiskundig Genootschap* (Dutch for: 'Mathematical Society') offered a prize for the solution of certain mathematical problems. In 1927, one of the prize problems was the formalisation of Brouwer's set theory, a theme suggested by Mannoury. Solutions had to be handed in anonymously, with a motto to identify the author later. There was one entry on this subject, under the motto '*Steenen voor brood*' ('Stones for bread').³⁵⁴ As the motto indicates, the author thought formalisation a good thing, but not the central issue in mathematics. The essay was crowned in the beginning of 1928, and its author turned out to be Heyting. In its report, the committee, consisting of Mannoury, Wolff and Schuh, praised the work as 'a formalisation fulfilled in a most knowledgeable way and with admirable perseverance'.³⁵⁵

Even though Brouwer was, in general, not in favour of formalisation, he did appreciate Heyting's work. He even went so far as to suggest him to formalise more than what he had done already.³⁵⁶

The word about Heyting's work spread quickly, even before it was published. Bernays was told about it by Łukasiewicz and Tarski at the 1928 Bologna conference,³⁵⁷ and Church wrote to Errera about it in 1929.³⁵⁸

Heyting's work was meant to appear in the *Mathematische Annalen*, where it had already been accepted for publication. However, because of the fight between Hilbert and Brouwer in its editorial board,³⁵⁹ Brouwer withdrew the paper for publication, an action to which Heyting consented. Instead, Brouwer suggested to publish it in the *Sitzungsberichte der Preussischen Akademie der Wissenschaften*, with mediation of Bieberbach.³⁶⁰ Publication in that journal, however, was not without problems either. As becomes clear from a letter Hurewicz wrote to Brouwer, the Berlin Academy questioned whether the level of Heyting's paper was higher than that of papers they could otherwise publish in its place.³⁶¹ Despite the publication

---

[352] Presumably because his wife was much attached to the region *Twente*, as his son Arend Heyting later told Troelstra; oral communication from Troelstra to the author, 26/9/1998.
[353] [Troelstra 1981, pp. 1–6]
[354] An expression derived from the Bible, meaning 'the useless for the useful'.
[355] 'een met grote kennis van zaken en bewonderenswaardige volharding ten einde gebrachte formaliseering', report on the answer to prize question number 9, 1927; [TLI Heyting, P 30-d]
[356] Letter from Brouwer to Heyting, 17/07/1928; [TLI Heyting, B bro-280717]
[357] Letter from Bernays to Heyting, 5/11/1930; [TLI Heyting, B ber1-301105]
[358] Letter from Church to Errera, 29/5/1929, [AC Church]
[359] See 2.8.
[360] Letter from Brouwer to Heyting, 28/9/1929, letter from Heyting to Brouwer, 7/10/1929; [TLI Heyting, B bro-290928; 291007*]
[361] Letter from Hurewicz to Brouwer, 25/11/1929, [MI Brouwer]

## 5.4. LATER REACTIONS

in this much lesser known journal, the papers drew substantial attention. Since no version of the original Heyting essay has survived, it is not clear how much was changed from the prize winning essay to the published papers in 1930.[362]

Halfway the second of the three papers, Heyting starts formalising intuitionistic arithmetic, choice sequences and set theory, which are not relevant to us. Also, these parts played a far less important role in the reception of intuitionism. Therefore, they are not treated here. What is exposed below is the contents of the first paper, devoted to intuitionistic propositional logic, and the first part of the second paper, which extends the formalisation to predicate logic. These parts drew most of the attention.

Heyting opens the first paper fully in Brouwerian style by playing down the value of language:[363]

> Die intuitionistische Mathematik ist eine Denktätigkeit, und jede Sprache, auch die formalistische, ist für sie nur Hilfsmittel zur Mitteilung. Es ist prinzipiell unmöglich, ein System von Formeln aufzustellen, das mit der intuitionistische Mathematik gleichwertig wäre, denn die Möglichkeiten des Denkens lassen sich nicht auf eine endliche Zahl von im voraus aufstellbaren Regeln zurückführen.[364]

The only justification for the use of a formalised language, Heyting claims, lies in its conciseness and definiteness, which make it easier to understand intuitionistic mathematics. This already makes it clear that Heyting, contrary to Brouwer, attached great importance to a didactical exposition of intuitionism.

Furthermore, Heyting continues, for the construction of mathematics one does not need to formulate general logical rules, since they will each time be discovered anew while building the mathematical system. However, Heyting maintains, if he would have described the language that follows the intuitionistic construction of mathematics, the result would be so different from the normal form of mathematics that one would again lose the advantages of conciseness and definiteness. Therefore, he does start his formalisation of intuitionistic mathematics with a logical calculus.[365]

Such an 'intuitionist's apology' may have seemed somewhat strange to the uninitiated reader, but it is completely understandable from the intuitionistic point of view. To Brouwer, logic and language (especially a formalised language) came second, after mathematics. Thus, to start the presentation of intuitionistic mathematics with a formalisation of logic could easily have been seen as running contrary

---

[362][Troelstra 1990, p. 2]
[363][Heyting 1930A, p. 191]
[364]'Intuitionistic mathematics is a thought activity, and for it every language, including the formalistic one, is only an aid for communication. It is fundamentally impossible to draw up a system of formulas which would be equivalent to intuitionistic mathematics, for the possibilities of thought cannot be reduced to a finite number of rules which can be drawn up in advance.'
[365][Heyting 1930A, p. 191]

to Brouwer's intention. In fact, however, very few people seem to have bothered about Heyting's introduction, and all attention was drawn to the formal system.

For Heyting, the introduction was not meant as a mere apology towards Brouwer. The regret he later showed about the impression the papers had left reinforces the value he attached to the remarks made there:[366]

> I regret that my name is known to-day mainly in connection with these papers, which were very imperfect and contained many mistakes. They were of little help in the struggle to which I devoted my life, namely a better understanding and appreciation of Brouwer's ideas. They diverted the attention from the underlying ideas to the formal system itself.

Indeed, the fact that this formalisation is sometimes taken as the core of intuitionism can be seen as 'Hilbert's ultimate victory'.[367] Nevertheless, I think Heyting is too pessimistic. Although it is true that the formal system he presented drew all attention, one must not forget that to many people this was the first time they could read about intuitionism in an understandable language, namely the language of formalised mathematics. There was a great need for such a contribution.

Heyting continues the introduction to his paper by remarking that, if one makes sure that the formulas in the calculus can only be interpreted in a meaningful way, then the consistency of the axiomatic system is guaranteed automatically. Here, consistency is taken in the sense that it is not possible to have a formula appearing in the formal system which, if interpreted, would yield a contradictory proposition. Consistency in the sense of a certain formula (nowadays indicated by $\bot$) not being provable is, Heyting maintains, less important. Heyting remarks that the system he presents is not complete in the sense of Hilbert-Ackermann,[368] since one can always add the formula $\neg\neg a \to a$.[369]

Heyting uses four basic connectives, namely $\to$, $\wedge$, $\vee$ and $\neg$.[370] He remarks that a first difference between intuitionistic and classical logic is that in the former, none of these concepts can be defined by means of the others.[371] Furthermore, Heyting gives some examples of implications which are true only one way in intuitionistic logic, whereas they go both ways in classical logic, such as: $\vdash (\neg a \vee b) \to (a \to b)$ and $\vdash (a \vee b) \to \neg(\neg a \wedge \neg b)$.[372]

---

[366][Heyting 1978, p. 15]
[367][Posy 1998, p. 301]
[368]One of the ways in which Hilbert and Ackermann had defined completeness was: if one adds a formula to a system which is not derivable in the system, then the system becomes inconsistent, [Hilbert & Ackermann 1928, p. 33].
[369]I modernised the symbolism used by Heyting.
[370]The nowadays usual notation $\neg$ seems to come from this paper of Heyting. Heyting remarks that he chose a new sign for negation, instead of the usual $\sim$, since negation is interpreted differently in intuitionistic logic, [Heyting 1930A, p. 192].
[371]In classical propositional logic, for example $\vee$ and $\neg$ form a functionally complete set of connectives, cf. [Van Dalen 1997, pp. 24–25].
[372][Heyting 1930A, pp. 192–193]

## 5.4. LATER REACTIONS

The section on negation, Heyting remarks, differs most from classical logic. He adds:[373]

> Den Schein, daß die dort hervortretenden Unterschiede eigentlich den wichtigsten Streitpunkt zwischen Intuitionisten und Formalisten bilden (...), konnte ich hier nicht vermeiden, weil der Formalismus ungeeignet ist, die mehr fundamentalen Streitpunkte auszudrücken.[374]

As to literature on logic, Heyting refers to the works of Russell and Whitehead, Hilbert and Ackermann, Bernays, and Glivenko as the ones he used most for his papers.[375]

Heyting next presents his axiomatic system. Contrary to Kolmogorov's earlier formalisation,[376] Heyting treats all the connectives. The axioms he uses are the following:

(1) $a \to a \wedge a$
(2) $a \wedge b \to b \wedge a$
(3) $(a \to b) \to ((a \wedge c) \to (b \wedge c))$
(4) $((a \to b) \wedge (b \to c)) \to (a \to c)$
(5) $b \to (a \to b)$
(6) $(a \wedge (a \to b)) \to b$
(7) $a \to (a \vee b)$
(8) $(a \vee b) \to (b \vee a)$
(9) $((a \to c) \wedge (b \to c)) \to (a \vee b) \to c$
(10) $\neg a \to (a \to b)$
(11) $((a \to b) \wedge (a \to \neg b)) \to \neg a$

Heyting had obtained these axioms by going through the axioms of the *Principia Mathematica* and keeping only those which remain valid intuitionistically.[377] On the basis of the axioms, Heyting proves a number of mostly rather elementary propositions. As Heyting noted in the introduction, the section on the negation is the one that deviates most from classical logic, with the use of double negations. Heyting formally proves the equivalence between single and triple negation,[378] which Brouwer (and Kolmogorov for part of the propositional calculus) had proved before.

Regarding the principle of the excluded middle, Heyting first proves that

$$(a \vee \neg a) \to (\neg \neg a \to a),$$

---

[373][Heyting 1930A, p. 193]
[374]'Here I could not avoid giving the impression that the differences which emerge there in fact constitute the most important point of conflict between intuitionists and formalists (...), since formalism is unsuited to express the more fundamental points of conflict.' Translation based on the English translation in [Mancosu 1998, p. 313].
[375][Heyting 1930A, p. 193]
[376]See 5.3.1.
[377]Letter from Heyting to Becker, 23/9/1933; [TLI Heyting, B bec-330923]
[378][Heyting 1930A, p. 198]

that is: if the principle of the excluded middle holds for a certain proposition $a$, then one can use *reductio ad absurdum* for $a$; the contrary, Heyting remarks, does not hold. However, Heyting continues, as Brouwer had pointed out before, the assertion that the principle of the excluded middle holds for any proposition and for any mathematical system and the one saying that *reductio ad absurdum* holds for any proposition and any mathematical system are equivalent.[379] Following Glivenko, Heyting proves that in intuitionistic logic one can prove the double negation of the principle of the excluded middle.[380]

In the appendix, Heyting follows a method indicated by Bernays to prove the independence of the axioms. Using (except for one case) not more than natural numbers, Heyting gives what we would nowadays call models in which each time ten of the eleven given axioms are true, the eleventh one false. Since in intuitionistic logic not every proposition is either true or false, Heyting uses the truth value 0 for a proposition for which $\vdash p$ holds, 1 for $\vdash \neg p$ and 2 for $\vdash \neg\neg p$. By varying the eleventh axiom and presenting a different model to which the new combination of axioms applies, he proves that all the axioms are independent of each other.[381]

In the second paper, Heyting moves on to what we now call predicate logic. He interprets the expression $\exists x a(x)$ as 'one can point out an object $x$ for which the proposition $a$ holds'.[382] He gives axioms for equality and substitution (indicated here by $[p/x]$ for: substitute $p$ for $x$), which I do not treat here, and then moves on to quantifiers. Heyting gives a full axiomatisation of intuitionistic predicate logic, which, however, has some flaws. For example, the axioms alone do not guarantee that, in the case of the introduction of a universal quantifier, the variable which is being quantified does not occur free in any of the hypotheses on which the proposition in question depends. Furthermore, Heyting's definition of $a(\neg x)$ as $(\forall x g[p/x] = g) \vee (\forall x g[p/x] \leftrightarrow g)$ does not conform to the intended meaning of '$a$ does not contain $x$'. Nevertheless, since Heyting's work presents the first axiomatisation of intuitionistic predicate logic, the full list of axioms is given below, in Heyting's formulation:

(12) $\quad \forall x a(x) \to a[p/x]$
(13) $\quad \forall x a(x) \to \forall(y) a[y/x]$
(14) $\quad (\forall y (y = y)^{383} \to a[y/x]) \to \forall(x) a(x)$
(15) $\quad a[p/x] \to \exists(x) a(x)$
(16) $\quad a(\neg x) \to (\exists x a(x) \to \forall x a(x))$
(17) $\quad \forall(x)(a(x) \to b(x)) \to (\exists(x) a(x) \to \exists(x) b(x))$

---

[379][Heyting 1930A, p. 199]. Formally: $\vdash \forall a(a \vee \neg a) \Leftrightarrow \vdash \forall a(\neg\neg a \to a)$.
[380][Heyting 1930A, p. 201]
[381][Heyting 1930A, pp. 202–205]
[382]'es kann ein Gegenstand $x$ angegeben werden, für welchen der Satz $a$ gilt', [Heyting 1930A, p. 207]
[383]Heyting uses a '$\stackrel{=}{=}$' here, to indicate strict identity; he reserves the single '=' for cardinal equality, [Heyting 1930B, p. 206]. The expression $\forall y(y = y)$ would nowadays be expressed by means of the existence predicate $E$ as $E(y)$, meaning '$y$ exists'.

## 5.4. LATER REACTIONS

On the basis of these axioms, Heyting again proves a number of theorems, some of which are weaker than their classical counterparts, such as $\neg \forall x \neg a(x) \to \neg\neg \exists x a(x)$ and $\neg \exists x \neg a(x) \to \forall x \neg\neg a(x)$.[384]

When Heyting's work became known, he was soon asked to present it at different occasions. For instance, he spoke on intuitionism at the 1930 Königsberg conference[385] and the following year he lectured in Münster and Göttingen, upon the invitation of Heinrich Scholz and Otto Neugebauer respectively.[386] Neugebauer asked Heyting to write a monograph on the foundations of mathematics for the *Ergebnisse der Mathematik*, of which he was an editor. Heyting agreed, but he asked for a co-author to write on logicism. It was agreed that Gödel should do so, who accepted the offer. On Saturdays, Heyting went to Münster to work in the library, which Scholz had put at his disposal. However, despite some correspondence between Heyting and Gödel, Gödel never finished his part, and in the end Heyting's part was published separately.[387] Therefore it only deals with intuitionism and proof theory, and not with logicism.[388]

**Kolmogorov** In 1932, Kolmogorov[389] published a second paper on intuitionistic logic, in which he presented an original interpretation of the intuitionistic calculus. The paper is signed 'Göttingen, 15. Januar 1931', an indication that intuitionistic research had even spread into the centre of resistance against intuitionistic mathematics.

Kolmogorov starts the paper with the pragmatic remark that his work can be read in two ways: one for those who do not recognize the intuitionistic epistemology, and one for those who do. For the former, it presents a calculus of solving problems, different from theoretical logic. Then, the following 'remarkable'[390] fact holds: in form, the problem calculus coincides with Brouwer's intuitionistic logic, as formalised by Heyting. For those who do adhere to intuitionistic epistemology, Kolmogorov claims that his paper shows that intuitionistic logic[391]

> durch die Aufgabenrechnung ersetzt werden sollte, denn ihre Objekte sind in Wirklichkeit keine theoretischen Aussagen, sondern vielmehr Aufgaben.[392]

---

[384][Heyting 1930B, pp. 210–213]. Links between Heyting's papers and modern results are given in [Troelstra 1978].
[385]See 4.4.1.
[386]Letter from Scholz to his colleagues, 1/2/1931; [TLI Heyting, B sch5-310201**]; [Troelstra 1981, p. 4]
[387][Heyting 1934]
[388][Troelstra 1981, pp. 4–5]
[389]For biographical information on Kolmogorov, see 5.3.1.
[390]'merkwürdig', [Kolmogorov 1932, p. 58]
[391][Kolmogorov 1932, p. 58]
[392]'should be replaced by the calculus of problems, since its objects are in reality not theoretical propositions, but rather problems.'

## 280  CHAPTER 5. REACTIONS: LOGIC AND THE EXCLUDED MIDDLE

In good Brouwerian tradition, Kolmogorov leaves no room for other interpretations but maintains that the one presented in his paper is the correct one. It was, however, never accepted as such and nowadays lives on as an alternative interpretation of intuitionistic logic.[393]

Kolmogorov does not define what a problem is, but explains it by means of some examples, such as 'to give four natural numbers $x, y, z, n$ for which the relation $x^n + y^n = z^n$ for $n > 2$ holds' or 'to disprove Fermat's theorem'.[394] Then, if $a$ and $b$ are two problems, $a \vee b$ means the problem 'to solve at least one of the problems $a$ and $b$', whereas $a \to b$ means 'to reduce a solution for $b$ to a solution for $a$'. $\neg a$ means the problem: 'assuming that a solution for $a$ is given, produce a contradiction'. Similarly, Kolmogorov defines the interpretation of 'and' and of expressions with problems as free variables.[395]

Next, Kolmogorov argues that we have to assume certain problems solved in order to use them in the mechanical application of problem solving rules. He remarks that the list of problems which he assumes solved is identical to Heyting's axioms and inference rules (for the propositional calculus). Indeed, it is a literal copy of Heyting's axioms.[396] He notes that the problem interpretation of the principle of the excluded middle $a \vee \neg a$ is the problem 'to give a general method to find for any problem $a$ either a solution, or to derive from the assumption of such a solution a contradiction'. Unless the reader considers himself to be omniscient, Kolmogorov argues, it is clear that this principle cannot be on the list of solved problems.[397]

Kolmogorov devotes the final section to a discussion on the interpretation of negation and existence. He correctly argues that Brouwer gave a new interpretation to negation by turning $\neg a$ into an existence statement: 'there exists a chain of logical inferences, starting from the assumption that $a$ is correct and concluding with a contradiction'. This is the first time that someone described the intuitionistic negation is such clear terms. But in intuitionistic logic, Kolmogorov continues, also existence statements are interpreted differently. He claims to be following Brouwer in interpreting the intuitionistic existence statement, but in fact the interpretation is his own:[398]

> (...) ist das, was Brouwer unter einer Existenzaussage versteht, vollständig in zwei Elemente zerlegt: das objektive Element (die Aufgabe) und das subjektive (ihre Lösung). Somit findet man keinen Gegenstand übrig, den man als Existenzaussage im eigentlichen Sinne zu bezeichnen hätte.[399]

---

[393] The standard interpretation is the so-called Brouwer-Heyting-Kolmogorov interpretation, going back to Heyting's proof interpretation in [Heyting 1934]; cf. [Troelstra & Van Dalen 1988, vol. 1, pp. 9–10; 31–32].
[394] At that time, there was no proof of Fermat's theorem.
[395] [Kolmogorov 1932, pp. 59–60]
[396] With only a misprint in the last axiom, [Kolmogorov 1932, p. 61].
[397] [Kolmogorov 1932, pp. 61–63]
[398] [Kolmogorov 1932, p. 65]
[399] '(...) what Brouwer understand by an existence statement is completely split into two ele-

## 5.4. LATER REACTIONS

Brouwer, however, would single out the latter as the intuitionistic existence statement.

Heyting and Kolmogorov had a small discussion on Kolmogorov's paper, from which only one letter (from Kolmogorov to Heyting) seems to have survived. From this, it becomes clear that Heyting wanted to include the 'hope'[400] to find a solution in Kolmogorov's interpretation. This is in accordance with one of Heyting's 1930 papers, where he described that, in intuitionism, a proposition expresses a certain 'expectation',[401] which can either be realised or disappointed. However, as Kolmogorov pointed out, hope (or expectation) to solve a problem is neither a problem nor a solution, and it is hard to work with if it does not come true.[402]

### 5.4.2 Gödel

'The most important logician of the 20th century' and perhaps even 'the greatest logician since Aristotle',[403] Kurt Gödel, was born in 1906 in Brünn (Brno), then part of the Austro-Hungarian monarchy. His family belonged to the community of Südeten Germans, a minority in the region. The young Kurt was especially interested in languages and mathematics. In 1924, Gödel graduated from the gymnasium and started his studies in Vienna. By that time, the monarchy had collapsed and Gödel found himself a Czechoslovakian citizen in Austria. He started as a physics student, but later switched to mathematics. Contrary to what was usual at the time, Gödel did not do part of his studies at other universities. He took mathematics courses from, among other persons, Furtwängler; he also attended philosophy courses, and he participated in Schlick's seminar on Russell's 'Introduction to Mathematical Philosophy' held in 1925-26. From 1926 to 1928, Gödel participated in meetings of the Schlick Circle, later called the *Wiener Kreis*,[404] discussing among other things Wittgenstein's *Tractatus*. Gödel later stressed that he did not agree with the Circle's main ideas, most notably not with Carnap's idea that mathematics should be regarded as a 'syntax of language'. He stated that he had arrived at a Platonistic point of view around 1925.[405] Gödel's most important discussion partners at the time were Carnap and Menger, and Gödel attended some of their lectures. In 1929, Gödel became an Austrian citizen.[406]

---

ments: the objective element (the problem) and the subjective element (its solution). Therefore, one finds no object left that one would have to indicate as an existence statement properly speaking.' Translation based on the English translation in [Mancosu 1998, p. 333].
[400]'Hoffnung', [TLI Heyting, B kol-32xx]
[401]'attente', [Heyting 1930D, p. 958]
[402]Letter from Kolmogorov to Heyting, 1932 (exact date unknown); [TLI Heyting, B kol-32xx]
[403]Feferman and Von Neumann respectively, in [Feferman 1986, p. 1; 8]
[404]On the Wiener Kreis, see 3.3.3.
[405][Wang 1987, p. 17-20]
[406][Dawson 1997, pp. 4-33]

**Completeness and incompleteness**  On the basis of request slips Gödel used in the libraries of the University of Vienna and of the *Technische Hochschule* in Brno, Dawson argued in his Gödel biography that '[t]he shift in Gödel's mathematical interests away from more classical mathematical fields toward logic and foundations thus seems to have occurred between the summer and fall of 1928.'[407] Dawson suggests that Gödel got interested in the topic by reading about it. However, he also notes that Hilbert and Ackermann's 1928 *Grundzüge der theoretischen Logik* was not among the library books Gödel requested. In this work the subject of Gödel's dissertation, the completeness of what we now call first order predicate logic, is posed as an open problem.[408] Gödel later revealed that that book was where he got the problem from.[409]

Another possibility to explain Gödel's shifted focus is that Brouwer was the one who drew Gödel's attention to logic. In March 1928, Brouwer had lectured in Vienna on intuitionism.[410] However, it is not clear whether Gödel attended one of the lectures. The closest indication we have is a passage in Carnap's diary from December 23, 1929, where he describes his meeting with Gödel that evening as follows:[411]

> Über Unerschöpflichkeit der Mathematik (...). Er ist durch Brouwers Wiener Vortrag zu diesen Gedanken angeregt worden. Die Mathematik ist nicht restlos formalisierbar. Er scheint recht zu haben.[412]

Carnap later maintained the same in one of his papers, where he, referring to Brouwer's first Vienna lecture, described Gödel's incompleteness theorem as 'the true kernel of Brouwer's conviction that mathematics cannot be formalized without remainder'.[413] This clearly shows the impact Brouwer's ideas had on Gödel regarding the incompleteness theorem.

The citations do not, however, solve the problem whether or not Gödel actually attended Brouwer's lecture. For by the end of December 1929, Brouwer's first Vienna lecture, *Mathematik, Wissenschaft und Sprache* ('Mathematics, science and language'), the one in which he treated the relationship between mathematics and language, had most probably appeared in print.[414] And would it not have been somewhat strange to note the influence of a lecture Brouwer delivered in March 1928 only in December 1929? The assumption that Gödel did not attend Brouwer's

---

[407][Dawson 1997, pp. 53–54]
[408][Hilbert & Ackermann 1928, p. 68]
[409][Wang 1987, p. 42]
[410]See 2.7.3.
[411]Diary passage Carnap, 23/12/1929; transcribed from Stolz-Schrey shorthand; cited from [Köhler 1991, p. 138].
[412]'About the inexhaustibility of mathematics (...). He was stimulated to his idea by Brouwer's Vienna lecture. Mathematics is not completely formalizable. He appears to be right.' English translation cited from [Wang 1987, p. 84].
[413]'der richtige Kern in der von Brouwer (...) ausgesprochenen Überzeugung, die Mathematik sei nicht restlos formalisierbar', [Carnap 1934A, p. 274]
[414]It was published in 1929. I do not know the exact date of publication, but chances are high that it lies before December 23.

## 5.4. LATER REACTIONS

lecture but read the published version is in accordance with other evidence. On the basis of a letter Gödel wrote to the American Philosophical Society, Dawson argued that Gödel probably did not attend Brouwer's lecture. In the letter, Gödel stated that he had 'seen Brouwer only at one occasion, in 1953, when he came to Princeton for a brief visit.'[415] There is one letter I know of in which Gödel explicitly refers to Brouwer's lecture by saying:[416]

> I only saw him [Wittgenstein, DH] once in my life when he attended a lecture. I think that it was Brouwer's.

It is not clear whether this means that Gödel attended the same lecture or not.

Summarizing, I think that the evidence we have suggests that Gödel did not attend Brouwer's lecture, but was influenced by his ideas by reading the published version of *Mathematik, Wissenschaft und Sprache*.

Dawson conjectures that Gödel started working at his dissertation in 1928 or early 1929.[417] If that is correct, Gödel obtained his results quickly, for his dissertation was formally approved by Hahn and Furtwängler in July 1929. In February 1930, Gödel received his degree.

In the dissertation,[418] Gödel proves the completeness[419] of what we now call first order predicate logic. In the original version of the dissertation, Gödel discusses the relation between his work and the views put forward by Brouwer and Hilbert in the introduction. In the published version, however, the introduction is skipped and there is no reference to the foundational debate.[420] I here follow the introduction, which is the most relevant part for our discussion.

As Gödel remarks, he makes use of the principle of the excluded middle in the proof, also for infinite totalities. He notes that it may seem that this makes the whole proof worthless, since what he sets out to prove is a certain kind of decidability (namely that every expression of first order predicate logic can either be proved by a finite number of inferences or be refuted by a counterexample).[421] And the principle of the excluded middle does not seem to express anything else than the decidability of every problem. However, Gödel argues, there are two things to be answered to this objection. In the first place, only intuitionists interpret the principle of the excluded middle in such a way. And secondly, even if one interprets the *tertium non datur* in this way, there is still a difference between decidability by any correct means, as the principle says in the intuitionistic interpretation, and decidability only by means of a specific set of inference rules.[422] The distinction

---

[415] Letter from Gödel to Corner, 19/1/1967; cited from: [Dawson 1997, p. 56]
[416] Letter from Gödel to Menger, 20/4/1972; cited from: [Köhler 1991, p. 138]
[417] [Dawson 1997, p. 53]
[418] For Gödel's remarks on the intuitionistic view on mathematical existence, see 4.4.3.
[419] Nowadays, completeness is defined as $\Gamma \models \varphi \Leftrightarrow \Gamma \vdash \varphi$. Gödel took completeness to mean what we would now describe as $\Gamma \models \varphi \Rightarrow \Gamma \vdash \varphi$.
[420] [Gödel 1930]
[421] Since Gödel intends to prove $\Gamma \models \varphi \Rightarrow \Gamma \vdash \varphi$, the two-valued logic on a semantical level transfers to the syntactical level of derivations.
[422] [Gödel 1929, pp. 60–64]

Gödel makes is correct and would also be recognized by an intuitionist. It shows that Gödel was well aware of Brouwer's criticism of the principle of the excluded middle and that he took it seriously.

Gödel first publicly announced the incompleteness theorem at the 1930 Königsberg conference.[423] At the same time and in the same city, Hilbert continued spreading his optimistic belief in the capacity of mathematicians to tackle all problems. In his lecture *Naturerkennen und Logik* ('Natural cognizing and logic'), delivered on the occasion of the honorary citizenship that his native city Königsberg offered him,[424] he denied that there could be any *ignorabimus* in mathematics.[425] His credo, the words with which he finished his lecture and which are now inscribed in his tombstone in Göttingen, was:[426]

Wir müssen wissen,
Wir werden wissen.[427]

Gödel's first incompleteness theorem states that for any formal system $S$ which is strong enough to include elementary arithmetic,[428] there exists a proposition $P$ which is true but formally undecidable in $S$, i.e., neither $P$ nor $\neg P$ can be proved in $S$.[429] Although this theorem already ran contrary to Hilbert's optimistic belief in the (formal) solvability of all mathematical problems, the consequences of Gödel's second incompleteness theorem were even more far-reaching. The second incompleteness theorem states that for any consistent formal system $S$ which is strong enough to include elementary arithmetic,[430] the formula which expresses the consistency of $S$ is unprovable in $S$. Gödel only sketches the proof; he planned to publish a full proof later on, but never did so.[431] Gödel points out that the theorem can be applied to classical analysis or set theory, which means that, if these systems are consistent, a consistency proof cannot be formalised in the system itself. He then continues:[432]

---

[423] On the conference, see 4.4.1.
[424] [Reid 1970, p. 192]
[425] The *'ignorabimus'*, Latin for 'we will not know', was an important theme in many academic discussions. It originated from Emile Du Bois-Reymond, who used it in a lecture delivered in 1872, [Du Bois Reymond, E. 1872].
[426] [Hilbert 1930A, p. 387]
[427] 'We must know, We shall know.'
[428] To be more precise: which is $\omega$-consistent, i.e., there is no (arithmetic) formula $A(x)$, containing only the free variable $x$, such that $A(x)$ is provable for $x = 0, 1, 2, ..$ and also $\neg \forall x A(x)$ is provable.
[429] The proof of Gödel's incompleteness theorem is explained in several places, e.g. in [Kleene 1986].
[430] To be more precise: which is $\omega$-consistent, i.e., there is no (arithmetic) formula $A(x)$, containing only the free variable $x$, such that $A(x)$ is provable for $x = 0, 1, 2, ..$ and also $\neg \forall x A(x)$ is provable.
[431] A proof of the part lacking in Gödel's publication was given in [Hilbert & Bernays 1934–1939, vol. 2, pp. 283–240]; cf. [Kleene 1986, p. 137].
[432] [Gödel 1931, p. 194]

## 5.4. LATER REACTIONS

Es sei ausdrücklich bemerkt, daß Satz XI [the second incompleteness theorem, DH] (...) in keinem Widerspruch zum Hilbertschen formalistischen Standpunkt steh[t]. Denn dieser setzt nur die Existenz eines mit finiten Mitteln geführten Widerspruchsfreiheitsbeweises voraus und es wäre denkbar, daß es finite Beweise gibt, die sich in $P$[433] *nicht* darstellen lassen.[434]

However, Gödel's second incompleteness theorem makes it clear that Hilbert's programme cannot be pursued by taking as finitary a selection of the usual arithmetic methods. Thus, one has to look for different means of proving the consistency, thereby substantially changing Hilbert's programme.

Von Neumann interpreted the incompleteness theorem in a stronger way: he considered it to have proved the impossibility of Hilbert's programme. He added that[435]

Kein Anlass zur Ablehnung des Intuitionismus mehr vorliegt (wenn man vom Ästhetischen absieht, der freilich in Praxi auch für mich der ausschlaggebende sein wird).[436]

From when he announced his incompleteness theorem in Königsberg on, Gödel interpreted the incompleteness result in terms of intuitionism, maintaining that for any formal system for which there exists a finitary consistency proof, there exist formally unprovable propositions which are intuitionistically provable.[437] As Gödel remarked in a discussion in the Schlick *Zirkel*, the incompleteness theorems do not apply to intuitionistic mathematics, since intuitionism, in Brouwer's view, does not want to be contained in a formal system. Gödel himself questioned whether all intuitionistic proofs could be contained in *one* formal system.[438]

Gödel's result reached some colleagues quickly via the informal circuit, even

---

[433] $P$ is the formal system obtained from that of the *Principia Mathematica* by omitting the ramification of types, taking the natural numbers as the lowest type and adding the Peano axioms, cf. [Kleene 1986, p. 129].

[434] 'I wish to note expressly that Theorem XI [the second incompleteness theorem, DH] (...) do[es] not contradict Hilbert's formalistic viewpoint. For this viewpoint presupposes only the existence of a consistency proof in which nothing but finitary means of proof is used, and it is conceivable that there exist finitary proofs that *cannot* be expressed in the formalism of $P$ [the formal system obtained from that of the *Principia Mathematica* by omitting the ramification of types, taking the natural numbers as the lowest type and adding the Peano axioms, [Kleene 1986, p. 129], DH].' Translation from [Gödel 1931, p. 195].

[435] Letter from Von Neumann to Carnap, 7/6/1931, [ASP Carnap, 029-08-03]; published in [Mancosu 1999A, pp. 39–40].

[436] 'There is no longer a reason to reject intuitionism (if one disregards the aesthetic issue, which in practice will also for me be the decisive factor).' Translation based on the English translation in [Mancosu 1999A, p. 40].

[437] Letter from Heyting to Gödel, 22/8/1931; letter from Gödel to Heyting, 15/11/1932; [TLI Heyting, B goe-310822\*; 321115].

[438] 'Schlick-Zirkel, Protokoll am 15.1.1931', [ASP Carnap, 081-07-07]; published in [Mancosu 1999A, pp. 36–37].

though the first messages were rather unclear.[439] For example, in January 1931 Freudenthal had heard from several sources the rumour that Artin had found a fundamental fact which proved Brouwer to be right against Hilbert in all aspects.[440] Heyting had heard similar rumours in Münster, where he had lectured in February 1931,[441] but there the result was ascribed to a member of the *Wiener Kreis*.[442] Brouwer later maintained that he was astonished about the amount of attention directed at Gödel's incompleteness theorems, claiming that the fact had been clear to him long before.[443]

Gödel only handed in his incompleteness paper as a *Habilitationsschrift* in 1932. It was accepted later that year, and the following year Gödel became *Privatdozent* at the University of Vienna.[444]

**Intuitionistic logic**  After his work on incompleteness, Gödel made some contributions to intuitionistic logic. In January 1932, he lectured twice in Hahn's seminar on mathematical logic on the subject of Heyting's formalisation of intuitionistic logic.[445] He published three short papers on intuitionistic logic, the first of which was published in 1932 and dealt with a question put forward by Hahn on the number of truth values in intuitionistic logic. Gödel proves that, firstly, intuitionistic propositional logic cannot be seen as a system of many-valued logic, refuting earlier efforts by Becker, Lévy and Barzin & Errera to interpret intuitionistic logic as a three-valued logic.[446] Secondly, Gödel shows that there is an infinite number of systems between intuitionistic and classical propositional logic, i.e., there is a monotone descending sequence of systems which all encompass intuitionistic propositional logic and which are all included in classical propositional logic.[447]

The two other papers appeared in 1933. In *Eine interpretation des intuitionistischen Aussagenkalküls* ('An interpretation of the intuitionistic propositional calculus'), Gödel adds a new concept to classical propositional logic, $B$ (*'beweisbar'*), meaning '$p$ is provable'. Here, as Gödel points out, 'provable' is to be taken in a sense wider than 'provable in a specific formal system'. He presents the axioms which $B$ has to obey, and uses $B$ to give a translation of intuitionistic into classical propositional logic. He remarks that the system for $B$ is equivalent to Lewis' system of strict implication, with a small addition.[448]

---

[439] Some of the early correspondence on Gödel's incompleteness theorems is collected in [Mancosu 1999A].
[440] Letter from Freudenthal to Heyting, 19/1/1931; [TLI Heyting, B fre1-310119]
[441] [TLI Heyting, B sch5-310201**]
[442] Letter from Heyting to Freudenthal, 6/3/1931; [TLI Heyting, B fre1-310306*]
[443] [Wang 1987, p. 57]
[444] [Dawson 1997, pp. 81–89]
[445] [Dawson 1997, p. 81]
[446] See 5.3.1 and 5.3.1.
[447] [Gödel 1932]
[448] [Gödel 1933F, pp. 300–302]. Lewis had developed his logic of strict implication in opposition to the interpretation of implication used by Russell and Whitehead. In Lewis' definition, $P$

## 5.4. LATER REACTIONS

The remaining paper represents Gödel's most important result regarding the clarification of the relationship between intuitionistic and classical logic. The paper contains Gödel's contribution to Menger's colloquium in June 1932, namely a translation from classical first-order arithmetic into its intuitionistic counterpart. Gödel opens the paper by presenting a translation ' from the classical propositional calculus $C$ into the intuitionistic propositional calculus $I$.[449] The translation is defined as follows:[450]

| $C$ | $I$ |
|---|---|
| $\neg p$ | $\neg p$ |
| $p \to q$ | $\neg(p \wedge \neg q)$ |
| $p \vee q$ | $\neg(\neg p \wedge \neg q)$ |
| $p \wedge q$ | $p \wedge q$ |

It should be pointed out that Gödel's description of the translation is incomplete, since he does not include atomic expressions. The remarks he makes point in contrary directions as to how we should interpret the full translation, making the paper even, in a sense, undecidable. On the one hand, Gödel states that every formula in $C$ which includes only $\wedge$ and $\neg$ has to be of the form $\neg A_1 \wedge \neg A_2 \wedge \ldots \wedge \neg A_n$. This is evidently not the case if the formula is atomic. Thus, one could interpret Gödel as saying that he only considers atomic formulas with negation. In that case, one does not have to extend the translation but should only remark that the translation does not apply to atomic formulas without negation. On the other hand, however, Gödel also states that he wants to show that 'the classical propositional calculus is (...) a subsystem of the intuitionistic one',[451] and that 'every [valid] classical formula also holds in $H$'.[452] But these remarks would have to be formulated in a stricter form if Gödel excluded atomic expressions without negation. Since I think that Gödel indeed wanted to prove that the full classical propositional logic is a subsystem of its intuitionistic counterpart, and since this is possible by extending the translation, I think the best option is to extend the translation he gives to include atomic expressions by defining for atomic $p$[453]

---

implies $Q$ if and only if it is *impossible* that $P$ is true and $Q$ is false; cf. [Kneale & Kneale 1964, pp. 548–549].

[449] Gödel uses the letters $A$ and $H$ (Heyting) respectively.

[450] Gödel uses different signs for both intuitionistic and classical negation and conjunction, but the distinction is not necessary for the present purpose.

[451] 'ist (...) der klasssische ein Teilsystem des intuitionistischen', [Gödel 1933E, p. 286]; translation in [Gödel 1933E, p. 287]

[452] 'ist jede klassische Formel auch in $H$ gültig', [Gödel 1933E, p. 286]; translation from [Gödel 1933E, p. 287]

[453] The introduction to Gödel's paper in the Collected Works does not solve the problem either, since it contains the misprint in Gödel's translation that $P' = P$ for atomic $P$, [Troelstra 1986, p. 282]; this should read $(\neg P)' = \neg P$. When I pointed the problem out to Troelstra, he opted for the first version, namely to restrict the domain of the translation (electronic communication to the author, 29/9/1998). Possibly, the cause of Gödel's slovenliness was that the paper was not published in one of the bigger journals.

288   CHAPTER 5. REACTIONS: LOGIC AND THE EXCLUDED MIDDLE

| C | I |
|---|---|
| $p$ | $\neg\neg p$ |

I now present Gödel's argument in a somewhat more explicit form, using the extended translation just given. Note that all classical expressions used in the tables above are classically equivalent to the expressions given in the table as their intuitionistic translation. Now let a classical formula $\varphi$ in propositional logic be given. Because of the equivalences just noted, $\varphi$ is equivalent to $\varphi'$, which is expressed by means of $\neg$ and $\wedge$ only. We have to show that $\varphi'$ is provable in $I$. Note that $\varphi'$ is of the form

$$\neg A_1 \wedge \neg A_2 \wedge \ldots \wedge \neg A_n.$$

Since $\varphi'$ holds in $C$, every single $\neg A_i$ has to hold. Now Glivenko has shown that then $\neg A_i$ is also provable in $I$.[454] Therefore, the conjunction of all $\neg A_i$, i.e., $\varphi'$, holds in $I$ as well, which was to be proved.[455]

The translatability of classical into intuitionistic propositional logic had already been proved by Kolmogorov (for part of it), in 1925, and by Glivenko, in 1928.[456] It seems that Kolmogorov's paper was unknown to Gödel. The difference is that Gödel uses a different translation from both of them.[457]

Gödel wants to extend the result to arithmetic and number theory,[458] that is, to show that by means of a translation, all classically valid theorems become provable in intuitionistic arithmetic and number theory. He therefore extends the translation to include the universal quantifier:[459]

| C | I |
|---|---|
| $\forall x A(x)$ | $\forall x' A'(x)$ |

He next proves that if a formula $\varphi$ is provable in Herbrand's system, then its translation $\varphi'$ is provable in its intuitionistic counterpart.

Gödel concludes[460]

> daß die intuitionistische Arithmetik und Zahlentheorie nur scheinbar enger ist als die klassische, in Wahrheit aber die ganze klassische, bloß mit einer etwas abweichenden Interpretation, enthält.[461]

---
[454]Gödel refers to [Glivenko 1929], which is treated in 5.4.1.
[455][Gödel 1933E, p. 286]
[456]See 5.3.1 and 5.4.1.
[457]Gödel's and Kolmogorov's translation, as well as modern results on the relationship between classical and intuitionistic logic, are treated in [Troelstra & Van Dalen 1988, pp. 56–68].
[458]as axiomatized by Herbrand in [Herbrand 1931].
[459]Gödel does not have to include the existential quantifier in his translation, since classically $\exists x \varphi(x)$ is equivalent to $\neg\forall(x)\neg\varphi(x)$. The full translation for predicate logic is nowadays known as the negative translation, since it only reaches a part of intuitionistic logic (one does not obtain intuitionistic expressions with $\vee$ or $\exists$). Later developments building on Gödel's translation are mentioned in [Troelstra 1986].
[460][Gödel 1933E, p. 294]
[461]'that the system of intuitionistic arithmetic and number theory is only apparently narrower than the classical one, and in truth contains it, albeit with a somewhat deviant interpretation.' Translation from [Gödel 1933E, p. 295].

## 5.4. LATER REACTIONS

Since most people saw intuitionistic mathematics as more restrictive than classical mathematics, this was a rather surprising result. The reason why it is possible, Gödel explains, is that the 'interdiction'[462] to use pure existence statements in intuitionistic mathematics is neutralised by the fact that one can, also in intuitionistic mathematics, negate universal statements, which leads formally to exactly the same results as in classical mathematics. Only in analysis and set theory, Gödel conjectures, does intuitionism put real restrictions on mathematics, not because of the rejection of the principle of the excluded middle, but of impredicative definitions.[463]

Finally, Gödel points out that his translation provides an intuitionistic consistency proof for classical arithmetic and number theory, albeit not finite in Hilbert's sense.[464]

Independently of Gödel, Gentzen and Bernays proved the same result on the translatability between classical and intuitionistic mathematics, using a somewhat different translation. Since Gödel's paper was already published, Gentzen withdrew the paper from publication.[465] Kolmogorov was optimistic about Gödel's result. As he wrote to Heyting, he thought that one could go much further in this direction, proving the consistency of a large part of classical mathematics by intuitionistic means. He refers to his 1925 paper, where he had made a similar remark.[466] The weak point in Kolmogorov's argumentation there, however, is that he assumes that the *-translation of all mathematical axioms we know is true in pseudo-mathematics.[467]

### 5.4.3 Barzin and Errera

Despite the papers that Church and Glivenko published in reaction to Barzin and Errera's 1927 paper, pointing out what was wrong in their argumentation,[468] Barzin and Errera did not give up. In fact, they continued putting forward their point of view in a flood of papers, lectures and published correspondence, trying to conquer ground from Belgium to Roumania.[469] Errera seems to have been the one most eager to fight Brouwer's views, since he also published papers on the subject on his own. Barzin and Errera's notorious contribution to the offshoots of the foundational crisis is treated below. I should like to make clear in advance that

---

[462]'Verbot', [Gödel 1933E, p. 294]
[463][Gödel 1933E, p. 294]
[464][Gödel 1933E, p. 294]
[465]Letter from Gentzen to Heyting, 28/11/1933; [TLI Heyting, B gen-331128]. An English translation of Gentzen's paper can be found in [Szabo 1969, pp. 53–67]. Although the paper was on Gentzen's name, Gentzen himself stated in a footnote to the original paper that Bernays should be considered as the co-author of the paper, [Szabo 1969, p. 313].
[466]Letter from Kolmogorov to Heyting, 1934 (exact date unknown); [TLI Heyting, B kol-34xx]
[467]See 5.3.1.
[468]See 5.3.1 and 5.4.1.
[469]In fact, they even continued the discussion into the 1950s, still putting forward views on intuitionism which Heyting had shown to be erroneous some twenty years before, cf. [Errera 1951].

## 290   CHAPTER 5. REACTIONS: LOGIC AND THE EXCLUDED MIDDLE

I did not aim at completeness in pointing out the mistakes, misinterpretations and wrong conclusions that are present in their contribution. The reader interested in such an account should consult the original publications.

The sequence starts with a paper published in a new Belgian journal in philosophy, with Barzin as the only editor. The paper, *Sur le principe du tiers exclu* ('On the principle of the excluded middle'), was published in 1929 and is relatively long, with several appendices in which proofs are given. The quotes Barzin and Errera use show that they were able to read Dutch,[470] which means that they were well-quipped to understand Brouwer's intuitionism – in principle, at least.

Barzin and Errera open their paper with the following words:[471]

> Une proposition est ou bien vraie ou bien fausse. C'est là le *principe du tiers exclu* qui passait depuis Aristote pour une des lois de tout raisonnement et qui conserve un caractère d'extrême évidence au regard du sens commun.[472]

Note that they thus do not take up a formalistic position, but a classical one, defending the principle of the excluded middle on the grounds that it has a certain contents and even a very strong one, namely that it belongs to the laws of thought, as Boole had characterised them almost a century earlier. This conviction of theirs helps to explain why they fought so strongly against Brouwer's views.[473]

They continue by stating that they found a contradiction in the foundations of Brouwer's mathematics, which has caused 'uneasiness'[474] among his followers. Barzin and Errera claim to have received 'an avalanche of letters'[475] from them. They now present their reaction to five papers which appeared in reaction to their 1927 publication.[476] I only treat the ones that formed a reaction to intuitionism as well, namely the papers by Lévy, Church and Glivenko.

Barzin and Errera resume their earlier argumentation that they added to Brouwer's logic the postulate that there exists at least one proposition which is neither true nor false, thus, in their words, which is *tiers* ('third'). For if there is no such proposition, they argue, then the principle of the excluded middle has to hold.[477] They are mistaken here already, since they do not consider the possibility that there are propositions for which one can neither prove that they are true or false, which is not a third truth value as they suggest.

---

[470] Cf., e.g., [Barzin & Errera 1929, p. 6].

[471] [Barzin & Errera 1929, p. 3]

[472] 'A proposition it either true or false. That is the *principle of the excluded middle* which since Aristotle has passed for one of the laws of all reasoning and which retains a character of extreme evidence for common sense.'

[473] Errera had expressed a similar point of view in a letter to Dresden; letter from Errera to Dresden, 17/8/1928; [ULB Errera].

[474] 'émoi', [Barzin & Errera 1929, p. 3]

[475] 'une avalanche de lettres', [Barzin & Errera 1929, p. 3]

[476] [Barzin & Errera 1929, p. 3]

[477] [Barzin & Errera 1929, pp. 4–5]

## 5.4. LATER REACTIONS

They first turn to Lévy's paper.[478] Lévy interprets 'true' as 'there exists a demonstration', which, Barzin and Errera claim, is a 'radical divergence'[479] from Brouwer's ideas. For Brouwer, they argue, 'true' means 'originating from the primary intuition',[480] which, in their view, has nothing to do with a proof.[481] We see here a prime example of how confused people could be about what role intuition played in intuitionism. The interpretation Lévy gives comes closer to the Brouwerian one.

Barzin and Errera next turn to Church's paper. They indicate that Church states that one can reject the principle of the excluded middle without admitting any 'third' proposition. But then they do not see why one should reject the principle:[482]

Pourquoi sauter à cloche-pied quand il est possible de marcher?[483]

Furthermore, they argue, Brouwer gave many examples of 'third' propositions, referring to some weak counterexamples in Brouwer's *'Über die Bedeutung des Satzes vom ausgeschlossenen Dritten in der Mathematik'*.[484] This makes it clear where the misunderstanding on the 'third' propositions comes from: Barzin and Errera do not differentiate between weak and strong counter-examples. Later, they were to modify their point of view on the Brouwerian counter-examples radically.

Regarding Glivenko's paper, finally, Barzin and Errera remark that what he showed, namely that there are no 'third' propositions, is exactly what they wanted to show. Unfortunately, they continue, they discovered some strange things in Glivenko's paper. For example, they consider Glivenko's proof that $\neg\neg(\neg p \vee p)$ holds a 'rather liberal interpretation'[485] of Brouwer's thought. They easily dismiss Glivenko's work.[486] In fact, however, Glivenko's conclusion was totally in accordance with Brouwer's ideas.

Barzin and Errera summarize that none of the works they discussed affected the validity of their proof. They conclude that the logical principles are not isolated truths, but that they belong together. In their view, Brouwer should have defined as 'true' anything following from the intuition of time, and as 'false' anything else. Then the principle of the excluded middle could have been kept.[487]

This again shows Barzin and Errera's dogmatic point of view concerning the status of the logical principles. Besides that, they confuse Brouwer's philosophical ideas on intuition with his demands for mathematical rigour, they do not see the

---

[478] For Lévy's paper, see 5.3.1.
[479] 'divergence radicale', [Barzin & Errera 1929, p. 5]
[480] 'tout ce qui découle de l'intuition originelle', [Barzin & Errera 1929, p. 7]
[481] [Barzin & Errera 1929, pp. 5–7]
[482] [Barzin & Errera 1929, p. 12]
[483] 'Why limp if it is possible to walk?'
[484] [Brouwer 1924N]; [Barzin & Errera 1929, pp. 9–12]
[485] 'une interprétation assez libre', [Barzin & Errera 1929, p. 13]
[486] [Barzin & Errera 1929, pp. 12–14]
[487] [Barzin & Errera 1929, pp. 17–18]

value of giving a positive interpretation to the words 'true' and 'false', and they do not understand the difference between weak and strong counterexamples.

De Donder, who had presented Barzin and Errera's paper in the *Académie Royale de Belgique*, asked Brouwer to react to Barzin and Errera's paper and the discussion on their earlier paper in the *Académie*. Brouwer intended to do so,[488] but he was 'pleasantly surprised'[489] to find that Heyting had already reacted to Barzin and Errera by publishing a paper,[490] and he felt that Heyting had said everything he had wanted to say. Apparently, Brouwer and Heyting did not consult each other on how to react to such papers. In his letter to De Donder, Brouwer did add that Barzin and Errera's ideas, 'besides the great interest they represent, are intenable in their essential tendency.'[491]

Heyting's paper that 'pleasantly surprised' Brouwer was published in the Bulletin of the *Académie Royale de Belgique* in 1930 under the title *'Sur la logique intuitionniste'* ('On intuitionistic logic'). In it, Heyting sets out to answer some of the questions that were raised in the discussion following on Barzin and Errera's first paper.[492] Heyting fully approves of Glivenko's axiomatization of intuitionistic logic, which had been published before his own.[493]

Heyting explains that in intuitionism, the affirmation of a proposition $p$ should be understood as meaning not, as Lévy thought,[494] '$p$ is provable', but as 'one can prove $p$'. In this way, Heyting wants to dispose of all reference to transcendental facts, and keep only empirical ones. The intuitionistic negation, Heyting maintains, means '$p$ implies contradiction'. Then, the third case is that one can neither prove $p$ nor $\neg p$. One could, Heyting continues, designate this case by $p'$, but one should understand that $p'$ can almost never be stated definitely. For, in general, it remains possible that one day either $p$ or $\neg p$ will be proved. The double negation of $p$, finally, means that one can derive a contradiction from the assumption that $p$ implies a contradiction. This is the case of which Brouwer and Wavre have given several examples.[495]

**Discussion with Heyting**  The most interactive part of Barzin and Errera's contribution was the public discussion they had with Heyting in the didactics

---

[488] Letter from De Donder to Brouwer, 26/10/1929; letter from Brouwer to De Donder, 13/6/1930, [MI Brouwer]
[489] 'agréablement surpris', letter from Brouwer to De Donder, 9/10/1930; [MI Brouwer]
[490] [Heyting 1930D]
[491] 'à part du grand intérêt qu'elles présentent, sont néanmoins intenables dans leur tendance essentielle', letter from Brouwer to De Donder, 9/10/1930, [MI Brouwer]
[492] See 5.3.1.
[493] See 5.4.1 and 5.4.1; [Heyting 1930D, pp. 957–960].
[494] See 5.3.1.
[495] [Heyting 1930D, pp. 958–961]

## 5.4. LATER REACTIONS

journal *L'Enseignement mathématique*. The discussion lasted from 1931 to 1933[496] and consisted of six letters altogether.

Barzin and Errera open the series with a short discussion of Heyting's formalisation of intuitionistic logic. They maintain that Heyting's work has shown that there is not a big difference between Brouwerian and classical logic. Now, it has become clear that the only difference is that intuitionistic logic uses one axiom less. They continue:[497]

> Demandons-nous maintenant quelle serait notre attitude, si un beau jour, un mathématicien nous enjoignait de proscrire la géométrie métrique et d'abandonner toute la partie de cette science qui ne découle pas de l'*analysis situs* [topology, DH]?[498]

We would, Barzin and Errera argue, ask him to prove this to be necessary. This could either be done by proving a contradiction, but Brouwer himself said that illegitimately using the principle of the excluded middle does not lead to a contradiction. (Note that Barzin and Errera now implicitly acknowledge that Glivenko was right when he proved that, in intuitionistic logic, $\neg\neg(\neg p \vee p)$ holds.) Or it could be done by showing that the principle of the excluded middle is in contradiction with our experience. But the Brouwerians never tried to do so. For, Barzin and Errera contend, one could not consider the Brouwerian counter-examples as such. (Note the radical change that has taken place in their appreciation of the counter-examples.) Finally, one could argue from a philosophical basis, and this, Barzin and Errera maintain, is what Brouwer does. However, one should leave philosophical considerations out of mathematics, and only look at mathematical necessities.[499]

Heyting responds by pointing to Kolmogorov's interpretation of the intuitionistic calculus as problems, which is independent of any intuitionistic idea.[500] He continues by stating that the supposed evidence of the principle of the excluded middle rests on metaphysical assumptions, namely that propositions have properties independently of us knowing so, and that Brouwer agrees with Barzin and Errera that one should keep philosophical considerations away from mathematical proofs. Brouwer does not reject the principle of the excluded middle, but he refuses to

---

[496] The last letters appeared in *L'Enseignement mathématique* 31, which is dated 1932. The letters themselves, however, are clearly and consistently dated in 1933. Therefore, it seems that the journal appeared too late, and that the dates of the letters should be taken as indicating the real time.

[497] [Barzin & Errera 1931, p. 249]

[498] 'Let us now ask what our attitude would be if, one day, a mathematician would order us to abolish metric geometry and to abandon every part of that science which does not result from *analysis situs* [topology, DH]?'

[499] [Barzin & Errera 1931]

[500] See 5.4.1.

# 294 CHAPTER 5. REACTIONS: LOGIC AND THE EXCLUDED MIDDLE

admit it, since he has not seen a demonstration. Changing the burden of proof, Heyting finishes his letter by stating:[501]

> En retournant un argument de MM. Barzin et Errera on peut dire que l'attitude des partisans de la logique classique ressemble à celle du mathématicien imaginaire qui soutiendrait que tout espace abstrait admet une métrique et qui reprocherait à ceux qui exigeraient une démonstration de ce théorème de vouloir attaquer la liberté de la science.[502]

Indeed, Brouwer had not maintained that the principle of the excluded middle *never* holds, but only that it does not hold universally.

In their reaction, Barzin and Errera express in a clearer form an argument which they had already used in their first contribution. They maintain that there is a contradiction in intuitionistic logic. For, they argue, on the one hand Heyting considers the principle of the excluded middle not to be true. However, Heyting formally proves that the principle is not false either. Since it is not true and not false, it has to be 'third'. But if there is a third proposition, the principle of the excluded middle is false, which leads to a contradiction. They finally argue that one cannot ask them to prove the principle of the excluded middle, since then metaphysics would enter into mathematics.[503] (Note that Barzin and Errera have thus moved from a classical to a formalistic position.)

Heyting has no difficulty in pointing out what is wrong in Barzin and Errera's argumentation. The various negations they use have different meanings. The negation used in saying that the principle of the excluded middle is not true is the classical negation, simply meaning that we have no proof for it. The one used in Heyting formal's proof that the negation of the principle of the excluded middle is not true is the strong, intuitionistic 'negation as absurdity'. Therefore, the contradiction noted by Barzin and Errera does not appear. Furthermore, as Heyting points out after Lévy and Dresden, how can one obtain a contradiction by simply leaving out an axiom? Regarding the question of providing a proof for the principle of the excluded middle, Heyting points out that his remark was not directed at formalists, but at classical mathematicians, those who adhere to some meaning in non-intuitionistic mathematics. He finishes his letter with an application of the principle of the excluded middle:[504]

---

[501][Heyting 1932A, p. 122]
[502]'Countering an argument of Mr. Barzin and Errera, one could say that the attitude of the partisans of classical logic resembles that of the imaginary mathematician who would maintain that every abstract space admits a metric and who would reproach those who would demand a proof of that theorem of wanting to attack the liberty of science.'
[503][Barzin & Errera 1932A]
[504][Heyting 1932B, p. 272]

## 5.4. LATER REACTIONS

> Des deux choses l'une: ou bien les mathématiques consistent de pensées humaines, ou bien elles sont purements formelles. Le but de M. Brouwer est de tirer toutes les conséquences de la première alternative.[505]

Barzin and Errera, in their final part in the correspondence, admit that the different interpretations Heyting gives to the negation make the contradiction disappear. But, they ask, at what price? If 'true' means 'provably true', 'false' means 'provably false' and 'third' that which is neither proved true nor false, thus incertain, then the only thing that takes place is a change in language. Furthermore, they argue, the intuitionistic point of view introduces the subjective into science. (It is not clear how a mere change of language can bring the subjective into science, but let me not start arguing with Barzin and Errera.) They do not see why Brouwer wanted to forbid the principle of the excluded middle, and mock at the meagre results of the big revolution.[506]

In his final answer, Heyting indicates that, now that agreement has been reached on the essential point, the discussion can be closed. He just wants to add a word on the value of the intuitionistic revolution. The revolution does not stop, Heyting argues, because one cannot forbid formalists to write down signs on paper. The essential thing is that, once one tries to interpret their results, these turn out to be erroneous. For formalists themselves, the consequences are also revolutionary: they have to renounce from all meaning when doing mathematics.[507]

> Qui, avant M. Hilbert, songeait à considérer les mathématiques comme un jeu arbitraire?[508]

Already at the time, there was little appreciation for the quality of Barzin and Errera's work on intuitionistic logic. Becker could 'hardly understand that it is still possible at this moment to misunderstand intuitionism in such a naive way'.[509] Scholz admired Heyting's patience with Barzin and Errera.[510]

**Later reactions** Three final contributions of Barzin and Errera fall within the period covered here. One is a lecture Errera delivered at the second conference of Roumanian mathematicians in Turnu-Severin in May 1932. The other two are papers, both from 1933, one published by Barzin and Errera in the mathematical

---

[505]'It is one or the other: either mathematics consists of human thoughts, or it is purely formal. Mr. Brouwer's goal is to draw all the consequences of the former alternative.'
[506][Barzin & Errera 1932B]
[507][Heyting 1932C, p. 275]
[508]'Who, before Mr. Hilbert, would have dreamt about considering mathematics as an arbitrary game?'
[509]'ich kann kaum verstehen, dass es heute noch möglich ist, den Intuitionismus in diese naiven Weise miszuverstehen, wie das diese Herren tun', letter from Becker to Heyting, 19/9/1933; [TLI Heyting, B bec-330919]
[510]Letter from Scholz to Heyting, 24/9/33; [TLI Heyting, B sch5-330924]

bulletin of the Roumanian Society of Sciences,[511] the other a paper by Errera in the *Revue de Métaphysique et de Morale*. Since the last two papers are not dated by month, it is hard to place them in exact relation to the correspondence that was going on at the same time in *L'Enseignement mathématique*.

Errera's lecture at the Roumanian conference is mostly a repetition of Barzin and Errera's earlier arguments concerning Lévy's paper and Heyting's formalisation. He again maintains that intuitionistic logic leads to a contradiction. He does add, paradoxically, that it can be of interest to investigate what one can prove without making use of the principle of the excluded middle.[512]

The paper written by Barzin and Errera is, again, mostly a repetition. They maintain that intuitionistic logic leads to a contradiction, and that Brouwer's reasons for rejecting the principle of the excluded middle are metaphysical.[513]

Errera is even more polemic in his 1933 paper than in most of their other contributions, claiming that[514]

> les intuitionnistes ont déclaré la guerre au genre humain, ou, tout au moins, à la *gens mathematica* tout entière.[515]

He again mentions the contradiction in intuitionistic logic, and grossly overrates the contribution he and Barzin made to the development of intuitionistic logic by presenting Heyting's formalisation of intuitionistic logic as a reaction to their work. There is, however, a new and even coherent argument in the paper. Following Hilbert's 1925 lecture, Errera now reasons that one should stick to the use of the principle of the excluded middle because of the simplicity of language, even if the results one gets do not always represent a concrete solution to a given problem.[516]

So finally, Barzin and Errera seem to have reached some kind of understanding of the intuitionistic criticism. Perhaps the most interesting part of the whole discussion is to see how their own position changed. They began as staunch supporters of the 'evident' principles of logic which represented the laws of thought. They ended by remarking that one should stick to classical logic in order to keep the language simple. In fact, that is probably the route most people followed. The merit of Barzin and Errera is that they did it slowly and publicly.

## 5.5 Conclusion

The debate on the principle of the excluded middle started with Weyl's 1921 paper *Über die neue Grundlagenkrise der Mathematik*. Weyl presented his doubts

---

[511] It is not clear what the Roumanian connection is.
[512] [Errera 1935A]
[513] [Barzin & Errera 1933]
[514] [Errera 1933, p. 27]
[515] 'the intuitionists have declared war on the human race, or, at least, on the whole *gens mathematica*.'
[516] [Errera 1933]

## 5.5. CONCLUSION

concerning the validity of this principle and concerning the correctness of Brouwer's criticism. Finally, he concluded in favour of Brouwer because, in Weyl's view, an existence statement did not constitute a proper proposition. Therefore, Weyl reasoned, the argument 'it is the case or it is not the case' did not apply.

Hilbert reacted the same year in lectures delivered in Copenhagen and Hamburg. He labelled the *tertium non datur* as a principle which was 'forbidden' by Brouwer and Weyl. One could, he claimed, prove the validity of using this principle for infinite totalities by means of a consistency proof (without, however, presenting one). Bernays showed more sympathy for intuitionism and maintained that, from Brouwer's point of view, rejection of the principle of the excluded middle was nothing but logical.

**Formalism vs. contents** The following year, Hilbert extended his argumentation. Regarding the question of transfinite statements in contentual mathematics, Hilbert used some combination of arguments used by Brouwer and Weyl before. He reasoned that, although the principle of the excluded middle is valid in finite totalities, this does not solve the problem of its general validity, since one cannot simply use arguments which are valid for finite totalities in infinite ones. In infinite totalities, Hilbert maintained, the negation of a universal or existential proposition does not have a precise meaning. Therefore, the meaningful statements $\forall x \varphi(x)$ and $\exists x \neg \varphi(x)$ are not opposed as contradictories, and thus the *tertium non datur* does not apply. Note that Hilbert's argumentation was about meaningful mathematics and not merely about a formal system. This lines up with his idea that formalised mathematics should represent mathematical thoughts. Hilbert's solution was to add certain transfinite axioms and to prove the resulting system consistent. He did not explicate what this procedure meant for the character of the mathematical objects involved.

Weyl quickly noted what happened. In 1924, he pointed out that Hilbert's negative reaction to intuitionism was not even justified from his own point of view, since Hilbert, too, respected the bounds of meaningful argumentations, including those on the issue of the excluded middle. Fraenkel later made the same comment. Weyl did modify his own view on the absolutistic claim he had before defended for intuitionism; he now maintained that one should do both Brouwer's meaningful mathematics and Hilbert's symbolic one.

A general debate developed in which many people paid attention to the intuitionistic criticism of the unrestricted use of the principle of the excluded middle. This criticism was based on a *contentual* interpretation of the excluded middle. All the people who were sensitive to the intuitionistic criticism shared with them a contentual interpretation of logic. Hardly anybody used the formalistic argument *par excellence* against intuitionism that the principle of the excluded middle holds simply because *we have defined the formal system that way*. Several arguments were used against the criticism, such as the one that not all pairs of propositions which are opposed to each other are opposed as contradictories, whereby the principle of the excluded middle does not apply (but this does not make the principle invalid); or the one that 'it is the case or not, even if we do not know so'. Also, the idea that logic represented the laws of thought was used against Brouwer's criticism. All these arguments rested on a contentual, rather than a purely formal, interpretation of logic. Thus, the people who used them were classical rather than formalistic mathematicians (or philosophers with a classical view on mathematics). The main exceptions were, in chronological order, Dingler, Hölder, Dresden and Church. They took up the pragmatic position that one can simply choose which logical axioms to use.

In 1925, Hilbert publicly changed to a different position. Now, he no longer claimed that all of formalised mathematics represented mathematical thoughts, but only the finite part of it. The rest was meaningless. He simply stated that we do not want to give up the simple rules of classical logic. Therefore, we have to add ideal elements, to make the mathematical system obey the rules of logic. This was exactly the reason which Wavre had given in his 1924 paper for the formalistic adherence to the rules of classical logic. However, in 1927 Hilbert still presented logic as representing the rules of thought, indicating that he had not moved towards a purely formalistic position.

Thus, the debate evoked by the intuitionistic criticism of classical logic reveals that there were actually very few formalistic mathematician at the beginning of the debate. People who were sensitive to the intuitionistic critique and who used a contentual interpretation of logic in arguing about intuitionism were classical mathematicians rather than formalists. Formalism only developed and spread because of the untenability of the claims of classical logic.

**Intuitionistic logic**  Over time, intuitionistic logic was formalised. First, Kolmogorov presented a formalisation of parts of intuitionistic propositional logic in 1925. He maintained that Brouwerian logic was always valid, Hilbertian, which made unrestricted use of the excluded middle, not. However, by means of a translation, Kolmogorov managed to prove that finite statements proved by means of the excluded middle in infinite totalities are also generally valid. Furthermore, Kolmogorov gave a relative consistency proof for classical logic. However, since Kolmogorov's paper was written in Russian, it remained almost completely unnoticed.

In 1927, others like Becker and Lévy tried to formalise intuitionistic logic. Glivenko did so in 1928. The final answer came in 1930, when Heyting published a revised version of his prize-winning essay on the formalisation of intuitionistic logic and mathematics. The papers soon became the standard formalisation of intuitionistic logic. From then on everybody could see that, in formal terms, all that happened in intuitionistic logic was the deletion of a classical axiom; it was not replaced by anything. The use of double negations provided one of the most visible differences with classical logic.

## 5.5. CONCLUSION

In 1927, Barzin and Errera claimed to have found a contradiction in Brouwer's logic. In their argumentation, they made use of what they called the truth value 'third'. Several people, amongst whom Glivenko with a very good contribution, pointed out that such a truth value is not part of intuitionistic logic. This, however did not stop them. In a flood of publications, they kept on putting forward their point of view. The only thing that changed during this debate was their own position. Started as defenders of the meaningful character of classical logic, they silently moved to a purely formalistic position, where the laws of classical logic were only kept for reasons of simplicity.

Gödel, who was inspired for his incompleteness theorem by one of Brouwer's Vienna lectures, worked on intuitionistic logic, too. He provided a translation of classical into intuitionistic arithmetic, thereby showing that intuitionistic arithmetic was only apparently narrower than its classical counterpart. In fact, it contained all of classical arithmetic, only with a different interpretation. This was one of the most illuminating contributions to the debate.

The overview just presented leads me to the following conclusion. The debate on logic and the principle of the excluded middle centered on questions of formalisation and contents. On the one hand, intuitionistic logic was formalised, first (partially) by Kolmogorov, later by Heyting. This clarified to many people what the relationship between intuitionistic and classical logic was, and it enabled Gödel to prove more exact results between the two in terms of translatability. On the other hand, the debate makes it clear that most of the participants to the foundationale debate adhered to a contentual rather than a purely formal interpretation of logic. This explains why they were sensitive to the intuitionistic criticism, which was directed against a contentual interpretation of classical logic. Over time, more people admitted that it was not the contentual interpretation of logic which was decisive, but the wish to adhere to the simple rules of classical logic, even if this meant (partially) abandoning the claim that mathematics represented some kind of contents.

# Chapter 6

# The foundational crisis in its context

Und wird mir das ganze
Getu hier zu trist,
Dann kauf ich mir'ne Kanone
Und werde Putschist.[1]

Hubert Cremer[2]

## 6.1 Introduction

It is often acknowledged that cultural circumstances in post-World War Germany may have effected the attitude of mathematicians in the foundational debate. However, usually no *specific* influences are pointed out. Thus, in his recent *Moderne - Sprache - Mathematik*, which drew considerable attention among historians of mathematics, Mehrtens writes:[3]

> Das Gefühl der großen Krise nach dem Schock von Weltkrieg und Revolution, damit die unausweichliche Frage, wie Sinn und Ordnung wiederherzustellen seien, erfaßten auch die Mathematiker.[4]

---

[1] 'And if the whole
Show here becomes too sad
Then I buy a gun
And become *Putschist*.'

[2] [Cremer 1927]; *'Putschists'*, people who participated in a coup, was a nickname used in Berlin for intuitionists.

[3] [Mehrtens 1990, p. 290]

[4] 'The sense of a big crisis after the shock of world war and revolution, carrying with it the unavoidable question how to restore meaning and order, got hold of mathematicians, too.'

Hermann Weyl, in a 1955 postscript to his 1921 *Grundlagenkrise* paper that started the foundational crisis, only maintains that the *style* in which the article was written reflects the excitement of the post-war period.[5] The most specific conjecture comes from Forman, who investigated the case of physicists in Weimar Germany and came to the conclusion that they used an accommodationist strategy to adapt to the hostile intellectual environment. He conjectured that mathematicians reacted in the same way. In this section, I treat the case of mathematicians in more detail and investigate if Forman's conjecture holds.

There are four areas of the cultural environment outside mathematics which possibly influenced the reception of Brouwer's intuitionism: philosophy, physics, art, and politics. Of these areas, I focus on those parts that may have influenced the foundational debate. Furthermore, I address Mehrtens' *Moderne – Gegenmoderne* thesis, by which he embeds the foundational debate (and more) into its cultural environment. In the final section, all collected insights are weighted against each other to provide a comprehensive conclusion. But to start with, I analyse the use of metaphors in the debate, since they provide a possible link as seen by contemporaries between the foundational debate in mathematics and its cultural environment.

## 6.2 Metaphors

Metaphors provide one of the most visible links between a discussion in a certain area and its cultural environment. As Mehrtens puts it:[6]

> Sie schaffen und strukturieren alte und neue Perspektiven; sie stellen Positionen fest oder verschieben sie; sie bestätigen oder verwandeln Identitäten von Mathematikern, von Arbeitsfeldern und von übergreifenden sozialen Ordnungen.'[7]

By referring to subjects outside the proper domain of discussion, the participants themselves place the debate in a wider context. In this way, popular metaphors in the debate tell us something about the self-perception of the debate. Whereas Mehrtens focuses on metaphors that are relatively close to the core of the mathematical language, such as 'intuition' and 'system',[8] I restrict myself to the more striking, politics-inspired metaphors used in the foundational debate.

### 6.2.1 Crisis and revolution

**Weyl's introduction of metaphors**  The polemic terms with which Hermann Weyl opened the foundational debate were 'crisis' and 'revolution'. The former only

---

[5][Weyl 1956, p. 247]
[6][Mehrtens 1990, p. 509]
[7]'They create and structure old and new perspectives; they fix or move positions; they confirm or alter identities of mathematicians, of fields of work and of over-arching social orders.'
[8][Mehrtens 1990, p. 499–509]

## 6.2. METAPHORS

figures in the title of his 1921 paper *Über die neue Grundlagenkrise der Mathematik* ('On the new foundational crisis in mathematics').[9] In the opening lines of the paper, Weyl implicitly justifies the 'crisis' label by arguing that the antinomies of set theory should not be seen as a matter remote to the core of mathematics, but rather as symptoms of the inner 'untenability'[10] of the foundations of mathematics of the time. The 'revolution' metaphor appears later, when Weyl proclaims to have found the solution:[11]

> So gebe ich also jetzt meinen eigenen Versuch preis und schliesse mich Brouwer an. In der drohenden Auflösung des Staatswesens der Analysis, die sich vorbereitet, wenn sie auch erst von wenigen erkannt wird, suchte ich festen Boden zu gewinnen, ohne die Ordnung, auf welcher es beruht, zu verlassen, indem ich ihr Grundprinzip rein und ehrlich durchführte;[12] und ich glaube, das gelang – soweit es gelingen konnte. Denn *diese Ordnung ist nicht haltbar in sich*, wie ich mich jetzt überzeugt habe, und Brouwer – das ist die Revolution![13]

From the citation only, it is not clear if Weyl was referring to any specific revolution. However, the opening line of Weyl's paper gives a further indication of the context in which the quote just given should be seen. There, he speaks about the antinomies of set theory which are normally seen as unrest in the 'remotest provinces of the mathematical Empire',[14] a hint at the situation in the eastern parts of the German Empire, where the war was smouldering on.[15] The Weimar republic was far from stable at the time, and the whole atmosphere was highly political. In these circumstances, I think Weyl's revolution metaphor was not merely an appeal to overthrow the existing order, but also obtained a more political connotation.

It is impossible to say if Weyl had any specific revolution in mind when we wrote these lines. However, one can try to re-create the impression these words may have had on readers in 1921. What did it mean to them to read an important mathematician proclaim the 'revolution'? They could have been reminded of the German 1918 revolution, by which the republic had been declared. If this was the case, it means that they saw Weyl's intuitionism as linked to republican,

---

[9]The contents of the paper, as well as Weyl's conversion to intuitionism, is treated in 4.2.1 and 5.2.1.
[10]*'Haltlosigkeit'*, [Weyl 1921, p. 211]
[11][Weyl 1921, p. 226]
[12]Weyl is referring to his book *Das Kontinuum*, [Weyl 1918A].
[13]'So I now abandon my own attempt and join Brouwer. In the threatening dissolution of the state of analysis, which is in preparation even though still only few recognise so, I tried to find solid ground without leaving the order upon which it rests, by carrying out its fundamental principle purely and honestly. And I believe this succeeded – as far as it could succeed. For *this order is in itself untenable*, as I have now convinced myself, and Brouwer – that is the revolution!' Translation based on the English translation in [Mancosu 1998, pp. 98–99].
[14]'entlegensten Provinzen des mathematischen Reichs', [Weyl 1921, p. 211]
[15]Cf. [Van Dalen 1995, p. 146].

304    CHAPTER 6. THE FOUNDATIONAL CRISIS IN ITS CONTEXT

democratic Germany. By 1921, however, the German revolution was generally considered 'stolen', and the most likely revolution was a communist one.

The threat of such a revolution seemed all too real. Russia had had its revolution in 1917, Germany in 1918, and by the spring of 1919 Austria and Hungary had had theirs, too.[16] With the view of spreading the revolution, the Third International (Comintern) was set up in March 1919 in Moscow. At the second congress of the Comintern, in July 1920, the executive committee aimed at becoming the general staff of the revolution. The Red Army reached the gates of Warsaw in 1920. By the end of that year, the mass communist party KPD emerged in Germany, persisting in the idea of a revolution, and similar parties were created shortly afterwards in other Western European countries.[17] In the light of these circumstances, I think the association Weyl's revolution metaphor created for the reader was that of a communist revolution. Thus, Weyl's call could be seen as much more radical, and the old order to be overthrown would include not only the Wilhelminian empire, but also that of parliamentary democracy.

Weyl used radical terms such as 'revolution' and 'crisis' on purpose. In a letter to Brouwer written before the publication of the 1921 paper, Weyl explained that the article was meant as a propaganda pamphlet, not as a scientific publication. Its purpose was to rouse the sleepers[18] – which it certainly did.

There is another aspect of the citation just given which hints at a popular current in society. Weyl's talk of getting rid of the old order reminds one of the Dada movement, originated in Zürich, Weyl's university city, in 1916. Dada aimed at destroying all art and replacing it by a new (social, intellectual, artistic) order.[19] It is known that Weyl was deeply interested in art and literature, especially poetry. Furthermore, he took actively part in the spiritual life in Zürich, an activity which was reinforced by his wife Hella's many contacts with journalists, theatre actors and the like. Thus, it is most likely that Weyl was *au courant* with Dada, although his appreciation of classical German literature makes it unlikely that he would wholly agree with them.[20]

**Political positions suggested by metaphors**   One of the first people to react to the revolution metaphor was Hilbert. In his 1921 Copenhagen and Hamburg lectures, Hilbert's first reaction to the intuitionistic attack,[21] he countered Weyl's proclaimed revolution in the following way:[22]

> Was Weyl und Brouwer tun, kommt im Prinzip darauf hinaus, daß sie die einstigen Pfade von Kronecker wandeln; sie suchen die Mathematik

---

[16] However, only the first of these led to a communist takeover.
[17] [Joll 1990, pp. 247–253]
[18] Letter from Weyl to Brouwer, 6/5/1920; cited from: [Van Dalen 1995, pp. 147–148]
[19] [Joll 1990, p. 307]
[20] [Chandrasekharan 1986B, p. 96; 105]
[21] See 4.2.2 and 5.2.2.
[22] [Hilbert 1922, pp. 159–160]

## 6.2. METAPHORS

dadurch zu begründen, daß sie alles ihnen unbequem Erscheinende über Bord werfen und eine Verbotsdiktatur à la Kronecker errichten. Dies heißt aber, unsere Wissenschaft zerstückeln und verstümmeln, und wir laufen Gefahr, einen großen Teil unserer wertvollsten Schätze zu verlieren, wenn wir solche Reformatoren folgen. (...) nein, Brouwer ist nicht, wie Weyl meint, die Revolution, sondern nur die Wiederholung eines Putschversuches mit alten Mitteln, der seinerzeit, viel schneidiger unternommen, doch gänzlich mißlang und jetzt zumal, wo die Staatsmacht durch Frege, Dedekind und Cantor so wohl gerüstet und befestigt ist, von vornherein zur Erfolglosigkeit verurteilt ist.[23]

Hilbert tries to discredit Weyl's revolution by renaming it a Putsch. Again, it is not *a priori* clear with what kind of Putsch readers would associate Hilbert's reference. The most well-known Putsch from that period was the Kapp-Putsch, which discontented army officers had staged in 1920. If this was the link Hilbert tried to establish, it means he wanted to alter the image of intuitionism into a conservative, old-fashioned current rather than something modern. On the other hand, a Putsch could just as well come from the (extreme) left. This interpretation would be in accordance with Hilbert's labeling of Brouwer and Weyl as 'reformers'. Had Hilbert wanted to position Brouwer and Weyl as right-wing, the word 'reactionaries' would have been more appropriate. Even Hilbert's reference to Kronecker is a reference to a former attempt at a reform. Hilbert's warning for the danger of loosing one's precious treasures referred to a danger all too real for intellectuals, since many of them belonged to the upper class (even in economical terms, until the inflation) and were highly impopular among the masses.[24] The threat, then, came from the left-wing, Spartakist side.[25] I think the more plausible interpretation of Hilbert's text for readers at the time would be to see Brouwer and Weyl depicted

---

[23]'What Weyl and Brouwer do amounts in principle to following the former path of Kronecker: they seek to ground mathematics by throwing overboard everything that makes them uneasy and by establishing a repressive dictatorship *à la* Kronecker. But this means to dismember and mutilate our science, and we run the danger of loosing a large number of our most valuable treasures if we follow such reformers. (...) no: Brouwer is not, as Weyl believes, the revolution, but only the repetition, with old tools, of an attempted coup which, at the time, was undertaken much more vigorously, but nevertheless failed completely; and especially now that the state power has been so well armed and strengthened by Frege, Dedekind and Cantor, is doomed in advance to failure.' Translation based on [Ewald 1996, vol. II, p. 1119].

[24]In his extensive study on the German academic community from 1890 to 1933, Ringer describes the situation in Germany after the revolution as follows: 'The lower classes allowed themselves to be led by the Social Democratic Party; but their anger was directed less against capitalism than against the bureaucratic monarchy and its traditional ruling castes. Apparently, they sensed that the universities and the gymnasiums were important parts of the old social and political system; for they showed almost as much resentment toward the institutions of higher learning as they did toward the officer corps.', [Ringer 1969, p. 200].

[25]The Spartakists had been on the extreme left side of the SPD (*Sozialdemokratische Partei Deutschlands*) and joined the USPD (*Unabhängige Sozialdemokratische Partei Deutschlands*) when it broke of from the SPD in 1917. In 1919, the USPD was transformed into the KPD (*Kommunistische Partei Deutschland*). The Spartakists strived for completion of the revolution and wanted to give all power to the councils (*Raten*).

as extreme leftists. Thus, it seems that Hilbert and Weyl agreed that intuitionism was an extreme left-wing current which wanted to overthrow the existing order, but that they valued it differently. Whereas Weyl could be read as depicting it as a broad, popular movement, Hilbert's description was more that of a small group that wanted to take over power. Hilbert's defense against the attack was to posit himself as the champion of the state, of the ones in power – thus, the republic. Hilbert, of course, *was* the man in power of the mathematical 'empire'.

Now that Hilbert had reacted, it was clear that something was going on in the foundations of mathematics.[26] Had Hilbert remained silent, one could imagine that Weyl would have been left as the Dutch Social Democrat Troelstra, who declared the revolution, but later had to apologize as it turned out that nobody joined in. Now that the counterrevolution was there, it was clear that there had been a revolution, too.

It is not clear whether the positions sketched by Hilbert and Weyl by means of metaphors are in accordance with their political preferences. Although neither of them was politically active, it is said that Weyl had sympathy for the republic, at least in its early years.[27] Hilbert sympathised with Leonard Nelson's *Internationale Sozialistische Kampfbund* ('International Socialist Fighting Union'), a non-marxist, left-wing break-of party from the SPD.[28] Bernays later reported that Hilbert supported the Weimar coalition right after the war, then, due to the communist troubles and the inflation, moved to the conservative opposition, to finally support the republic after about 1925.[29] Progressive positions such as Weyl's and, at times, Hilbert's, were not very common among German academics in the Weimar period, though it seems to have been less rare in the Göttingen mathematics and physics community.[30] It should be pointed out that neither of them appealed to a 'restoration' in mathematics, even though such a term must have been popular at the time among German academics. The name *Putschists* lived on as a nickname for intuitionists, at least in Berlin.[31]

The Dutch mathematician Wolff countered Hilbert calling Brouwer's actions a coup in the following way:[32]

---

[26] In fact, Hilbert's public reaction meant a break with his usual attitude to avoid open polemical debates, cf. [Rowe 2000, p. 78].

[27] [Forman 1971, pp. 113–114]

[28] [Dahms 1987A, pp. 16–17]

[29] Letter from Bernays to Reid, 27/11/1968; [ETH Bernays, HS 975-3775]

[30] Ringer wrote on the Weimar period: 'The [German, DH] academic community as a whole did everything in its power to resist the new regime', [Ringer 1969, pp. 200–201]. However, Siegmund-Schultze pointed out that in Göttingen as in Berlin 'liberal and republic-friendly dispositions were at least not untypical among mathematics and physics students', [Siegmund-Schultze 1998, p. 62].

[31] [Van Stigt 1990, p. 95]

[32] [Wolff 1922, p. 17]

## 6.2. METAPHORS

> Rustverstoring moge den hervormers verweten worden, erkend zij, dat zij niet alleen afbreken, maar ook opbouwen (...).[33]

Recognition of the positive side of the reform, however, was exactly what was lacking.[34]

There were some others who used political metaphors in the foundational debate. These, like Weyl and Hilbert, agree that intuitionism is more left-wing than formalism. For example, Brouwer's friend the mathematician Barrau, who lectured in Groningen, said in 1926:[35]

> Vrij-axiomatici [formalists, DH] echter zijn er velen en velerlei, een uiterste rechtervleugel van fanatieke logistici en hartstochtelijke verzamelaars; een linkervleugel, intuitionisten bijna, die, historisch gesproken, door hunne afwijzing van en hunne critiek op het extremisme in het eigen kamp, den weg bereiden voor het autonome intuitionisme; en verder een breed, massaal centrum, bovenal gehecht aan de verruiming van arbeidsveld, die het vrij-axiomatisme bracht.[36]

Similarly, Ramsey spoke of 'the Bolshevik menace of Brouwer and Weyl',[37] whereas Skolem pointed out the conservative character of Hilbert's programme because of its name 'new foundations of mathematics',[38] implying that mathematics itself should be kept, and only its foundations changed.

Thus, for those contemporaries who used political metaphors to speak out, intuitionism was the more progressive force, while formalism was more conservative. This runs contrary to the modern view, recently presented by Mehrtens, that modernism, represented in this period by formalism, was the more progressive type of mathematics.[39]

**'Revolution' and 'crisis' in a broader context**  In his standard work *The decline of the German mandarins*, Ringer described the period from the 1890s to the 1920s as one of increased self-doubt about the functioning of the German academic community, reaching its height in the early years of the Weimar republic. 'By 1920, no German professor doubted that a profound 'crisis of culture' was at hand.'[40] In literary and artistic circles, it was not uncommon to link an artis-

---

[33]'One may reproach the reformers of breaching the peace, one should recognize that they not only demolish, but also build up (...).'
[34]See 3.3.1.
[35][Barrau 1926, p. 19]
[36]'But free-axiomaticians [formalists, DH] are manifold and exist in many forms, an extreme right-wing of fanatic and passionate set-theoreticians; a left-wing, almost intuitionists, who, historically seen, pave the way for autonomous intuitionism by their rejection and criticism of the extremism in their own group. And further there is a broad, massive centre, which above all is attached to the broadening of the field of work which free-axiomatism brought.'
[37][Ramsey 1926A, p. 380]
[38][Skolem 1926, p. 12]
[39][Mehrtens 1996, p. 522]. For Mehrtens on modernism, see 6.7.
[40][Ringer 1969, p. 254]

308     CHAPTER 6. THE FOUNDATIONAL CRISIS IN ITS CONTEXT

tic revolution to the Russian October revolution.[41] Furthermore, the revolution metaphor was often used in the debate on the theory of relativity.[42] It seemed to appeal to the longing for something new, whether it be in politics, science or art, after the atrocities of the Great War.

The same terms, revolution and crisis, had been used by prominent scientists before the war, to describe the situation in their discipline.[43] In the early years of the Weimar republic, one could easily find the crisis metaphor in more or less popular presentations of science. Even though most academics did not feel the need to define what the crisis consisted of, it existed, if only because almost every educated German believed it did.[44] After 1921, the theme of the crisis had become 'a ritual and an obsession'.[45] There was a crisis in politics, a cultural crisis, a crisis of learning, a crisis in linguistics, a crisis in mechanics, a crisis in the concept of causality, a crisis in theoretical physics and a crisis in German physics.[46] Regarding the exact sciences, Forman argued that mathematicians and physicians were quite ready to use the crisis rhetoric in order to show that they were part of the anti-rational tendencies of their time.[47]

The crisis situation in the early years of the Weimar republic was all too real. The war had caused enormous losses. A civil war was going on in the streets, there were political assassinations and *Putsch* attempts. In 1919, the Versailles treaty was imposed upon Germany, which among other things proclaimed Germany to be the only offender of the war, made it lose some 70,000 square kilometres of its home country and all of its colonies, and announced a compensation to be paid,

---

[41][Joll 1990, p. 302]

[42]See 6.4.1.

[43]For example, in his 1912 inaugural lecture, Paul Ehrenfest spoke the following words: *'Gestatten Sie mir über eine Krise zu sprechen, die gegenwärtig eine fundamentale Hypothese der Physik – die Aetherhypothese – schwer bedroht. Diese Krise gibt, wie mir scheint, ein lebendiges Bild von der eigenthümlichen revolutionären Stimmung, welche augenblicklich die theoretische Physik beherrscht'*, ('Allow me to speak of a crisis, which at the moment severely threatens a fundamental hypothesis of physics, the ether hypothesis. I think that this crisis provides a lively picture of the remarkable revolutionary atmosphere which at the moment controls theoretical physics.'), [Ehrenfest 1912, p. 3], cited from [Klomp 1997, p. 47]; I corrected for various linguistic and typing errors in the quote given there.

[44][Ringer 1969, p. 245]

[45][Ringer 1969, p. 385]

[46][Ringer 1969, p. 384]; *Über die gegenwärtigen Krise der Mechanik* ('On the present crisis in mechanics'), [Von Mises 1921], cited from [Forman 1971, p. 62]; *Zur Krisis des Kausalitätsbegriffs* ('On the crisis of the concept of causality'), [Petzoldt 1922]; *Über die gegenwärtigen Krise der theoretischen Physik* ('On the present crisis in theoretical physics'), [Einstein 1922], cited from [Forman 1971, p. 62]; *Die gegenwärtige Krisis in der deutschen Physik* ('The present crisis in German physics'), [Stark 1922], cited from [Goenner 1993, p. 132]. Forman has more examples of the use of the crisis metaphor later in the 1920s.

Weyl's foundational crisis paper does not fit into the general pattern of the crisis papers presented by Ringer, in which an attack upon over-specialisation and positivism plays an important role, [Ringer 1969, p. 385].

[47][Forman 1971, pp. 58–59]. Forman's theses and his suggestions for the crisis in mathematics are treated in 6.4.2.

## 6.2. METAPHORS

the amount of which was to be specified. In 1921, the amount was fixed at an enormous 132 billion *Goldmark*. Inflation rose to astronomical heights.

**Metaphors as taken over by others** Both the crisis and the revolution metaphor were taken over by others in their contributions to the foundational debate. Even though these two words were among the most prominent metaphors used, they did not appear in more than some ten percent of the contributions. Fraenkel, the most important reporter of the debate, was an enthusiastic user of the terms; he employed at least one of the metaphors in most of his contributions. A comparison between the two metaphors learns that the crisis metaphor was used somewhat more often, and that it was much more prominent because of its appearance in the title of papers and lectures, a way in which the revolution metaphor was not used. A striking difference is that the revolution metaphor was used about as often in German-language contributions as in non-German ones, whereas the crisis metaphor was utilized almost exclusively in German.[48] Thus, it seems that, while the revolution was felt as an international phenomenon, the crisis was felt mainly in the German speaking areas, in accordance with Ringer's remarks. The fact that people used the crisis metaphor does not necessarily mean that they agreed there was a crisis; some used it to question the crisis or to argue against it.[49]

It is worth paying attention to the chronology of the use of both metaphors.[50] If we take the appearance of the metaphor 'revolution' as a measure for the development of the foundational debate, we see an *uninterrupted* debate from 1923 on, rising until 1927, slightly falling back in 1928, whereafter the revolution metaphor only appears occasionally. A similar development can be seen in the use of the crisis metaphor. Again, 1923 is the year in which the *uninterrupted* development starts, and the peak is in 1927. In 1928, there is a marked drop to only one contribution. In 1929, its use rises again, but in a different way. In an unprecedented contribution, the geometer Study, who was at the time working on a book about the foundational crisis[51] and who knew Brouwer personally, mocks at the use of

---

[48] The main non-German contribution in which the crisis metaphor figured prominently was Wavre's '*Y a-t-il une crise des mathématiques?*', ('Is there a crisis in mathematics?'), [Wavre 1924]. Wavre's answer was that no such crisis existed.

[49] For instance, Wavre put the question in the title, '*Y a-t-il une crise des mathématiques?*' ('Is there a crisis in mathematics?'), and his answer was negative, [Wavre 1924, p. 469], whereas Hölder spoke about the 'so-called foundational crisis' (*'die sogenannte Grundlagenkrise'*) in the title of his paper [Hölder 1926].

[50] As in 3.2.2, published lectures were classified according to the year in which they were given.

[51] The book, *Prolegomena zu einer Philosophie der Mathematik. Der Streit um die Grundlagen der Analysis* ('Prolegomena to a philosophy of mathematics. The battle on the foundations of analysis'), remains unpublished until today. In it, Study is dismissive of the foundational crisis, calling it *'eine Art von geistiger Infektionskrankheit'* ('a sort of spiritual infectious disease'). He describes intuitionism as *die Mathematik der Gespensterfurcht und der Polizeiverbote* ('the mathematics of fear for ghosts and of police-forbiddings'), referring to Hilbert's metaphors; Brouwer as 'Allah' and Weyl as his prophet.[52] The main problem Study has with intuitionism is what he calls its psychological character. He is, however, equally negative about Hilbert and formalism, and defends instead the contentual character of classical mathematics. On the persons

slogans:[53]

> Freilich kämpfen nur zwei noch kleinen Parteien diesen grimmen Kampf. Ihre Bannerträger führen Fahnen mit merkwürdig vielen übereinstimmenden Devisen daher:
> Transzendentaler Idealismus! Antinomien! Unwissenschaftlichkeit der Analysis! Cavete numeros infinitos![54] Tertium datur! Schafft euch neue Scheuklappen an! Tragt farbige Brillen!
> Um unliebsamen Verwechslungen vorzubeugen, sind aber auch Banner mit abweichenden Inschriften da:
> Intuition!     |     Figuren auf Papier!
> Reifgewordene Logik!     |     Metamathematik!
> Man nennt dies geräuschvolle Treiben Grundlagenstreit, auch reden einige von einer Krise der Mathematik. Zu der Stärke der Sprache scheint indessen die Überzeugungskraft der von beiden Seiten vorgebrachten Gründe nicht im Verhältnis direkter Proportionalität zu stehen.[55]

After 1931, finally, the crisis metaphor almost completely disappeared; even though a political and economic crisis, again, was all too real. When Hahn and Menger, together with other Viennese scientists, organised a lecture series in 1932–33, they called it *Krise und Neuaufbau in den exakten Wissenschaften* ('Crisis and new construction in the sciences') and argued that the crises had been followed by an erection of simpler but better grounded scientific edifices.[56]

The chronological development of the use of these metaphors provides us, I think, with the best possible answer to the question where to draw the end of the

---

involved most in the foundational debate Study wrote: '*Z.B. halte ich Brouwer, Weyl und Hilbert für geistig nicht normal. (In Bezug auf Hilbert hat mir das ein Psychiater bestätigt, der früher in Göttingen war und bei H. gehört hat.)*' ('For example, I take Brouwer, Weyl and Hilbert not to be mentally normal. (With regard to Hilbert, a psychiatrist, who formerly was in Göttingen and who attended lectures with H., has confirmed this to me.)' (letter from Study to Engel, 12/1/1929, [BMIG Study]).

Study's paper in the *Sitzungsberichte* in 1929 was presumably based on research done for the book. From his correspondence with Engel, one can see that Study worked on the *Prolegomena* at least from halfway 1929. He died in January 1930. The manuscript of the book can be found in library of the Mathematical Institute of the *Westfälische Wilhelms-Universität Münster*. I am grateful to David Rowe and Yvonne Hartwich for providing me with this information.

[53][Study 1929, p. 255]

[54]I have no idea why Study inserted a Latin slogan into the sequence.

[55]'Admittedly only two still small parties parties engage in this fierce fight. Their standard bearers bear forward banners with remarkably many corresponding slogans:
Transcendental idealism! Antinomies! Unscientific nature of analysis! Beware of infinite numbers! *Tertium datur*! Purchase new blinkers! Wear colourful glasses!
In order to avoid unpleasant mistakes, there are banners with deviating inscriptions, too:
    Intuition!     |     Figures on paper!
    Ripened logic!     |     Meta-mathematics!
One calls this noisy stir foundational battle, some even speak of a crisis of mathematics. Yet, the cogency of the reasons put forward from both sides does not seem to be in direct proportionality to the vigour of the language used.'

[56][Mark et al. 1933, Vorwort]

foundational crisis. In chapter three, I argued on the basis of other considerations that the end should be situated not later than 1930. Now, we can refine the picture. One of the characteristics of the foundational debate was its emotional character and the use of (political) metaphors like 'revolution' and 'crisis'. The use of both metaphors dropped markedly in 1928. At the same time, the first contribution mocking the metaphors appeared. This provides the best indication that after 1928 the situation had substantially changed. Therefore, the end of the foundational crisis is best set in 1928.[57]

It is not clear what may have caused this change in attitude. It seems that some kind of sobriety came over the debate. A rather speculative explanation might be the clarifying influence of Heyting's formalisation of intuitionistic logic, which spread already before its publication.[58]

## 6.3 Philosophy

In this section, I treat the relationship between philosophy and the foundational crisis in mathematics. In order to better understand the reactions to Brouwer's intuitionism, it is important to realize that in philosophy proper a number of developments took place in the Interbellum which, because of its similarities with mathematical intuitionism, may have influenced the way in which people perceived 'intuitionism' or pleas for 'intuition' in mathematics. The most important philosophical current in this respect was the so-called *Lebensphilosophie* ('philosophy of life').[59] I indicate different ways in which the philosophy of life may have influenced the foundational debate. First of all, it should be recognized that, instead of one (mathematical) intuitionism, there were in fact two, the other being a sub-current of *Lebensphilosophie*. The occurrence of the same name for different views may have misled some participants to the debate. There were only few contemporaries who explicitly differentiated between the two. Furthermore, I point out that Hilbert was directly involved in a fight against some representatives of *Lebensphilosophie* in the philosophical faculty in Göttingen. Finally, I discuss the popular life philosopher Spengler and look for Spenglerian presentations used in the foundational debate.

---

[57]This means that Fraenkel made a very good judgement, when stating in 1929 that he thought that intuitionism was over its height, [Fraenkel 1930A, p. 297]. For when the foundational crisis was over, the chance was gone that intuitionism could become the dominant current in mathematics.
[58]See 5.4.1.
[59]Many representatives of logical positivism, as it was adhered to in the *Wiener Kreis* and the Berlin *Gesellschaft für empirische Philosophie*, were directly involved in the foundational debate. Therefore, they are not treated in this chapter, which is devoted to the *context* of the foundational crisis, but in 3.3.3.

### 6.3.1 Lebensphilosophie

Lebensphilosophie was an important intellectual but above all popular current in Weimar Germany. In Lukács' well-known analysis *Die Zerstörung der Vernunft* ('The destruction of reason'), in which the author investigates Germany's road to Hitler in philosophy,[60] Lukács characterised *Lebensphilosophie* as the dominant ideology in postwar Germany.[61]

**Rickert** An important exposition of German *Lebensphilosophie* as perceived in the 1920s is provided by the Heidelberg scholar Heinrich Rickert (1863–1936). Rickert was a neo-Kantian philosopher who was active from the end of the 19th century until halfway the 1930s. He is seen as the most important student of Windelband;[62] Heidegger did his Habilitation with Rickert. The Freiburger school, to which Rickert and Windelband belonged, represented together with the Marburger school the dominant neo-Kantian current at the time.

In 1920, Rickert published a book called *Die Philosophie des Lebens. Darstellung und Kritik der philosophischen Modeströmungen unserer Zeit* ('The philosophy of life. Presentation and criticism of the fashionable philosophical currents of our time'),[63] in which he presents a characterisation and a critical analysis of *Lebensphilosopie* in its various forms. I based my analysis of *Lebensphilosophie* on Rickert's book since it presents an extensive treatment of the philosophy of life by a prominent philosopher at the time, in which the term 'intuitionism' figures frequently. I restrict myself to Rickert's description of what *Lebensphilosophie* in general was, and more specifically the version which he called *intuitive Lebensphilosophie*.[64]

---

[60][Lukács 1960, p. 10]. It is a widely held thesis that *Lebensphilosophie* functioned as fertile soil for the national socialist ideology. As Dahms pointed out, however, the result of the Nazi university cleansings meant, at least in the case of Göttingen, the *end* of the academic philosophy of life, [Dahms 1987B, p. 169].

[61]Lukács even claims that virtually all of the widely read bourgeois *Weltanschauungsliteratur* was *lebensphilosophisch* ([Lukács 1960, p. 351]). However, I have not seen any evidence that Lukács did extensive research to substantiate this specific point. In general, Lukács' book is loaded with dialectic materialist terminology, which adds more to its value as a time document written in Hungary in (originally) 1952 than to the power of its analysis. (*'Die Lebensphilosophie, wie sie in der imperialistischen Periode als philosophische Richtung auftritt und sich entfaltet, ist ein spezifisches Produkt dieser Zeit: der Versuch, vom Standpunkt der imperialistischen Bourgeoisie und ihrer parasitären Intelligenz jene Fragen philosophisch zu beantworten, die von der gesellschaftlichen Entwicklung, von den neuen Formen des Klassenkampfes gestellt wurden.'* etc., [Lukács 1960, p. 352] – 'The philosophy of life as it appears and develops itself as a philosophical current in the imperialistic period is a specific product of the time: the attempt to answer philosophically those questions which were posed by the development of society, by the new forms of the class-struggle, from the point of view of the imperialistic bourgeoisie and its parasitical intelligentsia.').

[62][Brown, Collinson & Wilkinson 1996, p. 665]

[63][Rickert 1920]

[64]All following information on *Lebensphilosophie* was taken from [Rickert 1920], unless stated otherwise. The description given here departs from Rickert in that it highlights those aspects of the philosophy of life which are relevant for a comparison with Brouwer's intuitionism.

## 6.3. PHILOSOPHY

It should be pointed out that Rickert himself was definitely not a follower of *Lebensphilosophie*.[65] The reason why he devoted a book to it was that he saw it as the most widely spread philosophy of the time to be taken seriously. Rickert was, at least later, recognised as one of the foremost critics of the philosophy of life.[66]

**Life and the lifeless**  The name *Lebensphilosophie* is explained in the following way. According to Rickert, the most modern view was the one holding that philosophy should occupy itself with life. The view on life is to be taken as the basis from which to formulate a complete philosophy. As the main predecessors of *Lebensphilosophie* Rickert mentions Schopenhauer, Nietzsche, Henri Bergson and William James.

Whereas life is to be taken as the basis for philosophy, it is opposed to everything that has become rigid and lifeless. These things should therefore be avoided. More specifically, Rickert maintains that adherents of *Lebensphilosophie* refuse to be restricted to fixed rules of grammar and to given sets of words.

Similarly, *Lebensphilosophie* rejects systems because of their rigidity and lifelessness. If one then reproaches the philosophy of life to be without principles, its adherents will answer that one cannot expect it to have a firm ground of related concepts, since it wants to be a living philosophy.

In Rickert's view, the fight of *Lebensphilosophie* against 'the system' brought it many adherents. For, so he maintains, systematic thinking is highly unsympathetic to many people.[67]

**Mathematics and science**  In looking for the essence of nature, the philosophy of life wants to keep away from matter seen as mere atom complexes ruled by mathematically formulated laws. Therefore, Rickert tells us, even materialists in disguise join the current of *Lebensphilosophie* by using organic reasonings against materialistic ones.[68] Classical sciences are seen as occupying themselves with mere appearances, whereas the only thing real is that which is sensed intuitively. In some of these views, scientific theories are described as dead 'constructions'.[69] It will not surprise that one of the main characteristics of *Lebensphilosophie* is its anti-rationalism and anti-intellectualism.

The life philosophical view on the universe is different from the scientific one. Referring to Bergson, Rickert describes it as acknowledging that everything that is is a creating act, an ever becoming and happening. Living can only unfold itself in becoming, not in being.

---

[65] '*Die Modephilosophie des Lebens wird bisweilen zum Lebenssumpf, und darin gibt es dann nur noch Froschperspektiven*', ('Sometimes, the fashionable philosophy of life becomes a life swamp, in which there are then only frog perspectives') [Rickert 1920, p. 155].

[66] [Ringer 1969, p. 337]

[67] The Nazi's, too, campaigned against what they called 'the system', meaning the Weimar republic, [Siegmund-Schultze 1998, p. 61].

[68] [Rickert 1920, p. 9]. Rickert's remark resembles the weaker version of Forman's thesis, see 6.4.2.

[69] [Rickert 1920, p. 153]

# 314 CHAPTER 6. THE FOUNDATIONAL CRISIS IN ITS CONTEXT

In discussing mathematics, Rickert makes the following side remark which is of interest to us:[70]

> Von einer lebendigen Mathematik war allerdings bisher wohl noch nicht die Rede, und diese Disziplin ist bei den meisten Vertretern der 'lebendigen' Wissenschaft auch wenig geliebt. Doch lassen sich sogar hier Richtungen konstatieren, die darauf ausgehen, die Welt des Mathematischen beweglicher und insofern lebendiger zu gestalten.[71]

Unfortunately, he does not say whom he sees as attempting to make mathematics more alive. Therefore, it is hard to say whether Rickert was aware of the foundational crisis in mathematics.[72]

**Intuitive *Lebensphilosophie*** Rickert divides *Lebensphilosophie* into two currents: the intuitive and the biological one. The first is the one that is of interest to us here. At several occasions[73] Rickert also calls this view the intuitionistic one.

Rickert characterizes intuitive *Lebensphilosophie* by its allegation that real life as given to us by intuition is opposed to dead forms:[74]

> (...) die 'Formen', in welche der Verstand die Welt bringt, machen das Leben 'unlebendig'. Es gilt daher, vorzudringen zum ungeformten, unverfälschten, reinen Inhalt, wie er sich der unmittelbaren Intuition als 'echtes' Leben darbietet.[75]

Rickert refers to Simmel, who used a special argumentation for the opposition between form and life. Life is, in Simmel's view, an unlimited continuity, and

---

[70] [Rickert 1920, p. 9]

[71] 'Until now there has been no talk yet of mathematics which is alive, and this discipline is not very much loved among most followers of 'living' academic research. But even here one can note schools which aim at forming the mathematical world more changeable and in that sense more alive.'

[72] In his *Allgemeine Grundlegung der Philosophie* ('General foundation of philosophy'), which he finished half a year after *Die Philosophie des Lebens*, Rickert wrote that even though a mathematical object is something different from the psychological act of experiencing it, one would not say that the object does not exist. '*Darüber besteht heute wohl kaum noch Streit.*' ('There is hardly any fight about that these days.', [Rickert 1921, p. 107]) This makes his earlier reference to 'living' currents within mathematics all the more puzzling. A further indication is given in a treatise he published in 1924. In a postscript to *Das Eine, die Einheit und die Eins* ('The singular, the unit and the one'), first published in 1911, Rickert used the terms 'intuitionism' and 'logicism' without even once mentioning Brouwer or Russell. Even though he mentioned that he had not included foreign literature in order not to make the paper too long, [Rickert 1924, pp. 83–84], this gives the impression that he was not aware of the foundational debate in mathematics.

[73] For example, [Rickert 1920, p. 36; p. 46; p. 52; p. 55]

[74] [Rickert 1920, p. 41]

[75] '(...) the 'forms', in which reason puts the world, make life 'unliving'. Therefore, the important thing is to penetrate to the unformed, unadulterated, pure content, the way it appears to the immediate intuition as 'real' life.'

## 6.3. PHILOSOPHY

continuity and form are opposed to each other, since form itself is a border and cannot be changed. In this view, intuition is often seen as opposed to reason.

Although Rickert does not point out so, the stress on intuition is often related to the works of the popular French philosopher Henri Bergson. In Bergson's view, there is an ontological dualism between living and unliving, between life and matter. The latter can be known by the intellect; the former only by intuition. Therefore, Bergson demanded that intuition be recognized as a new metaphysical method.[76]

Rickert notes that one often speaks of a 'hunger for intuition'[77] in his times. This is understandable in areas where certain forms, which life adopted and which thus made sense in a certain period, have survived themselves and thus lost the connection with intuition.

It is a thankless task, Rickert complains, to fight intuitionism these days, because the words 'intuition' and 'living' have such an appeal to many people. Furthermore, one can only reason in favour of the rational by departing from a rational basis.

**Rickert in 1923** In 1923, Rickert published a paper in which he was even more outspoken on the popularity of philosophical intuitionism. By then he no longer speaks of *Lebensphilosophie* in general as being widespread, but only of intuitionism, and he declares:[78]

Der Intuitionismus gilt vielen als die Philosophie der Zukunft.[79]

Again, Rickert gives some of the main characteristics of intuitionism. In the first place intuitionism adheres to the idea of substituting constructions by intuition. For constructions are abstract, forgetting about real life, and they disturb the unity of the world. Secondly, intuitionists maintain that 'the immediate'[80] can only be grasped immediately. Therefore, they refuse to obey to logical structures and intermediate methods.

On this basis, intuitionists demand that philosophy be shaped intuitionistically. Rickert disagrees, since he thinks that pure intuitionism would make philosophy vulnerable to arbitrariness.

**Logicism and formalism in philosophy** The term 'intuitionism' was not the only one familiar from the foundational debate in mathematics that appeared in Rickert's work. In another publication, he also used the terms 'logicism' and 'formalism'.[81] However, they were used less frequently and the stress put on them was less than with intuitionism.

---

[76] [Hentschel 1990, pp. 441–442]
[77] *'Hunger nach Anschauung'*, [Rickert 1920, p. 42]
[78] [Rickert 1923, p. 53]
[79] 'Many see intuitionism as the philosophy of the future.'
[80] *'das Unmittelbare'*, [Rickert 1923, p. 53]
[81] Both terms were used in a wider context, too. In Litt's *Die Philosophie der Gegenwart*, it is noted that *Lebensphilosophie* is opposed to the 'formalism' of a 'logical systematization',

# 316  CHAPTER 6. THE FOUNDATIONAL CRISIS IN ITS CONTEXT

Rickert discusses logicism in the context of the question to what extent philosophy should be bound by logic. To Rickert, this 'restriction' does not pose a problem, because he sees the bond to logic as the core of philosophy, i.e., of thinking about the world. Instead, he maintains that logicism should only be opposed if it would want to shape not only the form of the philosophical world according to purely logical rules, but also its contents.[82]

Rickert uses the term 'formalism' without explaining what is meant. He states that for any systematic philosophy the distinction between form and contents is indispensable, even though many philosophers may 'denounce'[83] formalism. Interestingly, Rickert remarks that only content changes, not form. This is exactly what happened in Hilbertian formalism in mathematics, too.

One of the main ideas of mathematical formalism, the consistency demand, is discussed by Rickert independent of the topic of formalism. He claims that many people see the unification of all scientific knowledge into one consistent philosophical system as the main task of philosophy. But, Rickert maintains, the formal criterion of consistency does not get one any further, and nobody, so he claims, seriously tried to achieve this goal.[84]

## 6.3.2 Mathematical and philosophical intuitionism: a comparison

I now turn to the question if and how intuitionism as described by Rickert is related to Brouwer's intuitionism. In order to distinguish between the two, I call the latter mathematical, the former philosophical intuitionism.

To start with, I should point out that the basis from which possible conclusions can be drawn is rather small. Apart from Rickert, I know of few philosophers who spoke of philosophical intuitionism.[85] On the other hand, Rickert was one of the well-known philosophers of his days, therefore his writings must have had some influence.[86]

**Similarities** The similarities between mathematical and philosophical intuitionism as described above are too obvious and too many not to take them into ac-

---

[Litt 1925, p. 32]. For further remarks on formalism, see 2.5. Logicism was, for instance, mentioned as the current around Natorp's *Sozialpädogogik*, which is hardly related to mathematical logicism, [Litt 1925, pp. 20–25]. In philosophy in general, it seems to have had a negative connotation. Cf. Ringer, who describes that Cohen and his followers were accused of 'logicism', [Ringer 1969, pp. 306–307].

[82][Rickert 1921, pp. 62–63]

[83]*'schelten auf'*, [Rickert 1921, p. 351]

[84][Rickert 1921, pp. 165–166]

[85]Bergson's intuitionism is also mentioned in [Messer 1927, p. 18]. Pacotte managed to mention Bergsonian intuitionism in mathematics without referring to Brouwer even once, cf. [Pacotte 1925].

[86]It is hard to say to what extent the interpretation given here can be extended to include other works on the philosophy of life from the period. More research would be needed to evaluate this.

## 6.3. PHILOSOPHY

count in evaluating the reception of Brouwer's intuitionism. First of all, it should be pointed out that it is quite unlikely that philosophical intuitionism influenced its mathematical namesake. It is, however, very well possible that some of the participants in the mathematical debate spoke about mathematical intuitionism, but actually meant its philosophical namesake (or did not clearly distinguish between the two).[87]

The first similarity one notes is the name: both are called intuitionism. The use of this name, however, is not new, neither in philosophy nor in mathematics.

Regarding philosophical intuitionism, the name was used in connection with the philosophy of Henri Bergson.[88] Bergson gained an international reputation with the publication of his *L'évolution créatrice* ('The creative evolution') in 1907,[89] and his views formed a popular topic among academics. In Germany, the use of 'intuitionism' for Bergson's philosophy can be traced back to a paper by Kroner in 1910. Kroner characterized Bergon's philosophy by the three isms anti-rationalism, biologism and intuitionism.[90] However, in France the term seems to have been used less frequently. For instance, Benda, in his 1914 *Sur le Succès du Bergsonisme* ('On the success of Bergsonism'),[91] constantly speaks of 'Bergsonisme' instead of intuitionism.

In the case of mathematical intuitionism, the first mentioning I know of dates from 1893. In that year, Felix Klein delivered a series of lectures in English at the Evanston Colloquium in Chicago. In the beginning of the first lecture, he states the following:[92]

> Among mathematicians in general, three main categories may be distinguished; and perhaps the names *logicians, formalists,* and *intuitionists* may serve to characterize them.

The description Klein gives of these groups is rather different from what we are used to nowadays: logicians are strong in giving strict definitions, formalists excel in devising algorithms, and intuitionists primarily resort to geometrical intuition.[93]

---

[87]For example, Barzin and Errera maintain that Brouwer's idea of truth has nothing to do with a proof, but should be understood as meaning 'originating from the primary intuition' (*'tout ce qui découle de l'intuition originelle'*), [Barzin & Errera 1929, p. 7]. Although they do not mention Bergson, their idea seems to have been influenced by Bergson's intuitionism rather than by Brouwer's. There are probably more examples to be found if one checks how people reacted to the use of the word 'intuition' in Brouwer's writings. I did not do so.
[88][Meyer 1982, p. 24]
[89][Kerszberg 1997, p. 56]
[90][Kroner 1910/11, p. 139]
[91][Benda 1914]
[92][Klein 1911, p. 2]
[93]In this respect, Paul Du Bois-Reymond's 1882 characterisation of formalism is much closer to the modern view. Du Bois-Reymond does not give a description of formalism, but notes that pure formalism would lead to a 'mere sign game' (*'ein blossen Zeichenspiel'*), comparable to chess, [Du Bois-Reymond, P. 1882, pp. 53–54].

The structuring of mathematicians along these names, however, dominated during a large part of the 20th century.

The use of words as 'perhaps' and 'may serve' seem to indicate that at the time of speaking, the division was not generally accepted or maybe not even well-known. Whether Klein was the one who started this grouping and whether it was taken by others from him is not known.

In the context of the foundational debate in the 1920s, the term 'intuitionism' became known by Brouwer's inaugural address *Intuitionisme en formalisme*, delivered in 1912 when he became professor at the university of Amsterdam. In the talk, he did not explain where the names came from, but merely indicated that intuitionism was largely French, whereas formalism was predominantly German. Also, he traced intuitionism back to Kant.[94] Although intuition played an important role in Kant's philosophy, Kant does not seem to have used the term 'intuitionism'.

The main characteristic of *Lebensphilosophie* in general, its adherence to life as opposed to the lifeless, can be found with Brouwer as well. In 1906, in a letter to Korteweg, Brouwer defended the direction his dissertation had taken with the following words:[95]

> (...) wat ik nu heb gebracht, behandelt uitsluitend, hoe de wiskunde in het leven wortelt, en hoe dus de uitgangspunten der theorie behooren te zijn (...).[96]

In the same letter, Brouwer acknowledged that cultural influences may have played a role, for he explains the difficulties Korteweg had in understanding him by saying that he [Brouwer] was a 'child of a different time'.[97] In his presentation of intuitionism, however, Brouwer did not use such considerations.

Furthermore, Brouwer and the life philosophers share a dislike of fixed grammatical rules, fixed logical structures, fixed systems (for example, axiomatic systems) and fixed forms. In 1906, Brouwer wrote:[98]

> Men heeft bij filosofie niets aan gebouwen en systemen (men wil direct worden aangeslagen met inzicht), want systemen hebben alleen waarde

---

[94] See 2.5, where also the question is discussed how Brouwer might have come to use the term 'intuitionism'.

[95] Letter from Brouwer to Korteweg, 5/11/1906; cited from [Van Dalen 1981, p. 9]; a similar remark can be found in [Brouwer 1905, p. 47].

[96] '(...) what I have presented now only treats how mathematics is rooted in life, and thus what the points of departure of the theory should be (...).'

[97] '*omdat ik een kind van een anderen tijd ben dan u*', letter from Brouwer to Korteweg, 5/11/1906; cited from [Van Dalen 1981, p. 9].

[98] [MI Brouwer, BMS3A, C. VIII.73]

## 6.3. PHILOSOPHY

> in toepassing in strijd tegen een vijand; maar filosofie <u>moet</u> niet worden toegepast. – Wijsbegeerte kan niet wiskundig in elkaar zitten.[99]

Also, Weyl's presentation of the intuitionistic continuum as a *Medium freien Werdens*, as opposed to the classical atomistic continuum, fits in very well with the life philosophical idea that living can unfold itself only in becoming:[100]

> Die Brouwersche Bemerkung ist einfach, aber tief: hier ersteht uns ein 'Kontinuum', in welches wohl die einzelnen reellen Zahlen hineinfallen, das sich aber selbst keineswegs in eine Menge fertig seiender reeller Zahlen auflöst, vielmehr ein *Medium freien Werdens*.[101]

In general, choice sequences could be combined nicely with the life philosophical opposition against causality and determinism, although neither Brouwer nor Weyl did so.

Also, *Lebensphilosophie*'s opposition against the intellect can be found in Brouwer's more mystical views expressed in *Leven, Kunst en Mystiek*.[102]

Finally, both mathematical and philosophical intuitionism make an appeal to intuition, a characteristic which at least superficially makes them comparable.

**Differences**  Besides the similarities between mathematical and philosophical intuitionism, there are two clear differences, too. The first one is the latter's anti-rationalism. Brouwer did not share this. Once one accepts his basis, i.e., the demand that mathematical concepts are meaningful, intuitionistic mathematics is worked out in a rational way.

The other and more striking difference is the philosophical intuitionism's dislike of constructions. As Rickert stressed repeatedly, philosophical intuitionists saw constructions as dead things that one should get rid of.[103] Brouwer, however, deliberately used constructions in order to create intuitionistic mathematics. This point is related to the former, since the creation of intuitionistic mathematics by means of constructions is a rational process.

---

[99]'In philosophy, buildings and systems are useless (one wants to be struck by direct insight), since systems are only valuable if applied in the fight against an enemy; but philosophy <u>must</u> not be applied. – Philosophy cannot be constituted mathematically.'

[100][Weyl 1921, p. 221]

[101]'Brouwer's remark is simple but deep: we have here the arising of a 'continuum' into which the individual real numbers do fall, but which itself in no way dissolves into a set of real numbers which are completed; we rather have a *medium of free becoming*.' Translation based on [Mancosu 1998, p. 94].

[102][Brouwer 1905, pp. 17–26]; see also 2.2.4.

[103]'*An die Stelle der Konstruktion habe die Intuition zu treten*' ('Constructions must be replaced by intuition'), [Rickert 1923, p. 53]. Cf. also [Rickert 1923, pp. 61–64].

### 6.3.3 Contemporaries' remarks

There is no abundance of Brouwer's contemporaries who mention the similarity between philosophical and mathematical intuitionism in their publications. Actually, I could only find three: Vollenhoven, Menger, and Fraenkel.[104]

**Vollenhoven** In his dissertation in theology *De wijsbegeerte der wiskunde van theïstisch standpunt* ('The philosophy of mathematics from a theistic point of view') published in 1918, the Dutch Ph.D. student Dirk Vollenhoven[105] aimed at giving a theistic foundation of mathematics. In order to do so, he divides the foundational currents in mathematics into three: empiricism, formalism and intuitionism. However, the definition he uses for these movements are quite different from the ones we are used to. Vollenhoven starts from the distinction between a material and a mental world, and next distinguishes between two monisms and a dualism. Dualists hold that there is a qualitative difference between mind and matter; for monists, no such difference exists. The two possible monisms are the one that holds that everything is matter, while the other sees everything as mind. Vollenhoven then labels material monism as empiricism, mental monism as formalism, and dualism as intuitionism.[106] In this way, the currents obtain a much wider meaning than we are used to, and Vollenhoven can extend intuitionism to include such various people as Bolzano, Kant, Leibniz, Newton, Pascal, Descartes, Galilei, Augustinus, Euclid, and even Aristotle, Plato and Socrates.

However, when discussing modern intuitionism, Vollenhoven comes closer to mathematical intuitionism. He devotes a large section to Poincaré, who 'more than anyone else'[107] has the right to serve as representing intuitionism in modern mathematics, and Brouwer is discussed in a separate section, too. Vollenhoven mostly agrees with Brouwer's construction of mathematics, but there is a fundamental difference regarding the question of mathematical truth, which, in Vollenhoven's view, is an absolute norm, independent of the act of thinking.[108] All in all, Vollenhoven judges Brouwer's 'vitalistic'[109] intuitionism dangerous for mathematics and dismisses it for being mystical and pan-theistic.[110] He does want to recon-

---

[104]Wavre does mention Bergson and his view on intuition in a paper on the foundational crisis and remarks that the intuitionistic view of the continuum seems to escape from Bergson's criticism. However, he does not label Bergson's view 'intuitionistic', [Wavre 1924, p. 455]. A similar remark applies to Van Os, [Van Os 1933, p. 215]. Study mentions both Bergsonian and Brouwerian intuitionism in his unpublished book *Prolegomena zu einer Philosophie der Mathematik* ('Prolegomena to a philosophy of mathematics'), and conjectures that the main similarity between them is that they both depart from psychological theses. On Study's book, see 6.2.1.

[105]Vollenhoven (1892–1978) was to become professor of philosophy at the *Vrije Universiteit Amsterdam* in 1926. Together with Dooyeweerd, he is seen as the father of reformational philosophy, a current in orthodox protestant philosophy. More information on Vollenhoven's early work may be found in [Kok 1992].

[106][Vollenhoven 1918, p. 3]
[107]'meer dan iemand', [Vollenhoven 1918, p. 352]
[108][Vollenhoven 1918, p. 387; p. 393]
[109]'vitalistisch', [Vollenhoven 1918, p. 402]
[110][Vollenhoven 1918, p. 351]

## 6.3. PHILOSOPHY

cile mathematical intuitionism and theistic philosophy in a 'theistic intuitionistic construction'.[111]

The part in Vollenhoven's dissertation where he pays most attention to comparing mathematical and philosophical intuitionism is the introductory section to modern intuitionism. There, Vollenhoven discusses the different forms of intuition used by Poincaré, Brouwer and Bergson. However, they are each treated separately, and no influences of the one on the other are mentioned.[112]

Apart from the resemblance between Brouwer and Bergson in using, at least in general terms, the same word 'intuition', Vollenhoven mentions another feature of Brouwer's intuitionism that coincides with Bergson's views. This is an aspect of mathematical intuitionism that is hardly ever mentioned at all, namely its link to life.[113] In general terms, Vollenhoven states that 'intuitionistic epistemology has to start from life'.[114] More specifically regarding Brouwer, he mentions the relation between mathematics and life as one of the first characteristics of Brouwer's intuitionism.[115]

It may be evident that the influence of a theological dissertation in Dutch on the foundational debate at large is negligable. However, Vollenhoven's work makes it clear that at least some people at the time of the debate were aware of a possible lumping together of Brouwer's and Bergson's intuitionism.[116]

In the dissertation, Vollenhoven expresses his gratitude to Brouwer for his initial help.[117] Therefore, it is reasonable to assume that Brouwer received a copy of the dissertation and thus saw himself in one class not only with Poincaré, but also with Bergson.

**Menger** In his 1930 paper *Der Intuitionismus*, Menger opened with a description of intuitionism in philosophy and in mathematics. He characterises the former current by its recognition of the intuition as a source of knowledge, and mentions Bergson and Husserl as two of its representatives. Mathematical intuitionism, Menger maintains, is different in that it does not extend our possible knowledge by adding intuition, but by a similar appeal aims at decreasing our knowledge, since not all mathematics is in accordance with our intuition. In fact, this is not true, and it is yet another example of the one-sided stress on negative aspects of intuitionism. A similarity pointed out by Menger is that both Brouwer and Bergson use a primordial intuition based on time.[118]

---

[111]'theistisch-intuitionistische constructie', [Vollenhoven 1918, p. 351]
[112][Vollenhoven 1918, pp. 349–351]
[113]Brouwer, too, did not stress this point.
[114]'De intuitionistische kenleer heeft (...) uit te gaan van het leven', [Vollenhoven 1918, p. 73]
[115][Vollenhoven 1918, p. 388]
[116]In later publications, Vollenhoven used the term 'intuitionism' in a broad sense, without reference to Brouwer, cf. [Vollenhoven 1919A], [Vollenhoven 1919B] and [Vollenhoven 1921].
[117][Vollenhoven 1918, p. VII]
[118][Menger 1930A, p. 311; 318]

**Fraenkel** Fraenkel only makes a remark about mathematical and philosophical intuitionism in one of his later contributions. In the paper 'On modern problems in the foundations of mathematics', published in 1933, Fraenkel simply remarks that Brouwerian intuitionism has 'nothing in common' with philosophical intuitionism, for instance of the Bergsonian brand.[119]

The fact that the first remarks about the differences between mathematical and philosophical intuitionism only appeared so late in the debate can mean two things. Either, it took a long time before philosophical intuitionism became known in mathematical circles, or it took a long time before people realised that the two were different. Unfortunately, with the present state of knowledge it is not possible to determine which of the two applies.

### 6.3.4 Göttingen and Hilbert

A factor which makes a linking between the two intuitionisms more likely is the fact that the philosophical seminar in Göttingen was one of the academic strongholds of *Lebensphilosophie*. Georg Misch and Herman Nohl lectured in Göttingen, both students of Dilthey, who is seen as one of the main exponents of the philosophy of life.[120] Unfortunately, I do not know if they used the name 'intuitionism' to indicate their point of view. They did oppose Nelson's supposed 'formalism' in ethics, when fighting with Hilbert in 1917 about the succession of Husserl after he had left Göttingen for Freiburg.[121]

Hilbert was one of the persons who was aware of the fact that intuitionism did not descend solely from Brouwer. In rough notes on a book by Messer, he mentions Bergsonian intuitionism. Presumably, the author meant was August Messer, a professor in Giessen who published on philosophy and pedagogy. Unfortunately, I have not been able to find out which book it was Hilbert made notes of.

In the notes, Hilbert writes that Bergson 'looks down on formalism',[122] and he makes the loose remark 'against Bergson's intuitionism'.[123] The rest of the notes contain a characterisation of irrational philosophy. Unfortunately, it is not clear when Hilbert made the notes. Therefore, it is hard to say if the association with Bergsonian intuitionism played a role in Hilbert's reaction to Brouwer's intuitionism.

---

[119] [Fraenkel 1933, p. 225]
[120] [Dahms 1987B, p. 169]
[121] The philosophical faculty put Misch on top of the list of proposed successors. A group of notable faculty members from the mathematical-physical department protested and proposed Nelson, [Dahms 1987B, pp. 170–171]. By that time, the mathematical-physical department of the Göttingen university was still part of the philosophical faculty.
[122] 'Bergson verachtet den Formalismus', [NSUB Hilbert, 607]
[123] 'Gegen den Intuitionismus von Bergson', [NSUB Hilbert, 607]

## 6.3.5 Spengler

Doubtlessly the most influential life philosopher was Oswald Spengler. *Der Untergang des Abendlandes* ('The decline of the West')[124] was an instant success, despite its size and its use of mathematical examples which were incomprehensible to a large audience.[125] The first edition appeared in 1918 and went through thirty printings within five years. Spengler's thesis was that cultures behave as autonomous organisms, each going through the same life cycle, but at different times and in different ways. Every cultural manifestation, whether it be art, mathematics or something completely different, is an expression of that particular culture and is, as such, completely different from manifestations which bear the same name but stem from different cultures.[126]

Spengler, a mathematician and scientist by training, devoted substantial attention to (the history of) mathematics and the sciences. Maybe for this reason, his book was widely read among Weimar mathematicians and physicists, too.[127] Even if they did not read the whole book, they must have hit upon themes and ideas that appear in Brouwer's intuitionism as well – for there are many. I mention some of them.

Spengler and Brouwer shared an aversion of logical argumentations. In Spengler's words:[128]

> Denn es handelt sich nach meiner Überzeugung nicht um eine neben andern mögliche und nur logisch gerechtfertigte, sondern um *die*, gewissermaßen natürliche, von allen dunkel vorgefühlte Philosophie der Zeit.[129]

Brouwer could have said the same about intuitionism, perhaps even somewhat more intolerantly towards other currents.

Like Brouwer, Spengler attacks the restrictions of mathematical laws, calling them a means for understanding dead forms.[130] In this way, Brouwer's choice sequences would fit in much better with what Spengler wanted.

---

[124] We find most of the characteristics of *Lebensphilosophie* in Spengler's book: '*Der Verstand, das System, der Begriff töten, indem sie 'erkennen'. (...) Das Anschauen beseelt. Es verleibt das Einzelne einer lebendigen, innerlich gefühlten Einheit ein.*' ('The intellect, the system, the comprehension kill, because they 'recognize'. (...) Intuiting animates. It absorbs the single thing into a living, innerly felt unity.') etc., [Spengler 1918, vol. 1, p. 147], [Spengler 1923, p. 137]. The 1918 quote is without '*das System*'. The book was revised in the 33rd edition, 1923; therefore, I each time give two references.

[125] Each of the two volumes of the book consisted of more than 600 pages, and Spengler freely speaks of, for example, convergence criteria for infinite sequences and of elliptic integrals, without further explanation, [Spengler 1918, vol. 1, p. 123], [Spengler 1923, p. 116].

[126] [Spengler 1923, p. 29]

[127] Forman found explicit references to Spengler with, among others, Max Born, Albert Einstein, Philipp Frank, Richard von Mises, Hermann Weyl and Wilhelm Wien, [Forman 1971, p. 56]; Weyl's reference to Spengler is somewhat joking, [Weyl 1922B, p. 209].

[128] [Spengler 1918, vol. 1, p. VII], [Spengler 1923, p. X]

[129] 'For the issue at stake is, in my view, not a philosophy which is possible besides others and which is only logically justified, but about *the*, let us say natural, philosophy of the time, which is darkly anticipated by all.'

[130] [Spengler 1918, vol. 1, p. 4], [Spengler 1923, p. 4]

Spengler wants his 'morphology of world history'[131] to catch all world forms and movements in a different order,[132]

> nicht zum Gesamtbilde alles Erkannten, sondern zu einem Bilde des Lebens, nicht des Gewordenen, sondern des Werdens (...).[133]

Weyl's presentation of the intuitionistic continuum as a *'Medium freien Werdens'* thus fitted in well with the general atmosphere created by Spengler.[134]

Already in his defense of intuitionism in the discussion with Pólya after his 1920 lectures, Weyl described the de-linking of mathematics from 'spiritual life' as 'the killing of mathematics'.[135] I have the impression, though I cannot give hard proof for it, that Weyl let himself be inspired by Spengler's *Untergang des Abendlandes* when writing some of the more popular parts of his 1921 paper. The following quote is one of the passages which suggests so. In Spengler we find:[136]

> Das Wort Europa sollte aus der Geschichte gestrichen werden. Es gibt keinen 'Europäer' als historischen Typus. (...) Das sind Worte, die aus einer banalen Interpretation der Landkarte stammen und dessen nichts Wirkliches entspricht. (...) Hier hat (...) eine bloße Abstraktion zu ungeheuren realen Konsequenzen geführt. (...) Orient und Okzident sind Begriffe von echtem historischen Gehalt. 'Europa' ist leerer Schall.[137]

One may compare it to Weyl's words, which I quote here once more, when he attacked a similarly established concept:

> *Ein Existentialsatz* –etwa 'es gibt eine gerade Zahl'– *ist überhaupt kein Urteil*[138] *im eigentlichen Sinn, das einen Sachverhalt behauptet;* Existential-Sachverhalte sind eine leere Erfindung der Logiker. '2 ist eine gerade Zahl': das ist ein wirkliches, einem Sachverhalt Ausdruck

---

[131]*'Morphologie der Weltgeschichte'*, in the full title of [Spengler 1923]

[132][Spengler 1918, vol. 1, p. 7], [Spengler 1923, p. 7]

[133]'not into an overall picture of everything accredited, but into a picture of life, not of what has become, but of what is becoming (...).'

[134][Weyl 1921, p. 221]. Spengler was apparently not aware of intuitionistic mathematics, and he maintained that there was no contact between becoming and any area of mathematics, [Spengler 1918, vol. 1, p. 178], [Spengler 1923, p. 165].

[135]See 4.2.1.

[136][Spengler 1918, vol. 1, pp. 21–22], [Spengler 1923, p. 22]

[137]'The word Europe should be crossed out of history. There are no 'Europeans' as a historical type. (...) These are words that were derived from a banal interpretation of the map and that do not correspond to anything real. (...) Here, (...) a mere abstraction has led to enormous real consequences. (...) Orient and occident are concepts of real historical contents. 'Europe' is empty sound.'

[138]As Van Dalen pointed out, Weyl's use of *'Urteil'* does not coincide with what we presently call 'judgement', [Van Dalen 1995, p. 157]. Rather, it is probably better translated as 'proposition', as can be seen from [Weyl 1918A, p. 1].

## 6.3. PHILOSOPHY

gebendes Urteil; 'es gibt eine gerade Zahl' ist nur ein aus diesem Urteil gewonnenes *Urteilsabstrakt*.[139]

Regarding mathematics proper, however, Spengler and Brouwer held different opinions. Spengler takes a relativist position and maintains that there is not a single mathematic, but several mathematics, linked to different cultures;[140] for Brouwer, there is only one. Furthermore, Spengler disagrees with the Kantian idea that the numbers originate from time as an a priori form of intuition.[141]

In line with Brouwer's thought, Spengler maintains that the truth of a philosophy does not require a consistency proof. Furthermore, Spengler claims a special place for intuition as a method.[142]

Spengler's view on constructions seems to be different from other life philosophers as characterised by Rickert. Spengler singles out constructions as the main characteristic of mathematics in antiquity, and opposes it to modern mathematics of the infinite, which treats classes of formal possibilities. Without making explicit his own opinion, Spengler remarks that there is nothing more impopular than modern mathematics, whereas literary works from Antiquity are all very popular.[143] Spengler is somewhat unclear here, but he seems to imply that mathematics which makes use of constructions is more popular than the one working with formal possibilities.

### 6.3.6 Summary

There are three possible ways in which philosophical intuitionism or, more broadly speaking, *Lebensphilosophie*, may have influenced the reception of Brouwer's intuitionism. The first is via Spengler, its most well-known representative at the time. Spengler's plea for intuition and becoming and against logic and being may have given mathematicians involved in the foundational debate the idea that what Spengler wanted was something like mathematical intuitionism (or, the other way round, that what Brouwer wanted was something like Spenglerian science).

The second way is via Hilbert. In the first place, there were life philosophers represented at the philosophical faculty in Göttingen. Since Hilbert was, until the faculty split in 1922 into a philosophical and a mathematical-physical part, a member of the same faculty and since he quarreled with some of the life philosophers about appointments, Hilbert's associations with such philosophers may have been

---

[139]'*An existence statement* —e.g., 'there is an even number'— *is not a proper proposition at all, that expresses a state of affairs*; existence states of affairs are an empty invention of logicians. '2 is an even number': that is a real proposition, expressing a state of affairs; 'there is an even number' is only a *proposition abstract*, obtained from the proposition.' (For the translation of 'Urteil', see the preceding footnote.)
[140][Spengler 1918, vol. 1, pp. 88–89], [Spengler 1923, p. 82]
[141][Spengler 1918, vol. 1, p. 94], [Spengler 1923, p. 87]
[142][Spengler 1918, vol. 1, p. 58; p. 81], [Spengler 1923, p. 58; p. 75]
[143][Spengler 1918, vol. 1, pp. 122–124], [Spengler 1923, pp. 115–117]

less than positive. Also, Hilbert was aware of the fact that Bergson's philosophy was called intuitionism, and that it was opposed to formalism. These factors may have strengthened his negative opinion on mathematical intuitionism, though it is unclear when Hilbert knew about the other meaning of intuitionism.

Finally, there is the link between mathematical intuitionism and philosophical intuitionism as described by Rickert. Not only the names coincide, also their adherence to concepts linked with life, their dislike of systems and their appeal to intuition are to a large degree comparable. However, Brouwer explicitly and repeatedly advocated the use of rational constructions in intuitionistic mathematics, whereas philosophical intuitionism rejected precisely this.

Whether Brouwer's contemporaries who discussed intuitionism were aware of the difference is another question. Judging from Rickert's description, philosophical intuitionism was a most popular current those days. Therefore, it seems reasonable to assume that most participants in the foundational debate in mathematics must at least have had some association with the term 'intuitionism' — associations maybe brought about more by philosophical intuitionism than by its mathematical namesake.

I can prove neither of these paths. But I think chances are that at least some of them did actually appear, and that thus philosophical intuitionism or *Lebensphilosophie* influenced the image participants in the foundational debate had of Brouwer's intuitionism. If so, it sheds more light on many of the misunderstandings in the foundational debate in mathematics.[144]

## 6.4 Physics

Traditionally, physics is the science closest to mathematics. At the time when the foundational debate raged in mathematics, similarly vigorous discussions took place in the domain of physics, most notably regarding quantum mechanics and the theory of relativity. In 1934, Mannoury suggested that there might be important similarities between these discussions:[145]

> (...) hat nicht die Arbeit Brouwers (...) eine ähnliche Bedeutung der klassischen Logik gegenüber als die relativistische und quantentheoretische Physik der Neuzeit, der klassischen Physik gegenüber?[146]

In this section, I investigate to what extent the discussions in physics had similar characteristics as or were linked to the discussions on Brouwer's intuitionism, so that they can help us improve our understanding of the latter's reception. Following the chronological development in physics, we start with Einstein's theory.

---

[144] The conjecture is reinforced by Heyting, who remarked that the denomination 'intuitionism' was 'the cause of much misunderstanding', [Heyting 1978, p. 7].

[145] [Mannoury 1934B, p. 330]

[146] '(...) does not Brouwer's work (...) have a similar importance towards classical logic as relativistic and quantum-theoretical modern physics has towards classical physics?'

## 6.4.1 Theory of relativity

Albert Einstein developed his special and general theory of relativity in the first two decades of the 20th century. A (relatively spoken) broad public discussion about the theories arose when measurements in 1919 showed that the deflection of light at the sun performed as the theory of relativity had predicted. Several arguments were raised against Einstein's theory, such as the idea that the theory of relativity was a mathematical rather than a physical theory, that it was not confirmed sufficiently by experiments, that one cannot discard the ether, and that the theory of relativity contradicts fundamental postulates about space and time.[147] Others were more demagogic in their attacks on the theory of relativity and labelled Einstein a revolutionary and scientific Bolshevik,[148] as one did with Brouwer.[149]

Thus, the debate on the theory of relativity arose at about the same time as the controversy on intuitionism, just after the Great War. Furthermore, some of the labels applied in popular presentations to the proponents of both new currents were the same. The question is whether such rather superficial similarities between the two debates suffice for drawing parallels between the two discussions, as Forman has suggested for the case of quantum mechanics.[150] This turns out not to be the case.

First, the 'revolution' similarity does not go all the way. Contrary to what was the case with intuitionism, the first instances in scientific literature calling the theory of relativity 'revolutionary' can already be found before the war started.[151] This means that, at least in scientific circles, the revolutionary character of the theory of relativity had already been proclaimed before the Russian and German revolutions had taken place. This lends the label a different context than in the case of the mathematical foundational debate, where it presumably carried a more political connotation.[152]

The one prominent person who figured in both the debates, Hermann Weyl, used the revolution metaphor in both cases. Also in the case of physics, Weyl opted for the new theory. In 1918, he described Einstein's theory of relativity as revolutionary, just like he later was to do with intuitionism.[153] However, his position is much less polemic than in the case of intuitionism:[154]

(...) kam in unsern Tagen der revolutionären Sturm zum Ausbruch,

---

[147][Goenner 1993, p. 109]. Goenner's main thesis in the paper is rather weakly supported, but this does not affect other interesting information presented there.

[148][Sticker 1922], cited from [Goenner 1993, p. 108]. Examples of the proclaimed revolutionary character of the theory of relativity abound and may be found in [Hentschel 1990, pp. 108–119], an extensive study which focuses on the reception of Einstein's theory of relativity in philosophical circles.

[149]See 4.2.1 and 3.3.2.

[150]See below.

[151][Hentschel 1990, pp. 108–119]

[152]See 6.2.

[153]See 4.2.1.

[154][Weyl 1918B, pp. 1–2]

## 328 CHAPTER 6. THE FOUNDATIONAL CRISIS IN ITS CONTEXT

> der jene Vorstellungen über Raum, Zeit und Materie, welche bis dahin als die festesten Stützen der Naturwissenschaften gegolten hatten, stürtzte; doch nur, um Platz zu schaffen für eine freiere und tiefere Ansicht der Dingen.[155]

Three years later, in a report written for the German Mathematical Society in 1921 on the 'most dramatic' part of the Bad Nauheim meeting on the theory of relativity the year before, the discussion between Einstein and Lenard, Weyl comments matter-of-fact-like and does not preach the revolution.[156]

Second, there were important characteristics that differed in the discussions on the theory of relativity and on intuitionism. In the first place, anti-Semitism played an important role in the fight against the theory of relativity, whereas it was completely absent in the debate on intuitionism. Anti-Semitism is often even seen as the main factor explaining the opposition to the theory of the Jewish scientist Einstein. Moreover, no *theoretical* physicist of any standing publicly opposed the theory of relativity. Paul Weyland, one of the more prominent early antagonists, was an engineer with strong anti-Semitic feelings, who mostly contributed by organising anti-Einstein activities.[157] The most able physicists who opposed the theory of relativity, Ernst Gehrcke, Philipp Lenard and Johannes Stark, were all three experimental physicists who complained about the growing influence of theoretical physics because of Einstein's theory of relativity.[158] This was an important difference with the mathematical foundational debate, where two of the greatest mathematicians of their time who had both contributed substantially to the field of the foundations of mathematics opposed each other. Finally, the discussion on the theory of relativity peaked much earlier, around 1921, whereas the foundational debate in mathematics was at its height around 1927.[159]

Thus, despite some similarity in time occurrence and some of the labels used, the fights against the theory of relativity and against intuitionism differ substantially, mostly regarding the role of anti-Semitism and the absence or presence of a big name in the field opposing the new theory involved. Let us now turn to possible links between the debates on intuitionism and on that other highly disputed part of physics at the time, quantum mechanics.

---

[155] '(...) in our days the revolutionary storm had broken out, which overthrew those conceptions of space, time and matter which until then had been considered the most secure pillars of the sciences; but only to make place for a freeer and deeper view on things.'
[156] [Weyl 1922A]; the term *'dramatischste'* figures on page 51.
[157] [Goenner 1993, p. 108; 120–128]
[158] [Hentschel 1990, pp. 135–136]. Lenard was Nobel laureat and the later champion of *Deutsche Physik*.
[159] See 3.2.2.

## 6.4.2 Quantum mechanics

Quantum mechanics was developed as a theory of atomic and subatomic physics around the years 1925–1926 by Heisenberg, in terms of matrix mechanics, and by Schrödinger, using wave mechanics. In 1926, Schrödinger proved the formal equivalence of both types of quantum mechanics. With Born's statistical interpretation of the wave function in the same year, an a-causal element was introduced right into the heart of quantum mechanics.[160]

**The Forman theses** As the historian Paul Forman pointed out, talk about a-causal elements in physics had already appeared in popular lectures given by physicists in the Weimar period before the development of quantum mechanics. In his well-known *Weimar Culture, Causality, and Quantum Theory*, Forman analysed the relations between the Weimar intellectual milieu on the one hand and mathematics and physics on the other, with a special focus on quantum theory. The remarkable Forman thesis holds that physicists used an accommodationist strategy to adapt not only the presentation but even the contents of their theories to the hostile cultural environment of Weimar Germany:[161]

> (...) suddenly deprived by a change in public values of the approbation and prestige which they had enjoyed before and during World War I, the German physicists were impelled to alter their ideology and even the content of their science in order to recover a favorable public image.

Forman primarily points to *Lebensphilosophie* in the form of Spengler's *Untergang des Abendlandes*[162] as the dominant cultural factor with which physicists wanted to be associated. The idea of causality was the one that was most affected by the adaptation. Following the intellectual current of the day, physicists plead for the disposing of causality in physics.[163] In the 1920s, Becker already made a similar remark in a letter to Weyl, whom he wrote about influence of the *Zeitgeist* on scientific research:[164]

> es liegt doch sicher nicht an das Wachstum unserer objektiven Erkenntnis der physischen Welt, dass wir heute auf dem Sprunge sind selbst in der Physik Indeterministen zu werden!'[165]

As Klomp pointed out,[166] Forman altered his thesis substantially in a later publication. There, Forman maintains that 'physicists sought to make the most

---
[160][Forman 1971, pp. 100–104], [Forman 1984, p. 335]. Causality is here taken in the sense as it was used in the 1920s, namely as lawfulness, [Forman 1971, p. 65].
[161][Forman 1971, pp. 109–110]
[162]'Decline of the West', see 6.3.1.
[163][Forman 1971, pp. 109–110]
[164]Letter from Becker to Weyl, 10/10/1924; [ETH Weyl, HS 91-473]
[165]'it will surely not be a consequence of the growth of our objective knowledge of the physical world that we at the moment are about to become indeterminists even in physics!'
[166][Klomp 1997, p. 227]

of a-causality in quantum mechanics, exaggerating and trumpeting it'.[167] The proclaimed relationship between the hostile cultural environment and the *contents* of a physical theory has disappeared. Rather, the contents of the theory are almost completely eliminated from the story:[168]

> (...) there was little connection between quantum mechanics and the philosophical constructions placed on it, or the world-view implications drawn from it. The physicists allowed themselves, and were allowed by others, to make the theory out to be whatever they wanted it to be – better, whatever their cultural milieu obliged them to want it to be.

Thus, we in fact have two Forman theses. In the original one, Forman maintains that physicists adapted the contents of physical theories, especially on the subject of causality, to their *lebensphilosophische* intellectual milieu. In the later version, he only states that physicists gave a rather free interpretation of physical theories, exaggerating the role of a-causality, to bring it in accordance with popular intellectual preferences. I call the old one the strong Forman thesis, the new one the moderate version.[169]

Forman only touches upon the case of mathematics, but he does suggest a link with the situation in physics. Referring to the accommodationist attitude of physicists in Weimar Germany, Forman conjectures that 'the intuitionist movement in mathematics, which won so many adherents and created so much furor in Germany in this period, was primarily an expression of just such inclinations and aims.'[170] Forman's characterisation of the foundational crisis in mathematics, however, is something of a caricature. He maintains that Weyl precipitated the crisis 'virtually out of thin air', whereupon 'considerable numbers of German mathematicians rallied to L.E.J. Brouwer's standard calling for a complete reconstruction of mathematics', while they seem almost to have welcomed the destructive impact of intuitionism 'in a spirit of abnegation and resignation'.[171] In fact, Weyl's crisis arose out of problems that are important enough for anyone who takes foundational matters seriously. Most support Brouwer got came from philosophers, not from mathematicians, and most of the mathematicians who supported him only paid lip-service to intuitionism but did not contribute anything to intuitionistic mathematics. Finally, I have hardly anywhere found a spirit of resignation in the whole discussion.

---

[167][Forman 1984, p. 343]

[168][Forman 1984, p. 344]. The way in which Forman connects his two works is rather curious. In his 1984 publication he refers to the earlier one by simply stating that, regarding causality, he 'argued the point at length in an earlier publication', [Forman 1984, p. 335]. However, the point argued there was quite different.

[169]To put it in Mehrtens' terms (see 6.7): in the strong thesis, Forman maintains that both the discourse *of* and the discourse *on* physics changed; in the moderate version he only argues for a change in the discourse *on* physics.

[170][Forman 1971, p. 7]

[171][Forman 1971, pp. 60–61]

## 6.4. PHYSICS

The question to answer here is not which of the two Forman theses is the better one, but whether one of them can be applied to the foundational debate in mathematics, as Forman suggests.

**Application to the foundational crisis**  Let us first consider the strong Forman thesis. Applied to the foundational crisis, this would mean that intuitionists adapted not only the presentation but also the contents of intuitionistic mathematics to the hostile Weimar cultural environment. This thesis is easily refuted. For there were only few intuitionists who really contributed to intuitionistic mathematics and logic, and all of them were outside the (direct) influence of Weimar culture: Brouwer and Heyting were Dutch, while Kolmogorov and Glivenko were (Byelo)Russian.[172]

If we apply the moderate Forman thesis to the foundational crisis, the statement would be that intuitionists made most out of the presentation of, and implications drawn from, intuitionistic mathematics in order to conform to the dominant *lebensphilosophische* current of the day. The only thing that can be said in support of this thesis is that Weyl, who had a perfect sense for the *Zeitgeist*, made use of descriptions which fitted into *Lebensphilosophie* in his 1921 paper.[173] However, there is an important restriction to the application of Forman's moderate thesis to mathematics. One of the labels that occur frequently in the foundational debate is 'constructive'. Apparently, the label was so popular that first Brouwer and Weyl on the one hand and Hilbert and Bernays on the other fought for it, then it was granted to intuitionism for some time, while in 1930 the discussion opened again with intuitionism, formalism and logicism all three claiming to be 'constructive'.[174] However, the term 'constructive' is definitely one which is *opposed* by *Lebensphilosophie*.[175] Thus, even if mathematicians used some terms which were popular at the time, they did not hesitate to use distinctly anti-*lebensphilosophische* terms if they preferred these.

**D'Abro's conjecture**  Finally, I touch upon another suggested similarity between the discussion on quantum mechanics and that on the foundations of mathematics. In his 1939 book on quantum mechanics, D'Abro suggested that 'the quantum theorists occupy the position of the intuitionists while Einstein and Planck occupy that of the formalists.' D'Abro loosely supports the idea by pointing out that, where Brouwer rejects the principle of the excluded middle in infinite totalities while retaining it in finite ones, a similar attitude can be found with quantum

---
[172]Weyl, despite all his propaganda, made almost no contribution to intuitionism proper. Furthermore, he lived in Switzerland during most of the time the debate raged.
[173]See 6.3.5.
[174]See 4.2.1, 4.2.2 and 4.4.1.
[175]Cf., for instance, Litt's description of *Lebensphilosophie*: '*Es [Das Leben, DH] fühlt sich (...) begnadet mit einer inneren Produktivität, die die Berechnungen des kausalen wie die Konstruktionen des logischen Denkens Lügen straft*', ('It [life, DH] feels (...) gifted with an inner productivity that belies the computations of causal as well as the constructions of logical thinking') [Litt 1925, p. 36]. See also 6.3.1.

theorists who reject determinism for microscopic processes, while leaving it intact on more ordinary levels of physical processes.[176] If we continue to stress the importance of the philosophy of life for scientists' attitude, D'Abro's suggestion is supported by Litt's description of *Lebensphilosophie* in 1925, where he pointed out that the philosophy of life is opposed to both formalism and determinism.[177]

I superficially examine D'Abro's suggestion by looking at persons who were involved in both the debates. As far as I know, there were only two: Hermann Weyl and Joseph Petzoldt. In these cases, D'Abro's conjecture indeed holds. In fact, both of them linked quantum mechanics to the foundational crisis in mathematics.

In the 1920s, Petzoldt was a philosopher in his sixties. He was influenced by Ernst Mach, the example of the *Wiener Kreis*, and was specialised in epistemology, *Naturphilosophie* and methodology of the sciences. From 1922 on, he held an extraordinary professorship at the *Technische Universität Berlin-Charlottenburg*. He was the founder and first president of the Berlin *Gesellschaft für positivistische Philosophie*, which existed from 1912 to 1921, and co-founder of the *Internationale Gesellschaft für empirische Philosophie*[178] in 1927.[179] In a lecture given before the latter society in Berlin in May 1927, Petzoldt points to the joint crisis in mathematics and physics:[180]

> Es ist nicht übertrieben, wenn man geradezu von einer Krise spricht, in die (...) zunächst die exaktesten Wissenschaften geraten sind, die Physik und die Mathematik, aber auch von einer Bedrohung unserer gesamten Wissenschaft, denn man bezweifelt das *Gesetz der Kausalität* und den *Satz vom ausgeschlossenen Dritten*.[181]

The lecture was given just after Brouwer had lectured on intuitionism in Berlin.[182] I do not know whether Petzoldt attended these lectures, but I would say chances are that he at least heard about them. Petzoldt's stand towards quantum mechanics and intuitionism is the same: he opposes both.[183]

Weyl did exactly the opposite: he supported both new currents. Even before quantum mechanics had been developed and labeled 'a-causal', Weyl had already

---

[176][D'Abro 1939, p. 212]. The second argument D'Abro uses, the insistence by intuitionists on definitions by a finite number of words and by quantum theorists on (at least in principle) observable notions, is, I think, too far-fetched.

[177][Litt 1925, p. 36]; the characterisation is repeated in the 1930 third edition, [Litt 1930, p. 32].

[178]See 3.3.3.

[179][Hentschel 1990, pp. 401–403]

[180][Petzoldt 1927, p. 154]

[181]'It is not exaggerated if one speaks of an outright crisis, in which (...) in the first place the most rigorous sciences have ended up, physics and mathematics, but also of a threat to all of our science, for one doubts the *law of causality* and the *theorem of the excluded middle*.'

[182]See 2.7.1.

[183]Petzoldt's argumentation is not very strong, cf. [Petzoldt 1927, pp. 155–158]. In a paper published two years earlier, Petzoldt had not opposed intuitionism. Some of Petzoldt's ideas, like those on the intuitve component of mathematical signs, seem intuitionistic rather than formalistic, cf. [Petzoldt 1925, p. 354].

linked intuitionism (without naming it) to an a-causal view on physics. In his 1920 paper *Das Verhältnis der kausalen zur statistischen Betrachtungsweise in der Physik* ('The relationship between the causal and statistical view in physics'), Weyl argued that statistics should obtain an independent position in physics alongside the standard, causal way of thinking. This could be done, among other things, by seeing the continuum as something that is infinitely becoming, instead of already being.[184]

Thus, the two persons who were active in both debates do not refute D'Abro's suggestion; they actually reinforce it by both establishing a more or less explicit link between intuitionism and quantum mechanics. However, the number is too small to draw further conclusions.

## 6.5 Art

'Constructive' was a popular label with both Brouwer and Weyl on the one hand, and Hilbert and Bernays on the other, in the beginning of the debate, and later also with Carnap, speaking for logicism.[185] Apparently, constructivity was not seen as something inherent to one of the mathematical currents, but it made appeal to some broader popularity. The question is where this popularity came from. The answer may lay in art.

German culture in the Weimar period is often characterised as synonymous with modernity, a revolutionary break with the past, that created hope for a new society. At the same time, to some contemporaries such dominant cultural movements in Weimar Germany as Expressionism and Dadaism meant a violent disturbance of the established values.[186] The area of art in which, at least in the beginning of the Weimar period, tension was highest, was that of visual art. An abundance of new isms came up, characterised above all by a political and artistic radicality, striving for a true revolution.[187] One of these currents was constructivism. Below, we will take this constructivism as a starting point to investigate the link between art and the discussions around Brouwer's intuitionism.

---

[184][Weyl 1920, p. 121]. Weyl made a similar claim in a letter to Pauli (letter from Weyl to Pauli, 9/12/1919; in: [Pauli 1979, pp. 5-6]). The statement is a clear, though implicit reference to the intuitionistic continuum, which Weyl put forward a year later as a medium of free becoming. In the final paragraph, Weyl even links statistical methods in organic matters to an organising power wrested from causality: life, [Weyl 1920, p. 122].
[185]See 4.2.1, 4.2.2 and 4.4.1.
[186][Joll 1990, p. 300],[Peukert 1987, pp. 166-167]
[187][Hermann & Trommler 1978, pp. 353-355]

## 6.5.1 Constructivism

The name 'constructivism' gained popularity in art before mathematics.[188] It originated from Russian artists who saw themselves as revolutionaries in art after the 1917 Revolution. In March 1921, the so-called First Working Group of Constructivists was set up in Moscow. Their members included Aleksey Gan and Aleksandr Rodchenko, who were seen as part of the revolutionary avant-garde. The former wrote the first constructivist programme, which appeared in April 1921. In the manifest, he stated as one of the aims of constructivism to fight art and to promote technique. The following year, Gan's book *Konstruktivizm* was published, in which he further propagated the constructivist ideas.

More important for our story than Russian constructivism, however, is the international version of constructivism, which was inspired both artistically and politically by the Russian current. International constructivism was established at the Düsseldorf *Internationale Kongreß fortschrittlicher Künstler* ('Congress of International Progressive Artists'), held in May 1922. Here, the Dutchman Theo van Doesburg organised the *Internationale Sektion konstruktivistischer Künstler* ('International Section of Constructivistic Artists'), in which also Hans Richter and El Lissitzky participated. The section published a declaration in the journal *De Stijl* ('The Style'), in which they stressed their opposition to subjectivity and their dedication to the systematization of the means of expression. In September 1922, the Manifesto of International Constructivism was released. The centre of activity of the international constructivists during the 1920s was in Germany. Thus, Theo van Doesburg promoted De Stijl aesthetics in Berlin and at the Bauhaus. The Bauhaus managed to combine the different influences, and became the western centre of international constructivism. By the mid-1920s, constructivism included important groups in the Netherlands, Germany, Czechoslovakia and Poland; Piet Mondriaan and Walter Gropius are among the most famous persons associated to it.[189]

Thus, a possible explanation for the popularity of the constructivity concept in mathematics lies in the rise of constructivism in art, which became well-known at the same time. The coinciding popularity of the constructivity concept in both mathematics and art makes it tempting to see constructivity as something which David Auben called a cultural connector. A cultural connector, in Auben's definition, is a convenient tool allowing actors to connect different spheres of culture.[190] In this case, the connection may have taken place especially by way of associating the cultural spheres of mathematics and art in the context of the foundational debate in mathematics. Admittedly, there are no explicit references showing that

---

[188] In the beginning of the 1920s, the term 'constructivism' seems to have been unknown in philosophy (of mathematics). For example, is does not appear in the *Systematisches Wörterbuch der Philosophie* ('Systematic dictionary of philosophy'), [Clauberg & Dubislav 1923].

[189] [Loddur 1996], [Honisch & Prinz 1977, pp. 1/102–1/103]; Hermann and Trommler claim that constructivism was over its height after 1923, [Hermann & Trommler 1978, p. 359].

[190] [Auben 1997, p. 299]

mathematical (self-proclaimed) constructivists saw themselves linked in any way to constructivism in art. In the beginning of the foundational debate, the meaning of the word 'constructive' was rather vague. Therefore, it is hard to judge to what extent its use was metaphorical or served as a cultural connector. Another possible explanation for the simultaneous occurrence is that there was some broader background which was common to both constructivism in mathematics and in art. Further research would be needed to check whether there is any such connection.[191]

## 6.6 Politics

The political circumstances in Weimar Germany influenced the foundational activities especially in the beginning and the end of the republic. The former period is covered in the section on metaphors.[192] Here, only the later period of Weimar history is treated, its decline and the rise of the Third *Reich*.

### 6.6.1 Mathematics and the rise of the Third *Reich*

In general, German universities provided fertile ground for Nazi activists. Already in 1930, the Nazi's obtained majorities at many of the universities' student elections. National Socialist students demonstrated against Jewish, liberal or internationalist professors. Some of the university cities were among the first to be ruled by the national socialists. Thus, the *Nationalsozialistische Deutsche Arbeiterpartei* already had an absolute majority in the Göttingen city council by 1931.[193]

**Purges** On January 30, 1933, Hitler became *Kanzler* of the German *Reich*. On April 7, the *Gesetz zur Wiederherstellung des Berufsbeamtentums* ('Act for the Restoration of the Professional Civil Service') provided the basis for the dismissal of civil servants, including scientists, disliked by the new regime.[194] The third paragraph ordered that all 'non-Aryan' civil servants had to retire; the fourth opened the possibility to also send civil servants on forced retire whose unconditional loyalty to the national state could be doubted.[195] There were some exception clauses, however, whereby the number of mathematicians affected was limited at first: in Göttingen, with its relatively high number of 'non-Aryan' scientists in the Mathematical Institute, the law applied to none of the ordinary professors. It did apply

---
[191] In modern expositions on constructivism, it is maintained that the link between the constructivisms in various fields is purely coincidental, see [Thiel 1984, p. 449]. However, I have not seen any evidence that the subject was researched thoroughly.
[192] See 6.2.
[193] [Dahms 1987A, p. 16], [Ringer 1969, pp. 436–437]
[194] The case of mathematicians emigrating from nazi Germany is treated in much more detail in [Siegmund-Schultze 1998].
[195] The relevant paragraphs are reprinted in [Brüning, Ferus & Siegmund-Schultze 1998, p. 5].

to Emmy Noether, who, being a woman, was not allowed to hold a full chair anyway, and to Paul Bernays, who worked as an assistant.[196] Landau and Bernstein, although 'non-Aryan', were allowed to stay since they had been a professor under the Wilhelminian empire and thus fell under one of the exception clauses. The same applied to Courant, whose 'non-Aryan-ness' was overruled by him having fought as a soldier during the war.

Within three weeks, a new law was approved which limited the number of new 'non-Aryan' students to 1.5 percent, and the number of women to 10 percent.[197]

The situation on the floor soon radicalised, too. The physicist and Nobel laureate James Franck staged a public protest against the non-Aryan paragraph by resigning his Göttingen professorship. Franck, who was himself not affected by the new law, motivated his move in the local Göttingen newspaper.[198] A popular campaign against the Göttingen mathematicians and physicists followed, which was primarily directed against Courant, as head of the Mathematical Institute. Presumably, this caused the ministry to react. On April 25, it sent a telegram announcing the immediate and forced retirement of, among other people, Bernstein, Courant and Noether.[199] Courant was picked for being a Jew and being the head of the Mathematical Institute, while Bernstein and Noether were not only 'non-Aryan', but also politically left-wing.[200] On top of that, Noether was a woman. Still, Landau continued lecturing for some time, and Bernays stayed as Hilbert's assistant at his expense.

Also at other universities, important mathematicians were expelled: Fraenkel (Kiel), Reidemeister (Königsberg), Baer (Halle), Schur (Berlin), and Hopf (Aachen) were all told to leave.[201] The mathematical centres in Berlin and Göttingen were hit hardest.[202] Reidemeister was one hundred percent 'Aryan'; he was dismissed because of his liberal persuasion and because his sister worked at the socialist municipality in Vienna.[203] To make it worse, Reidemeister had publicly disapproved of the new regime by spending the full hour of a mathematical lecture proving that the national-socialist students' action against the rector of the university of Königsberg was irreconcilable with logical thought. Apparently, this had thoroughly upset the new authorities, since he was dismissed while three 'non-Aryan' colleagues of his were left to lecture for at least one more semester.[204] However, Reidemeister fought his dismissal, claiming among other things that he had re-

---

[196] The full list of Göttingen mathematicians eventually affected by the law is much longer and can be found in [Becker, H. 1987, pp. 493–498].
[197] [Brüning, Ferus & Siegmund-Schultze 1998, p. 5]
[198] [Dahms 1987A, p. 27]
[199] [Schappacher 1987, p. 349]
[200] Noether had been a member of the USPD and, later, of the SPD; Bernstein had been president of the Göttingen section of the left-wing liberal DDP, [Schappacher 1987, p. 347, p. 362].
[201] [Nazi 'purge' 1933]
[202] [Siegmund-Schultze 1998, p. 60]
[203] [Anonymous 1933], cited from [Siegmund-Schultze 1998, p. 310]
[204] [Artzy 1972, pp. 97–98].

## 6.6. POLITICS

jected the 'propagandistic representation of the logistic philosophy'.[205] Helped by a petition initiated by Blaschke, he was eventually reinstalled.[206] Schur managed to stay for two years more.[207]

Student boycotts put more pressure on 'undesired' professors. Such actions were staged against, among other persons, Blumenthal, Landau, Reichenbach and Reidemeister. In this way, Landau was forced to leave 'voluntarily'.[208]

The same year, Hermann Weyl, who had a Jewish wife, accepted the third offer he got from Princeton and left for the United States. Emmy Noether did the same. By 1934, the situation had deteriorated to such an extent that both Bernays and Courant decided to leave the country, too.[209]

In 1935, more pressure was put on Jews. The Nürnberg laws were accepted, which withheld German citizenship to Jews and forbade marriages between Jews and 'citizens of German blood'.

The changed political climate was reflected in mathematical journals, too. For example, the 1935 volume of the *Jahresbericht der deutschen Mathematiker-Vereinigung* contains reports of a lecturer paying tribute to the *Führer*,[210] a military man writing on mathematics and soldiers,[211] and a Göttingen contribution proclaiming that what is needed is a mathematics 'rooted in irrationality'.[212] It goes without saying that by that time the *Jahresbericht* had been transformed substantially.

However, still not all mathematical activities were affected. There were some exceptions to the publications supporting the new regime. Thus, when Emmy Noether died in 1935, Van der Waerden published a memorial article in the *Mathematische Annalen*.[213] In his extensive study on mathematical emigration because of Hitler, Siegmund-Schultze characterises Van der Waerden's paper as 'about the maximum of a *public* stand made against the regime among mathematicians',[214] even though it contained no political comments. The publication did not cause any problems. In the same way, the Jewish mathematician Harald Bohr could still publish a paper in 1937,[215] and Blumenthal's name could even be kept on the cover of the *Mathematische Annalen* as one of the editors, despite the fact that he was Jewish, until 1939.

---

[205]'Die propagandistische Vertretung der logistischen Philosophie habe ich (...) abgelehnt', letter from Reidemeister to the ministry, 13/5/1933, cited from: [Siegmund-Schultze 1998, p. 78]
[206][Siegmund-Schultze 1998, pp. 64–67]
[207][Fletcher 1986, p. 16]
[208][Siegmund-Schultze 1998, p. 58]
[209][Reid 1970, pp. 203–206]
[210][Schmeidler 1935, p. 4]
[211][Lechner 1935]
[212]'im Irrationalen verwurzelte', [Siegert 1935, p. 19]
[213][Van der Waerden 1935]
[214]'ungefähr das Maximum an öffentlicher Stellungnahme gegen das Regime unter den Mathematikern', [Siegmund-Schultze 1998, p. 64]
[215][Jesser & Bohr 1937]

In the early years of the Nazi regime, altogether some 1700 faculty members and young scholars lost their jobs, an estimated 80 percent on racial grounds.[216] Of these, some 150 were mathematicians.[217] In this way, Germany managed to lose much of its intellectual capacity within a few years. When the freedom of learning and the idea of objectivity in scholarship were officially rejected and a certificate of political reliability was required in order to enter the university, it was clear that the universities had been transformed completely.[218]

### 6.6.2 Bieberbach's racial interpretation of the foundational debate

Even though the rise of the Nazi government destroyed much of the existing mathematical world, some saw opportunities, too. Becker, who had been pessimistic on the chances of an intuitionistic breakthrough in 1926,[219] wrote in September 1933:[220]

> Ich glaube, daß man bei uns in Deutschland, in Gefolge der nationalen Revolution, für den Intuitionismus wieder Sinn gewinnen wird. Denn wir wenden uns jetzt ja auf allen Gebieten gegen den 'leeren Konstruktivismus' und die 'reine intellektualistische Dialektik'. Es wäre also wohl gerade jetzt eine Ausbeutung intuitionistischer Ideen bei uns möglich.[221]

The argument Becker uses in support of his reasoning, namely the turning against the 'empty constructivism', is somewhat curious, since most people, Becker included, regarded constructivity as a characteristic of intuitionism. Perhaps the adjective 'empty' makes the difference. However, much more famous than Becker, in this respect, was Bieberbach.

The 1934 volume of the *Jahresbericht der deutschen Mathematiker-Vereinigung* contained the famous open letter by Bieberbach[222] to Harald Bohr in which Bieberbach defended the existence of a relationship between race and the style of mathematical creation. Such a position came as a surprise to many contemporaries, who had regarded Bieberbach as left-wing, liberal, and a good republican.

---

[216][Ringer 1969, p. 440]
[217][Brüning, Ferus & Siegmund-Schultze 1998, p. 4]. Full lists of mathematicians emigrated during the whole Nazi period because of the Nazi regime or persecuted or killed by it are given in [Siegmund-Schultze 1998, pp. 292–298; 301–303].
[218][Ringer 1969, pp. 439–440]
[219]See 4.3.1.
[220]letter from Becker to Heyting, 19/09/1933; [TLI Heyting, B bec-330919]
[221]'I think that here in Germany, because of the national revolution, one will attach value to intuitionism again. For in all areas we are now turning away from 'empty constructivism' and 'purely intellectual dialectics'. So right now an exploitation of intuitionistic ideas could be possible here.'
[222]For biographical information on Bieberbach, see 5.3.1.

## 6.6. POLITICS

Furthermore, Bieberbach had until then always behaved loyally towards Jewish colleagues and students.[223] Now, Bieberbach cited the *Führer* in support of his reasoning. At the same time, he also pleaded for international cooperation between the different peoples and races.[224] Bieberbach repeated his arguments in the *Unterrichtsblätter für Mathematik und Naturwissenschaften* ('Educational magazines for mathematics and the sciences'), in which he defended the Göttingen students' boycott against the Jew Landau. In Bieberbach's view, the students had rightly felt that there was something un-German in Landau's teaching. They wanted to be educated in their own *völkische Geist* ('popular spirit'). Bieberbach's reasoning was more subtle than the standard Nazi propaganda, for he explicitly stated that this detracted nothing from Landau's 'undisputed merits' in mathematics.[225]

Bieberbach promoted his new ideas outside mathematics, too. Together with his four sons, he had marched along with the storm troopers (SA) from Potsdam to Berlin in a highly publicised march, and in November 1933 he had joined the SA. In the beginning of 1934 Bieberbach was appointed deputy to the Nazi rector of the Berlin university.[226]

German mathematics, Bieberbach proclaimed, was rooted in *Anschaulichkeit* ('intuition'). Whereas some years before Bieberbach had given a well-founded interpretation of the foundational debate without linking it to questions of race, now the foundational crisis was included in his racial polemic:[227]

> Überhaupt bin ich der Meinung, daß der ganze Grundlagenstreit der Mathematik zu erklären ist als ein Streit gegensätzlicher psychologischer Typen, also in erster Linie als ein Rassenstreit. Das Aufkommen des Intuitionismus scheint mir nur eine Bestätigung dieser Auffassung zu sein.[228]

Bieberbach repeated his view in a paper in the *Sitzungsberichte der Preussischen Akademie der Wissenschaften*, in which he used the formalistic and intuitionistic stands on the excluded middle and on mathematical existence as examples of the types they supposedly represented.[229] Bieberbach and others went on to promote *Deutsche Mathematik* with a separate journal founded in 1936, but in general they found few supporters. More specifically, Bieberbach's interpretation of the foundational crisis in racial terms did not lead to a new debate.[230]

---

[223][Biermann 1988, pp. 198–199]
[224][Bieberbach 1934B]
[225]'*unbestrittenen Verdiensten*', [Bieberbach 1934A, p. 237]. It may be interesting to note that Bieberbach's reasoning is similar to the one used by later white South African governments in defence of the apartheid regime.
[226][Mehrtens 1987, pp. 219–220]
[227][Bieberbach 1934A, p. 241]; his non-racial account is in [Bieberbach 1926], see also 5.3.1.
[228]'More generally I am of the opinion that the whole foundational battle in mathematics can be explained as a fight between opposite psychological types, thus in the first place as a race fight. It seems to me that the rise of intuitionism is only a confirmation of this view.'
[229][Bieberbach 1934C, p. 9]
[230]The case of Bieberbach's *Deutsche Mathematik* is dealt with in much more detail in [Mehrtens 1987].

## 6.7 *Moderne* and *Gegenmoderne*

'One of the most provocative and original contributions to the history of mathematics (...) in recent decades'[231] is Herbert Mehrtens' *Moderne – Sprache – Mathematik* ('Modernism – Language – Mathematics').[232] Mehrtens presents an overview of 19th and 20th century foundations of mathematics, not only as an intellectual development, but also in terms of social and cultural history, introducing semiotics and Foucault into the historiography of mathematics.[233] His thesis is that the main developments in mathematics in that period can be ordered around a dialectic process between two currents, a modernist movement and its opposition, countermodernism.[234] The difference between modernism and countermodernism is to be found in what Mehrtens calls the discourse *on* mathematics, where the character and value of mathematics is discussed, rather than in the discourse *of* mathematics. Thus, it is a question of opposing views on the self-understanding of mathematics.

**Modernism and countermodernism** Mathematical modernism, which arose around the turn of the century, can be characterised by its claim that truth and contents of a mathematical text are determined by working at the text itself. In the modernist view, mathematics is nothing else than mathematical texts, and these texts do not represent anything outside the texts. In the text, signs can be manipulated following certain rigorous rules, as long as the system is free from contradictions. In this way, mathematics becomes the language of pure possibilities. Thus, modernists plea for disciplinary autonomy and conceptual self-reference. Countermodernists oppose this view and maintain that there is some primordial basis in which mathematics is rooted, like intuition.[235] Thus, countermodernists preserve a certain external truth for mathematical texts, at the expense of restricting the mathematician's autonomy. It should be pointed out that modernism was not a deliberate choice in favour of freedom in mathematics made at a certain moment, but rather the result of a slow and laborious historical process.[236]

Mehrtens places the origins of modernism in Dedekind and Riemann, of countermodernism in Klein.[237] Countermodernism is modern in the sense that, like modernism, it abandons the idea of a transcendent reality for mathematics.[238]

---

[231] [Rowe 1997, p. 534]
[232] A most valuable review of (among other things) Mehrtens' book is [Rowe 1997], in which counterweight is given to some of Mehrtens' claims.
[233] [Scholz, E. 1992, p. 92]
[234] Mehrtens chose the name 'modernism', which is known from cultural history, in order to embed the history of mathematics into its cultural context, [Mehrtens 1996, p. 520].
[235] 'Anschauung', [Mehrtens 1990, p. 76]
[236] [Mehrtens 1990, pp. 13], [Scholz, E. 1992, p. 93]
[237] [Mehrtens 1990, pp. 67–68]
[238] The primordial intuition, from which mathematics is constructed in the intuitionistic viewpoint, is not transcendental, since it is essentially linked to the mathematician. See 4.4.3 and 4.4.1, where Heyting explicitly rejects the assumption of a transcendental existence of mathematical objects as a means of proof.

## 6.7. MODERNE AND GEGENMODERNE

That is why Mehrtens calls it countermodernism (*'Gegenmoderne'*) and not antimodernism. Countermodernism arises with modernism and is part of the modern world.[239] Thus, for countermodernists there are certain ways in which the mathematician has to work, whereas modernists put themselves forward as the champions of freedom.[240] Where countermodernists strive for some kind of reality and eternal truth,[241] modernists prefer freedom of development and consistency.[242] The main metaphors for the countermodernists are 'intuition' (in the form of both German words *Anschauung* and *Intuition*) and 'construction', whereas the modernist discourse is characterised by 'freedom', 'creator', 'sign' and 'system'. There are even radical modernists, such as Hausdorff and Zermelo, who refrain from a binding metaphor altogether and restrict themselves to the 'working mathematician'.[243]

**The foundational crisis**  Mehrtens places the foundational crisis in the dualism of modernism and countermodernism, where Hilbert and his folllowers assume the role of modernists, the Brouwerians that of countermodernists. This leads to a new interpretation of the foundational crisis. If it was a crisis at all, Mehrtens maintains, then it was so because the relationship between language and speaker changed drastically. There was no crisis of 'the' foundations of mathematics, for the foundations were uncertain before and after the crisis, too. Rather, Mehrtens maintains, it was a crisis regarding the concept of truth, on which mathematicians had come to disagree. Furthermore, the crisis was most productive in foundational and logical terms, stimulating new developments such as intuitionism and metamathematics. Mehrtens' final assertion is that the crisis faded away after about 1925, when the Weimar republic was consolidated.[244]

Mehrtens' analysis is one of the most complete ones given of this part of the history of mathematics. On the basis of the material gathered in this book, I evaluate it as follows.

First of all, I think Mehrtens' *Moderne – Gegenmoderne* distinction applies well to the debate around intuitionism. The fundamental difference between the parties involved indeed lies in the fact that intuitionists want to maintain a bond between mathematics and the human mind, whereas their opponents are willing to give this up to a large degree and do mathematics in a more formalistic way.

I agree with Mehrtens' judgement that the foundational crisis did not give certainty to mathematical foundations, but was nevertheless most fruitful. It pro-

---

[239] [Mehrtens 1996, p. 521]
[240] [Mehrtens 1990, pp. 7–10]
[241] It is questionable how 'eternal' the intuitionistic truth is. Strictly spoken, from the intuitionistic point of view truth is linked to the individual mathematician who has completed certain mathematical constructions.
[242] [Mehrtens 1990, p. 237]. Of course, the consistency demand does restrict the mathematician's freedom somewhat, but it is needed for the obvious reason that one can derive any proposition from a contradictory system.
[243] [Mehrtens 1990, pp. 511–512]
[244] [Mehrtens 1990, p. 13; 294–299]

vided an impetus for new lines of thought, both inside and outside mathematics. One need only think of the examples of Gödel, Heyting and Wittgenstein to notice the broad and positive influence it had.

I disagree, however, with Mehrtens' link between the consolidation of the Weimar republic and the fading away of the foundational crisis after 1925. Looking at the public contributions to the foundational debate, it becomes clear that the debate continued expanding after 1925 and that the zenith of the discussion was around 1927.[245] Thus, the relationship which Mehrtens posits between the stabilisation of the Weimar republic and the decline of the foundational debate is inappropriate.

Finally, I think Mehrtens is correct in judging that the crisis focused on the changed relation between speaker and language, and on the concept of mathematical truth. It should be stressed, however, that the formalistic position was only developed in opposition to the intuitionistic one, after intuitionists had criticised classical mathematics. The foundational debate, which focused on more concrete themes such as the question of mathematical existence and the validity of the principle of the excluded middle,[246] over time gave rise to a new, formalistic view on mathematics. Thus, in this period, to put it in Mehrtens' language, countermodernism gave rise to modernism, rather than the other way round.

## 6.8 Conclusion

The question we started this chapter with was if any specific characteristics of the cultural context could be pointed out which may have influenced the reception of Brouwer's intuitionism. Having now analysed the various possible interactions between the foundational debate in mathematics and its cultural context, I come to the following conclusion.

The most fruitful link between the foundational crisis and its environment lies in the use of metaphors, which provides us with a most appropriate way of demarcating the foundational debate in chronological terms. In his 1921 paper *Über die neue Grundlagenkrise der Mathematik*, Weyl introduced the term 'foundational crisis' and labelled Brouwer the 'revolution', evoking associations with a communist revolution. Hilbert tried to oppose it by re-labeling Brouwer's reform attempts as a *Putsch*, but Weyl's terms were the ones that echoed through the debate. The revolution metaphor was used throughout the various languages employed in the debate; the crisis metaphor mostly in German. Despite the popularity of Wilhelminian Germany in academic circles, none of the currents associated itself with the old order, thus reflecting their progressive attitude. The metaphors provide an excellent way of measuring the development of the debate, since they reflect the emotional character of the foundational crisis. In 1928, the use of them dropped markedly, indicating the end of the foundational crisis.

---

[245] See 3.2.2.
[246] See 3.3.1.

## 6.8. CONCLUSION

A valuable contribution in terms of placing the foundational debate in mathematics in a wider context is provided by Mehrtens'. In Mehrtens' interpretation, the controversy focused around a modernist current and its opposition, countermodernism. Countermodernists see mathematics as rooted in some kind of primordial basis; modernists maintain that the truth of a mathematical text is determined only by the text itself. Mehrtens identifies Hilbertians as modernists, whereas intuitionists belong to countermodernism. The main reason for the foundational crisis, Mehrtens argued, was that the relation between language and speaker changed. This is certainly true. It should be added, however, that there were very few formalists in the beginning of the debate. The change which took place regarding the meaning of the mathematical language was one away from a classical and towards a more formalistic view, evoked by the rise of intuitionism and its justified criticism of the claims of classical mathematics. Thus, in mathematics counter-modernism contributed substantially to the development of modernism.

Two subjects touched upon in this chapter are open to more research. The first one are the various possible links between intuitionism in mathematics and its *lebensphilosophische* namesake. *Lebensphilosophie*, one of the dominant philosophies in Weimar Germany, had several sub-currents, one of which was called 'intuitionism'. This philosophical intuitionism shared some characteristics with mathematical intuitionism, such as its adherence to intuition and its dislike of logic and fixed systems. However, whereas philosophical intuitionism was an irrational philosophy which considered constructions to be dead, Brouwer's intuitionism was based on rational constructions. In academic circles, especially Bergson's form of intuitionism was a popular subject. It is not clear to what extent this influenced the reactions to Brouwer's intuitionism. Contributions in which both are mentioned are rare, and most of them appeared later in the debate. It is known that Hilbert opposed Bergson's intuitionism, but he did not explicitly link it with or separate it from Brouwerian intuitionism. Similarly, Hilbert quarrelled with life philosophers in the Göttingen faculty about university appointments, but it is not known if these presented themselves as intuitionists. Finally, Weyl used Spengler-like passages in his 1921 paper, but also here hard proof of a link is absent. Thus, more research would be needed to establish a more definite relationship both intuitionisms.

The same applies, although the number of possible links is smaller, to the constructivity label in mathematics and the rise of constructivism in art. It is hard to explain the popularity of the constructivity label in the foundational debate. The fact that it was claimed by different currents, and that the first criticism about its use as a slogan without clear meaning only arose in 1928, indicate that it could draw on some general popularity. A tempting explanation would be to link it with the rise of constructivism in art, which gained popularity at about the same time. However, there are no explicit references to substantiate this suggestion.

One suggestion about how to embed the foundational crisis in its cultural context is largely refuted. Forman's conjecture that mathematicians applied an accommodationist strategy to adapt to the hostile Weimar cultural environment does not hold in general, especially because of the popularity of the constructivity label in mathematical circles. Forman's well-known thesis holds that physicists adapted both the presentation and the contents of their theories to the hostile cultural environment of Weimar Germany, most notably to *Lebensphilosophie*. A later and weaker version of the thesis focuses on the presentation and leaves the contents of physical theories out of the adaptation. Forman suggested that similar attitudes among mathematicians played a role in the popularity of intuitionism in mathematics. However, almost all intuitionists were outside the direct influence of Weimar Germany. Forman's thesis definitely does not apply to the contents of intuitionistic mathematics, which was created almost singlehandedly by Brouwer. There may have been some influence on the presentation in Weyl's 1921 paper, where passages appear which fit well into the life philosophical discourse. However, the 'constructivity' label was popular throughout the debate, something which runs contrary to *Lebensphilosophie*. Thus, even the weaker version of Forman's thesis does not hold generally.

Similarly, the similarities between the debate on Einstein's theory of relativity and that on intuitionism are only superficial. Furthermore, the importance of a factor like anti-Semitism in the former discussion makes it clear that there is more that divides the two debates than that joins them.

Finally, Hitler's rise to power in 1933 had serious implications for the mathematical world. Many mathematicians were dismissed from their position, and publications supporting the new regime appeared in mathematical journals. Not surprisingly, 1933 is the year in which the number of German contributions to the foundational debate dropped significantly. Bieberbach put forward a racial interpretation of the foundational debate, but this did not lead to a revived interest in the foundational controversy.

# Conclusion

Most mathematicians are wroth, rather justifiably so I think, at the suggestion that mathematics has no content and that proofs of the impossibility of certain combinations or of the indemonstrability of certain theorems are a sort of meta-chess. Between wordless thought and thoughtless words, the decision would possibly rest with the former.

Alice Ambrose[247]


Only very few mathematicians were willing to accept the new, exigent standards for their own daily use. Very many, however, admitted that Weyl and Brouwer were prima facie right, but they themselves continued to trespass, that is, to do their own mathematics, in the old, 'easy' fashion – probably in the hope that somebody else, at some other time, might find the answer to the intuitionistic critique and thereby justify them *a posteriori*.

John von Neumann[248]

The key questions for this research were how people reacted to Brouwer's intuitionism, and why they did so. The analysis performed in the preceding chapters leads me to the following conclusions.

**Beginning of the debate**   The foundational crisis started with the publication of Weyl's paper *Über die neue Grundlagenkrise der Mathematik* in 1921. Weyl had been converted to intuitionism after talks he had had with Brouwer in the Swiss Alps in the Summer of 1919. Weyl was impressed by Brouwer's personality as well as by his ideas. In the paper, Weyl presented both the intuitionistic criticism, focusing on pure existence statements versus constructions and on the unlimited

---
[247][Ambrose 1933, p. 598]
[248][Von Neumann 1947, p. 188]

use of the principle of the excluded middle, and the alternative it offered, based on choice sequences. Weyl's paper was polemic, written with the purpose to rouse the sleepers. Evoking associations with a communist uprising, Weyl declared Brouwer to be the revolution, which would overthrow the existing order. Weyl joined it.

Weyl had chosen the right tone for his paper. After the Great War, Germany's defeat and the ensuing crisis, old beliefs were shattered. The Russian revolution had shown that it was possible to overthrow the existing order. If an entire society could be overthrown and replaced by something new, then why not mathematics? And if anybody could do so, this had to be the leading mathematicians of the new generation: Brouwer and Weyl.

Brouwer had developed intuitionistic mathematics from 1907 onwards. Even though he changed his view on certain issues along the way, he remained faithful to his basic idea, inspired by Mannoury, of mathematics as constructions carried out in the human mind. Intuitionism as developed by Brouwer knew two sides. One was its radical criticism of uses and concepts of classical mathematics, such as Cantorian set theory and the unlimited use of the principle of the excluded middle. On the positive side, intuitionism offered an alternative set theory and function theory based on choice sequences. There was hardly any resonance to Brouwer's intuitionistic work until Weyl entered the scene in 1921. Whereas Brouwer's papers were usually hard to read, Weyl's paper was clear and, moreover, polemic. In addition, more attention was created by Weyl's public conversion to intuitionism.

**General characterisation** The foundational crisis which followed the publication of Weyl's 1921 paper increased steadily during the years 1922–1927. In 1928, the number of contributions to the debate declined, whereafter it started to fluctuate. The emotional character of the reactions is clearly noticeable in the use of the 'crisis' and 'revolution' metaphors. In 1928, their use dropped markedly. Furthermore, the first reactions criticising the use of these popular slogans appeared in that year. Therefore, the end of the foundational crisis is best set in 1928. When Brouwer, at the end of that year, was dismissed from the editorial board of the *Mathematische Annalen*, this confirmed an existing trend rather than created a new situation.

The changed tone in the debate was clearly noticeable at the 1930 Königsberg conference, where participants representing the different currents in the debate showed a reconciliatory attitude. Instead of fighting each other, they mentioned their own weaknesses and the strengths of other currents. The common and explicit search for a synthesis had started. Carnap and Fraenkel were among the most prominent persons who promoted the new direction.

Altogether, there were more than 250 public reactions to intuitionism up to the year 1933. Even though this only represents a fraction of the total number of mathematical and philosophical publications during the period, it is a large amount compared to how much attention foundational matters normally attracted. The fact that such a comprehensive discussion could develop shows that mathematicians and others felt that there was a need to discuss the foundations of mathematics. Apparently, there was unclarity or insecurity about the status of the foundations of mathematics in the beginning of the 1920s, and people cared about it.

The foundational crisis that followed Weyl's paper was almost completely a European affair. This reinforces the argument that the specific circumstances after the first World War fostered the sometimes emotional debate on the foundations of mathematics. In the United States, where no trench war had been fought and where no threat of overthrowing society existed, reactions to intuitionism were late and few. Moreover, the tone in which Americans reacted to the new current was much more pragmatic.

Within Europe, the use of the characteristic metaphors was not spread uniformly. Whereas the revolution metaphor can be found throughout Europe, the crisis metaphor was largely German. This indicates that the German cultural sphere was at the center of the debate, a conclusion which is in accordance with the fact that more than half of the contributions were in German. Whereas German and Dutch were the languages in which the first part of the debate was carried out, the discussion spread to the French and English speaking world in 1924, by papers by Wavre and Dresden respectively.

In the debate that developed, most contributors did not clearly differentiate between Brouwer's and Weyl's contributions to intuitionism. Even though Brouwer was the one who had developed intuitionistic mathematics while Weyl was its most important protagonist, people tended to refer to 'Brouwer's and Weyl's intuitionism'. Moreover, most people learned about Brouwer's intuitionism through Weyl's presentation of it. Apart from the Brouwerian counter-examples, Brouwer's own intuitionistic publications during the foundational crisis had surprisingly little influence on the course of the debate.

Logicism, which is sometimes seen as a third current in the debate, played hardly any role of importance. It had few supporters: Russell and Wittgenstein had stopped contributing to the foundations of mathematics, Ramsey was relatively isolated in Britain, and Carnap only became active later in the debate. The image of the tripartite debate seems to have been fostered by the 1930 Königsberg conference, in which all three currents were represented. This image was then projected backwards onto the foregoing part of the debate.

Another common misunderstanding is that the foundational debate arose because of the set-theoretical paradoxes. In fact, the paradoxes were mostly only used as an introduction to foundational problems, and they were nowhere near the center of the debate.

By the beginning of the 1930s the foundational debate was definitely over. Neither Gödel's second incompleteness theorem, which destroyed Hilbert's original programme, nor Bieberbach's racial interpretation of the foundational debate was enough to lead to its resurrection.

**Reactions**   Hilbert reacted immediately to Weyl's 1921 paper. In the same year, he delivered lectures in Copenhagen and Hamburg, in which he answered Weyl's attack with more polemics. If anyone would have felt attacked by the criticism of the state of mathematics, it was Hilbert, the *Generaldirektor*.[249] Apparently, he felt the need to defend himself and the mathematical heritage. The political metaphors Hilbert used to counter Weyl's attack identified him with the current state power of Weimar Germany. Neither Hilbert nor Weyl, thus, associated himself with Wilhelminian Germany, even though that must have been a popular position among German academics. Hilbert tried to rename the intuitionistic revolution a *Putsch*, suggesting that it had little support. However, Weyl's metaphors were the ones that echoed through the debate.

During the debate, Hilbert remained the most polemic participant. It is not clear if Hilbert had by that time made an association between Brouwerian and Bergsonian intuitionism, which may have added to the fierceness of his reaction. The fact that Hilbert took over most of the intuitionistic demands, as long as mathematics was to be considered meaningful, did not temper his desire to argue fiercely against intuitionism.

Now that Hilbert had reacted, it was clear that something was rotten in the state of mathematics. Since there was a counter-revolution, it was indisputable that there had been a revolution, too.

Two themes dominated the debate: mathematical existence and the principle of the excluded middle. Thus, the stress came to lay on critical aspects of intuitionism, where intuitionists 'forbade' mathematicians the use of pure existence statements and of the principle of the excluded middle in infinite domains. This happened despite the fact that Weyl had devoted as much attention to the positive aspects of intuitionism, namely its use of choice sequences and the alternative theory of the continuum. In this way, a debate developed that, for intuitionists, was one-sided.

It is in this context that the frequently used argument that intuitionism had catastrophic consequences for mathematics should be seen. In general, intuitionism was seen as a threat, not as an opportunity. The argument was reinforced by pointing out Kronecker as one of Brouwer's predecessors. Kronecker was well-known for his dictum that God had created the natural numbers, all the rest was man-made. This, too, was seen as a restrictive stand. It is hard to say whether the stress on the negative aspects of intuitionism was put there on purpose by its opponents, or that they reacted to these parts because these were simply aspects of intuitionism that affected them most. Also, the fact that only a few people understood intuitionistic set theory based on choice sequences must have played a role.

---

[249]The term 'General Director' comes from Minkowski, [Minkowski 1973, p. 130].

It was mainly philosophers who paid proper attention to the positive aspects of intuitionism (without, mostly, going into the mathematical details). However, the majority of the participants in the foundational debate were mathematicians. Mathematicians and philosophers generally moved around in different circuits (in terms of journals and conferences), and only the odd person crossed the border. This isolated someone like Becker, who was among the first contributors to the debate and who tried to establish a link between Brouwer's intuitionism and Husserl's phenomenology. Apart from Weyl, very few mathematicians were able to understand his contributions, because of the technical vocabulary one had to master in order to understand phenomenology.

**Mathematical existence** The debate on mathematical existence centered on the subjects of consistency and constructivity, where the former was in fact directed at the existence of mathematical systems, the latter at that of mathematical objects. Weyl had put forward the intuitionistic demand that mathematical existence statements had to be constructive. Most of the contributions after Weyl aimed at clarification; people tried to explain what existence meant or should mean in mathematics, rather than arguing for or against a certain point of view. Wavre was the first to give a clear interpretation of pure existence statements. In 1924, he claimed (in modern terms) that formalists saw $\exists x \neg \varphi(x)$ as $\neg \forall x \varphi(x)$. Fraenkel, the most influential commentator of the debate, used the issue of mathematical existence to give the first coherent presentation of intuitionism by a non-intuitionist, maintaining in 1927 that all intuitionistic ideas could be derived from the 'sharp distinction' between pure and constructive existence statements.

The fact that the discussion was mostly about the meaning and use of the word 'to exist' means that a terminological solution was possible, simply by using different terms for the different meanings. Such a solution was indeed suggested by Dingler in 1926, and later again by Menger. However, Dingler's contribution remained largely unnoticed, and the idea was not taken over by the mathematical community at large during the debate.

The reactions to the intuitionistic view on mathematical existence show that there was at the time no clearly established meaning for the term 'to exist' in mathematics. What we now see as the formalistic demand of consistency was linked by a number of mathematicians, most notably Hilbert, to some kind of existence. But what would existence then mean? It became clear to most of the participants that it could not mean constructivity in the intuitionistic sense, since that was a narrower concept. The fact that consistency did not imply any other form of existence was already pointed out in Brouwer's dissertation. Perhaps, some had a Platonistic idea of existence in mind, but few defended it openly. In the end, most people ended up recognizing both the contentual, constructive view and the purely formal view on existence, as Weyl had done in 1924. This means that the debate on mathematical existence, provoked by intuitionism, not only brought clarification about what existence meant in mathematics and when an existence statement could be seen as constructive, but also contributed to the establishment of the purely formalistic view on mathematical existence, where existence meant nothing more than consistency.

**Constructivity** A special feature of the debate on mathematical existence was the fight for the label 'constructive'. In the beginning of the debate, both Brouwer and Weyl on the one hand, Hilbert and Bernays on the other claimed the term for their current. It is not clear why the label was so popular. It is tempting to see a link with constructivism in art, which became popular at the same time. However, no explicit references from one field to the other are known.

It soon became clear that most participants to the debate saw intuitionistic existence as 'constructive', whereupon Hilbert and Bernays did not repeat their claim. The first criticism of the vagueness of this label was voiced only in 1928, by Menger. Whereas Fraenkel had in the same year described the most characteristic element of intuitionism as the 'sharp distinction' between pure and constructive existence statements, Menger pointed out that nobody had clearly defined what constructive meant. Menger was soon joined by others. Nevertheless, at the 1930 Königsberg conference, the debate about the label revived, with both formalism and logicism, and intuitionism to a lesser degree, claiming to be 'constructive'.

The sub-debate on the constructivity label enables us to judge a suggestion made by Forman. Forman suggested that mathematicians had a similar accommodationist attitude as physicists showed towards the hostile Weimar cultural environment, especially towards *Lebensphilosophie*. The popularity of the constructivity label in circles of mathematicians makes it difficult to defend this thesis. In most contemporary descriptions of the philosophy of life, it becomes clear that *Lebensphilosophie* was actually very much opposed to constructions, which it considered to be 'dead'. The only thing that may be true of Forman's suggestion is that Weyl seems to have used some Spenglerian descriptions in the presentation of intuitionism in his 1921 paper.

The discussion on mathematical existence contributed substantially to the clarification of the meaning of existence and to the acceptance of the formalistic view. A very different effect is that it may have played an important role in the views of the 'second Wittgenstein'. It is known that Wittgenstein started philosophising again after he visited one of Brouwer's lectures in Vienna in 1928. After that, Wittgenstein went through two phases regarding his views on mathematical subjects such as infinity and existence. In the first phase, he basically agreed with Brouwer's intuitionistic point of view. In the second, however, he moved to a more descriptive position, simply pointing out that different uses of mathematical terms implied different meanings. This 'meaning is use' argument applies very well to the differences between the intuitionistic and formalistic conception of mathematical existence, and Wittgenstein applied it to them. What remains unclear, however, is whether he formulated his general insight thanks to the discussion on mathematical existence, or that it was a mere application of an idea that he had obtained in a different way.

**Excluded middle** In the discussion on the principle of the excluded middle, arguments played a more important role than in the one on mathematical existence. Several reasonings were used against Brouwer's criticism of the general validity of the principle of the excluded middle, such as the Platonistic idea that one out of two possibilities had to be the case, whether or not mathematicians could determine which one it was; the idea that some expressions were not opposed as contradictories and therefore the principle of the excluded middle simply did not apply; and the idea that the rules of logic represented the laws of thought and thus had to be true. All these were contentual arguments, based on the assumption that logic had some kind of meaning. Only very few people, most notably Dingler and Church, took up the purely formalistic position that one can simply choose which logical rules to use.

The discussion on the excluded middle was more fruitful than the one on mathematical existence. Kolmogorov, Heyting, Gödel and Glivenko, men of the younger generation, contributed substantially to the development of intuitionistic logic. It seemed to have been easier for them than for people from older generations to work in both intuitionistic and formalistic systems. Kolmogorov already published a formalisation of a part of intuitionistic propositional logic, together with what can be interpreted as a translation of classical logic into its intuitionistic counterpart, in 1925. However, since the paper was written in Russian, it remained unnoticed. The discussion on the excluded middle inspired people such as Becker, Lévy, and Barzin & Errera to make positive contributions (to logic), something which was much scarcer in the discussion on mathematical existence. In 1930, Heyting's formalisation of intuitionistic logic appeared, which soon became the standard formalisation and clarified to many non-intuitionists in what sense intuitionistic logic differed from classical logic. Gödel, who had been inspired by Brouwer for his incompleteness theorems, published a translation of classical into intuitionistic arithmetic in 1932. Thereby, he showed that intuitionistic arithmetic contained all of its classical counterpart, albeit in a different interpretation.

The effect of the discussion was that classical logic lost its absolute status. The change is best noted in Barzin and Errera's reactions. They started as staunch supporters of classical logic as representing the laws of thought. After their discussion with Heyting, they only maintained that classical logic should be conserved because of its simplicity. At the beginning of the 1930s, classical logic had become one out of various logics, and was thus seen more as a formal and somewhat arbitrary system. Also in this respect, the discussion about the intuitionistic point of view in fact contributed to a more formalistic view on mathematics.

**Intuitionism, formalism and classical mathematics** The foundational crisis is often characterised as a discussion between formalists and intuitionists, both by contemporaries of the debate and by modern commentators. However, the discussion on logic and the principle of the excluded middle, and to a lesser degree that on mathematical existence and constructivity, makes it clear that most arguments used in the debate, either in favour or against intuitionism, were based on a contentual (rather than a formalistic) interpretation of mathematics and logic. Hardly anybody used the purely formalistic argument that, e.g., the principle of the excluded middle applies simply because *we have defined the formal system that way*. Instead, what Weyl attacked in his 1921 paper was classical, contentual mathematics; and the vast majority of those who reacted to the intuitionistic critique were, judged by the way they reacted, classical mathematicians. They saw mathematics as possessing some kind of contents, and thus were sensitive to the intuitionistic arguments. This explains why they reacted so massively and sometimes emotionally to intuitionism: as classical mathematicians, they felt attacked by the intuitionistic criticism – and rightly so.

It is even unclear to what extent Hilbert was a formalist. To be sure, during the first years of the debate Hilbert did not describe his own position as 'formalistic', nor did Brouwer or Weyl label him as such in widely read publications. Baldus was to first to characterise Hilbert as a formalist during the foundational crisis, in 1923, but his description of formalism differed from Brouwer's. In Baldus' version of formalism, the essential feature of seeing mathematics as a formal system without contents was lacking. In his earlier contributions on the foundations of mathematics, Hilbert always mentioned 'objects of thought' as constituting the contents of mathematics. In the beginning of the foundational crisis, he maintained that the formulas of formalised mathematics represented the thoughts of ordinary mathematics and that the consistency of a mathematical system implied the truth of the mathematical propositions of that system, thus holding on to a contentual interpretation. Only in 1925 did he explicitly and publicly drop the contentual character of *some* of the statements of formalised mathematics, arguing that these elements did not mean anything, but were needed to adhere to the rules of classical logic. This was exactly the reason which Wavre had given in 1924 for the formalistic position. Thus, in response to the intuitionistic criticism of classical logic, Hilbert moved towards a more, but not purely, formalistic position over time. This may help in explaining Hilbert's ambiguous reaction to intuitionism. While he took over the intuitionistic demands regarding constructive existence and the use of the principle of the excluded middle on what he called the 'finitary level', he argued fiercely against them in 'ordinary' mathematics. Therefore, Hilbert was not so much arguing against the intuitionistic criticism, for, as Von Neumann, Weyl, Fraenkel and Becker already observed, he actually agreed with it. He was fighting for the right to work with formalised mathematics, which he at least during part of the debate considered to have some kind of contents.

Hence, Mehrtens' interpretation of the foundational debate as one in which a modernist and a counter-modernist view opposed each other is essentially correct. It should be pointed out, however, that modernism, represented in this period by formalism, only developed and spread because of the counter-modernist (intuitionistic) criticism of classical mathematics. Thus, contrary to what Mehrtens' labels suggest, in the case of the foundational crisis in the 1920s, counter-modernism gave rise to modernism.

For all the sympathy for the intuitionistic point of view that one can find in some of the reactions, only very few authors actually contributed to intuitionistic mathematics. The power of their daily practice, of tradition, of their educational background was stronger than the desire to build up something new. Even though the discourse *on* mathematics changed, this had little, if any, effect upon the discourse *of* mathematics. The old order, as Weyl had remarked correctly in 1921, could indeed not be maintained. But instead of adapting to the demands of contentual mathematics as intuitionists wanted, the claim of contents was (partially) dropped.

If Brouwer had any strategy for getting intuitionism accepted by the mathematical community, it clearly failed. However, most likely Brouwer did not. One gets the impression that he was more interested in the development of his ideas and the value he attached to them, than in their presentation and general acceptance. Furthermore, Brouwer overrated the importance of rational arguments in discussions. He fought an ever growing number of battles during his life, and it does not seem to have hindered his faith if persons of name opposed him. Brouwer's outspoken personality and the rigidity with which he defended his point of view contributed to splitting the community of mathematicians into those who supported him and those who did not.

It is much more likely that Weyl had a strategy when he entered the debate. As he wrote to Brouwer, he intended his 1921 paper to rouse the sleepers. Measured by the number of awakenings, it certainly worked. It did not, however, establish intuitionism as the main current in the foundations of mathematics. Weyl himself soon dropped the exclusive claim he had defended for intuitionism and tried to reconcile intuitionism, which he thought correct from a philosophical point of view, with formalism, which fitted more into the need for symbolic systems in the sciences. Weyl remained occupied with the choice between intuitionism and formalism for the rest of his life.

For all its opposition to the formalisation of mathematics, Brouwer's intuitionism actually contributed, by its justified criticism of classical mathematics and by the debate which it provoked, to its development and spreading. In the course of the debate, it became clear that the laws of classical logic did not have an absolute status, based on a contentual interpretation. If one wanted to adhere to them, one had to allow meaningless statements in mathematics. Similarly, mathematicians came to realise that consistency did not imply existence in some other sense, at least not in terms of constructivity. Thus, the debate on intuitionism led to the separation between mathematics and contents, and to the dilution of what absolute views there had been left of mathematics. The fact that hardly anybody used the purely formalistic argument that we are simply working at a formal system and that therefore, by definition, certain logical rules apply, makes it clear that there were actually very few formalists in the beginning of the foundational debate. Formalism only arose because of the intuitionistic criticism of classical mathematics, and the fight was mostly between intuitionism and classical mathematics. The result of the debate was totally opposite to what Brouwer had fought for.

In his inaugural lecture in 1912, Brouwer introduced the dichotomy between intuitionism and formalism to a large audience. The image he created there has remained as the characterisation of the currents involved in the foundational crisis until today. In fact, however, very few mathematicians were formalists at the beginning of the foundational debate. What Brouwer sketched in 1912 was not so much a situation that existed then or even ten years later, but the positions that resulted if one thought the matter through to its end. Brouwer was way ahead of his time.

# Appendix A

# Chronology of the debate

**1897**

Brouwer starts his mathematics study in Amsterdam; attends lectures by Mannoury.

**1899**

Hilbert publishes *Grundlagen der Geometrie*;[1] axiomatic foundations of geometry.

**1904**

Weyl starts his studies in Göttingen; attends lectures by Hilbert.
Brouwer starts his Ph.D. in Amsterdam under Korteweg.
Hilbert lectures at the International Congress of Mathematicians in Heidelberg; presents consistency proofs.

**1905**

Brouwer publishes *Leven, kunst en mystiek*; first criticism of language.

**1907**

Brouwer's dissertation *Over de grondslagen der wiskunde* appears; first act of intuitionism: separation of mathematics and mathematical language; mathematics is a free creation and is independent of logic; consistency does not imply existence.

**1908**

Brouwer's *Onbetrouwbaarheid der logische principes* appears; rejection of the principle of the excluded middle.
Weyl writes his dissertation under Hilbert.
Beginning of Brouwer's topological work, which brings him world fame.

---

[1]Some titles have been abbreviated. For full bibliographical details of published works, see the bibliography.

## APPENDIX A. CHRONOLOGY OF THE DEBATE

**1909**

    Brouwer becomes *privaat-docent* in Amsterdam.

    Mannoury's *Methodologisches und Philosophisches* appears; first public reaction to Brouwer's intuitionism.

Summer     Brouwer and Hilbert meet in Scheveningen; Brouwer tells Hilbert about the difference between mathematics and meta-mathematics.

**1910**

    Weyl becomes *Privatdozent* in Göttingen; holds a Cantorian view on the foundations of mathematics.

**1911**

    Brouwer's first visit to Göttingen; meets Weyl.

**1912**

    Brouwer delivers his inaugural address *Intuitionisme en formalisme* in Amsterdam; introduces the names 'intuitionism' and 'formalism' to a large audience.

    End of Brouwer's topological period.

    Bernays becomes *Privatdozent* in Zürich.

**1913**

    Weyl becomes professor in Zürich.

**1914**

    Bieberbach delivers his inaugural lecture in Basel; takes up a formalistic position; first public reaction to the intuitionistic view on mathematical existence.

April     Brouwer becomes chairman of the *Wiskundig Genootschap*.

August     Beginning of the First World War.

**1915**

    Brouwer becomes an editor of the *Mathematische Annalen*.

**1916**

October 31     De Haan delivers his inaugural lecture in Amsterdam on legal significs; first public reaction to the intuitionistic criticism of the excluded middle.

**1917**

April     Mannoury succeeds Brouwer as chair of the *Wiskundig Genootschap*.

September     Hilbert lectures on axiomatic thinking in Zürich.

Autumn     Bernays comes to Göttingen as Hilbert's assistant.

**1918**

    Second act of intuitionism: Brouwer publishes a constructive set theory based on choice sequences.

    Hilbert's *Axiomatisches Denken* appears in the *Mathematische Annalen*.

    Weyl's *Das Kontinuum* appears; rejects impredicative definitions.

|   |   |
|---|---|
| November | First edition of Spengler's *Untergang des Abendlandes*. Revolution in Germany: the republic is declared. End of the First World War. |

**1919**

|   |   |
|---|---|
| Summer | Brouwer's *Intuitionistische Mengenlehre* appears in the *Jahresbericht*. Brouwer and Weyl meet in Engadin; Weyl is converted to intuitionism. Hilbert offers Brouwer a chair in Göttingen; at the same time, Brouwer receives an offer from Berlin; Brouwer declines both. |
| December | Weyl lectures in Zürich on the new (intuitionistic) foundations of mathematics; discussion with Pólya. |

**1920**

|   |   |
|---|---|
|  | (or 1921) Fraenkel meets Brouwer in Amsterdam. |
| March | Kapp-putsch. |
| May 11 | Weyl lectures on the continuum in Göttingen; Hilbert misses the lecture. |
| July 28–30 | Weyl lectures on the foundations of analysis in Hamburg. |
| Fall | Weyl visits Brouwer in the Netherlands. |
|  | Brouwer offers Weyl a chair in Amsterdam; Weyl declines |
| September 22 | Brouwer lectures at the Bad Nauheim *Naturforscherversammlung*; presents first weak (mathematical) counter-examples against the principle of the excluded middle. |

**1921**

|   |   |
|---|---|
| April | Weyl's *neue Grundlagenkrise* appears in the *Mathematische Zeitschrift*; stress on construction vs. pure existence, rejection of the excluded middle, and new continuum by means of choice sequences; Weyl declares Brouwer to be the revolution, and publicly joins intuitionism; beginning of the foundational crisis. |
| Spring | Hilbert lectures in Copenhagen on the new foundations of mathematics. |
| July 25–27 | Hilbert lectures in Hamburg on the new foundations of mathematics. |
| September 23 | Bernays lectures in Jena on Hilbert's thoughts on the foundations of arithmetic. |

**1922**

|   |   |
|---|---|
|  | Hilbert's lectures appear in the *Abhandlungen* as *Neubegründung der Mathematik*; claims that consistency implies the truth of the propositions; recognizes that the excluded middle does not apply universally in contentual mathematics; introduces proof theory: distinction between mathematics and meta-mathematics. |
|  | Hilbert offers Weyl a position in Göttingen; Weyl declines. |
|  | Bernays' *Über Hilberts Gedanken* appears in the *Jahresbericht*. |
| September | Hilbert lectures on the logical foundations of mathematics before the *Deutsche Naturforscher-Gesellschaft* in Leipzig; claims that formalised mathematics represents mathematical thoughts. |

**1923**

|   |   |
|---|---|
|  | Second edition of Fraenkel's *Einleitung in die Mengenlehre* appears; stress |

|  |  |
|---|---|
|  | on principle of the excluded middle in the presentation of intuitionism. Brouwer lectures in Amsterdam; triple negation equals a single one. Hilbert's lecture *Die logische Grundlagen* appears in the *Mathematische Annalen*. Becker's *Habilitationsschrift 'Phänomenologische Begründung der Geometrie'* appears; first philosophical reaction to intuitionism during the foundational crisis. |
| September | At the *Jahresversammlung der Deutschen Mathematiker-Vereinigung* in Marburg, Brouwer lectures on the role of the principle of the excluded middle, and Fraenkel lectures on the new ideas on the foundations of analysis and set theory; Brouwer visits Fraenkel in Marburg. |
| November | Hitler-putsch. |
| December 1 | Baldus delivers his rector's address on intuitionism and formalism in Karlsruhe; first to characterise Hilbert as a formalist during foundational crisis. |

**1924**

|  |  |
|---|---|
|  | Brouwer proves the uniform continuity of every full function. Weyl publishes *Randbemerkungen zu Hauptprobleme*; points out that there is no difference between Brouwer and Hilbert on a contentual level. Beginning of the *Wiener Kreis*. Baldus' rector's address *Intuitionismus und Formalismus* appears in writing. Wavre's paper on the crisis of mathematics appears in the *Revue de Métaphysique et de Morale*; first reaction to intuitionism in French during the foundational crisis; stress on the excluded middle and existence. Dresden's paper on Brouwer's contributions to the foundations of mathematics appears in the *Bulletin of the American Mathematical Society*; first reaction to intuitionism in English during the foundational crisis. |
| January 8 | Bernays lectures in Berlin on new investigations on Hilbert's axiomatic method. |
| June 18 | Fraenkel lectures in Marburg on the crisis in the foundations of mathematics. |

**1925**

|  |  |
|---|---|
|  | Brouwer starts a series of expository papers on intuitionism in the *Mathematische Annalen*. Kolmogorov publishes a paper on the principle of the excluded middle (in Russian); first formalisation of intuitionistic logic. |
| June 4 | Hilbert lectures on the infinite at the *Weierstraß-Woche* in Münster; first admission that some formulas in mathematics are meaningless. |
| June 8-12 | Fraenkel lectures on the foundations of set theory in Kiel. |
| Fall | It becomes known that Hilbert suffers from pernicious anaemia, at that time generally a fatal disease. |

**1926**

|  |  |
|---|---|
|  | Discussion between Wavre and Lévy in the *Revue de Métaphysique et de Morale*. |

|   |   |
|---|---|
|  | Dingler's *Zusammenbruch der Wissenschaft* appears; suggestion to differentiate terminologically between the different views on existence. |
|  | Hilbert's *Über das Unendliche* appears in the *Mathematische Annalen*. |
|  | Gonseth's *Fondements des mathématiques* appears. |
| February 15 | Bieberbach lectures on the scientific ideal of mathematicians; takes up an intuitionistic position. |
| September | Germany becomes member of the League of Nations. |

**1927**

A partial reprint of Hilbert's 1926 paper appears in the *Jahresbericht*.

Fraenkel's Kiel lectures appear as *Zehn Vorlesungen über die Grundlegung der Mengenlehre*; central issue: the distinction between construction and pure existence; first coherent presentation of intuitionism by a non-intuitionist; remarks that Hilbert takes over the intuitionistic criticism of the excluded middle.

Becker's *Mathematische Existenz* and Heidegger's *Sein und Zeit* appear jointly in Husserl's *Jahrbuch*; Becker argues that, from a phenomenological point of view, intuitionism is right.

Weyl's paper on the present epistemological state in mathematics appears in *Symposion*.

Lietzmann's *Aufbau und Grundlage* appears; first high school book in which intuitionism figures.

Barzin and Errera claim that intuitionistic logic leads to a contradiction.

|   |   |
|---|---|
| Jan. – March | Brouwer lectures on intuitionism in Berlin. |
| May | First lecture organised by the Berlin *Gesellschaft für empirische Philosophie*, by Petzoldt; rejects intuitionism. |
| July | Hilbert lectures on the foundations of mathematics in Hamburg; characterises logic as representing the laws of thought; Weyl defends intuitionism by pointing out the difference between contentual and non-contentual mathematics. |
| September 19 | Hilbert lectures on *Das Auswahlaxiom in der Mathematik* at the *Jahresversammlung* in Leipzig. |
| December | Brouwer lectures on formalism in Amsterdam; formulates four insights which would suffice to end the debate. |

**1928**

For the first time since 1922, the number of reactions to intuitionism drops; also, the use of the 'revolution' and 'crisis' metaphors drops markedly; end of the foundational crisis.

Hilbert's *Die Grundlagen der Mathematik* appears.

Menger's first criticism of the constructivity concept appears.

Lily Herzberg's paper on Reichenbach's lecture appears; first female contribution to the foundational debate.

Glivenko reacts to Barzin and Errera's paper; presents a formalisation of intuitionistic logic; proves Barzin and Errera wrong.

Heyting receives a prize of the *Wiskundig Genootschap* for his formalisation of intuitionism.

# 360  APPENDIX A. CHRONOLOGY OF THE DEBATE

|            |                                                                                                                                                                                                                                                             |
|---|---|
|               | Church takes up a pragmatic position, claiming that one can use a logic with or without the principle of the excluded middle. |
|               | The third edition of Fraenkel's *Einleitung in die Mengenlehre* appears. |
| March 10, 14  | Brouwer lectures in Vienna on the philosophical aspects of intuitionism; Wittgenstein (and possibly Gödel) is among the public, and starts philosophising again. |
| April         | Husserl lectures in Amsterdam, meets Brouwer. |
| September 3   | International Congress of Mathematicians in Bologna; Brouwer and Bieberbach call for a boycott, but Hilbert leads a substantial German delegation; Hilbert lectures on the problems in the foundations of mathematics. |
|               | Hilbert dismisses Brouwer from the editorial board of the *Mathematische Annalen*. |

## 1929

|   |   |
|---|---|
|              | Hilbert's *Probleme der Grundlegung der Mathematik* appears in the conference notes. |
|              | Study mocks at the slogans of the foundational debate. |
| Sept. 15–17  | First *Tagung für Erkenntnislehre der exakten Wissenschaften*, Prague; Fraenkel lectures on the foundational debate, points out similarities between Hilbert's and Brouwer's views. |

## 1930

|   |   |
|---|---|
|              | Special issue of the *Blätter für Deutsche Philosophie* on the philosophical foundations of mathematics; Carnap and Fraenkel suggest a synthesis between intuitionism, formalism and logicism. |
|              | Heyting's formalisation of intuitionistic logic and mathematics appears. |
|              | Hilbert's 1928 lecture appears, in abbreviated form, in the *Mathematische Annalen*. |
| September 5  | Von Neumann, Heyting and Carnap present formalism, intuitionism and logicism respectively at the Königsberg conference, followed by a general discussion; conciliatory tone; Gödel announces his first incompleteness theorem. |
| October      | Weyl lectures on the levels of infinity in Jena; if one takes mathematics by itself, intuitionism is right, but if one includes science, one should do mathematics in Hilbert's way. |
| December     | Hilbert lectures in Hamburg on *Die Grundlegung der elementaren Zahlenlehre*. |

## 1931

|   |   |
|---|---|
|   | – 1933: discussion between Barzin and Errera and Heyting on intuitionistic logic. |
|   | Gödel publishes his incompleteness theorems, inspired by Brouwer's Vienna lecture. |
|   | Special issue of *Erkenntnis* on the foundational crisis in mathematics, including reports from the Königsberg conference. |

**1932**

Brouwer lectures on *Willen, weten, spreken* in Amsterdam.
Kolmogorov publishes the problem interpretation of intuitionistic logic.
First radio lecture on the foundational debate, by Fraenkel.

June  Gödel presents a translation of classical into intuitionistic arithmetic at Menger's colloquium.

# Appendix B

# Public reactions to Brouwer's intuitionism

The list below contains all public reactions to Brouwer's intuitionism until 1933 known to me. Reviews have only been included in case the review mentioned Brouwer and/or intuitionism, but the work reviewed not, in order to avoid 'double counting'.

Reactions have been classified according to the year in which they were first made public. Thus, a lecture given in year $Y$ and published in year $Y + 1$ is listed below under year $Y$, but in the bibliography under $Y + 1$. In case the year in which the contribution was made public is not known exactly, an 'informed guess' has been made. Within each year, the authors have been ordered alphabetically.

Some titles have been abbreviated; for full bibliographical data, see the bibliography.

| Year | Author | Contribution |
|---|---|---|
| 1909 | Mannoury | Methodologisches und Philosophisches |
| 1914 | Bieberbach | Über die Grundlagen der modernen Mathematik |
|  | Frizell | A non-enumerable well-ordered set |
|  | Voss | Über die mathematische Erkenntnis |
| 1916 | Van Eeden | Een machtig Brouwsel |
|  | De Haan | Rechtskundige significa en hare toepassing |
|  | De Haan | Wezen en taak der rechtskundige significa |
|  | Korselt | Auflösung einiger Paradoxien |
| 1917 | Mannoury | Over de sociale betekenis van de wiskundige denkvorm |
| 1918 | Vollenhoven | Wijsbegeerte der wiskunde van theïstisch standpunt |
| 1919 | Bernstein | Die Mengenlehre Cantors und der Finitismus |
|  | Fraenkel | Einleitung in die Mengenlehre |

## 364  APPENDIX B. PUBLIC REACTIONS TO BROUWER'S INTUITIONISM

|      | De Haan | Rechtskundige significa |
|------|---------|-------------------------|
| 1920 | Weyl | Über die Grundlagen der Mathematik |
| 1921 | Bernays | Hilberts Gedanken zur Grundlegung der Arithmetik |
|      | Brodén | Verschiedene Gesichtspunkte bei Grundlegung der Analysis |
|      | Courant, Bernays | Neue arithmetische Theorien von Weyl und Brouwer |
|      | Fraenkel | Zermelosche Begründung der Mengenlehre |
|      | Hellinger | Weyls Untersuchungen zu den Grundlagen der Mathematik |
|      | Hilbert | Neubegründung der Mathematik |
|      | Koschmieder | Über den Brouwer-Weylschen Zahlbegriff |
|      | Ostrowski | Hilbert, Vorträgen über die Grundlagen der Mathematik |
|      | Reidemeister | Bericht über die Hamburger Vorträge von Hilbert |
|      | Reidemeister | Hilbert über die Grundlagen der Mathematik |
| 1922 | Boomstra | Beteekenis der meetkundige axioma's |
|      | Fraenkel | Grundlagen der Cantor-Zermeloschen Mengenlehre |
|      | Fraenkel | Probleme der Mengenlehre |
|      | Hilbert | Die logischen Grundlagen der Mathematik |
|      | Schoenflies | Zur Erinnerung an Georg Cantor |
|      | Wolff | Over het subjectieve in de wiskunde |
| 1923 | Baldus | Formalismus und Intuitionismus in der Mathematik |
|      | Becker | Phänomenologische Begründung der Geometrie |
|      | Behmann | Algebra der Logik und Entscheidungsproblem |
|      | Finsler | Gibt es Widersprüche in der Mathematik? |
|      | Fraenkel | Einleitung in die Mengenlehre |
|      | Fraenkel | Die neueren Ideeen zur Grundlegung der Analysis |
|      | Fraenkel | Die Axiome der Mengenlehre |
|      | Levi | Sui procedimenti transfiniti |
|      | London | Bedingungen der Möglichkeit einer deduktiven Theorie |
| 1924 | Doetsch | Der Sinn der reinen Mathematik und ihre Anwendung |
|      | Dresden | Brouwer's contributions to the foundations of mathematics |
|      | Fraenkel | Über die gegenwärtige Grundlagenkrise der Mathematik |
|      | Fraenkel | Die gegenwärtige Krise in den Grundlagen der Mathematik |
|      | Grelling | Mengenlehre |
|      | Hölder | Die mathematische Methode |
|      | Mannoury | Mathesis en mystiek |
|      | Von Neumann | Eine Axiomatisierung der Mengenlehre |
|      | Wavre | Y a-t-il une crise des mathématiques? |
|      | Weyl | Randbemerkungen zu Hauptprobleme der Mathematik |
| 1925 | Bieberbach | Die Entwicklung der nichteuklidischen Geometrie |
|      | Brodén | Einige Worte über aktuelle mathematische Prinzipfragen |
|      | Chwistek | review Chwistek, Zasady czystej teorji typów |
|      | Fraenkel | Der Streit um das Unendliche in der Mathematik |

|      | Fraenkel | Untersuchungen über die Grundlagen der Mengenlehre |
|------|----------|-----|
|      | Fraenkel | Zehn Vorlesungen über die Grundlegung der Mengenlehre |
|      | Gonseth | Sur la logique intuitioniste |
|      | Heyting | Intuitionistische axiomatiek der projectieve meetkunde |
|      | Hilbert | Über das Unendliche |
|      | Horák | Sur les antinomies de la théorie des ensembles |
|      | Колмогоров | О принципе Tertium non datur |
|      | Lietzmann | Formalismus und Intuitionismus in der Mathematik |
|      | Lipps | Bemerkungen zur Theorie der Prädikation |
|      | Petzoldt | Beseitigung der mengentheoretischen Paradoxa |
|      | Rivier | A propos du principe du tiers exclu |
|      | Skolem | Litt om de vigtigste diskussioner i den senere tid |
|      | Weyl | Die heutige Erkenntnislage in der Mathematik |
|      | Zariski | Gli sviluppi più recenti della teoria degli insiemi |
| 1926 | Barrau | De onbemindheid der wiskunde |
|      | Betsch | Fiktionen in der Mathematik |
|      | Bieberbach | Vom Wissenschaftsideal der Mathematiker |
|      | Dingler | Der Zusammenbruch der Wissenschaft |
|      | Dubislav | Über das Verhältnis der Logik zur Mathematik |
|      | Finsler | Über die Grundlegung der Mengenlehre |
|      | Gonseth | Les fondements des mathématiques |
|      | Hadamard | Préface |
|      | Hölder | Der angebliche circulus vitiosus |
|      | Kohnstamm | Schepper en schepping I |
|      | Larguier d. Bancels | Logique d'Aristote et le principe du tiers exclu |
|      | Lévy | Sur le principe du tiers exclu |
|      | Lévy | Critique de la logique empirique |
|      | Löwy | Die Krisis in der Mathematik |
|      | Menger | Bericht über die Dimensionstheorie |
|      | Otto | West-östliche Mystik: Vergleich und Unterscheidung |
|      | Ramsey | The foundations of mathematics |
|      | Ramsey | Mathematical logic |
|      | Scholz | review Boutroux, L'Idéal scientifique |
|      | Wäsche | Grundzüge zu einer Logik der Arithmetik |
|      | Wavre | Logique formelle et logique empirique |
|      | Wavre | Sur le principe du tiers exclu |
| 1927 | Ackermann | Was ist Mathematik? |
|      | Anonymous | Mathematische Schnaderhüpfl |
|      | Barzin & Errera | Sur la logique de M. Brouwer |
|      | Becker | Mathematische Existenz |
|      | Burkamp | Begriff und Beziehung |
|      | Burkamp | Die Krisis des Satzes vom ausgeschlossenen Dritten |
|      | Carnap | Eigentliche und uneigentliche Begriffe |
|      | Courant | Über die allgemeine Bedeutung des mathematischen Denken |

# 366 APPENDIX B. PUBLIC REACTIONS TO BROUWER'S INTUITIONISM

|      |      |      |
|------|------|------|
|      | Dresden | Some philosophical aspects of mathematics |
|      | Fraenkel | review Buchholz, Das Problem der Kontinuität |
|      | Fraenkel | review Skolem, Einige Bemerkungen |
|      | Härlen | Über Vollständigkeit und Entscheidbarkeit |
|      | Hartmann | Der Satz vom ausgeschlossenen Dritten in der Mathematik |
|      | Heidegger | Sein und Zeit |
|      | Heiss | Der Mechanismus der Paradoxien |
|      | Hilbert | Die Grundlagen der Mathematik |
|      | Landry | Grundlagenkrisis der Logik |
|      | Lévy | Logique classique, Logique brouwerienne et Logique mixte |
|      | Lietzmann | Aufbau und Grundlage der Mathematik |
|      | Mahnke | Leibniz als Begründer der symbolischen Mathematik |
|      | Von Neumann | Zur Hilbertschen Beweistheorie |
|      | Petzoldt | Rationales und empirisches Denken |
|      | Pierpont | Mathematical Rigor, Past and Present |
|      | Reymond | L'Axiomatique logique et le principe du tiers exclu |
|      | Weyl | Diskussionsbemerkungen zum zweiten Hilbertschen Vortrag |
|      | Weyl | Philosophie der Mathematik und Naturwissenschaft |
|      | Winternitz | Bemerkungen zu Brouwers intuitionistischer Mathematik |
| 1928 | Becker | Das Symbolische in der Mathematik |
|      | Church | On the law of the excluded middle |
|      | Doetsch | review Enriques, Zur Geschichte der Logik |
|      | Fraenkel | Einleitung in die Mengenlehre |
|      | Glivenko | Sur la logique de M. Brouwer |
|      | Gonseth | Mathematik und Erkenntnis |
|      | Grelling | Philosophy of the exact sciences: its present status |
|      | Hamel | Über die philosophische Stellung der Mathematik |
|      | Hardy | Mathematical proof |
|      | Herzberg, A. | Möglichkeitsfragen Satz vom ausgeschlossenen Dritten |
|      | Herzberg, L. | Der mathematische Grundlagenstreit und die Philosophie |
|      | Heyting | Intuitionistischen Axiomatik der projektiven Geometrie |
|      | Hilbert | Probleme der Grundlegung der Mathematik |
|      | Lipps | Untersuchungen zur Phänomenologie der Erkenntnis |
|      | Marcus | Der Triumph uber die Logik |
|      | Menger | Bemerkungen zu Grundlagenfragen |
|      | Menger | Bemerkungen zu Grundlagenfragen II. Die Paradoxien |
|      | Orlov | Über die Theorie der Verträglichkeit von Aussagen |
|      | Reidemeister | Exaktes Denken |
|      | Stammler | Begriff Urteil Schluss |
| 1929 | Barzin, Errera | Sur le principe du tiers exclu |
|      | Becker | Über den sogenannten 'Anthropologismus' |
|      | Belinfante | Intuitionistischen Theorie der unendlichen Reihen |
|      | Brand, Deutschbein | Einführung philosophischen Grundlagen der Mathematik |
|      | Cassirer | Philosophie der symbolischen Formen III. Phänomenologie |
|      | Deutschbein | Die philosophische Bildungswert der Mathematik |

| | | |
|---|---|---|
| | Dresden | Mathematical certainty |
| | Dubislav | Zur Lehre von den sog. schöpferischen Definitionen |
| | Dubislav | Zur Philosophie der Mathematik und Naturwissenschaft |
| | Fraenkel | Der Streit um die Grundlagen der Mathematik |
| | Fraenkel | review Pasch, Mathematik am Ursprung |
| | Fraenkel | Heutige Gegensätze in der Grundlegung der Mathematik |
| | Glivenko | Sur quelques points de la Logique de M. Brouwer |
| | Gödel | Über die Vollständigkeit des Logikkalküls |
| | Hahn | Die Bedeutung der wissenschaftliche Weltauffassung |
| | Heyting | De telbaarheidspraedicaten van prof. Brouwer |
| | Hirsch | Neuaufbau der Mathematik |
| | Leśniewski | Grundzüge neuen Systems der Grundlagen der Mathematik |
| | Nagel | Intuition, consistency, and the excluded middle |
| | Reichenbach | Die Weltanschauung der exakten Wissenschaften |
| | Schmeidler | Neue Grundlagenforschungen in der Mathematik |
| | Schmidt | Über Gewissheit in der Mathematik |
| | Skolem | Über die Grundlagendiskussionen in der Mathematik |
| | Study | Die angeblichen Antinomien der Mengenlehre |
| | Weiss | The nature of systems |
| | Weyl | Consistency in mathematics |
| 1930 | Becker | Zur Logik der Modalitäten |
| | Belinfante | Eine besondere Klasse von non-oszillierenden Reihen |
| | Bell | Debunking Science |
| | Bernays | Philosophie der Mathematik und Hilbertsche Beweistheorie |
| | Carnap | Die logizistische Grundlegung der Mathematik |
| | Carnap | Die Mathematik als Zweig der Logik |
| | Droste | De eenheid der wiskunde |
| | Dubislav | Über den sogenannten Gegenstand der Mathematik |
| | Fraenkel | Das Problem des Unendlichen in der neueren Mathematik |
| | Hahn et al. | Diskussion zur Grundlegung der Mathematik |
| | Haupt | Existenzbeweise in elementaren und höheren Mathematik |
| | Herbrand | Les bases de la logique hilbertienne |
| | Herbrand | Badania nad teorją dowodu |
| | Heyting | Sur la logique intuitionniste |
| | Heyting | Die formalen Regeln der intuitionistischen Logik |
| | Heyting | Die formalen Regeln der intuitionistischen Mathematik |
| | Heyting | Die formalen Regeln der intuitionistischen Mathematik |
| | Heyting | Die intuitionistische Grundlegung der Mathematik |
| | Hirsch | Philosophie der Mathematik |
| | Kaufmann | Das Unendliche in der Mathematik und seine Ausschaltung |
| | Kynast | Logik und Erkenntnistheorie der Gegenwart |
| | Łukasiewicz | Philosophische Bemerkungen zu mehrwertigen Systemen |
| | Menger | Der Intuitionismus |
| | Menger | Über die sogenannte Konstruktivität |
| | Von Neumann | Die formalistische Grundlegung der Mathematik |
| | Rivier | L'empirisme dans les sciences exactes |

|      |                  |                                                                      |
|------|------------------|----------------------------------------------------------------------|
|      | Scholz           | Die Axiomatik der Alten                                              |
|      | Weyl             | Die Stufen des Unendlichen                                           |
| 1931 | Barzin, Errera   | Sur la logique de M. Heyting                                         |
|      | Belinfante       | Hardy-Littlewoodsche Umkehrung des Stetigkeitssatzes                 |
|      | Belinfante       | Elemente der Funktionentheorie und Picardschen Sätze                 |
|      | Bell             | Mathematics and speculation                                          |
|      | Dingler          | Philosophie der Logik und Arithmetik                                 |
|      | Dubislav         | Die sog. Grundlagenkrise der Mathematik                              |
|      | Freudenthal      | Qualität und Quantität in der Mathematik                             |
|      | Herbrand         | Sur la non-contradiction de l'arithmétique                           |
|      | Heyting          | Die intuitionistische Mathematik                                     |
|      | Hofmann          | Das Problem des Satzes vom ausgeschlossenen Dritten                  |
|      | Jørgensen        | A treatise of formal logic                                           |
|      | Kaufmann         | Bemerkungen zum Grundlagenstreit in Logik und Mathematik             |
|      | Mannoury         | Woord en gedachte                                                    |
|      | Menger           | Über den Konstruktivitätsbegriff. Zweite Mitteilung                  |
|      | Scholz           | Geschichte der Logik                                                 |
|      | Scholz           | Über das Cogito, ergo sum                                            |
|      | Scholz           | review Bolzano, Wissenschaftslehre                                   |
|      | Weyl             | The open world                                                       |
| 1932 | Barzin, Errera   | Note sur la logique de M. Heyting                                    |
|      | Belinfante       | Den intuitionistischen Beweis der Picardischen Sätze                 |
|      | Bentley          | Linguistic analysis of mathematics                                   |
|      | Burkamp          | Logik                                                                |
|      | Carnap           | Über Hilbert                                                         |
|      | Church           | A set of postulates for the foundation of logic                      |
|      | Van Dantzig      | Over de elementen van het wiskundig denken                           |
|      | Dubislav         | Die Philosophie der Mathematik in der Gegenwart                      |
|      | Errera           | Sur le principe du tiers exclu                                       |
|      | Fraenkel         | Über das Unendliche                                                  |
|      | Gödel            | Zum intuitionistischen Aussagenkalkül                                |
|      | Hedrick          | Tendencies in the logic of mathematics                               |
|      | Heyting          | Anwendung der intuitionistischen Logik auf die Definition            |
|      | Heyting          | A propos d'un article de MM. Barzin et Errera                        |
|      | Hirsch           | Der Streit um das Unendliche                                         |
|      | Jørgensen        | Über die Ziele und Probleme der Logistik                             |
|      | Kolmogorov       | Zur Deutung der intuitionistischen Logik                             |
|      | Menger           | Die neue Logik                                                       |
|      | Reymond          | La fonction propositionnelle en logique algorithmique                |
|      | Reymond          | Les principes de la Logique et la critique contemporaine             |
|      | Scholz           | Augustinus und Descartes                                             |
|      | Vollenhoven      | De noodzakelijkheid eener christelijke logica                        |
|      | Weyl             | Zu David Hilberts siebzigstem Geburtstag                             |
|      | Zawirski         | Les logiques nouvelles et le champ de leur application               |
|      | Zilsel           | Bemerkungen zur Wissenschaftslogik                                   |

1933 | Ambrose | A controversy in the logic of mathematics
| Barzin, Errera | Réponse à M. Heyting
| Barzin, Errera | La logique de M. Brouwer
| Bell | Remarks on the Preceding Note on Many-Valued Truths
| Black | The nature of mathematics
| Dassen | Réflexions sur antinomies et sur la logique empiriste
| Errera | Remarques sur les mathématiques intuitionnistes
| Fraenkel | On modern problems in the foundations of mathematics
| Garcia | Assaigs moderns per a la fonamentació
| Gödel | Zur intuitionistischen Arithmetik und Zahlentheorie
| Gödel | Interpretation des intuitionistischen Aussagenkalküls
| Gödel | The present situation in the foundations of mathematics
| Heyting | Sur la logique intuitionniste
| Heyting | Réponse à MM. Barzin et Errera
| Hirsch | Intuition und logische Form
| Löwy | Commentar zu Jozef Poppers Abhandlung
| Van Os | Wiskunde en wijsbegeerte
| Scholz | Über die Ableitbarkeit der Mathematik aus der Logik
| Vredenduin | De autonomie der wiskunde

# Appendix C

# Logical notations

The main logical notations used in this book are the following:

| Connective | Meaning |
|---|---|
| ∧ | and (conjunction) |
| ∨ | or (disjunction) |
| ¬ | not (negation) |
| → | if ... then (implication) |
| ∀ | for all (universal quantifier) |
| ∃ | there exists (existential quantifier) |

Brackets () are used to indicate the binding of connectives. If there are no brackets, binding is ruled by the convention that ¬ and the quantifiers bind strongest, followed by ∧ and ∨, and → binds weakest.

The notation ⊢ indicates derivability (a syntactical property), ⊨ indicates truth (a semantical property).[1]

---

[1] For more information on logic and its notations, cf., for example, [Van Dalen 1997].

# Glossary

In the glossary below, some of the more technical terms from mathematics, logic and philosophy are explained.

**Analytic judgement** An analytic judgement is a judgement that is true on purely logical grounds, because the meaning is already implicit in the subject. For example: a square has four sides. Analytic judgements are opposed to synthetic judgements.

**Aristotelian logic** See classical logic.

**Axiom of choice** The axiom of choice states that, for any collection of non-empty sets $S$, there exists a set $Z$ such that, for each $S$, $Z$ contains one and only one element in common with $S$.

**Burali-Forti paradox** The Burali-Forti paradox is connected to the theory of transfinite numbers. Assume that $\Omega$ is the set of all ordinals. Since $\Omega$ is a well-ordered set, it has an ordinal number $o$, which by definition is bigger than all the ordinals in $\Omega$. However, since $\Omega$ is the set of all ordinals, $o$ has to be in $\Omega$ as well, whereby $o$ is bigger than itself.

**Cardinal number** A cardinal number is an object that is assigned to a set, which indicates its number of elements irrespective of its ordering. Two sets $A$ and $B$ have the same cardinal number is there exists a bijection from $A$ to $B$. Set $A$ has a smaller cardinal number than set $B$ if there exists a bijection between $A$ and a proper part of $B$ and is there is no bijection between $A$ and $B$. For instance, the sets $\overline{\mathbb{N}}$ and $\overline{\mathbb{N}}+1*$, which is the set $\overline{\mathbb{N}}$ with the extra element $1*$ added at the end, have the same cardinal number by the bijection $i : x \mapsto \begin{cases} 1* & \text{if } x = 1 \\ x - 1 & \text{otherwise.} \end{cases}$
Similarly, the set $\overline{\mathbb{N}}$ has a smaller cardinal number than the set $\overline{\mathbb{N}}$ by the bijection $i : x \mapsto x$, with the observation that a bijection between $\overline{\mathbb{N}}$ and $\overline{\mathbb{N}}$ does not exist.

**Categorical** An axiom system is called categorical if any two models of the axiom system are isomorphic. For example, Hilbert's axioms for Euclidian geometry are categorical in second-order logic, but not in first-order logic.

**Classical logic** Classical logic is the traditional Aristotelian logic, satisfying the principle of the excluded middle. Classical logic was the dominant logic since Antiquity.

**Complete** An axiom system $\Gamma$ is called complete if anything that is provable in $\Gamma$ is true and vice versa. In symbolic form: $\Gamma \vdash \varphi \Leftrightarrow \Gamma \models \varphi$ for all propositions $\varphi$.

**Consistent** An axiom system is called consistent if it is not possible to derive a contradiction from the axioms.

**Constructive** A constructive existence proof of a sentence $\exists(x)\varphi(x)$ consists of the formulation of an algorithm to construct an object $a$ for which $\varphi(a)$ holds demonstrably.

**Decidable** A theory $T$ is called decidable if there exists an algorithm that, for each proposition $\varphi$, checks if $\varphi$ can be derived from $T$ (in other words: if $T \vdash \varphi$). For example, classical propositional logic is decidable by means of truth tables and the completeness theorem.

**Epistemology** Epistemology is the study of knowledge and the justification of belief. Basic epistemological questions are: What can we know? What is the difference between knowing and having a true belief?

**Formalism** Formalism is a current in the foundations of mathematics. In the formalistic view, mathematics consists of symbols which are manipulated according to certain rules. The only criterion which an axiom system has to fulfil is that of consistency. The consistency proof is given on the so-called meta-mathematical level, in which contentual reasonings are used. Hilbert is generally seen as the leading representative of formalism.

**Idealism** Idealism comes in two forms: an epistemological and a metaphysical version. The former originates from Kant and holds that all conceivable propositions deal with human experiences. In other words: there is no evidence-transcendent truth. The latter evolved out of the former by modifications which philosophers like Fichte and Hegel introduced to Kant's position. Metaphysical idealism considers the mind to be the only bearer of reality.

**Impredicative definition** An impredicative definition is a definition in which the entity to be defined is defined in terms of the collection of all the elements of a certain kind, of which the entity to be defined itself is one. In other words: an entity $E$ is defined by means of a collection $C$, to which $E$ belongs. The definition of the ordinal number of the set of all ordinals, used in the Burali-Forti paradox, is an example of an impredicative definition.

**Indirect definition** An indirect definition is a definition of a mathematical object by means of a set of postulates or axioms in which the object to be defined appears. The most well-known example is Hilbert's *Grundlagen der Geometrie*, in which Hilbert considered the concepts 'point', 'line' and 'plane' to be defined indirectly.

**Intuitionism** Intuitionism is a current in the foundations of mathematics. In the intuitionistic view, mathematics consists of mental constructions. This has several implications for mathematical practice. On the one hand, intuitionism allows for 'private' mathematical objects such as choice sequences, which are constructed by an individual mathematician. On the other hand, intuitionism rejects pure existence proofs and the use of the principle of the excluded middle in infinite totalities. Intuitionism was developed in the beginning of the 20th century by L.E.J. Brouwer.

**Law of the excluded middle** See principle of the excluded middle.

***Lebensphilosophie*** *Lebensphilosophie* ('life philosophy') is a philosophical current which takes life as the basis for philosophy. It is not possible to describe *Lebensphilosophie* in a system, since it refuses to be reduced to a fixed system, which it considers to be dead. Bergson and Spengler are well-known life philosophers. *Lebensphilosophie* was especially popular in Weimar Germany.

**Logical Positivism** Logical positivism is a philosophy that can be characterised by the following principles. The only genuine propositions, i.e., propositions that are meaningful in the sense that they are strictly true or false about the world, are those that are verifiable by the methods of science. Mathematical and logical propositions are also meaningful, but their truth is discovered by analysis and not by observation.

**Logicism** Logicism is a current in the foundations of mathematics. In the logicist view, mathematics consists of abbreviated logical reasonings. Logicism was developed by Frege and Russell.

**Meta-mathematics** See under formalism.

**Non-predicative definition** See impredicative definition.

**Ontology** Ontology is the logical-conceptual study of being as such. In ontology, one looks for the most general statements, which are valid for everything that exists. Ontology can also be seen as part of metaphysics; understood in this way, the focus of ontological research is on the meaning(s) of being. Basic ontological questions are: do objects exist, and if so in what way?

**Ordered set** A set $(S, \leq)$ is called (totally) ordered if for all $x, y, z \in S$ the following properties hold:

1. $x \leq y \leq z \Rightarrow x \leq z$;
2. $x \leq y \leq x \Leftrightarrow x = y$;
3. $x \leq y \vee y \leq x$.

**Ordinal number** An ordinal number is a number assigned to a well-ordered set that indicates its number of elements with respect to the ordering. Two well-ordered sets $A$ and $B$ have the same ordinal number if there is a bijection from $A$ to $B$ that respects the ordering. If there is a bijection between $A$ and a proper initial segment of $B$ respecting the ordering, then the ordinal number of $A$ is less than the ordinal number of $B$. For example, the ordinal numbers of $^-$ and $\{2n|n \in {}^-\}$ are the same, by the order-respecting bijection $i : x \mapsto 2x$. Equally, the ordinal number of $^-$ is less than that of $^-+1*$, which is the set $^-$ with the extra element $1*$ added at the end, by the order-respecting bijection $i : x \mapsto x$.

**Phenomenology** Phenomenology or, more specifically, transcendental phenomenology is a current in anti-realist philosophy which originated from Husserl. Phenomenology strives to arrive at pure consciousness by a procedure of transcendental reduction. This involves a reflection on consciousness, including the suspending of all beliefs regarding the real existence of objects. Objects are only taken as intentional objects, i.e., as objects of consciousness. Intentionality thus becomes the medium in which what counts as real is constituted. Transcendental phenomenology, one could say, is the description of the essential structures of the constitution of the world in transcendental subjectivity.[2]

**Principle of the excluded middle** The principle of the excluded middle is the logical principle according to which any proposition is either true or false, i.e., either the proposition itself or its negation has to hold. In symbolic form: $\varphi \vee \neg \varphi$ holds for any

---

[2][Brown, Collinson & Wilkinson 1996, pp. 890-891]

proposition $\varphi$. Once the principle of the excluded middle is accepted, it becomes possible to prove theorems by using *reductio ad absurdum*.

**Proof by contradiction** See *reductio ad absurdum*.

**Pure existence theorem** A pure existence theorem (German: *reine Existenzsatz*) is an existence theorem in which the existence is proved in a non-constructive way, i.e., no method is given in the proof by which the mathematical object can be constructed. Typically, this will be a proof using *reductio ad absurdum*. For example, the theorem 'there exist irrational numbers $a$ and $b$ such that $a^b$ is rational' is a non-constructive existence theorem, since it is proved in the following way. Suppose the theorem is not true, that is, suppose that for all irrational numbers $a$ and $b$, $a^b$ is also irrational. Take $a = b = \sqrt{2}$. Then, following the assumption, $\sqrt{2}^{\sqrt{2}}$ must be irrational as well. Again applying the assumption, so must $\sqrt{2}^{\sqrt{2}^{\sqrt{2}}}$. But $\sqrt{2}^{\sqrt{2}^{\sqrt{2}}} = \sqrt{2}^{\sqrt{2}\cdot\sqrt{2}} = \sqrt{2}^2 = 2$ and thus rational, contradiction. Thus, by reduction ad absurdum, there exist irrational numbers $a$ and $b$ such that $a^b$ is rational. However, the proof gives no answer to the question which irrational numbers $a$ and $b$ fulfil the demand that $a^b$ is rational.

**Realism** Realism is a current in metaphysics which originates from Plato. For a realist, there exists an outer reality which in its existence is independent of our experience of it. Thus, for a realist it is possible that propositions are true, even though we do not and maybe even cannot recognize so.

***Reductio ad absurdum*** A proof by *reductio ad absurdum* is a proof in which a proposition is proved to be true by assuming that it is false and deriving a contradiction. In symbolic form: $(\neg\varphi \to \bot) \Rightarrow \varphi$. For example, one can prove by *reductio ad absurdum* that there is an infinite number of prime numbers in the following way. Assume that the statement is false, i.e., there is a finite number of prime numbers, say $n$. Denote the primes by $p_1, \ldots, p_n$. Then $(p_1 \cdot \ldots \cdot p_n) + 1$ is not divisible by $p_i$ for any $i \in \{1, \ldots, n\}$. Hence, we have constructed a new prime number, in contradiction with the assumption that the number of primes was $n$. Therefore, by *reductio ad absurdum*, there is an infinite number of prime numbers.

**Synthetical judgement** A synthetical judgement is a judgement that is obtained by a synthesis between a subject and a predicate, i.e., the judgement gets its meaning from non-logical sources as experience. Synthetical judgements are opposed to analytic judgements.

**Tertium non datur** See principle of the excluded middle.

**Vienna Circle** See *Wiener Kreis*.

**Well-ordered set** A set $S$ is called well-ordered if it is ordered (see: Ordered set) and if every non-empty subset of $S$ has a first element. The set of natural numbers is an example of a well-ordered set.

***Wiener Kreis*** The *Wiener Kreis* ('Vienna Circle') was a group of philosophically interested academics, who started meeting in Vienna in 1924 under the direction of Schlick. Logical positivism was the philosophy that bound them together. Some of its more well-known members were Hahn, Feigl, Carnap, and Reidemeister. Gödel attended meetings of the Circle but was never a member. Wittgenstein only met with some of the individual members of the *Wiener Kreis*.

# Bibliography

**Archival sources:**

[AC Church] Nachlaß Alonzo Church, private archive of Alonzo Church jr., Hudson, USA

[ASP Carnap] Nachlaß Rudolf Carnap, Archives of Scientific Philosophy, Pittsburgh; copy at the Philosophisches Archiv, Universität Konstanz, Konstanz, Deutschland

[ASP Feigl] Nachlaß Herbert Feigl, Archives of Scientific Philosophy, Pittsburgh; copy at the Philosophisches Archiv, Universität Konstanz, Konstanz, Deutschland

[BAP Nelson] Nachlaß Leonard Nelson, Bundesarchiv Abteilung Potsdam, Potsdam, Deutschland

[BCUL Gonseth] Fonds Ferdinand Gonseth, Bibliothèque Cantonale et Universitaire de Lausanne, Lausanne, Suisse

[BF Fraenkel] Nachlaß Abraham A. Fraenkel, private archive of Benjamin Fraenkel, Jerusalem, Israel

[BMIG Study] Korrespondenz Study–Engel, Bibliothek des mathematischen Instituts der Justus–Liebig–Universität Giessen, Deutschland

[BMIM Study] Manuskripte Study, Bibliothek des mathematischen Instituts der Westfälische Wilhelms-Universität Münster, Deutschland

[ETH Bernays] Nachlaß Paul Bernays, Eidgenössische Technische Hochschule–Bibliothek, Wissenschaftshistorische Sammlungen, Zürich, Schweiz

[ETH Weyl] Nachlaß Hermann Weyl, Eidgenössische Technische Hochschule-Bibliothek, Wissenschaftshistorische Sammlungen, Zürich, Schweiz

[IMLG Scholz] Nachlaß Heinrich Scholz, Institut für mathematische Logik und Grundlagenforschung, Westfälische Wilhelms-Universität, Münster, Deutschland

[JNUL Einstein] Nachlaß Albert Einstein, The Jewish National and University Library, Department of Manuscripts and Archives, Jerusalem, Israel

[JNUL Fraenkel] Nachlaß Abraham A. Fraenkel, The Jewish National and University Library, Department of Manuscripts and Archives, Jerusalem, Israel

[MI Brouwer] Brouwer Archive, Mathematisch Instituut, Universiteit Utrecht, Nederland

[NSUB Hilbert] Nachlaß David Hilbert, Niedersächsische Staats- und Universitätsbibliothek Göttingen, Abteilung für Handschriften und seltene Drucken, Göttingen, Deutschland

[NSUB Klein] Nachlaß Felix Klein, Niedersächsische Staats- und Universitätsbibliothek Göttingen, Abteilung für Handschriften und seltene Drucken, Göttingen, Deutschland

[TLI Heyting] Nachlaß Heyting, Instituut voor Taal, Logica en Informatie, Universiteit van Amsterdam, Nederland

[ULB Errera] Errera archive, Archives de l'Université Libre de Bruxelles, Belgique

**Other sources:**

[D'Abro 1939] D'Abro, A., *Decline of mechanism*, Van Nostrand, 1939; cited from the unabridged replication as *The rise of the new physics*, vol. 1, Dover, New York 1952

[Ackermann 1927] Ackermann, W., 'Was ist Mathematik?', *Zeitschrift für mathematischen und naturwissenschaftlichen Unterricht aller Schulgattungen* 58 (1927), pp. 449–455

[Ajdukiewicz 1935] Ajdukiewicz, K., 'Der logistische Antiirrationalismus in Polen', *Erkenntnis* 5 (1935), pp. 151-161; lecture delivered at the *Prager Vorkonferenz des Ersten Internationalen Kongresses für Einheit der Wissenschaft* on August 31, 1934

[Ambrose 1933] Ambrose, A., 'A Controversy in the logic of mathematics', *The philosophical review* 42 (1933), pp. 594–611

[Anonymous 1933] Anonymous, 'Bericht über die gegenwärtigen Verhältnisse an Deutschen Universitäten, soweit Mathematik und theoretische Physik dabei betroffen sind', cited from: [Siegmund-Schultze 1998, pp. 308–311]; note to the Rockefeller Foundation, written before June 26, 1933; the author may be Weyl or Courant

[Artzy 1972] Artzy, R., 'Kurt Reidemeister', *Jahresbericht der deutschen Mathematiker-Vereinigung* 74 (1972), pp. 96–104

[Aspray 1981] Aspray, Jr., W.F., *From mathematical constructivity to computer science: Alan Turing, John von Neumann, and the origins of computer science in mathematical logic*, Ann Arbor, Michigan etc. 1981; dissertation

[Aspray & Kitcher 1988] Aspray, W.F., Kitcher, P., (ed.) *History and Philosophy of Modern Mathematics*, University of Minneapolis Press, Minneapolis 1988

[Van Atten 1999] Atten, M.S.P.R. van, *Phenomenology of choice sequences*, Zeno, Utrecht 1999; dissertation

[Auben 1997] Auben, D., 'The Withering Immortality of Nicolas Bourbaki: A Cultural Connector at the Confluence of Mathematics, Structuralism, and the Oulipo in France', *Science in Context* 10 (1997), pp. 297-342

[Aull & Lowen 1997] Aull, C.E., Lowen, R. (eds.), *Handbook of the History of General Topology*, volume 1, Kluwer, the Netherlands 1997

[Bachmann, Behnke & Franz 1972] Bachmann, F., Behnke, H., Franz, W., 'In memoriam Kurt Reidemeister', *Mathematische Annalen* 199 (1972), pp. 1-11

[Baire et al. 1905] Baire, R., Borel, E., Hadamard, J., Lebesgue, H., 'Cinq lettres sur la theorie des ensembles', *Bulletin de la Société mathématique de France* 33 (1905), pp. 261-273; English translation in [Moore 1982, pp. 311-320] and [Ewald 1996, vol. II, pp. 1077-1086]

[Baldus 1924] Baldus, R., *Formalismus und Intuitionismus in der Mathematik*, Braun, Karlsruhe i.B. 1924; rector's address delivered at the *Technische Hochschule Karlsruhe* on December 1, 1923

[Barrau 1926] Barrau, J., *De onbemindheid der wiskunde*, Wolters, Groningen etc. 1926; lecture at the assignment of the rectorship of the *Rijks-universiteit te Groningen*, delivered on September 20, 1926

[Barzin & Errera 1927] Barzin, M., Errera, A., 'Sur la logique de M. Brouwer', *Académie Royale de Belgique, Bulletins de la Classe des Sciences* 5-13 (1927), pp. 56-71

[Barzin & Errera 1929] Barzin, M., Errera, A., 'Sur le principe du tiers exclu', *Archives de la Société Belge de Philosophie* 1 (1928/29), pp. 3-26

[Barzin & Errera 1931] Barzin, M., Errera, A., 'Sur la logique de M. Heyting', *L'Enseignement mathématique* 30 (1931), pp. 248-250

[Barzin & Errera 1932A] Barzin, M., Errera, A., 'Note sur la logique de M. Heyting', *L'Enseignement mathématique* 31 (1932), pp. 122-124

[Barzin & Errera 1932B] Barzin, M., Errera, A., 'Réponse à M. Heyting', *L'Enseignement mathématique* 31 (1932), pp. 273-274; letter dated 18/5/1933

[Barzin & Errera 1933] Barzin, M., Errera, A., 'La logique de M. Brouwer, état de la question', *Bulletin Mathématique de la Société Roumaine des Sciences* 35 (1933), pp. 51-52

[Bayer 1951] Bayer, R. (ed.), *Philosophie XVI. Congrès International de Philosophie des Sciences*, Paris, 1951

[Becker, H. 1987] Becker, H., 'Aufstellung der Professoren, Lehrbeauftragten und Nachwuchswissenschaftler, die infolge der nationalsozialistischen Maßnahmen die Universität Göttingen verlassen mußten', in: [Becker, H., et al. 1987, pp. 489-499]

[Becker, H., et al. 1987] Becker, H., Dahms, H.-J., Wegeler, C., *Die Universität Göttingen unter dem Nationalsozialismus. Das verdrängte Kapitel ihrer 250jährigen Geschichte*, Saur, München etc. 1987

[Becker, O. 1923] Becker, O., 'Beiträge zur phänomenologischen Begründung der Geometrie und ihrer physikalischen Anwendungen', *Jahrbuch für Philosophie und phänomenologische Forschung* 6 (1923), pp. 385–560; Habilitationsschrift Freiburg, January 1922

[Becker, O. 1927] Becker, O., 'Mathematische Existenz. Untersuchungen zur Logik und Ontologie mathematischer Phänomene', *Jahrbuch für Philosophie und phänomenologische Forschung* 8 (1927), pp. 439–809

[Becker, O. 1927/28] Becker, O. , 'Das Symbolische in der Mathematik', *Blätter für Deutsche Philosophie* 1 (1927/28), pp. 329–348

[Becker, O. 1929] Becker, O., 'Über den sogenannten 'Anthropologismus' in der Philosophie der Mathematik', *Philosophischer Anzeiger* 3 (1928–29), pp. 369–387

[Becker, O. 1930] Becker, O., 'Zur Logik der Modalitäten', *Jahrbuch für Philosophie und phänomenologische Forschung* 11 (1930), pp. 497–548

[Becker, O. 1964] Becker, O., *Grundlagen der Mathematik*, Karl Alber, Freiburg/ München 1964; cited from the 4th. edition (identical with the 2nd.), *Grundlagen der Mathematik in geschichtlicher Entwicklung*, Suhrkamp, Frankfurt am Main 1990

[Behmann 1924] Behmann, H., 'Algebra der Logik und Entscheidungsproblem', *Jahresbericht der deutschen Mathematiker-Vereinigung* 32 (1924), pp. 66–67; report of a lecture delivered at the *Jahresversammlung* at Marburg in September 1923

[Behnke 1978] Behnke, H., *Semesterberichte. Ein Leben an deutschen Universitäten im Wandel der Zeit*, Vandenhoeck & Ruprecht, Göttingen 1978

[Belinfante 1929] Belinfante, H., 'Zur intuitionistischen Theorie der unendlichen Reihen', *Sitzungsberichte der Preussischen Akademie der Wissenschaften, physikalisch-mathematische Klasse* 29 (1929), pp. 639–660

[Belinfante 1930] Belinfante, H., 'Über eine besondere Klasse von non-oszillierenden Reihen', *Verhandelingen der Koninklijke Akademie van Wetenschappen, Afdeeling Natuurkunde, Sectie 1* 33 (1930), pp. 1170–1179

[Belinfante 1931A] Belinfante, H., 'Die Hardy-Littlewoodsche Umkehrung des Abelschen Stetigkeitssatzes in der intuitionistischen Mathematik', *Verhandelingen der Koninklijke Akademie van Wetenschappen, Afdeeling Natuurkunde, Sectie 1* 34 (1931), pp. 401–412

[Belinfante 1931B] Belinfante, H., 'Über die Elemente der Funktionentheorie und die Picardschen Sätze in der intuitionistischen Mathematik', *Verhandelingen der Koninklijke Akademie van Wetenschappen, Afdeeling Natuurkunde, Sectie 1* 34 (1931), pp. 1395–1397

[Belinfante 1932] Belinfante, H., 'Über den intuitionistischen Beweis der Picardischen Sätze', in: [Saxer 1932, pp. 345–346]

[Bell 1930] Bell, E.T., *Debunking Science*, University of Washington Book Store, Seattle 1930

[Bell 1931] Bell, E.T., 'Mathematics and speculation', *The Scientific Monthly* 32 (1931), pp. 193–209

[Bell 1933] Bell, E.T., 'Remarks on the Preceding Note on Many-Valued Truths', *The physical review* (2) 43 (1933), p. 1033

[Benacerraf & Putnam 1964] Benacerraf, P., Putnam, H. (eds.), *Philosophy of mathematics. Selected readings*, Basil Blackwell, Oxford 1964

[Benacerraf & Putnam 1983] Benacerraf, P., Putnam, H. (eds.), *Philosophy of mathematics. Selected readings*, Cambridge University Press, Cambridge etc. 1983; second edition

[Benda 1914] Benda, J., *Sur le Succès du Bergsonisme*, Mercure de France, Paris 1914; published originally in *Mercure de France*, numéros 1er et 16 juillet 1913 (the first part), and *Cahiers de la quinzaine* (2e cahier de la 15e serie) under the title 'Une philosophie pathétique' (the second part)

[Bentley 1932] Bentley, A.F., *Linguistic analysis of mathematics*, The Principia Press, Bloomington, Indiana, 1932

[Benz 1983] Benz, W., *Das Mathematische Seminar der Universität Hamburg in seinen ersten Jahrzehnten, Jahrbuch Überblick Mathematik 1983*, Bibliographisches Institut Mannheim, pp. 191–202; cited from: [Blaschke 1982–1986, Volume 6, pp. 275–283];

[Bernays 1922A] Bernays, P., 'Über Hilberts Gedanken zur Grundlegung der Arithmetik', *Jahresbericht der deutschen Mathematiker-Vereinigung* 31 (1922), pp. 10–19; English translation in: [Mancosu 1998, pp. 215–222]; lecture delivered at the *Mathematikertagung* at Jena, September 1921

[Bernays 1922B] Bernays, P., 'Die Bedeutung Hilberts für die Philosophie der Mathematik', *Die Naturwissenschaften* 10 Heft 4 (Sonderheft 'David Hilbert zur Feier seines sechzigsten Geburtstages') (1922), pp. 93–99; English translation in: [Mancosu 1998, pp. 189–197]

[Bernays 1923] Bernays, P., 'Erwiderung auf die Note von Herrn Aloys Müller: 'Über Zahlen als Zeichen' ', *Mathematische Annalen* 90 (1923), pp. 159–163; English translation in: [Mancosu 1998, pp. 223–226]

[Bernays 1926] Bernays, P., 'Axiomatische Untersuchung des Aussagen-Kalkuls der 'Principia Mathematica' ', *Mathematische Zeitschrift* 25 (1926), pp. 305–320

[Bernays 1927A] Bernays, P., 'Zusatz zu Hilberts Vortrag über 'Die Grundlagen der Mathematik' ', *Abhandlungen aus dem mathematischen Seminar der Hamburgischen Universität* V (1927), Leipzig 1926 [sic], pp. 89–92

[Bernays 1927B] Bernays, P., 'Probleme der theoretischen Logik', *Unterrichtsblätter für Mathematik und Naturwissenschaften* 33 (1927), pp. 369–377; cited from: [Bernays 1976, pp. 1–16]

[Bernays 1928] Bernays, P., 'Die Grundbegriffe der reinen Geometrie in ihrem Verhältnis zur Anschauung', *Die Naturwissenschaften* 12 (1928), pp. 197–203

[Bernays 1930] Bernays, P., 'Die Philosophie der Mathematik und die Hilbertsche Beweistheorie', *Blätter für Deutsche Philosophie* 4 (1930/31), pp. 326–367; English translation in: [Mancosu 1998, pp. 234–265]

[Bernays 1935] Bernays, P., 'Sur le platonisme dans les mathématiques', *L'Enseignement mathématique* 34 (1935), pp. 52–69

[Bernays 1954] Bernays, P., 'Zur Beurteilung der Situation in der beweistheoretischen Forschung', *Revue Internationale de Philosophie* VIII (1954), pp. 9–13

[Bernays 1971] Bernays, P., 'Zum Symposium über die Grundlagen der Mathematik', *Dialectica* 25 (1971), pp. 171–195

[Bernays 1976] Bernays, P., *Abhandlungen zur Philosophie der Mathematik*, Wissenschaftliche Buchgesellschaft, Darmstadt 1976

[Bernstein 1919] Bernstein, F., 'Die Mengenlehre Georg Cantors und der Finitismus', *Jahresbericht der deutschen Mathematiker-Vereinigung* 28 (1919), pp. 63–78

[Bertin, Bos & Grootendorst 1978] Bertin, E.M.J., Bos, H.J.M., Grootendorst, A.W., *Two decades of mathematics in the Netherlands · 1920–1940. A retrospection on the occasion of the bicentennial of the* Wiskundig Genootschap, Mathematical Centre, Amsterdam 1978; two volumes

[Beth 1935] Beth, E.W., *Rede en aanschouwing in de wiskunde*, Noordhoff, Groningen–Batavia 1935; dissertation

[Beth 1944] Beth, E.W., *Geschiedenis der Logica*, Servire, Den Haag 1944

[Betsch 1926] Betsch, Chr., *Fiktionen in der Mathematik*, Frommann, Stuttgart 1926

[Bieberbach 1914] Bieberbach, L., 'Über die Grundlagen der modernen Mathematik', *Die Geisteswissenschaften* 33 (1914), pp. 896–901

[Bieberbach 1925] Bieberbach, L., 'Über die Entwicklung der nichteuklidischen Geometrie im 19. Jahrhundert', *Sitzungsberichte der Preussischen Akademie der Wissenschaften* (1925), pp. 381–398

[Bieberbach 1926] Bieberbach, L., 'Vom Wissenschaftsideal der Mathematiker', lecture delivered on February 15, 1926

[Bieberbach 1934A] Bieberbach, L., 'Persönlichkeitsstruktur und mathematisches Schaffen', *Unterrichtsblätter für Mathematik und Naturwissenschaften* 40 (1934), pp. 236–243; lecture delivered at the *Hauptversammlung*

[Bieberbach 1934B] Bieberbach, L., 'Die Kunst des Zitierens. Ein offener Brief an Herrn Harald Bohr in København.', *Jahresbericht der deutschen Mathematiker-Vereinigung* 44 (1934), pp. 1–3

[Bieberbach 1934C] Bieberbach, L., 'Stilarten mathematischen Schaffens', *Sitzungsberichte der Preussischen Akademie der Wissenschaften, physikalisch–mathematische Klasse* 1934, pp. 351–360

[Biermann 1973] Biermann, K.-R., 'Leopold Kronecker', in: [Gillispie 1970–1978, Vol. VII, pp. 505–509]

[Biermann 1976] Biermann, K.-R., 'Karl Theodor Wilhelm Weierstrass', in: [Gillispie 1970–1978, Vol. XIV, pp. 219–224]

[Biermann 1988] Biermann, K.-R., *Die Mathematik und ihre Dozenten an der Berliner Universität, 1810–1933*, Akademie-Verlag, Berlin 1988

[Bishop 1967] Bishop, E., *Foundations of constructive analysis*, McGraw-Hill, New York etc. 1967

[Black 1933] Black, M., *The Nature of Mathematics*, 1933; cited from the 1950 reprint, The Humanities Press, New York

[Blaschke 1982–1986] Blaschke, W., *Gesammelte Werke*, Burau et al. (ed.), Thales Verlag, Essen 1982–1986, 6 volumes

[Blumenthal 1935] Blumenthal, O., 'Lebensgeschichte', in: [Hilbert 1932–1935, vol. 3, pp. 388–429]

[Bockstaele 1949] Bockstaele, P., *Het intuïtionisme bij de Franse wiskundigen*, Verhandelingen van de Koninklijke Vlaamse Academie voor wetenschappen, letteren en schone kunsten van België, Klasse der wetenschappen, Jaargang XI, No. 32, Brussel 1949

[Bolzano 1930/31] Bolzano, B., *Wissenschaftslehre*, Schultz, W. (ed.), vol. III and IV Felix Meiner, Leipzig 1930/31; 2nd., improved edition

[Boole 1854] Boole, G., *An investigation of the laws of thought, on which are founded the mathematical theories of logic and probabilities*, London 1854; cited from a later (undated) Dover Publications edition

[Boomstra 1922] Boomstra, W., *De beteekenis der meetkundige axioma's*, Maks & V.d. Klits, Bandoeng; lecture delivered on July 1, 1922, at Bandoeng

[Borel 1898] Borel, E., *Leçons sur la théorie des fonctions*, Paris, Gauthier-Villars et fils, 1898

[Borel 1905] Borel, E., 'Quelques remarques sur les principes de la théorie des ensembles', *Mathematische Annalen* 60 (1905), pp. 194–195; Engslish translation in [Ewald 1996, vol. II, pp. 1076–1077]

[Borel 1907] Borel, E., 'La logique et l'intuition en mathématiques', *Revue de Métaphysique et de Morale* 15 (1907), pp. 273–283

[Borel 1912] Borel, E., 'La philosophie mathématique et l'infini', *La Revue de Mois* 14 (1912), pp. 218–227; cited from: [Borel 1967, pp. 307–316]

[Borel 1914] Borel, E., *Leçons sur la théorie des fonctions*, Gauthier-Villars, Paris 1914; 2. édition

[Borel 1926] Borel, E., 'Correspondance à propos de la récente discussion sur les nombres irrationnels entre M. R. Wavre en M. P. Lévy', *Revue de Métaphysique et de Morale* 34 (1926), pp. 271–276; English translation in: [Mancosu 1998, pp. 296–300]

[Borel 1928] Borel, E., *Leçons sur la théorie des fonctions*, Paris 1928, 3. édition

[Borel 1967] Borel, E., *Philosophe et homme d'action*, Fréchet, M. (ed.), Gauthier-Villars, Paris 1967

[Borwein 1998] Borwein, J.M., 'Brouwer-Heyting Sequences Converge', *The Mathematical Intellligencer* 20 (1998), pp. 14–15

[Bos 1987] Bos, H.J.M., *Vanuit herkenning en verbazing*, inaugural lecture, Rijksuniversiteit Utrecht, Utrecht 1987; also in: *Euclides* 63 (1987), pp. 65–76; English translation: 'Recognition and wonder; Huygens, tractional motion and some thoughts on history of mathematics', *Tractrix, yearbook for the history of science, medicine, technology and mathematics* 1 (1989), pp. 3–20

[Boutroux 1920] Boutroux, P., *L'Idéal scientifique des Mathématiciens dans l'antiquité et dans les temps modernes*, Félix Alcan & Cie, Paris 1920

[Boyer & Merzbach 1989] Boyer, C.B., Merzbach, U.C., *A history of mathematics*, Wiley, New York etc. 1989; second, revised edition

[Braithwaite 1931] Braithwaite, R.B., 'Editor's introduction', in: [Ramsey 1931, pp. ix–xiv]

[Brand & Deutschbein 1929] Brand, W., Deutschbein, M., *Einführung in die philosophischen Grundlagen der Mathematik*, Moritz Diesterweg, Frankfurt am Main 1929

[Brodén 1921] Brodén, T., 'Über verschiedene Gesichtspunkte bei der Grundlegung der mathematischen Analysis', *Lunds Universitets Årsskrift* N.F. Avd. 2, Bd. 17, Nr. 7 (1921), 15 p.

[Brodén 1925] Brodén, T., 'Einige Worte über aktuelle mathematische Prinzipfragen', *Skandinaviska matematikerkongressen* 6 (1925), pp. 229–239; lecture delivered at the København conference, 31/8–4/9 1925

[Brouwer 1905] Brouwer, L.E.J.,[3] *Leven, Kunst en Mystiek*, J. Waltman Jr., Delft 1905; English translation in [Brouwer 1996]; lectures delivered in Delft in the first half of 1905

---

[3]The coding of Brouwer's publications follows the one used in the Brouwer biography [Van Dalen 1999A].

[Brouwer 1907] Brouwer, L.E.J., *Over de grondslagen der wiskunde*, Amsterdam 1907; commercial edition at Maas & Van Suchtelen, Amsterdam etc. 1907; cited from: [Van Dalen 1981]; dissertation

[Brouwer 1908] Brouwer, L.E.J., 'De onbetrouwbaarheid der logische principes', *Tijdschrift voor wijsbegeerte* 2 (1908), pp. 152–158; cited from: [Van Dalen 1981, pp. 253–259]

[Brouwer 1909] Brouwer, L.E.J., 'Het wezen der meetkunde', 23 p.; cited from: [Brouwer 1919B]; inaugural lecture as *privaatdocent*, delivered on October 12, 1909

[Brouwer 1910C] Brouwer, L.E.J., 'Zur Analysis Situs', *Mathematische Annalen* 68 (1910), pp. 422–434; cited from: [Brouwer 1975–1976, vol. 2, pp. 352–366]

[Brouwer 1910E] Brouwer, L.E.J., 'Beweis des Jordanschen Kurvensatzes', *Mathematische Annalen* 69 (1910), pp. 169–175; cited from: [Brouwer 1975–1976, vol. 2, pp. 377–383]

[Brouwer 1910F] Brouwer, L.E.J., 'Über eineindeutige, stetige Transformationen von Flächen in sich', *Mathematische Annalen* 69 (1910), pp. 176–180; cited from: [Brouwer 1975–1976, vol. 2, pp. 244–248]

[Brouwer 1911A] Brouwer, L.E.J., review of [Mannoury 1909], *Nieuw Archief voor Wiskunde* (2) 9 (1911), pp. 199–201

[Brouwer 1911C] Brouwer, L.E.J., 'Beweis der Invarianz der Dimensionszahl', *Mathematische Annalen* 70 (1911), pp. 161–165; cited from: [Brouwer 1975–1976, vol. 2, pp. 430–434]

[Brouwer 1912] Brouwer, L.E.J., *Intuitionisme en formalisme*, Noordhoff, Groningen 1912; cited from: [Brouwer 1919B]; inaugural address as *extra-ordinarius* delivered at the *Universiteit van Amsterdam* on October 14, 1912

[Brouwer 1913] Brouwer, L.E.J., 'Intuitionism and formalism', *Bulletin of the American Mathematical Society* 20 (1913), pp. 81–96; translation of [Brouwer 1912] by Arnold Dresden

[Brouwer 1914] Brouwer, L.E.J., review of Schoenflies, A., Hahn, H., *Die Entwicklung der Mengenlehre und ihrer Anwendungen. Erster Hälfte*, Teubner, Leipzig etc. 1913, *Jahresbericht der deutschen Mathematiker-Vereinigung* 23 (1914), pp. 78–83; cited from: [Brouwer 1975–1976, vol. 1, pp. 139–144]

[Brouwer 1917A] Brouwer, L.E.J., 'Addenda en corrigenda over de grondslagen van de wiskunde', *Verslagen Koninklijke Akademie voor Wetenschappen te Amsterdam* 25 (1917), pp. 1418–1423; also in: *Nieuw Archief voor Wiskunde* (2) 12 (1918), pp. 439–445; cited from: [Van Dalen 1981, pp. 261–267]

[Brouwer 1917D] Brouwer, L.E.J., 'Anti-Nationalistische Literatuur', *De Nieuwe Amsterdammer* 3/2/1917, p. 110

[Brouwer 1918] Brouwer, L.E.J., 'Begründung der Mengenlehre unabhängig vom logischen Satz vom ausgeschlossenen Dritten. Erster Teil: Allgemeine Mengenlehre', *KNAW Verhandelingen* eerste sectie 12 no. 5 (1918), 43 p.; cited from: [Brouwer 1975–1976, vol. 1, pp. 191-221]

[Brouwer 1919A] Brouwer, L.E.J., 'Begründung der Mengenlehre unabhängig vom logischen Satz vom ausgeschlossenen Dritten. Zweiter Teil, Theorie der Punktmengen', *KNAW Verhandelingen* 7 (1919), pp. 1-33

[Brouwer 1919B] Brouwer, L.E.J., *Wiskunde, waarheid, werkelijkheid*, Noordhoff, Groningen 1919; reprint of [Brouwer 1908], [Brouwer 1909] and [Brouwer 1912]

[Brouwer 1919D] Brouwer, L.E.J., 'Intuitionistische Mengenlehre', *Jahresbericht der deutschen Mathematiker-Vereinigung* 28 (1919), pp. 203-208; cited from: [Brouwer 1975–1976, vol. 1, pp. 230-245]

[Brouwer 1921A] Brouwer, L.E.J., 'Besitzt jede reelle Zahl eine Dezimalbruchentwickelung?', *Mathematische Annalen* 83 (1921), pp. 201-210; also in: *KNAW Verslagen* 29 (1921), pp. 803-812, and in: *KNAW Proceedings* 23 (1922), pp. 955-964; cited from: [Brouwer 1975–1976, vol. 1, pp. 236-245]; English translation in: [Mancosu 1998, pp. 28-35]

[Brouwer 1921H] Brouwer, L.E.J., 'Wis- en natuurkunde en wijsbegeerte', *De Nieuwe Kroniek* 18/6/1921, pp. 1-2

[Brouwer 1922] Brouwer, L.E.J., 'Wis- en natuurkunde en wijsbegeerte', *De Nieuwe Kroniek* 11/2/1922, pp. 3-5

[Brouwer 1923] Brouwer, L.E.J., 'Over de rol van het principium tertii exclusi in de wiskunde, in het bijzonder in de functietheorie', *Wis- en Natuurkundig Tijdschrift* II (1923), pp. 1-7

[Brouwer & De Loor 1924] Brouwer, L.E.J., De Loor, B., 'Intuitionistischer Beweis des Fundamentalsatzes der Algebra', *KNAW Proceedings* 27 (1924), pp. 186-188

[Brouwer 1924D2] Brouwer, L.E.J., 'Beweis, dass jede volle Function gleichmässig stetig ist', *KNAW Proceedings* 27 (1924), pp. 189-193; cited from: [Brouwer 1975–1976, vol. 1, pp. 286-290]

[Brouwer 1924N] Brouwer, L.E.J., 'Über die Bedeutung des Satzes vom ausgeschlossenen Dritten in der Mathematik, insbesondere in der Funktionentheorie', *Journal für die reine und angewandte Mathematik* 154 (1924), pp. 1-7; cited from: [Brouwer 1975–1976, vol. 1, pp. 268-274]

[Brouwer 1925A] Brouwer, L.E.J., 'Zur Begründung der intuitionistischen Mathematik. I', *Mathematische Annalen* 93 (1925), pp. 244-257

[Brouwer 1925E] Brouwer, L.E.J., 'Intuitionistische Zerlegung mathematischer Grundbegriffe', *Jahresbericht der deutschen Mathematiker-Vereinigung* 33 (1925), pp. 251-256; English translation in: [Mancosu 1998, pp. 290-292]

# BIBLIOGRAPHY

[Brouwer 1928] Brouwer, L.E.J., 'Intuitionistische Betrachtungen über den Formalismus', *KNAW Proceedings* 31 (1928), pp. 374–379; also in: *Sitzungsberichte der Preussischen Akademie der Wissenschaften* 1928, pp. 48–52; cited from: [Brouwer 1975–1976, vol. 1, pp. 409–414]; English translation in: [Mancosu 1998, pp. 40–44]

[Brouwer 1929] Brouwer, L.E.J., 'Mathematik, Wissenschaft und Sprache', *Monatshefte für Mathematik und Physik* 36 (1929), pp. 153–164; cited from: [Brouwer 1975–1976, vol. 1, pp. 417–428]; English translation in: [Mancosu 1998, pp. 45–53]; lecture delivered in Vienna on March 10, 1928

[Brouwer 1930] Brouwer, L.E.J., *Die Struktur des Kontinuums*, Wien, 1930; cited from: [Brouwer 1975–1976, vol. 1, pp. 429–440]; English translation in: [Mancosu 1998, pp. 54–63]; lecture delivered in Vienna on March 14, 1928

[Brouwer 1933] Brouwer, L.E.J., 'Willen, weten, spreken', *Euclides* 9 (1932/33), pp. 177–193; address delivered in Amsterdam on December 12, 1932

[Brouwer & Mannoury 1946] Brouwer, L.E.J., *Toespraak van Prof. Dr L.E.J. Brouwer en antwoord van Prof. Dr G. Mannoury*, Stadsdrukkerij Amsterdam, 1946

[Brouwer 1947A] Brouwer, L.E.J., Address delivered on September 16, 1946, on the conferment upon Professor G. Mannoury of the honorary degree of Doctor of Science, *Jaarboek der Universiteit van Amsterdam* 1946–1947, II; cited from the English translation in: [Brouwer 1975–1976, Vol. I, pp. 472–476]

[Brouwer 1947B] Brouwer, L.E.J., 'Richtlijnen der intuïtionistische wiskunde', *KNAW Proceedings* 50 (1947), p. 339; English translation in [Brouwer 1975–1976, vol. 1, p. 477]

[Brouwer 1952] Brouwer, L.E.J., 'Historical background, principles and methods of intuitionism', *South African Journal of Science* 49 (1952), pp. 139–146; cited from: [Brouwer 1975–1976, vol. 1, pp. 508–515]; lecture delivered to Section A of the South African Association for the Advancement of Science, Cape Town, July 1952

[Brouwer 1956] Brouwer, L.E.J., 'Voorgeskiedenis, beginsels en metodes van die intuïtionisme', *Tijdskrif vir wetenskap en kuns* (N.R.) 12 (1956), pp. 186–197; lecture delivered by Brouwer at several occasions during his visit to South Africa in 1952

[Brouwer 1975–1976] Brouwer, L.E.J., *Collected Works*, North Holland, Amsterdam etc., Elsevier, New York, 1975–1976; volume 1: Philosophy and foundations of mathematics, Heyting, A. (ed.); volume 2: Geometry, analysis, topology and mechanics, Freudenthal, H. (ed.)

[Brouwer 1992] Brouwer, L.E.J., *Intuitionismus*, Dalen, D. van (ed.), Bibliographisches Institut, Mannheim etc. 1992; posthumous publication of the 1927 *Berliner Gastvorlesungen* and of the first chapter of a planned book

[Brouwer 1996] Brouwer, L.E.J., 'Life, Art and Mysticism', Van Stigt, W.P. (ed.), *Notre Dame Journal of Formal Logic* 37 (1996), pp. 389–429; English translation of [Brouwer 1905]

[Brown, Collinson & Wilkinson 1996] Brown, S., Collinson, D. and Wilkinson, R., *Biographical Dictionary of Twentieth-Century Philosophers*, Routledge, London etc. 1996

[Brüning, Ferus & Siegmund-Schultze 1998] Brüning, J., Ferus, D., and Siegmund-Schultze, R., *Terror and Exile. Persecution and Expulsion of Mathematicians from Berlin between 1933 and 1945*, Deutsche Mathematiker-Vereinigung, Berlin 1998; text of the exhibition organised on the occasion of the International Congress of Mathematicians, Berlin, August 1998

[Buchholz 1927] Buchholz, H., *Das Problem der Kontinuität. Die Unmöglichkeit absoluter metrischer Präzision und die erkenntnistheoretischen Konsequenzen dieser Unmöglichkeit*, Krueger, München 1927

[Burali-Forti 1897] Burali-Forti, C., 'Una questione sui numeri transfiniti', *Rendiconti del Circolo matematico di Palermo* 11 (1897), pp. 154–164; cited from the English translation in [Van Heijenoort 1967, pp. 104–111]

[Burkamp 1927A] Burkamp, W., *Begriff und Beziehung. Studien zur Grundlegung der Logik*, Felix Meiner, Leipzig 1927

[Burkamp 1927B] Burkamp, W., 'Die Krisis des Satzes vom ausgeschlossenen Dritten', *Beiträge zur Philosophie des Deutschen Idealismus* 4 (1927), pp. 59–81

[Burkamp 1932] Burkamp, W., *Logik*, E.S. Mittler & Sohn, Berlin 1932

[Cairns 1973] Cairns, D., *Guide for translating Husserl*, Martinus Nijhoff, Den Haag 1973

[Cajori 1980] Cajori, F., *A history of mathematics*, Chelsea, New York 1980, 3rd edition; first edition 1893, second, revised edition 1919

[Calinger 1996] Calinger, R. (ed.), *Vita Mathematica. Historical Research and Integration with Teaching*, The Mathematical Association of America, Washington, D.C., 1996

[Canfield 1986] Canfield, J.V. (ed.), *The Philosophy of Wittgenstein*, Garland, New York etc. 1986

[Cantor 1883] Cantor, G., 'Ueber unendliche, lineare Punktmannigfaltigkeiten 5', *Mathematische Annalen* 21 (1883), pp. 545–591; part 5 in a series of 6 that appeared in the *Mathematische Annalen* from 1879 to 1884; cited from: Cantor, G., *Ueber unendliche, lineare Punktmannigfaltigkeiten*, Asser, G. (ed.), Teubner, Leipzig 1984

[Cantor 1932] Cantor, G., *Gesammelte Abhandlungen mathematischen und philosophischen Inhalts*, Zermelo, E. (ed.), Springer, Berlin 1932

[Cardano 1545] Cardano, H., *Artis magnae, sive de regulis algebraicis liber unus*, Nuremberg 1545

[Carnap 1927] Carnap, R., 'Eigentliche und uneigentliche Begriffe', *Symposion* 1 Heft 4 (1927), pp. 355–374

[Carnap 1930] Carnap, R., 'Die Mathematik als Zweig der Logik', *Blätter für Deutsche Philosophie* 4 (1930/31), pp. 298–310

[Carnap 1931] Carnap, R., 'Die logizistische Grundlegung der Mathematik', *Erkenntnis* 2 (1931), pp. 91–105; lecture delivered at the Königsberg Tagung in September 1930

[Carnap 1932] Carnap, R., 'Über Hilbert', lecture delivered at the Mathematical Circle in Prague, January 22, 1932; notes for the lecture, dated January 19, 1932, in [ASP Carnap, 110-07-16]

[Carnap 1934A] Carnap, R., *Logische Syntax der Sprache*, Springer, Wien 1934

[Carnap 1934B] Carnap, R., 'Die Antinomien und die Unvollständigkeit der Mathematik', *Monatshefte für Mathematik und Physik* 41 (1934), pp. 263–284

[Carnap 1963] Carnap, R., 'Intellectual Autobiography', in: [Schilpp 1963, pp. 3–84]

[Cassirer 1929] Cassirer, E., *Philosophie der symbolischen Formen. Dritter Teil: Phänomenologie der Erkenntnis*, Bruno Cassirer, Berlin 1929

[Castelnuovo 1909] Castelnuovo, G., (ed.), *Atti del IV Congresso Internazionale dei Matematici*, vol. I: Relazione sul Congresso – Discorsi e Conference, Accademia dei Lincei, Roma 1909; conference held in Rome on April 6-11, 1908

[Caygill 1995] Caygill, H., *A Kant Dictionary*, Blackwell, Oxford 1995

[Chandrasekharan 1986A] Chandrasekharan, K. (ed.), *Hermann Weyl 1885–1985*, Springer, Berlin etc. 1986

[Chandrasekharan 1986B] Chandrasekharan, K., 'Report on the Celebration', in: [Chandrasekharan 1986A, pp. 95–108]

[Chevalley & Weil 1957] Chevally, C., Weil, A., 'Hermann Weyl (1885–1955)', *L'Enseignement mathématique* (2) 3 (1957), pp. 157–187

[Church 1928] Church, A., 'On the law of the excluded middle', *Bulletin of the American Mathematical Society* 34 (1928), pp. 75–78

[Church 1932] Church, A., 'A set of postulates for the foundation of logic', *Annals of mathematics* second series 33 (1932), pp. 346–366, and 34 (1933), pp. 839–864

[Church 1938] Church, A., 'The constructive second number class', *Bulletin of the American Mathematical Society* 44 (1938), pp. 224–232

[Chwistek 1922] Chwistek, L., 'Zasady czystej teorji typów', *Przegląd Filozoficzny* 25 (1922), pp. 359–391

[Chwistek 1925] Chwistek, L., review of [Chwistek 1922], *Jahrbuch über die Fortschritte der Mathematik* 48 (1921–22; published 1925/28), p. 216

[Clauberg & Dubislav 1923] Clauberg, K.W., Dubislav, W., *Systematisches Wörterbuch der Philosophie*, Felix Meiner, Leipzig 1923

[Collingwood 1959] Collingwood, E.F., 'Émile Borel', *The Journal of The London Mathematical Society* XXXIV (1959), pp. 488–512

[Costabel 1970] Costabel, P., 'René Louis Baire', in: [Gillispie 1970–1978, Vol. I, pp. 406–408]

[Courant & Bernays 1921] Courant, R., Bernays, P., 'Über die neuen arithmetischen Theorien von Weyl und Brouwer', lectures delivered in Göttingen at February 1 and 8, 1921; mentioned in *Jahresbericht der deutschen Mathematiker-Vereinigung* 30 (1921), p. *32*

[Courant 1928] Courant, R., 'Über die allgemeine Bedeutung des mathematischen Denkens', *Die Naturwissenschaften* 16 (1928), pp. 89–94; lecture delivered at the conference of German philologists and pedagogues in Göttingen, September 1927

[Cremer 1927] Cremer, H., 'Mathematische Schnaderhüpfl', in: *Häufungspunkte. Mathematischer Reimsalat für die studierende Jugend von 3 bis 17 Semestern*, Mathematische Verein an der Universität Berlin, 1927; cited from: [Van Dalen 2001, p. 313]

[D'Abro] See under A

[Dahms 1987A] Dahms, H.-J., 'Einleitung', in: [Becker, H., et al. 1987, pp. 15–60]

[Dahms 1987B] Dahms, H.-J., 'Aufstieg und Ende der Lebensphilosophie: Das philosophische Seminar der Universität Göttingen zwischen 1917 und 1950', in: [Becker, H., et al. 1987, pp. 169–199]

[Van Dalen 1981] Dalen, D. van (ed.), *L.E.J. Brouwer - Over de grondslagen der wiskunde*, Stichting Mathematisch Centrum, Amsterdam 1981; reprint of [Brouwer 1907], including the unpublished parts, the Brouwer-Korteweg correspondence, and some papers by or about Brouwer

[Van Dalen 1984A] Dalen, D. van (ed.), *L.E.J. Brouwer, C.S. Adama van Scheltema - Droeve snaar, vriend van mij. Brieven*, De Arbeiderspers, Amsterdam 1984

[Van Dalen 1984B] Dalen, D. van (ed.), 'Four letters from Edmund Husserl to Hermann Weyl', *Husserl Studies* 1 (1984), pp. 1-12

[Van Dalen 1990] Dalen, D. van, 'The War of the Frogs and the Mice, or the Crisis of the *Mathematische Annalen*', *The Mathematical Intelligencer* 12 (1990), pp. 17–31

[Van Dalen 1992] Dalen, D. van, 'Einleitung', in: [Brouwer 1992, pp. 7–15]

[Van Dalen 1995] Dalen, D. van, 'Hermann Weyl's Intuitionistic Mathematics', *Bulletin of Symbolic Logic* 1 (1995), pp. 145–169

[Van Dalen 1997] Dalen, D. van, *Logic and Structure*, Springer, Berlin etc. 1997; third edition

[Van Dalen 1999A] Dalen, D. van, *Mystic, Geometer, and Intuitionist: The Life of L.E.J. Brouwer. Volume I: The dawning Revolution*, Clarendon Press, Oxford 1999

[Van Dalen 1999B] Dalen, D. van, 'From Brouwerian Counter-Examples to the Creating Subject', *Studia Logica* 62 (1999), pp. 305–314

[Van Dalen 2000] Dalen, D. van, 'Brouwer and Fraenkel on Intuitionism', *Bulletin of the Association of Symbolic Logic* 6 (2000), pp. 284–310

[Van Dalen 2001] Dalen, D. van, *L.E.J. Brouwer. Een biografie*, Bert Bakker, Amsterdam 2001

[Dancy 1989] Dancy, J., *Introduction to contemporary epistemology*, Basil Blackwell, Oxford 1989

[Danneberg et al. 1994] Danneberg, L., Kamlah, A., Schäfer, L. (eds.), *Hans Reichenbach und die Berliner Gruppe*, Vieweg, Braunschweig/Wiesbaden 1994

[Danneberg & Schernus 1994] Danneberg, L., Schernus, W., 'Die *Gesellschaft für wissenschaftliche Philosophie*: Programm, Vorträge und Materialien', in: [Danneberg et al. 1994, pp. 391–481]

[Van Dantzig 1932] Dantzig, D. van, *Over de elementen van het wiskundig denken*, Noordhoff, Groningen etc. 1932; inaugural lecture delivered at the *Technische Hoogeschool* Delft, October 4, 1932

[Dassen 1933] Dassen, C.C., 'Réflexions sur quelques antinomies et sur la logique empiriste', *Anales de la Sociedad Científica Argentina* 115 (1933), pp. 135–166; lectures delivered in Buenos Aires on October 19, 1929, June 21, 1930, and December 17, 1932

[Dauben 1979] Dauben, J.W., *Georg Cantor. His Mathematics and Philosophy of the Infinite*, Harvard University Press, Cambridge etc. 1979

[Davis 1982] Davis, M., 'Why Gödel Didn't Have Church's Thesis', *Information and Control* 54 (1982), pp. 3–24

[Dawson 1984] Dawson, J.W. Jr., 'Discussion on the Foundations of Mathematics', *History and Philosophy of Logic* 5 (1984), pp. 111-129; annotated English translation of [Hahn et al. 1931]

[Dawson 1997] Dawson, J.W. Jr., *Logical Dilemmas. The Life and Work of Kurt Gödel*, A K Peters, Wellesley, Massachusetts, 1997

[De Haan] See under H

[Dedekind 1887] Dedekind, R., *Was sind und was sollen die Zahlen?*, Vieweg & Sohn, Braunschweig 1887; cited from the 1969 re-edition

[Deppert et al. 1988] Deppert, W., Hübner, K., Oberschelp, A., Weidemann, V. (eds.), *Exact Sciences and their Philosophical Foundations/Exakte Wissenschaften und ihre philosophische Grundlagen. Vorträge des Internationalen Hermann-Weyl-Kongresses, Kiel 1985*, Peter Lang, Frankfurt am Main etc. 1988

[Detlefsen 1993] Detlefsen, M., 'Poincaré vs. Russell on the Rôle of Logic in Mathematics', *Philosophia Mathematica* 3, Vol. 1 (1993), pp. 24–49

[Detlefsen 1994] Detlefsen, M., 'Constructivism', in: [Grattan-Guinness 1994, Vol. 1, pp. 656–664]

[Deutschbein 1929] Deutschbein, M., 'Der philosophische Bildungswert der Mathematik', *Zeitschrift für Deutsche Bildung* (1929), pp. 326–333

[Dieudonné 1972] Dieudonne, J., 'Notice nécrologique sur Paul Lévy, Membre de la Section de Géométrie', *Comptes Rendues de l'Académie de Sciences de Paris* 274 (1972), pp. 137–144

[Dieudonné 1975] Dieudonné, J., 'Jules-Henri Poincaré', in: [Gillispie 1970–1978, Vol. XI, pp. 51–61]

[Dingler 1926] Dingler, H., *Der Zusammenbruch der Wissenschaft und der Primat der Philosophie*, Reinhardt, München 1926

[Dingler 1931] Dingler, H., *Philosophie der Logik und Arithmetik*, Reinhardt, München 1931

[Doetsch 1924] Doetsch, G., 'Der Sinn der reinen Mathematik und ihrer Anwendung', *Kantstudien* 29 (1924), pp. 439–459

[Doetsch 1928] Doetsch, G., review of [Enriques 1927], *Jahresbericht der deutschen Mathematiker-Vereinigung* 37 (1928), pp. 69–70

[Van Dooren 1983] Dooren, W. van, *Denkwegen in de geschiedenis van de nieuwere wijsbegeerte*, Van Gorcum, Assen 1983

[Dragalin 1988] Dragalin, A.G., 'Constructive propositional calculus', in: [Hazewinkel 1988–1993, Vol. 2, p. 361]

[Dreben & Van Heijenoort 1986] Dreben, B., Van Heijenoort, J., 'Introductory note to *1929*, *1930* and *1930a*', in: [Gödel 1986–1995, vol. 1, pp. 44–59]

[Dresden 1924] Dresden, A., 'Brouwer's contributions to the foundations of mathematics', *Bulletin of the American Mathematical Society* 30 (1924), pp. 31–40; presented to the Society in December 1923

[Dresden 1928] Dresden, A., 'Some philosophical aspects of mathematics', *Bulletin of the American Mathematical Society* 34 (1928), pp. 438–452; address presented at the joint meeting of the American Mathematical Society, the Mathematical Association of America and the American Association for the Advancement of Science, at Nashville, December 29, 1927

[Dresden 1929] Dresden, A., 'Mathematical certainty', *Scientia* 45 (1929), pp. 369–376

[Droste 1930] Droste, J., *De eenheid der wiskunde*, E.J. Brill, Leiden 1930; inaugural lecture delivered in Leiden on September 24, 1930

[Du Bois Reymond, E. 1872] Du Bois Reymond, E., 'Über die Grenzen des Naturerkennens', in: [Du Bois Reymond, E. 1974, pp. 54–77]

[Du Bois Reymond, E. 1974] Du Bois Reymond, E., *Vorträge über Philosophie und Gesellschaft*, Felix Meiner, Hamburg 1974

[Du Bois-Reymond, P. 1882] Du Bois-Reymond, P., *Die allgemeine Functionentheorie. Erster Theil. Metaphysik und Theorie der mathematischen Grundbegriffe: Grösse, Grenze, Argument und Function*, Laupp, Tübingen 1882

[Dubislav 1926] Dubislav, W., 'Über das Verhältnis der Logik zur Mathematik', *Annalen der Philosophie und philosophische Kritik* 5 (1925–1926), pp. 193–208

[Dubislav 1929A] Dubislav, W., 'Zur Philosophie der Mathematik und Naturwissenschaft', *Annalen der Philosophie und philosophische Kritik* 8 (1929), pp. 135–145

[Dubislav 1929B] Dubislav, W., 'Zur Lehre von den sog. schöpferischen Definitionen', *Philosophisches Jahrbuch der Görres-Gesellschaft* 42 (1929), pp. 42–53

[Dubislav 1930] Dubislav, W., 'Über den sogenannten Gegenstand der Mathematik', *Erkenntnis* 1 (1930/31), pp. 27–48

[Dubislav 1931] Dubislav, W., 'Die sog. Grundlagenkrise der Mathematik', *Unterrichtsblätter für Mathematik und Naturwissenschaft* 37 (1931), pp. 146–152

[Dubislav 1932] Dubislav, W., *Die Philosophie der Mathematik in der Gegenwart*, Junker und Dünnhaupt, Berlin 1932

[Dubucs 1988] Dubucs, J-P, 'L.E.J. Brouwer: topologie et constructivisme', *Revue d'histoire des sciences* 41 (1988), pp. 133–155

[Dummett 1964] Dummett, M., 'Wittgenstein's philosophy of mathematics', in: [Benacerraf & Putnam 1964, pp. 491–509]

[Dummett 1977] Dummett, M., *Elements of intuitionism*, Clarendon Press, Oxford 1977

[Edwards 1989] Edwards, H.M., 'Kronecker's Views on the Foundations of Mathematics', in: [Rowe & McCleary 1989, vol. I, pp. 67–77]

[Van Eeden 1916] Eeden, F. van, 'Een machtig Brouwsel' (1916); cited from: [Van Eeden 1925, pp. 75–108]

[Van Eeden 1925] Eeden, F. van, *Langs den weg: verspreide opstellen*, Roermond 1925

[Ehrenfest 1912] Ehrenfest, P., *Zur Krise der Lichtaetherhypothese*, Berlin 1912; inaugural lecture

[Einstein 1922] Einstein, A., 'Über die gegenwärtigen Krise der theoretischen Physik', *Kaizo* 4 (1922), pp. 1-8

[Embree et al. 1997] Embree, L., et al. (eds.), *Encyclopedia of Phenomenology*, Kluwer, Dordrecht etc. 1997

[Enderton 1995] Enderton, H., 'In memoriam: Alonzo Church', *Bulletin of Symbolic Logic* 1 (1995), pp. 486-488

[Enriques 1927] Enriques, F., *Zur Geschichte der Logik. Grundlagen und Aufbau der Wissenschaft im Urteil der mathematischen Denker*, Teubner, Leipzig und Berlin 1927; German translation by Bieberbach

[Errera 1933] Errera, A., 'Quelques remarques sur les mathématiques intuitionnistes', *Revue de Métaphysique et de Morale* 40 (1933), pp. 27-39

[Errera 1935A] Errera, A., 'Sur le principe du tiers exclu', *Mathematica* 9 (1935), pp. 73-79; lecture delivered at the second conference of Roumanian Mathematicians, Turnu-Severin, May 5-9, 1932

[Errera 1935B] Errera, A., 'Réponse à quelques objections', *L'Enseignement mathématique* 34 (1935), pp. 103-111

[Errera 1951] Errera, A., 'Observations de M. Errera sur la communication du professeur Heyting', in: [Bayer 1951, pp. 87-89]

[Ewald 1996] Ewald, W., *From Kant to Hilbert: A Source Book in the Foundations of Mathematics*, Clarendon Press, Oxford 1996; two volumes

[Feferman 1986] Feferman, S., 'Gödel's life and work', in: [Gödel 1986-1995, vol. I, pp. 1-36]

[Feferman 1988] Feferman, S., 'Weyl vindicated: 'Das Kontinuum' 70 years later', *Atti del Congresso* Temi e prospettive della logica e della filosofia della scienza contemporaneanee, Vol. 1, CLUEB, Bologna 1988

[Fehr, Flournoy & Claparède 1912] Fehr, H., Flournoy, Th., Claparède, Ed., *Enquête de l'Enseignement Mathématique sur la méthode de travail des mathématiciens*, Paris, Gauthier-Villars etc., 1912; deuxième édition

[Feigl 1969] Feigl, H., 'The Wiener Kreis in America', in: [Flemming & Bailyn 1969, pp. 630-673]

[Fenstad 1970] Fenstad, J.E., 'Thoralf Albert Skolem in Memoriam', in: [Skolem 1970, pp. 9-15]; English translation of a Norwegian paper in the *Norsk Matematisk Tidsskrift* in 1963

[Finsler 1926A] Finsler, P., 'Gibt es Widersprüche in der Mathematik?', *Jahresbericht der deutschen Mathematiker-Vereinigung* 34 (1926), pp. 143-155; inaugural lecture at the *Universität Köln*, 1923

[Finsler 1926B] Finsler, P., 'Über die Grundlegung der Mengenlehre. Erster Teil. Die Mengen und ihre Axiome', *Mathematische Zeitschrift* 25 (1926), pp. 683-713

[Fischer et al 1990] Fischer, G., Hirzebruch, F., Scharlau, W., Törnig, W. (eds.), *Ein Jahrhundert Mathematik 1890–1990, Festschrift zum Jubiläum der DMV*, Braunschweig, Vieweg 1990

[Flemming & Bailyn 1969] Flemming, D., Bailyn, B. (eds.), *The intellectual migration. Europe and America, 1930–1960*, Harvard University Press, Cambridge, Massachusetts, 1969

[Fletcher 1986] Fletcher, C., 'Refugee Mathematicians: A German Crisis and a British Response, 1933–1936', *Historia Mathematica* 13 (1986), pp. 13–27

[Fogelin 1968] Fogelin, R.J., 'Wittgenstein and intuitionism', *American Philosophical Quarterly* 5 (1968), pp. 267–274; cited from: [Canfield 1986, Vol. 11, pp. 153–160]

[Forman 1971] Forman, P., 'Weimar Culture, Causality, and Quantum Theory, 1918–1927: Adaptation by German Physicists and Mathematicians to a Hostile Intellectual Environment', in: [McCormmach 1971, pp. 1–115]

[Forman 1984] Forman, P., '*Kausalität, Anschaulichkeit*, and *Individualität*, or How Cultural Values Prescribed the Character and the Lessons Ascribed to Quantum Mechanics', in: [Stehr & Meja 1984, pp. 333–347]

[Fowler 1987] Fowler, D.H., *The Mathematics of Plato's Academy*, Clarendonn Press, Oxford 1987

[Fraenkel 1919] Fraenkel, A., *Einleitung in die Mengenlehre. Eine gemeinverständliche Einführung in das Reich des unendlichen Größen*, Springer, Berlin 1919; first edition

[Fraenkel 1921] Fraenkel, A., 'Über die Zermelosche Begründung der Mengenlehre', lecture delivered at the *Mathematikertagung* at Jena at September 22, 1921; mentioned in *Jahresbericht der deutschen Mathematiker-Vereinigung* 30 (1921), pp. 97–98

[Fraenkel 1922A] Fraenkel, A., 'Zu den Grundlagen der Cantor-Zermeloschen Mengenlehre', *Mathematische Annalen* 86 (1922), pp. 230–237

[Fraenkel 1922B] Fraenkel, A., 'Probleme der Mengenlehre', *Unterrichtsblätter für Mathematik und Naturwissenschaften* 28 (1922)

[Fraenkel 1923A] Fraenkel, A., *Einleitung in die Mengenlehre. Eine elementare Einführung in das Reich des Unendlichgrossen*, Springer, Berlin 1923; second edition

[Fraenkel 1923B] Fraenkel, A., 'Die Axiome der Mengenlehre', *Scripta Universitatis atque Bibliothecae hierosolymitanarum*, Mathematica et physica vol. 1 (1923); bilingual paper (German and Hebrew)

[Fraenkel 1924A] Fraenkel, A., 'Die neueren Ideen zur Grundlegung der Analysis und Mengenlehre', *Jahresbericht der deutschen Mathematiker-Vereinigung* 33 (1924), pp. 97–103; lecture delivered at the *Deutsche Mathematikertagung* in Marburg in September 1923

[Fraenkel 1924B] Fraenkel, A., 'Über die gegenwärtige Grundlagenkrise der Mathematik', *Sitzungsberichte der Gesellschaft zur Beförderung der gesamten Naturwissenschaften zu Marburg* 6 (1924), pp. 117–132

[Fraenkel 1925A] Fraenkel, A., 'Über die gegenwärtige Krise in den Grundlagen der Mathematik', *Unterrichtsblätter für Mathematik und Naturwissenschaften* 31 (1925), pp. 249–254, 270–274; lecture delivered before the *Gesellschaft zur Beförderung der gesamten Naturwissenschaften* in Marburg in June 1924

[Fraenkel 1925B] Fraenkel, A., 'Der Streit um das Unendliche in der Mathematik', *Scientia* 38 (1925), pp. 141–152, 209–218; French translation 'Les discussions sur l'infini en mathématiques' in the supplement, pp. 49–58; 81–88

[Fraenkel 1925C] Fraenkel, A., 'Untersuchungen über die Grundlagen der Mengenlehre', *Mathematische Zeitschrift* 22 (1925), pp. 250–273

[Fraenkel 1927A] Fraenkel, A., *Zehn Vorlesungen über die Grundlegung der Mengenlehre*, Teubner, Leipzig etc. 1927; lectures held before the Kant-Gesellschaft in Kiel on June 8–12, 1925

[Fraenkel 1927B] Fraenkel, A., review of [Buchholz 1927], *Jahresbericht der deutschen Mathematiker-Vereinigung* 36 (1927), pp. *78–79*

[Fraenkel 1927/28A] Fraenkel, A., review of [Skolem 1923], *Jahrbuch über die Fortschritte der Mathematik* 49 (1923; published 1927/28), pp. 138–139

[Fraenkel 1927/28B] Fraenkel, A., review of [Becker, O. 1923], *Jahrbuch über die Fortschritte der Mathematik* 49 (1923; published 1927/28), p. 391

[Fraenkel 1928] Fraenkel, A., *Einleitung in die Mengenlehre*, Springer, Berlin 1928; third edition

[Fraenkel 1929A] Fraenkel, A., 'Der Streit um die Grundlagen der Mathematik', lecture delivered at the *Versammlung Deutscher Philologen und Schulmänner in Salzburg, 25.-28. September 1929*; report in [Lietzmann 1930, p. 35]

[Fraenkel 1929B] Fraenkel, A., review of [Pasch 1927], *Jahresbericht der deutschen Mathematiker-Vereinigung* 38 (1929), pp. *93–94*

[Fraenkel 1930A] Fraenkel, A., 'Die heutige Gegensätze in der Grundlegung der Mathematik', *Erkenntnis* 1 (1930/31), pp. 286–302; lecture delivered at the first *Tagung für Erkenntnislehre der exakten Wissenschaften*, Praha, September 1929

[Fraenkel 1930B] Fraenkel, A., 'Das Problem des Unendlichen in der neueren Mathematik', *Blätter für Deutsche Philosophie* 4 (1930/31), pp. 279–297

[Fraenkel 1930C] Fraenkel, A., 'Der Streit um die Grundlagen der Mathematik', *Zeitschrift für mathematische Unterricht* 61 (1930), p. 35

[Fraenkel 1932] Fraenkel, A., 'Über das Unendliche', radio lecture delivered on *RAVAG (Radio Wien)* in March 1932; written version available in [JNUL Fraenkel]

[Fraenkel 1933] Fraenkel, A., 'On modern problems in the foundations of mathematics', *Scripta Mathematica* 1 (1933), pp. 222–227

[Fraenkel 1947] Fraenkel, A., 'The recent controversies about the foundation of mathematics', *Scripta Mathematica* 13 (1947), pp. 17–36

[Fraenkel 1951] Fraenkel, A., 'On the crisis of the principle of the excluded middle', *Scripta Mathematica* 17 (1951), pp. 5–16

[Fraenkel 1959] Fraenkel, A., 'Philosophie der Mathematik', in: [Heinemann 1959, pp. 334–359]

[Fraenkel 1967] Fraenkel, A., *Lebenskreise. Aus den Erinnerungen eines jüdischen Mathematikers*, Deutsche Verlags-Anstalt, Stuttgart 1967

[Fréchet 1965] Fréchet, M., 'La vie et l'oeuvre d'Émile Borel', *L'Enseignement mathématique* (IIe série) XI (1965), pp. 1–97

[Frege 1893] Frege, G., *Grundgesetze der Arithmetik*, Jena 1893; cited from the 1962 reprint, Georg Olms, Hildesheim

[Frege 1906] Frege, G., 'Antwort auf die Ferienplauderei des Herrn Thomae', *Jahresbericht der deutschen Mathematiker-Vereinigung* 15 (1906), pp. 586–590

[Frege 1976] Frege, G., *Wissenschaftliche Briefwechsel*, Gabriel, G., Hermes, H., Kambartel, F., Thiel, C., Veraart, A. (eds.), Felix Meiner, Hamburg 1976

[Freudenthal 1932] Freudenthal, H., 'Qualität und Quantität in der Mathematik', *Euclides* 8 (1932), pp. 89–98; inaugural address, delivered at the university of Amsterdam on May 28, 1931

[Freudenthal & Heyting 1967] Freudenthal, H., Heyting, A., 'Levensbericht van L.E.J. Brouwer', *Jaarboek der Koninklijke Nederlandse Akademie van Wetenschappen* 1966–1967, pp. 335–340; English translation in: [Brouwer 1975–1976, part II, pp. X–XV]

[Freudenthal 1981] Freudenthal, H., 'L.E.J. Brouwer – Topoloog, intuïtionist, filosoof', *Nieuw Archief voor Wiskunde* (3) XXIX (1981), pp. 249–253

[Freytag 1937] Freytag, Baron B. von, *Zum Problem der 'mathematischen Existenz'*, Greifswald 1937; dissertation

[Frizell 1914] Frizell, A.B., 'A non-enumerable well-ordered set', *Bulletin of the American Mathematical Society* 2 (1914–1915), pp. 404–405

[Galavotti 1991] Galavotti, M.C., 'Introduction', in: [Ramsey 1991A, pp. 11–29]

[Garcia 1933] Garcia, D., 'Assaigs moderns per a la fonamentació de les matemàtiques', *Societat Catalana de Ciencies físiques, químiques i matemàtiques, Memòries* 1 (1933), pp. 225–273

[Garciadiego 1986] Garciadiego, A., 'On Rewriting the History of the Foundations of Mathematics at the Turn of the Century', *Historia Mathematica* 13 (1986), pp. 39–41

[Gay 1969] Gay, P., *Weimar Culture. The outsider as insider*, Penguin, London etc. 1969; cited from the 1992 edition

[Geiger 1928] Geiger, M., review of [Becker, O. 1927], *Göttingische gelehrte Anzeigen unter der Aufsicht der Gesellschaft der Wissenschaften* 190 (1928), pp. 401–419

[Gentzen 1969] Gentzen, G., 'On the relation between intuitionist and classical arithmetic', in: [Szabo 1969, pp. 53–67]; English translation of the paper originally intended to be published in the *Mathematische Annalen* in 1933, but withdrawn by Gentzen when he found out that Gödel had already proved its main result

[Gethmann 1995] Gethmann, C.F., 'Realismus (erkenntnistheoretisch)', in: [Mittelstraß 1980–96, Vol. 3, pp. 500–502]

[Gillies 1982] Gillies, D.A., 'Wittgenstein and revisionism', *British Journal for the Philosophy of Science* 33 (1982), pp. 422–433

[Gillispie 1970–1978] Gillispie, L.C. (ed.), *Dictionary of Scientific Biography*, Charles Scribner's Sons, New York 1970–1978; 15 volumes

[Glivenko 1928] Glivenko, M.V., 'Sur la logique de M. Brouwer', *Académie Royale de Belgique, Bulletin de la Classe des Sciences* (5) 14 (1928), pp. 225–228

[Glivenko 1929] Glivenko, M.V., 'Sur quelques points de la Logique de M. Brouwer', *Académie Royale de Belgique, Bulletin de la Classe des Sciences* (5) 15 (1929), pp. 183–188; English translation in: [Mancosu 1998, pp. 301–305]

[Godeaux 1960] Godeaux, L., 'Alfred Errera (1886–1960)', *Bolletino della Unione matematica italiana* 1960, pp. 575–578

[Gödel 1929] Gödel, K.,[4] 'Über die Vollständigkeit des Logikkalküls', dissertation, Universität Wien 1929; cited from: [Gödel 1986–1995, vol. I, pp. 60–101], German original with English translation; dissertation defended on February 6, 1930

[Gödel 1930] Gödel, K., 'Die Vollständigkeit der Axiome des logischen Funktionenkalküls', *Monatshefte für Mathematik und Physik* 37 (1930), pp. 349–360; cited from: [Gödel 1986–1995, vol. I, pp. 102–123], German original with English translation

[Gödel 1931] Gödel, K., 'Über formal unentscheidbare Sätze der *Principia Mathematica* und verwandter Systeme I', *Monatshefte für Mathematik und Physik* 38 (1931), pp. 173–198; cited from: [Gödel 1986–1995, vol. I, pp. 144–195], German original with English translation

---

[4]The coding of Gödel's publications follows the one used in Gödel's Collected Works, [Gödel 1986–1995].

[Gödel 1932] Gödel, K., 'Zum intuitionistischen Aussagenkalkül', *Anzeiger der Akademie der Wissenschaften in Wien* 69 (1932), pp. 65–66; cited from: [Gödel 1986–1995, vol. I, pp. 222–225], German original with English translation

[Gödel 1933E] Gödel, K., 'Zur intuitionistischen Arithmetik und Zahlentheorie', *Ergebnisse eines mathematischen Kolloquiums* 4 (1933), pp. 34–38; cited from: [Gödel 1986–1995, vol. I, pp. 286–295], German original with English translation

[Gödel 1933F] Gödel, K., 'Eine Interpretation des intuitionistischen Aussagenkalküls', *Ergebnisse eines mathematischen Kolloquiums* 4 (1933), pp. 39–40; cited from: [Gödel 1986–1995, vol. I, pp. 300–303], German original with English translation

[Gödel 1933O] Gödel, K., 'The present situation in the foundations of mathematics', lecture delivered in December 1933 in Cambridge, Massachusetts; published posthumously in and cited from [Gödel 1986–1995, vol. III, pp. 45–53]

[Gödel 1938A] Gödel, K., 'Vortrag bei Zilsel', lecture delivered in Vienna in January 1938; posthumously published in and cited from [Gödel 1986–1995, vol. III, pp. 86–113]

[Gödel 1986–1995] Gödel, K., *Collected Works*, Volumes I–III, Feferman, S. (ed. in chief), Oxford University Press etc., New York etc. 1986–1995

[Goenner 1993] Goenner, H., 'The reaction to relativity theory I: The Anti-Einstein Campaign in Germany in 1920', *Science in Context* 6 (1993), pp. 107-133

[Goldfarb 1988] Goldfarb, W., 'Poincaré against the Logicists', in: [Aspray & Kitcher 1988, pp. 61–81]

[Goldstein et al. 1996] Goldstein, C., Gray, J., Ritter, J. (eds.), *L'Europe Mathématique: Histoires, Mythes, Identités*, Éditions de la Maison de l'homme, Paris 1996

[Gonseth 1925] Gonseth, F., 'Sur la logique intuitioniste', lecture delivered before the *Société mathématique Suisse* at Aarau, August 9, 1925; reported in *L'Enseignement mathématique* 24 (1924–25), p. 299

[Gonseth 1926] Gonseth, F., *Les Fondements des Mathématiques. De la Géometrie d'Euclide à la Relativité générale et à l'Intuitionisme*, avec une préface de M. Jacques Hadamard, Albert Blanchard, Paris 1926

[Gonseth 1929] Gonseth, F., 'Mathematik und Erkenntnis', *Jahresheft des Bernschen Gymnasiallehrer-Vereins* 1929; lecture delivered at the *Jahresversammlung des Bernschen Gymnasiallehrervereins*, February 25, 1928

[Goodstein 1951] Goodstein, R.L., *Constructive formalism*, University College Leicester, 1951

[Gottwald, Ilgauds & Schlote 1990] Gottwald, S., Ilgauds, H.-J. und Schlote, K.-H., *Lexikon bedeutender Mathematiker*, Bibliographisches Institut Leipzig, Leipzig 1990

[Grattan-Guinness 1981] Grattan-Guinness, I., 'On the development of logic between the two world wars', *The American Mathematical Monthly* 88 (1981), pp. 495–509

[Grattan-Guinness 1984] Grattan-Guinness, I., 'Notes on the Fate of Logicism from *Principia Mathematica* to Gödel's Incompletability Theorem', *History and Philosophy of logic* 5 (1984), pp. 67–78

[Grattan-Guinness 1994] Grattan-Guinness, I. (ed.), *Companion Encyclopedia of the History and Philosophy of the Mathematical Sciences*, Routledge, London etc. 1994

[Grattan-Guinness 1997] Grattan-Guinness, I., 'A Retreat from Holism: Carnap's Logical Course, 1921-43', *Annals of Science* 54 (1997), pp. 407–421; essay review of Cirera, R., Ibarra, A., Mormann, Th. (eds.), *El Programa de Carnap: Ciencia, lenguaje, filosofia*, De Bronce, Barcelona 1996

[Gray 1991] Gray, J., 'Did Poincaré Say 'Set Theory Is a Disease'?', *The Mathematical Intelligencer* 13 (1991), pp. 19–22

[Grelling 1924] Grelling, K., *Mengenlehre*, Teubner, Berlin 1924

[Grelling 1928] Grelling, K., 'Philosophy of the exact sciences: its present status in Germany', *The Monist* 38 (1928), pp. 97–119; translated from German

[Gruenberg 1988] Gruenberg, K.W., 'Obituary: Kurt August Hirsch', *Bulletin of the London Mathematical Society* 20 (1988), pp. 350–358

[De Haan 1916A] Haan, J.I. de, *Rechtskundige significa en hare toepassing op de begrippen: 'aansprakelijk, verantwoordelijk, toerekeningsvatbaar'*, W. Versluys, Amsterdam 1916; dissertation

[De Haan 1916B] Haan, J.I. de, *Wezen en taak der rechtskundige significa*, Van Kampen & Zoon, Amsterdam 1916; inaugural lecture at the *Hoogeschool van Amsterdam*, delivered on October 31, 1916

[De Haan 1919] Haan, J.I. de, *Rechtskundige significa*, Johannes Müller, Amsterdam 1919

[Hacker 1975] Hacker, P.M.S., *Insight and illusion. Wittgenstein on philosophy and the metaphysics of experience*, Oxford University Press, London etc. 1975

[Hadamard 1926] Hadamard, J., 'Préface', in: [Gonseth 1926, pp. V–XI]

[Hahn 1930] Hahn, H., 'Die Bedeutung der wissenschaftliche Weltauffassung, insbesondere für Mathematik und Physik', *Erkenntnis* 1 (1930/31), pp. 96–105; lecture delivered at the first *Tagung für Erkenntnislehre der exakten Wissenschaften*, Praha, September 1929

[Hahn et al. 1931] Hahn, H., et al., 'Diskussion zur Grundlegung der Mathematik', *Erkenntnis* 2 (1931), pp. 135–155; English in *History and Philosophy of Logic* 5 (1984); discussion held on September 7, 1930, at the Königsberg conference

[Hallett 1995] Hallett, M., 'Logic and Mathematical Existence', in: [Krüger & Falkenburg 1995, pp. 33–82]

[Halmos 1950] Halmos, P.R., *Measure Theory*, Van Nostrand, Toronto etc. 1950

[Hamel 1928] Hamel, G., *Über die philosophische Stellung der Mathematik*, Gesellschaft von Freunden der Technischen Hochschule Berlin, Berlin 1928; lecture delivered upon taking over the rectorship of the *Technische Hochschule Berlin* on June 30, 1928

[Hardy 1929] Hardy, G.H., 'Mathematical proof', *Mind* 38 n.s. (1929), pp. 1–25; Rouse Ball Lecture in Cambridge University, 1928

[Härlen 1928] Härlen, H., 'Über Vollständigkeit und Entscheidbarkeit', *Jahresbericht der deutschen Mathematiker-Vereinigung* 37 (1928), pp. 226–230; lecture delivered at the fourth German *Mathematikertagung*, Bad Kissingen, September 1927

[Hartmann 1927] Hartmann, E., 'Der Satz vom ausgeschlossenen Dritten in der Mathematik', *Philosophisches Jahrbuch der Görres-Gesellschaft* 40 (1927), pp. 127–128

[Haupt 1930] Haupt, O., 'Existenzbeweise in der elementaren und höheren Mathematik', *Unterrichtsblätter für Mathematik und Naturwissenschaften* 36 (1930), pp. 224–226; extract from a lecture delivered at a meeting of the *Deutsche Verein zur Förderung des mathematischen und naturwissenschaftlichen Unterrichts* in Würzburg, April 14, 1930

[Hausdorff 1904] Hausdorff, F., 'Der Potenzbegriff in der Mengenlehre', *Jahresbericht der deutschen Mathematikervereinigung* 13 (1904), pp. 569–571

[Hawkins 1973] Hawkins, T., 'Henri Léon Lebesgue', in: [Gillispie 1970–1978, Vol. VIII, pp. 110–112]

[Hazewinkel 1988–1993] Hazewinkel, M. (ed.), *Encyclopaedia of Mathematics*, 9 volumes, Kluwer Academic Publishers, Dordrecht etc. 1988–1993; an updated and annotated translation of the Soviet 'Mathematical Encyclopaedia', I.M. Vinogradov (ed.)

[Hedrick 1933] Hedrick, E.R., 'Tendencies in the logic of mathematics', *Science* 77 (1933), pp. 335–343; address at a meeting of the American Association for the Advancement of Science, Atlantic City, December 28, 1932

[Heidegger 1927] Heidegger, M., 'Sein und Zeit. (Erster Hälfte)', *Jahrbuch für Philosophie und phänomenologische Forschung* 8 (1927), pp. 1–438

[Van Heijenoort 1967] Heijenoort, J. van, *From Frege to Gödel. A Source Book in Mathematical Logic 1879–1931*, Harvard University Press, Cambridge, Massachusetts, 1967

[Heinemann 1959] Heinemann, F. (ed.), *Die Philosophie im XX. Jahrhundert*, Ernst Klett, Stuttgart 1959

[Heinzmann 1986] Heinzmann, G. (ed.), *Poincaré, Russell, Zermelo et Peano*, Blanchard, Paris 1986

[Heiss 1928] Heiss, R., 'Der Mechanismus der Paradoxien und das Gesetz der Paradoxienbildung', *Philosophischer Anzeiger* 2 (1927–1928), pp. 403–433

[Hellinger 1921] Hellinger, E., 'Über Weyls Untersuchungen zu den Grundlagen der Mathematik', lecture delivered at Frankfurt at June 24, 1921; mentioned in *Jahresbericht der deutschen Mathematiker-Vereinigung* 30 (1921), p. *106*

[Hentschel 1990] Hentschel, K., *Interpretationen und Fehlinterpretationen der speziellen und der allgemeinen Relativitätstheorie durch Zeitgenossen Albert Einsteins*, Birkhäuser, Basel etc. 1990

[Herbrand 1930A] Herbrand, J., 'Les bases de la logique hilbertienne', *Revue de Métaphysique et de Morale* 37 (1930), pp. 243–255; cited from: [Herbrand 1968, pp. 155–166]

[Herbrand 1930B] Herbrand, J., *Badania nad teorją dowodu. Recherches sur la théorie de la démonstration*, Nakładem towarzystwa naukowego Warszawskiego, Warszawa 1930

[Herbrand 1931] Herbrand, J., 'Sur la non-contradiction de l'arithmétique', *Journal für die reine und angewandte Arithmetik* 166 (1931), pp. 1–8

[Herbrand 1968] Herbrand, J., *Écrits logiques*, Presses Universitaires de France, Paris 1968

[Hermann & Trommler 1978] Hermann, J., Trommler, F., *Die Kultur der Weimarer Republik*, Nymphenburger, Darmstadt 1978

[Hersch 1979] Hersch, R., 'Some Proposals for Reviving the Philosophy of Mathematics', *Advances in Mathematics* 31 (1979), pp. 31–50

[Herzberg, A. 1928] Herzberg, A., 'Möglichkeitsfragen betreffend den Satz vom ausgeschlossenen Dritten, Kausalität und Telepathie', *Annalen der Philosophie* 7 (1928), pp. 338–343

[Herzberg, L. 1928] Herzberg, L., 'Der mathematische Grundlagenstreit und die Philosophie', *Annalen der Philosophie* 7 (1928), pp. 47–48

[Hesseling 1999] Hesseling, D.E., *Gnomes in the fog. The reception of Brouwer's intuitionism in the 1920s*, Utrecht 1999; dissertation

[Heyerman 1981] Heyerman, E., *Intuition and the intellect. On the relation between mathematics, philosophy and mysticism in the work of L.E.J. Brouwer, including a comparison with Nicolas of Cusa*, Utrecht University, Department of Mathematics, preprint 208, Utrecht 1981

[Heyting 1925] Heyting, A., *Intuitionistische axiomatiek der projectieve meetkunde*, Noordhoff, Groningen 1925; dissertation

[Heyting 1928] Heyting, A., 'Zur intuitionistischen Axiomatik der projektiven Geometrie', *Mathematische Annalen* 98 (1928), pp. 491–538

[Heyting 1929] Heyting, A., 'De telbaarheidspraedicaten van prof. Brouwer', *Nieuw Archief voor wiskunde* (2) 16 (1929), pp. 47–58; lecture delivered before the Wiskundig Genootschap in January 1929

[Heyting 1930A] Heyting, A., 'Die formalen Regeln der intuitionistischen Logik', *Sitzungsberichte der Preussischen Akademie von Wissenschaften, physikalisch-mathematische Klasse* (16/1/1930), pp. 42–56; cited from: [Heyting 1980, pp. 191–205]; English translation in: [Mancosu 1998, pp. 311–327]

[Heyting 1930B] Heyting, A., 'Die formalen Regeln der intuitionistischen Mathematik', *Sitzungsberichte der Preussischen Akademie von Wissenschaften, physikalisch-mathematische Klasse* (16/1/1930), pp. 57–71; cited from: [Heyting 1980, pp. 206–220]

[Heyting 1930C] Heyting, A., 'Die formalen Regeln der intuitionistischen Mathematik', *Sitzungsberichte der Preussischen Akademie von Wissenschaften, physikalisch-mathematische Klasse* (20/3/1930), pp. 158–169; cited from: [Heyting 1980, pp. 221–232]

[Heyting 1930D] Heyting, A., 'Sur la logique intuitionniste', *Académie Royale de Belgique, Bulletin de la Classe des Sciences* S. 5, 16 (1930), pp. 957–963; English translation in: [Mancosu 1998, pp. 306–310]

[Heyting 1931A] Heyting, A., 'Die intuitionistische Grundlegung der Mathematik', *Erkenntnis* 2 (1931), pp. 106–115; lecture delivered at the Königsberg Tagung in September 1930

[Heyting 1931B] Heyting, A., 'Die intuitionistische Mathematik', *Forschungen und Fortschritte* 7 (1931), pp. 38–39; cited from: [Heyting 1980, pp. 250–251]

[Heyting 1932A] Heyting, A., 'A propos d'un article de MM. Barzin et Errera', *L'Enseignement mathématique* 31 (1932), pp. 121–122

[Heyting 1932B] Heyting, A., 'Sur la logique intuitionniste', *L'Enseignement mathématique* 31 (1932), pp. 271–272; letter dated 11/2/1933 [sic]

[Heyting 1932C] Heyting, A., 'Réponse à MM. Barzin et Errera', *L'Enseignement mathématique* 31 (1932), pp. 274–275; letter dated 19/6/1933 [sic]

[Heyting 1932D] Heyting, A., 'Anwendung der intuitionistischen Logik auf die Definition der Vollständigkeit eines Kalküls', in: [Saxer 1932, vol. II, pp. 344–345]; cited from: [Heyting 1980, pp. 261–262]

[Heyting 1934] Heyting, A., *Mathematische Grundlagenforschung · Intuitionismus · Beweistheorie*, Springer, Berlin 1934

[Heyting 1956] Heyting, A., *Intuitionism, an introduction*, North-Holland, Amsterdam 1956; cited from the 1971 3rd edition

[Heyting 1962] Heyting, A., 'After thirty years', in: [Nagel, Suppes & Tarski 1962, pp. 194–197]

[Heyting 1967] Heyting, A., 'L.E.J. Brouwer (27 februari 1881 – 2 december 1966), vernieuwer op de toppen en in de diepste gronden der wiskunde', *De Gids* 130 (1967), pp. 287–294; cited from: [Heyting 1980, pp. 674–681]

[Heyting 1975] Heyting, A., 'Introduction', [Brouwer 1975–1976, part I, pp. XIII–XV]

[Heyting 1978] Heyting, A., 'History of the foundation [sic] of mathematics', *Nieuw Archief voor Wiskunde* (3) XXVI (1978), pp. 1–21

[Heyting 1980] Heyting, A., *Collected Papers*, Amsterdam 1980

[Heyting 1983] Heyting, A., 'The intuitionist foundations of mathematics', in: [Benacerraf & Putnam 1983, pp. 52–61]; English translation of [Heyting 1931A]

[Heywood 1947] Heywood, R.B. (ed.), *The Works of the Mind*, The University of Chicago Press, Chicago 1947

[Hilbert 1899] Hilbert, D., *Grundlagen der Geometrie*, Teubner, Berlin etc. 1899; cited from [Hilbert 1909]

[Hilbert 1900] Hilbert, D., 'Über den Zahlbegriff', *Jahresbericht der deutschen Mathematiker-Vereinigung* 8 (1900), pp. 180–184; English translation in [Ewald 1996, vol. II, pp. 1089–1095]

[Hilbert 1901] Hilbert, D., 'Mathematische Probleme', *Archiv für Mathematik und Physik* 3. Reihe, Band 1 (1901), pp. 44–63; 213–237; cited from: [Hilbert 1932–1935, vol. III, pp. 290–329]; partial English translation in [Ewald 1996, vol. II, pp. 1096–1105]; address delivered at the International Mathematicians' Conference in Paris, 1900

[Hilbert 1905A] Hilbert, D., 'Über die Grundlagen der Logik und Arithmetik', *Verhandlungen des 3. internationalen Mathematiker-Kongresses in Heidelberg*, Leipzig 1905, pp. 174–185; lecture delivered in 1904; cited from: [Hilbert 1909, pp. 263–279]; English translation in: [Van Heijenoort 1967, pp. 129–138]

[Hilbert 1905B] Hilbert, D., 'Über das Dirichletsche Prinzip', *Journal für die reine und angewandte Mathematik* 129 (1905), pp. 63–67; cited from: [Hilbert 1932–1935, vol. III, pp. 10–14]

[Hilbert 1909] Hilbert, D., *Grundlagen der Geometrie*, Teubner, Leipzig etc. 1909; third edition

[Hilbert 1918] Hilbert, D., 'Axiomatisches Denken', *Mathematische Annalen* 78 (1918), pp. 405–415; cited from: [Hilbert 1932–1935, vol. 3, pp. 146–156]; English translation in [Ewald 1996, vol. II, pp. 1105–1115]

[Hilbert 1922] Hilbert, D., 'Neubegründung der Mathematik. Erste Mitteilung', *Abhandlungen aus dem Mathematischen Seminar der Hamburgischen Universität* I (1922), pp. 157-177; English translation in [Ewald 1996, vol. II, pp. 1115-1134]; reprinted in: [Mancosu 1998, pp. 198-214]; lectures delivered in spring 1921 in Copenhagen and in July 1921 in Hamburg

[Hilbert 1923] Hilbert, D., 'Die logischen Grundlagen der Mathematik', *Mathematische Annalen* 88 (1923), pp. 151-165; cited from: [Hilbert 1932-1935, vol. 3, pp. 178-191]; English translation in [Ewald 1996, vol. II, pp. 1134-1148]; lecture delivered in September 1922 before the *Deutsche Naturforscher-Gesellschaft* in Leipzig.

[Hilbert 1926] Hilbert, D., 'Über das Unendliche', *Mathematische Annalen* 95 (1926), pp. 161-190; English translation in: [Van Heijenoort 1967, pp. 367-392]; lecture delivered at the Weierstraß memorial meeting of the *Westfälische Mathematische Gesellschaft* in Münster on June 4, 1925

[Hilbert 1927] Hilbert, D., 'Die Grundlagen der Mathematik', *Abhandlungen aus dem Mathematischen Seminar der Hamburgischen Universität* V (1927), pp. 65-85; English translation in: [Van Heijenoort 1967, pp. 464-479]; lecture delivered in July 1927 in Hamburg

[Hilbert & Ackermann 1928] Hilbert, D., Ackermann, W., *Grundzüge der theoretischen Logik*, Springer, Berlin 1928

[Hilbert 1930A] Hilbert, D., 'Probleme der Grundlegung der Mathematik', *Mathematische Annalen* 102 (1930), pp. 1-9; English translation in: [Mancosu 1998, pp. 227-233]; lecture delivered at the International Mathematicians' Conference in Bologna on September 3, 1928

[Hilbert 1930B] Hilbert, D., 'Naturerkennen und Logik', *Die Naturwissenschaften* 18 (1930), pp. 959-963; cited from: [Hilbert 1932-1935, vol. 1, pp. 378-387]; English translation in [Ewald 1996, vol. II, pp. 1157-1165]

[Hilbert & Bernays 1934-1939] Hilbert, D., Bernays, P., *Grundlagen der Mathematik*, Springer, Berlin 1934-1939; two volumes

[Hilbert 1932-1935] Hilbert, D., *Gesammelte Abhandlungen*, Springer, Berlin 1932-1935; three volumes

[Hilbert 1992] Hilbert, D., *Natur und mathematisches Erkennen. Nach der Ausarbeitung von Paul Bernays*, Rowe, D.E. (ed.), Birkhäuser, Basel etc.; lectures delivered in Göttingen in 1919-1920

[Hirsch 1929] Hirsch, K., 'Neuaufbau der Mathematik', *Das Unterhaltungsblatt der Vossischen Zeitung* 8/10/1929

[Hirsch 1930] Hirsch, K., 'Philosophie der Mathematik. Tagung für Erkenntnislehre in Königsberg', source and date unknown; presumably published in the *Vossische Zeitung* in 1930; in: [JNUL Fraenkel]

[Hirsch 1932] Hirsch, K., 'Der Streit um das Unendliche', source and date unknown; presumably published in 1932; in: [JNUL Fraenkel]

[Hirsch 1933] Hirsch, K., *Intuition und logische Form. Zur gegenwärtigen Philosophie der Mathematik*, 1933; dissertation

[Hjelmslev 1916] Hjelmslev, J., *Elementær Geometri*, København 1916

[Hoffmann 1994] Hoffmann, D., 'Zur Geschichte der Berliner 'Gesellschaft für empirische/wissenschaftliche Philosophie' ', in: [Danneberg et al. 1994, pp. 21–31]

[Hofmann 1931] Hofmann, P., *Das Problem des Satzes vom ausgeschlossenen Dritten*, Pan-Verlaggesellschaft, Berlin 1931; also in *Kant-Studien* 36 (1931), pp. 84–125

[Hölder 1924] Hölder, O., *Die mathematische Methode. Logisch erkenntnistheoretische Untersuchungen im Gebiete der Mathematik Mechanik und Physik*, Springer, Berlin 1924

[Hölder 1926] Hölder, O., 'Der angebliche circulus vitiosus und die sogenannte Grundlagenkrise in der Analysis', *Berichte über die Verhandlungen der Sachsischen Akademie der Wissenschaften, Mathematisch-physikalische Klasse* 78 (1926), pp. 243–250; English translation in: [Mancosu 1998, p. 143–148]

[Honisch et al. 1977] Honisch, D., et al., *Tendenzen der Zwanziger Jahre. 15. Europäische Kunstausstellung Berlin 1977*, Dietrich Reimer, Berlin 1977

[Honisch & Prinz 1977] Honisch, D., Prinz, U. (eds.), *Dokumente*, in: [Honisch et al. 1977, pp. 1/84–1/109]

[Horák 1925] Horák, J.M., 'Sur les antinomies de la théorie des ensembles', *Bulletin internationale de l'Académie des Sciences de Bohême* 26 (1925), pp. 38–44

[Hurwitz 1898] Hurwitz, A., 'Über die Entwickelung der allgemeinen Theorie der analytischen Funktionen in neuerer Zeit', in: [Rudio 1898, pp. 91–112]

[Husserl 1913] Husserl, E., 'Ideen zu einer reinen Phänomenologie und phänomenologischen Philosophie. I. Allgemeine Einführung in die reine Phänomenologie', *Jahrbuch für Philosophie und phänomenologische Forschung* 1 (1913), pp. 1–323; cited from the 1976 edition, Schuhmann, K. (ed.), Martinus Nijhoff, Den Haag

[Husserl 1970] Husserl, E., *Philosophie der Arithmetik*, Eley, L. (ed.), mit ergänzenden Texten (1890–1901), Martinus Nijhoff, Den Haag 1970

[Husserl 1994] Husserl, E., *Briefwechsel*, Schuhmann, K. und E. (eds.), Kluwer, Dordrecht etc. 1994; neun Bände; the ones cited from are: Band III, Die Göttinger Schule; Band IV, Die Freiburger Schule

[Jeltsch-Fricker 1988] Jeltsch-Fricker, R., 'In memoriam Alexander M. Ostrowski', *Elemente der Mathematik* 43 (1988), pp. 33–38

[Jesser & Bohr 1937] Jesser, B., Bohr, H., 'Über die Umkehrung von analytischen fastperiodischen Funktionen', *Mathematische Annalen* 113 (1937), pp. 461–488

[Johnson 1987] Johnson, D.M., 'L.E.J. Brouwer's coming of age as a topologist', in: [Phillips 1987, pp. 61-97]

[Joll 1990] Joll, J., *Europe since 1870. An international history*, Penguin, London etc. 1990, fourth edition

[Jørgensen 1931] Jørgensen, J., *A treatise of formal logic. Its evolution and main branches, with its relations to mathematics and philosophy*, vol. 3 , Levin & Munksgaard and Humprey Milford, Copenhagen and London 1931

[Jørgensen 1932/33] Jørgensen, J., 'Über die Ziele und Probleme der Logistik', *Erkenntnis* 3 (1932/33), pp. 73-100

[Juvet 1927] Juvet, G., 'Les fondements des mathématiques de la géométrie d'Euclide à la relativité générale et à l'intuitionisme', *Revue générale des sciences pures et appliquées* 38 (1927), pp. 133-140

[Kahn 1975] Kahn, D.W., *Topology. An introduction to the point-set and algebraic areas*, The Williams & Wilkins Company, Baltimore 1975

[Kant 1912-1918] Kant, I., *Werke*, Cassirer, E., et al. (eds.), Bruno Cassirer, Berlin 1912-1918, XI Bände

[Kant 1781/1787] Kant, I., *Kritik der reinen Vernunft*, Kehrbach, K., Philipp Reclam jun., zweite verbesserte Auflage, Leipzig 1787; first edition 1781; cited from: [Kant 1912-1918, Band III], Görland, A. (ed.), Bruno Cassirer, Berlin 1913

[Kant 1998] Kant, I., *Critique of pure reason*, Guyer, P., Wood, A.W. (eds. and translators), Cambridge University Press, Cambridge 1998; English translation of [Kant 1781/1787]

[Kass 1996] Kass, S., 'Karl Menger', *Notices of the AMS* 43 (1996), pp. 558-561

[Kaufmann 1930] Kaufmann, F., *Das Unendliche in der Mathematik und seine Ausschaltung*, Franz Deuticke, Leipzig und Wien 1930; second, unchanged edition, Wissenschaftliche Buchgesellschaft, Darmstadt 1968

[Kaufmann 1931] Kaufmann, F., 'Bemerkungen zum Grundlagenstreit in Logik und Mathematik', *Erkenntnis* 2 (1931), pp. 262-290

[Keirsey & Bates 1984] Keirsey, D., Bates, M., *Please understand me. Character & temperament types*, Prometheus Nemesis, Del Mar 1984

[Kerszberg 1997] Kerszberg, P., 'Henri Bergson', in: [Embree et al. 1997, pp. 56-61]

[Kienzler 1994] Kienzler, W., *Wittgensteins Wende zu seiner Spätphilosophie 1930-1932. Eine historische und systematische Darstellung*, Konstanz 1994; dissertation

[Kitcher & Aspray 1988] Kitcher, P., Aspray, W., 'An Opiniated Introduction', in: [Aspray & Kitcher 1988, pp. 3-57]

[Kleene 1936] Kleene, S.C., '$\lambda$-definability and recursiveness', *Duke Mathematical Journal* 2 (1936), pp. 340-353

[Kleene 1986] Kleene, S.C., 'Introductory note to *1930b*, *1931* and *1932b*', in: [Gödel 1986–1995, vol. I, pp. 126–141]

[Klein 1911] Klein, F., *Lectures on Mathematics*, republished by the American Mathematical Society, New York 1911; lectures delivered at the Evanston Colloquium from August 28 to September 9, 1893, before members of the Congress of Mathematics held in connection with the World's Fair in Chicago; reported by Alexander Ziwet

[Kline 1972] Kline, M., *Mathematical Thought from Ancient to Modern Times*, Oxford University Press, New York 1972

[Klomp 1997] Klomp, H., *De Relativiteitstheorie in Nederland. Breekijzer voor democratisering in het Interbellum*, Epsilon, Utrecht 1997; dissertation

[Kneale & Kneale 1964] Kneale, W., Kneale, M., *The development of logic*, Clarendon Press, Oxford, second edition 1964

[Kneser 1925] Kneser, A., 'Leopold Kronecker', *Jahresbericht der deutschen Mathematiker-Vereinigung* 33 (1925), pp. 210–228

[Knorr 1983] Knorr, W.R., 'Construction as Existence Proof in Ancient Geometry', *Ancient Philosophy* 3 (1983), pp. 125–148

[Koetsier & Van Mill 1997] Koetsier, T., Van Mill, J., 'General Topology, in Particular Dimension Theory, in The Netherlands: the Decisive Influence of Brouwer's Intuitionism', in: [Aull & Lowen 1997, pp. 135–180]

[Köhler 1991] Köhler, E., *Gödel und der Wiener Kreis*, in: [Kruntorad 1991, pp. 127–158]

[Kohnstamm 1926] Kohnstamm, Ph., *Schepper en schepping. Een stelsel van personalistische wijsbegeerte op bijbelschen grondslag*, vol. 1: Het waarheidsprobleem. Grondleggende kritiek van het christelijk waarheidsbewustzijn, Tjeenk Willink & Zoon, Haarlem 1926

[Kok 1992] Kok, J.H., *Vollenhoven. His early development*, Dordt College Press, Iowa 1992; dissertation

[Kolmogorov 1925] Колмогоров, А. Н., 'О принципе tertium non datur', Математический сборник 32 (1925), pp. 646–667; cited from the English translation 'On the principle of the excluded middle' in: [Van Heijenoort 1967, pp. 414–437]

[Kolmogorov 1932] Kolmogorov, A.N., 'Zur Deutung der intuitionistischen Logik', *Mathematische Zeitschrift* 35 (1932), pp. 58–65; English translation in: [Mancosu 1998, pp. 328–334]

[Kolmogorov & Forin 1975] Kolmogorov, A.N., Forin, S.V., *Reelle Funktionen und Funktionalanalysis*, Deutscher Verlag der Wissenschaften, Berlin 1975; translated from Russian

[König 1956] König, R., 'Hermann Weyl 9/11/1885 – 9/12/1955', *Bayerische Akademie der Wissenschaften, Jahrbuch 1956*, pp. 236–248

[Korselt 1916] Korselt, A., 'Auflösung einiger Paradoxien', *Jahresbericht der deutschen Mathematiker-Vereinigung* 25 (1916), pp. 132–138

[Koschmieder 1921] Koschmieder, L., 'Über den Brouwer-Weylschen Zahlbegriff', lecture delivered in Breslau in (presumably) 1921; reported in *Jahresbericht der deutschen Mathematiker-Vereinigung* 30 (1921), p. 106

[Krämer 1988] Krämer, S., *Symbolische Maschinen. Die Idee der Formalisierung in geschichtlichem Abriß*, Wissenschaftliche Buchgesellschaft, Darmstadt 1988

[Kreisel 1962] Kreisel, G., 'Foundations of Intuitionistic Logic', in: [Nagel, Suppes & Tarski 1962, pp. 198–210]

[Kreisel 1964] Kreisel, G., 'Hilbert's programme', in: [Benacerraf & Putnam 1964, pp. 157–180]; revised version of a paper in *Dialectica* 12 (1958), pp. 346–372

[Kreisel 1968] Kreisel, G., 'Lawless sequences of natural numbers', *Compositio Mathematica* 20 (1968), pp. 222–248

[Kreisel & Newman 1969] Kreisel, G., Newman, M.H.A., 'Luitzen Egbertus Jan Brouwer', *Biographical memoirs of fellows of the Royal Society* 15 (1969), pp. 39–68

[Kronecker 1887] Kronecker, L., 'Über den Zahlbegriff', *Crelle, Journal für die reine und angewandte Mathematik* 101 (1887), pp. 337–355; revised version of the *Philosophische Aufsätze, Eduard Zeller zu seinem fünfzigjährigen Doctor-Jubiläum gewidmet* VIII, Leipzig 1887, pp. 261–274; English translation in [Ewald 1996, vol. II, pp. 947–955]

[Kroner 1910/11] Kroner, R., 'Henri Bergson', *Logos. Internationale Zeitschrift für Philosophie der Kultur* I (1910/11), pp. 125–150

[Krüger & Falkenburg 1995] Krüger, L., Falkenburg, B. (eds.), *Physik, Philosophie und die Einheit der Wissenschaft*, Spektrum, Heidelberg etc. 1995

[Kruntorad 1991] Kruntorad, P. (ed.), *Jour fixe der Vernunft. Der Wiener Kreis und die Folgen*, Hölder-Pichler-Tempsky, Wien 1991

[Kundera 1993] Kundera, M., *Les testaments trahis*, Gallimard, 1993

[Kundera 1995] Kundera, M., *Testaments Betrayed*, Faber and Faber, London and Boston 1995; English translation of [Kundera 1993] by Linda Asher

[Kushner 1988] Kushner, B.A., 'Constructive mathematics', in: [Hazewinkel 1988–1993, Vol. 2, pp. 356–359]

[Kynast 1930] Kynast, R., *Logik und Erkenntnistheorie der Gegenwart*, Junker und Dünnhaupt, Berlin 1930

[Landry 1927] Landry, H., 'Grundlagenkrisis der Logik', newpaper article, presumably in a 1927 issue of the *Vossische Zeitung*; in: [JNUL Fraenkel]

[Larguier des Bancels 1926] Larguier des Bancels, J., 'La logique d'Aristote et le principe du tiers exclu', *Revue de théologie et de philosophie* (nouvelle série) 14 (1926), pp. 120–124

[Lebesgue 1904] Lebesgue, H., *Leçons sur l'intégration et la recherche des fonctions primitives*, Gauthier-Villars, Paris 1904; cited from: [Lebesgue 1972–73, vol. II]

[Lebesgue 1972–73] Lebesgue, H., *Œuvres scientifiques*, L'Enseignement mathématique, 1972–1973; five volumes

[Lechner 1935] Lechner, G., 'Soldaten und Mathematik', *Jahresbericht der deutschen Mathematiker-Vereinigung* 45 (1935), pp. *15–17*

[Leclerc 1927] Leclerc, M. (ed.), 'L'Axiomatique logique et le principe du tiers exclu', *Bulletin de la Société française de Philosophie* 1927, pp. 1–23; session of January 29, 1927; lecture by Reymond, followed by a general discussion

[Leśniewski 1929] Leśniewski, S., 'Grundzüge eines neuen Systems der Grundlagen der Mathematik', *Fundamenta Mathematicae* 14 (1929), pp. 1–81

[Leupold 1961] Leupold, R., *Die Grundlagenforschung bei Hermann Weyl*, Mainz 1961; Inaugural-Dissertation

[Levi 1923] Levi, B., 'Sui procedimenti transfiniti', *Mathematische Annalen* 90 (1923), pp. 164–173

[Lévy 1926A] Lévy, P., Sur le principe du tiers exclu et sur les théorèmes non susceptibles de démonstration', *Revue de Métaphysique et de Morale* 33 (1926), pp. 253–258

[Lévy 1926B] Lévy, P., 'Critique de la logique empirique', *Revue de Métaphysique et de Morale* 33 (1926), pp. 545–551

[Lévy 1927] Lévy, P., 'Logique classique, Logique brouwerienne et Logique mixte', *Académie Royale de Belgique, Bulletins de la Classe des Sciences* 5-13 (1927), pp. 256–266

[Lévy 1964] Lévy, P., 'Remarques sur un théorème de Paul Cohen', *Revue de Métaphysique et de Morale* 69 (1964), pp. 88–94

[Lévy 1970] Lévy, P., *Quelques aspects de la pensée d'un mathématicien*, Blanchard, Paris 1970

[Lietzmann 1925] Lietzmann, W., 'Formalismus und Intuitionismus in der Mathematik', *Zeitschrift für mathematischen und naturwissenschaftlichen Unterricht* 56 (1925), pp. 355–358

[Lietzmann 1927] Lietzmann, W., *Aufbau und Grundlage der Mathematik*, Teubner, Leipzig und Berlin 1927

[Lietzmann 1930] Lietzmann, W., report on the *Versammlung Deutscher Philologen und Schulmänner in Salzburg, 25.-28. September 1929*, *Zeitschrift für mathematischen und naturwissenschaftlichen Unterricht aller Schulgattungen* 61 (1930), pp. 34–36

[Lietzmann 1949] Lietzmann, W., *Das Wesen der Mathematik*, Vieweg & Sohn, Braunschweig 1949

[Lietzmann 1960] Lietzmann, W., *Aus meinen Lebenserinnerungen*, Fladt, K. (ed.), Vandenhoeck & Ruprecht, Göttingen 1960

[Lindner 1980] Lindner, H., ' 'Deutsche' und 'gegentypische' Mathematik. Zur Begründung einer 'arteigenen' Mathematik im 'Dritten Reich' durch Ludwig Bieberbach', in: [Mehrtens & Richter 1980, pp. 88–115]

[Lipps 1925] Lipps, H., 'Bemerkungen zur Theorie der Prädikation', *Philosophischer Anzeiger* 1 (1925–1926), pp. 57–71

[Lipps 1927–1928] Lipps, H., *Untersuchungen zur Phänomenologie der Erkenntnis*, Zwei Teile, Friedrich Cohen, Bonn 1927–1928

[Litt 1925] Litt, Th., *Die Philosophie der Gegenwart und ihr Einfluss auf das Bildungsideal*, Teubner, Leipzig etc. 1925

[Litt 1930] Litt, Th., *Die Philosophie der Gegenwart und ihr Einfluss auf das Bildungsideal*, dritte abermals verbessert Auflage, Teuber, Leipzig etc. 1930

[Loddur 1996] Loddur, C., 'Constructivism', in: [Turner 1996, vol. 7, pp. 767–772]

[London 1923] London, F., 'Über die Bedingungen der Möglichkeit einer deduktiven Theorie', *Jahrbuch für Philosophie und phänomenologische Forschung* 6 (1923), pp. 335–384

[Lopshitz & Rashevskii 1969] Lopshitz, A.M., Rashevskii, P.K., *Veniamin Fedorovich Kagan*, Moscow State University, Moscow 1969

[Lorenz 1994] Lorenz, Chr., *De constructie van het verleden. Een inleiding in de theorie van de geschiedenis*, Boom, Amsterdam etc. 1994, vierde, herziene druk

[Lorenzen 1950] Lorenzen, P., 'Konstruktive Begründung der Mathematik', *Mathematische Zeitschrift* 53 (1950), pp. 162–202

[Löwy 1926] Löwy, H., 'Die Krisis in der Mathematik und ihre philosophische Bedeutung', *Die Naturwissenschaften* 30 (1926), pp. 706–708

[Löwy 1932/22] Löwy, H., 'Commentar zu Jozef Poppers Abhandlung 'Über die Grundbegriffe der Philosophie und die Gewißheit unserer Erkenntnisse' ', *Erkenntnis* 3 (1932/33), pp. 324–347

[Lukács 1960] Lukács, G., *Die Zerstörung der Vernunft*, Luchterhand, Darmstadt etc. 1960; cited from the 1974 edition as volume 9 in Lukács *Werke*

[Łukasiewicz 1930] Łukasiewicz, J., 'Uwagi filozoficzne o wielowartościowych systemach rachunku zdań/Philosophische Bemerkungen zu mehrwertigen Systemen des Aussagenkalküls', *Sprawozdania z posiedzen towarzystwa naukowego Warszawskiego/Société des sciences et des lettres de Varsovie* 23 (1930), pp. 51–77

[Mahnke 1927] Mahnke, D., 'Leibniz als Begründer der symbolischen Mathematik', *Isis* 30 (1927), pp. 279–293

[Majer 1988] Majer, U., 'Zu einer bemerkenswerten Differenz zwischen Brouwer und Weyl', in: [Deppert et al. 1988, pp. 543–552]

[Majer 1989] Majer, U., 'Ramsey's conception of theories: an intuitionistic approach', *History of Philosophy Quarterly* 6 (1989), pp. 233–258

[Mancosu 1998] Mancosu, P., *From Brouwer to Hilbert. The debate on the foundations of mathematics in the 1920s*, Oxford University Press, New York etc. 1998

[Mancosu 1999A] Mancosu, P., 'Between Vienna and Berlin: The Immediate Reception of Gödel's Incompleteness Theorems', *History and Philosophy of Logic* 20 (1999), pp. 33–45

[Mancosu 1999B] Mancosu, P., 'Between Russell and Hilbert: Behmann on the foundations of mathematics', *The Bulletin of Symbolic Logic* 5 (1999), pp. 303–330

[Mancosu 2002] Mancosu, P., 'On the constructivity of proofs. A debate among Behmann, Bernays, Gödel, and Kaufmann', in: *Reflections on the foundations of mathematics. Essays in honor of Solomon Feferman*, Sieg, W., Sommer, R., Talcott, C. (eds.), Association for Symbolic Logic, Lecture Notes in Logic 15 (2002), pp. 346–368

[Mancosu & Ryckman 2002] Mancosu, P., Ryckman, T., 'Mathematics and Phenomenology: The correspondence between Oskar Becker and Hermann Weyl', *Philosophia Mathematica* 10 (2002), pp. 130–202

[Mannoury 1898–99] Mannoury, G., 'Lois cyclomatiques', *Nieuw Archief voor Wiskunde* Tweede reeks, deel III (1898–1899), pp. 126–152

[Mannoury 1900A] Mannoury, G., 'Sphères de seconde espèce', *Nieuw Archief voor Wiskunde* Tweede reeks, deel IV (1900), pp. 83–89

[Mannoury 1900B] Mannoury, G., 'Surfaces-images', *Nieuw Archief voor Wiskunde* Tweede reeks, deel IV (1900), pp. 112–129

[Mannoury 1909] Mannoury, G., *Methodologisches und Philosophisches zur Elementar-Mathematik*, P. Visser Azn., Haarlem 1909

[Mannoury 1917] Mannoury, G., *Over de sociale betekenis van de wiskundige denkvorm*, Noordhoff, Groningen 1917; address delivered at the occasion of accepting the extraordinary professorship at the *Universiteit van Amsterdam* on October 8, 1917

[Mannoury 1924] Mannoury, G., *Mathesis en mystiek. Een signifiese studie van kommunisties standpunt*, Maatschappij voor goede en goedkoope lectuur, Amsterdam 1924; cited from: reprint at Bohn, Scheltema & Holkema, Utrecht 1978

[Mannoury 1931] Mannoury, G., *Woord en gedachte. Een inleiding tot de signifika, inzonderheid met het oog op het onderwijs in de wiskunde*, Noordhoff, Groningen 1931

[Mannoury 1934A] Mannoury, G., 'Die signifischen Grundlagen der Mathematik', *Erkenntnis* 4 (1934), pp. 288–309

[Mannoury 1934B] Mannoury, G., 'Die signifischen Grundlagen der Mathematik', *Erkenntnis* 4 (1934), pp. 317–345

[Marcus 1928] Marcus, E., 'Der Triumpf über die Logik', article in unknown newspaper, 1928; in: [MI Brouwer]

[Mark et al. 1933] Mark, H., Thirring, H., Hahn, H., Nöbeling, G., and Menger, K., *Krise und Neuaufbau in den exakten Wissenschaften*, Franz Deuticke, Leipzig und Wien 1933; lectures delivered in Vienna in 1932–33

[Mason 1996] Mason, R., 'Wittgenstein', in: [Brown, Collinson & Wilkinson 1996, pp. 845–849]

[McCarty 1983] McCarty, C., 'Intuitionism: an introduction to a seminar', *Journal of Philosophical Logic* 12 (1983), pp. 105–149

[McCormmach 1971] McCormmach, R. (ed.), *Historical Studies in the Physical Sciences* Volume 3, University of Pennsylvania Press, Philadelphia 1971

[Mehrtens & Richter 1980] Mehrtens, H., Richter, S., *Naturwissenschaft, Technik und NS-Ideologie. Beiträge zur Wissenschaftsgeschichte des Dritten Reiches*, Suhrkamp, Frankfurt am Main 1980

[Mehrtens 1987] Mehrtens, H., 'Ludwig Bieberbach and 'Deutsche Mathematik'', in: [Phillips 1987, pp. 195–241]

[Mehrtens 1990] Mehrtens, H., *Moderne · Sprache · Mathematik. Eine Geschichte des Streits um die Grundlagen der Disziplin und des Subjekts formaler Systeme*, Suhrkamp, Frankfurt am Main 1990

[Mehrtens 1996] Mehrtens, H,. 'Modernism vs. Counter-Modernism, Nationalism vs. Internationalism: Style and Politics in Mathematics, 1900–1950', in: [Goldstein et al. 1996, pp. 518–529]

[Mehrtens 1998] Mehrtens, H., 'Wider den mathematischen Fundamentalismus', *Ethik und Sozialwissenschaften. Streitforum für Erwägungskultur* 9 (1998), pp. 467–469

[Menger 1926] Menger, K., 'Bericht über die Dimensionstheorie', *Jahresbericht der deutschen Mathematiker-Vereinigung* 35 (1926), pp. 113–150

[Menger 1928A] Menger, K., 'Bemerkungen zu Grundlagenfragen', *Jahresbericht der deutschen Mathematiker-Vereinigung* 37 (1928), pp. 213–226

[Menger 1928B] Menger, K., 'Bemerkungen zu Grundlagenfragen. II. Die Mengentheoretischen Paradoxien', *Jahresbericht der deutschen Mathematiker-Vereinigung* 37 (1928), pp. 298–302

[Menger 1928C] Menger, K., 'Bemerkungen zu Grundlagenfragen. III. Über Potenzmengen', *Jahresbericht der deutschen Mathematiker-Vereinigung* 37 (1928), pp. 303–308

[Menger 1928D] Menger, K., 'Bemerkungen zu Grundlagenfragen. IV. Axiomatik der endlichen Mengen und der elementargeometrischen Verknüpfungsbeziehungen', *Jahresbericht der deutschen Mathematiker-Vereinigung* 37 (1928), pp. 309–325

[Menger 1930A] Menger, K., 'Der Intuitionismus', *Blätter für Deutsche Philosophie* 4 (1930–1931), pp. 311-325

[Menger 1930B] Menger, K., 'Über die sogenannte Konstruktivität bei arithmetischen Definitionen', *Anzeiger der Akademie der Wissenschaften in Wien, mathematisch-naturwissenschaftliche Klasse* 67 (1930), pp. 257–258

[Menger 1931] Menger, K., 'Über den Konstruktivitätsbegriff. Zweite Mitteilung', *Anzeiger der Akademie der Wissenschaften in Wien, mathematisch-naturwissenschaftliche Klasse* 68 (1931), pp. 7–8

[Menger 1933] Menger, K., 'Die neue Logik', in: [Mark et al. 1933, pp. 93–122]

[Menger 1994] Menger, K., *Reminiscences of the Vienna Circle and the Mathematical Colloquium*, Golland, L, McGuinness, B., Sklar, A. (eds.), Kluwer, Dordrecht etc. 1994

[Meschkowski 1972] Meschkowski, H. (ed.), *Grundlagen der modernen Mathematik*, Wissenschaftliche Buchgesellschaft, Darmstadt 1972

[Messer 1927] Messer, A., *Einführung in die Erkenntnistheorie*, Felix Meiner, Leipzig 1927, dritte, umgearbeitete Auflage

[Meyer 1982] Meyer, R., 'Bergson in Deutschland. Unter besonderer Berücksichtigung seiner Zeitauffassung', *Phänomenologische Forschungen* 13 (1982), pp. 10–64

[Minkowski 1973] Minkowski, H., *Briefe an David Hilbert*, Rüdenberg, L., Zassenhaus, H. (eds.), Springer, Berlin 1973

[Mints 1988] Mints, G.E., 'Constructive logic', in: [Hazewinkel 1988–1993, pp. 354–356]

[Von Mises 1921] Mises, R. von, 'Über die gegenwärtige Krise der Mechanik', *Zeitschrift für angewandte Mathematik und Mechanik* 1 (1921), pp. 425–431

[Mittag-Leffler 1927] Mittag-Leffler, G., 'Zusätzliche Bemerkungen', *Acta Mathematica* 50 (1927), pp. 25–26

[Mittelstraß 1980–96] Mittelstraß, J. (ed.), *Enzyklopädie Philosophie und Wissenschaftstheorie*, B.I.-Wissenschaftsverlag, Mannheim etc. - Metzler, Stuttgart etc., 4 Bände, 1980–1996

[Molk 1885] Molk, J., 'Sur une notion qui comprend celle de la divisibilité et sur la théorie générale de l'élimination', *Acta Mathematica* 6 (1885), pp. 1–166

[Monk 1990] Monk, R., *Wittgenstein: the duty of genius*, Jonathan Cape, London 1990

[Monna 1975] Monna, A.F., *Dirichlet's Principle. A mathematical comedy of errors and its influence on the development of analysis*, Oosthoek, Scheltema & Holkema, Utrecht 1975

[Monnoyeur 1995] Monnoyeur, F. (ed.), *Infini des philosophes, infini des astronomes*, Belin, Paris 1995

[Moore 1978] Moore, G.H., 'The origin of Zermelo's axiomatization of set theory', *Journal of Philosophical Logic* 7 (1978), pp. 307–329

[Moore 1980] Moore, G.H., 'Beyond First-order Logic: The Historical Interplay between Mathematical Logic and Axiomatic Set Theory', *History and Philosophy of Logic* 1 (1980), pp. 95-137

[Moore 1982] Moore, G.H., *Zermelo's Axiom of Choice. Its Origins, Development, and Influence*, Springer, New York etc. 1982

[Moore 1987] Moore, G.H., 'A house devided against itself: the emergence of first-order logic as the basis for mathematics', in: [Phillips 1987, pp. 98–136]

[Mueller 1981] Mueller, I., *Philosophy of Mathematics and Deductive Structure in Euclid's Elements*, MIT Press, Cambridge, Massachusetts, etc., 1981

[Müller, C. 1986] Müller, C., 'Zum 100. Geburtstag von Hermann Weyl', *Jahresbericht der Deutschen Mathematiker-Vereinigung* 88 (1986), pp. 159–189

[Müller, H.M. 1996] Müller, H.M., *Schlaglichter der deutschen Geschichte*, Bundeszentrale für politische Bildung, Bonn 1996, dritte, überarbeitete und erweiterte Auflage

[Nagel 1929] Nagel, E., 'Intuition, consistency, and the excluded middle', *The Journal of Philosophy* 26 (1929), pp. 477–489

[Nagel, Suppes & Tarski 1962] Nagel, E., Suppes, P., and Tarski, A. (eds.), *Logic, methodology and philosophy of science. Proceedings of the 1960 International Congress*, Stanford University Press, Stanford 1962

[Nazi 'purge' 1933] 'Nazi 'purge' of the universities', *The Manchester Guardian Weekly* May 19, 1933 (no author mentioned); cited from: [Flemming & Bailyn 1969, p. 234]

[Van Nes 1901] Nes, H.M. van, *De nieuwe mystiek*, Bredée, Rotterdam 1901, second edition

[Von Neumann 1924] Neumann, J. von, 'Eine Axiomatisierung der Mengenlehre', *Journal für die reine und angewandte Mathematik* 154 (1924), pp. 219–240; English translation in: [Van Heijenoort 1967, pp. 393–413]

[Von Neumann 1927] Neumann, J. von, 'Zur Hilbertschen Beweistheorie', *Mathematische Zeitschrift* 26 (1927), pp. 1–46

[Von Neumann 1931] Neumann, J. von, 'Die formalistische Grundlegung der Mathematik', *Erkenntnis* 2 (1931), pp. 116–121; lecture delivered at the Königsberg Tagung in September 1930

[Von Neumann 1947] Neumann, J. von, 'The mathematician', in: [Heywood 1947, pp. 180–196]; lecture delivered at the University of Chicago in 1946

[Neurath 1930] Neurath, O., 'Historische Bemerkungen', *Erkenntnis* 1 (1930/31), pp. 311–314

[Nicholson 1997] Nicholson, G., 'Hermeneutic phenomenology', in: [Embree et al. 1997, pp. 304–308]

[Null 1997] Null, G.T., 'Formal and material ontology', in: [Embree et al. 1997, pp. 237–241]

[Oikonomou 1926] ⎯⎯⎯⎯⎯, '⎯⎯⎯⎯⎯⎯⎯⎯⎯⎯⎯⎯⎯⎯⎯⎯⎯⎯', *Bulletin de la Société Mathématique Grèce* 7 (1926), pp. 67–80; lecture delivered before the Greek Mathematical Society on November 22, 1925

[Ore 1939] Ore, O., 'James Pierpont – in memoriam', *Bulletin of the American Mathematical Society* 45 (1939), pp. 481–486

[Orlov 1928] Orlov, I.E., 'Über die Theorie der Verträglichkeit von Aussagen', *Bulletin de l'Association de Recherches scientifiques à la Faculté des Sciences de la première Université de Moscou* 35 (1928), pp. 263–286; Russian with a French abstract; I have not seen the original; review in *Fortschritte der Mathematik* 54 (1928), p. 54

[Van Os 1933] Os, Ch. H. van, 'Wiskunde en wijsbegeerte', *De Gids* 97-3 (1933), pp. 192–221

[Ostrowski 1921] Ostrowski, A., 'David Hilbert, Zu seinen Vorträgen über die Grundlagen der Mathematik', article in a Hamburg newspaper, July 1921; in: [NSUB Hilbert, 751]

[Otto 1926] Otto, R., *West-östliche Mystik: Vergleich und Unterscheidung zur Wesensdeutung*, 1926; cited from the second edition, Perthes, Gotha 1929

[Pacotte 1925] Pacotte, J., *La pensée mathématique contemporaine*, Librairie Félix Alcan, Paris 1925

[Palmer & Colton 1995] Palmer, R., Colton, J., *A history of the modern world*, McGraw-Hill, New York etc. 1995, 8th. edition

[Pasch 1927] Pasch, M., *Mathematik am Ursprung. Gesammelte Abhandlungen über Grundfragen der Mathematik*, Felix Meiner, Leipzig 1927

[Pauli 1979] Pauli, W., *Wissenschaftlicher Briefwechsel mit Bohr, Einstein, Heisenberg u.A./Scientific correspondence with Bohr, Einstein, Heisenberg a.o.*, Hermann, A., Von Meyenn, K., Weiskopf, V.F. (eds.), Springer, New York etc. 1979; Band I/Volume I: 1919–1929

[Peckhaus 1990] Peckhaus, V., *Hilbertprogramm und Kritische Philosophie*, Vandenhoeck und Ruprecht, Göttingen 1990

[Petkov 1990] Petkov, P.P. (ed.), *Mathematical logic*, Plenum Press, New York & London 1990

[Petzoldt 1922] Petzoldt, J., 'Zur Krisis des Kausalitätsbegriffs', *Die Naturwissenschaften* 10 (1922), pp. 693–695

[Petzoldt 1925] Petzoldt, J., 'Beseitigung der mengentheoretischen Paradoxa durch logisch einwandfreie Definition des Mengenbegriffs', *Kant-Studien* 30 (1925), pp. 346–356

[Petzoldt 1927] Petzoldt, J., 'Rationales und empirisches Denken', *Annalen der Philosophie* 6 (1927), pp. 145–160

[Peukert 1987] Peukert, D.J.K., *Die Weimarer Republik. Krisenjahre der Klassischen Moderne*, Suhrkamp, Frankfurt am Main 1987

[Phillips 1987] Phillips, E.R. (ed.), *Studies in the history of mathematics*, The Mathematical Association of America, USA 1987

[Pierpont 1928] Pierpont, J., 'Mathematical Rigor, Past and Present', *Bulletin of the American Mathematical Society* 34 (1928), pp. 23–53; lecture delivered at the Annual Meeting of the American Mathematical Society, held in connection with the meetings of the American Association for the Advancement of Science, at Nashville, December 28, 1927

[Pinl 1969] Pinl, M., 'Kollegen in einer dunklen Zeit', *Jahresbericht der deutschen Mathematiker-Vereinigung* 71 (1969), pp. 167–228

[Pöggeler 1969] Pöggeler, O., 'Oskar Becker als Philosoph', *Kant-Studien* 60 (1969), pp. 298–311

[Pöggeler 1996] Pöggeler, O., '"Eine Epoche gewaltigen Werdens" Die Freiburger Phänomenologie in ihrer Zeit', *Phänomenologische Forschungen* 30 (1996), pp. 9–32

[Poggendorff 1936–1940] *Poggendorffs biographisch-literarisches Handwörterbuch für Mathematik, Astronomie, Physik mit Geophysik, Chemie, Kristallographie und verwandte Wissensgebiete. Band VI: 1923 bis 1930*, Stobbe, H. (ed. for the first two parts), Chemie, Berlin 1936–1940

[Poincaré 1894] Poincaré, H., 'Sur la nature du raisonnement mathématique', *Revue de Métaphysique et de Morale* 2 (1894), pp. 371–384; English translation in [Ewald 1996, vol. II, pp. 972–982]

[Poincaré 1902] Poincaré, H., *La science et l'hypothèse*, Flammarion, Paris 1902; reedition Flammarion, Paris 1968

[Poincaré 1905] Poincaré, H., *La valeur de la science*, Flammarion, Paris 1970 (no edition mentioned); first edition 1905

[Poincaré 1908] Poincaré, H., *Science et méthode*, Flammarion, Paris 1908

[Poincaré 1909] Poincaré, H., 'L'avenir des mathematiques', in: [Castelnuovo 1909, pp. 167–182]

[Pólya 1972] Pólya, G., 'Eine Erinnerung an Hermann Weyl', *Mathematische Zeitschrift* 126 (1972), pp. 296–298

[Posy 1998] Posy, C.J., 'Brouwer versus Hilbert', *Science in Context* 11 (1998), pp. 291–325

[Ramsey 1926A] Ramsey, F.P., 'The foundations of mathematics', *Proceedings of the London Mathematical Society* Ser. 2 Vol. 25 (1926), pp. 338–384; paper read on November 12, 1925

[Ramsey 1926B] Ramsey, F.P., 'Mathematical logic', *The Mathematical Gazette* 13 (1926), pp. 185–194; paper read before the British Association, Section A, Oxford, in August 1926; cited from: [Ramsey 1978, pp. 213–232]

[Ramsey 1931] Ramsey, F.P., *The Foundations of Mathematics and other Logical Essays*, Braithwaite, R.B. (ed.), Routledge & Kegan Paul, London 1931; cited from the fourth impression, 1965

[Ramsey 1978] Ramsey, F.P., *Foundations. Essays in Philosophy, Logic, Mathematics and Economics*, Mellor, D.H. (ed.), Routledge & Kegan Paul, London etc. 1978

[Ramsey 1991A] Ramsey, F.P., *Notes on philosophy, probability and mathematics*, Galavotti, M.C. (ed.), Bibliopolis, Napoli 1991

[Ramsey 1991B] Ramsey, F.P., 'Principles of finitist mathematics', in: [Ramsey 1991A, pp. 197–202]

[Ramsey 1991C] Ramsey, F.P., 'The formal structure of intuitionist mathematics', in: [Ramsey 1991A, pp. 203–220]; presumably written in 1929

[Reeder 1997A] Reeder, H.P., 'Felix Kaufmann', in: [Embree et al. 1997, pp. 382–385]

[Reeder 1997B] Reeder, H.P., 'Ludwig Wittgenstein', in: [Embree et al. 1997, pp. 732–736]

[Reichenbach, H. 1924] Reichenbach, H., *Axiomatik der relativistischen Raum-Zeit-Lehre*, Vieweg, Braunschweig 1924

[Reichenbach, H. 1929] Reichenbach, H., 'Die Weltanschauung der exakten Wissenschaften', *Die Böttcherstraße* 1 (1929), pp. 44–46; I have not seen the original; review in *Fortschritte der Mathematik* 55 (1929), p. 28

[Reichenbach, H. 1935] Reichenbach, H., 'Wahrscheinlichkeitslogik', *Erkenntnis* 5 (1935), pp. 37–43; lecture delivered at the *Prager Vorkonferenz des Ersten Internationalen Kongresses für Einheit der Wissenschaft*, September 1, 1934

[Reichenbach, M. 1994] Reichenbach, M., 'Erinnerungen und Reflexionen', in: [Danneberg et al. 1994, pp. 9–17]

[Reid 1970] Reid, C., *Hilbert*, Springer, New York etc. 1970, cited from the fourth printing, 1983

[Reidemeister 1921A] Reidemeister, K., 'Bericht über die Hamburger Vorträge von D. Hilbert', *Jahresbericht der deutschen Mathematiker-Vereinigung* 30 (1921), pp. 106-107

[Reidemeister 1921B] Reidemeister, K., 'Hilbert über die Grundlagen der Mathematik', *Hamburgische Correspondent* 1921; a series of three articles, signed with 'R-r.', presumably Reidemeister; exact dates unknown; in: [NSUB Hilbert, 751]

[Reidemeister 1928-29] Reidemeister, K., 'Exaktes Denken', *Philosophischer Anzeiger* 3 (1928-29), pp. 15-47

[Reidemeister 1971] Reidemeister, K., *Hilbert Gedenkband*, Springer, Berlin etc. 1971

[Reymond 1932A] Reymond, A., *Les Principes de la Logique et la critique contemporaine*, Boivin & $C^{ie}$, Paris 1932; lectures delivered at the university of Paris in 1927 and 1930

[Reymond 1932B] Reymond, A., 'La fonction propositionnelle en logique algorithmique et le principe du tiers exclu', in: [Saxer 1932, pp. 198-199]; lecture delivered at the International Congress of Mathematicians in Zürich, September 4-12, 1932

[Richardson 1976] Richardson, J., *The grammar of justification: an interpretation of Wittgenstein's philosophy of language*, Sussex University Press, London etc. 1976

[Rickert 1920] Rickert, H., *Die Philosophie des Lebens. Darstellung und Kritik der philosophischen Modeströmungen unserer Zeit*, J.C.B. Mohr, Tübingen 1920

[Rickert 1921] Rickert, H., *System der Philosophie. Erster Teil: Allgemeine Grundlegung der Philosophie*, J.C.B. Mohr, Tübingen 1921

[Rickert 1923] Rickert, H., 'Die Methode der Philosophie und das Unmittelbare. Eine Problemstellung', *Logos* XII (1923), pp. 235-280; cited from: [Rickert 1939, pp. 51-96]

[Rickert 1924] Rickert, H., *Das Eine, die Einheit und die Eins. Bemerkungen zur Logik des Zahlbegriffs*, zweite, umgearbeitete Auflage, J.C.B. Mohr, Tübingen 1924

[Rickert 1939] Rickert, H., *Unmittelbarkeit und Sinndeutung. Aufsätze zur Ausgestaltung des Systems der Philosophie*, J.C.B. Mohr, Tübingen 1939

[Ridder 1931] Ridder, J., *De ontwikkeling van het intergraalbegrip*, Groningen etc. 1931; lecture delivered at the *Universiteit van Groningen* in September 1931

[Ringer 1969] Ringer, F.K., *The decline of the German Mandarins. The German Academic Community, 1890-1933*, Harvard University Press, Cambridge, Massachusetts, 1969

[Ritter & Grunder 1971-1995] Ritter, J., Grunder, K., *Historisches Wörterbuch der Philosophie*, Wissenschaftliche Buchgesellschaft, Darmstadt 1971-1995, 9 Bände (A-Sp)

[Rivier 1925] Rivier, W., 'A propos du principe du tiers exclu', *Revue de théologie et de philosophie* (nouvelle série) 13 (1925), pp. 215–221; lecture delivered before the Société romande de philosophie in the winter of 1924–25

[Rivier 1930] Rivier, W., 'L'empirisme dans les sciences exactes', *Archives de la Société Belge de Philosophie* 2 (1929/30), pp. 3–13

[Rowe & McCleary 1989] Rowe, D.E., McCleary, J. (ed.), *The History of Modern Mathematics*, Proceedings of the Symposium on the History of Modern Mathematics, Academic Press, Boston etc. 1989; volume I: Ideas and their perception; volume II: Institutions and Applications

[Rowe 1996] Rowe, D., 'New Trends and Old Images in the History of Mathematics', in: [Calinger 1996, pp. 3–16]

[Rowe 1997] Rowe, D., 'Perspective on Hilbert', *Perspectives on Science* 5 (1997), pp. 533–570; essay review of [Toepell 1986], [Mehrtens 1990] and [Peckhaus 1990]

[Rowe 2000] Rowe, D., 'The calm before the storm: Hilbert's early views on foundations', in: Hendricks, V. et al. (eds.), *Proof Theory: History and Philosophical Significance*, pp. 55-93, Kluwer, 2000

[Rudio 1898] Rudio, F. (ed.), *Verhandlungen des ersten internationalen Mathematiker-Kongresses in Zürich vom 9. bis 11. August 1897*, Leipzig 1898; reprint by Kraus, Nendeln/Liechtenstein 1967

[Rundschau 1930] 'Rundschau', *Erkenntnis* 1 (1930/31), pp. 80–88

[Russell 1937] Russell, B., *The Principles of Mathematics*, second edition 1937; cited from the ninth impression, George Allan & Unwin, London 1972

[Russell 1968] Russell, B., *The autobiography of Bertrand Russell*, George Allen and Unwin, London 1968, second impression

[Salz 1921] Salz, A., *Für die Wissenschaft. Gegen die Gebildeten unter ihren Verächtern*, München 1921

[Sarkar 1992] Sarkar, S., 'Rudolf Carnap, 1891–1970: The editor's introduction', *Synthese* 93 (1992), pp. 1–14

[Saxer 1932] Saxer, W., *Verhandlungen des Internationalen Mathematiker-Kongresses Zürich 1932. II. Band: Sektions-Vorträge*, cited from the 1967 reprint, Kraus, Nendeln/Liechtenstein

[Scanlon 1997] Scanlon, J., 'Eidetic method', in: [Embree et al. 1997, pp. 168–171]

[Schappacher 1987] Schappacher, N., 'Das Mathematische Institut der Universität Göttingen 1929–1950', in: [Becker, H., et al. 1987, pp. 345-373]

[Schappacher & Kneser 1990] Schappacher, N., unter Mitwirkung von M. Kneser, 'Fachverband — Institut — Staat, Streiflichter auf das Verhältnis von Mathematik zu Gesellschaft und Politik in Deutschland seit 1890 unter besonderer Berücksichtigung der Zeit des Nationalsozialismus', in [Fischer et al 1990, pp. 1–82]

[Scheler 1913] Scheler, M., *Der Formalismus in der Ethik und die materiale Wertethik. Neuer Versuch der Grundlegung eines ethischen Personalismus, Jahrbuch für Philosophie und phänomenologische Forschung* I und II (1913); cited from the second edition, Niemeyer, Halle a.d. S. 1921

[Schilpp 1963] Schilpp, P.A., *The Philosophy of Rudolf Carnap*, The Library of Living Philosophers Volume XI, Open Court, La Salle, Illinois, 1963

[Schmeidler 1929] Schmeidler, W., 'Neue Grundlagenforschungen in der Mathematik', *Unterrichtsblätter für Mathematik und Naturwissenschaften* 35 (1929), pp. 193–198

[Schmeidler 1935] Schmeidler, W., 'Der mathematische Unterricht im Dritten Reich', *Jahresbericht der deutschen Mathematiker-Vereinigung* 45 (1935), pp. *4–6*

[Schmidt 1929] Schmidt, E., *Über Gewissheit in der Mathematik*, Preussische Druckerei- und Verlags-Aktiengesellschaft, Berlin 1930; inaugural address as rector of the Friedrich-Wilhelms-Universität in Berlin, delivered on October 15, 1929; cited from: [Meschkowski 1972, pp. 56–64]

[Schmitz 1990] Schmitz, H.W., *De Hollandse Significa. Een reconstructie van de geschiedenis van 1892 tot 1926*, Van Gorcum, Assen 1990; Dutch translation of the German original

[Schoenflies 1922] Schoenflies, A., 'Zur Erinnerung an Georg Cantor', *Jahresbericht der deutschen Mathematiker-Vereinigung* 31 (1922), pp. 97–106

[Schoenflies 1927] Schoenflies, A., 'Die Krisis in Cantor's mathematischem Schaffen', *Acta Mathematica* 50 (1927), pp. 1–23

[Scholz, E. 1992] Scholz, E., review of [Mehrtens 1990], *Centaurus* 35 (1992), pp. 92–95

[Scholz, H. 1926] Scholz, H., review of [Boutroux 1920], *Deutsche Literaturzeitung* 46 (1926), pp. 2294–2302

[Scholz, H. 1930] Scholz, H., 'Die Axiomatik der Alten', *Blätter für deutsche Philosophie* 4 (1930/31), pp. 259–278; cited from: [Scholz, H. 1961, pp. 27–44]

[Scholz, H. 1931A] Scholz, H., *Geschichte der Logik*, Junker und Dünnhaupt, Berlin 1931

[Scholz, H. 1931B] Scholz, H., 'Über das Cogito, ergo sum', *Kantstudien* 36 (1931), pp. 126–147; cited from: [Scholz, H. 1961, pp. 75–94]

[Scholz. H. 1931C] Scholz, H., review of [Bolzano 1930/31], *Deutsche Literaturzeitung* 45 (1931), pp. 2152–2156

[Scholz, H. 1931/32] Scholz, H., 'Augustinus und Descartes', *Blätter für deutsche Philosophie* 5 (1931/32), pp. 405–423; cited from [Scholz, H. 1961, pp. 45–61]

[Scholz, H. 1933] Scholz, H., 'Über die Ableitbarkeit der Mathematik aus der Logik', radio lecture delivered on May 2, 1933; in [IMLG Scholz]

[Scholz, H. 1961] Scholz, H., *Mathesis universalis. Abhandlungen zur Philosophie als strenger Wissenschaft*, Hermes, H., Kambartel, F., Ritter, J. (eds.), Wissenschaftliche Buchgesellschaft, Darmstadt 1961

[Schopenhauer 1844] Schopenhauer, A., *Die Welt als Wille und Vorstellung II. Erster Teilband*, 1844; cited from the 1977 re-edition, Diogenes, Zürich

[Schuhmann 1977] Schuhmann, K., *Husserl-Chronik. Denk- und Lebensweg Edmund Husserls*, Martinus Nijhoff, Den Haag 1977

[Sepp 1988] Sepp, H.R., *Edmund Husserl und die phänomenologische Bewegung*, Karl Alber, Freiburg etc. 1988

[Serfati 1995] Serfati, M., 'Infini <<nouveau>>: principes de choix effectifs', in: [Monnoyeur 1995], pp. 207–238

[Sheldrake 1994] Sheldrake, R., *Seven experiments that could change the world. A Do-It-Yourself Guide to Revolutionary Science*, Fourth Estate, London 1994

[Sieg & Parsons 1995] Sieg, W., Parsons, C., 'Introductory note to *1938A', in: [Gödel 1986–1995, vol. III, pp. 62–85]

[Sieg 1999] Sieg, W., 'Hilbert's Programs: 1917–1922', *The Bulletin of Symbolic Logic* 5 (1999), pp. 1–44

[Siegert 1935] Siegert, K., 'Die Bedeutung der Mathematik für die Ausbildung der Juristen im Dritten Reich', *Jahresbericht der deutschen Mathematiker-Vereinigung* 45 (1935), pp. *17–19*

[Siegmund-Schultze 1993] Siegmund-Schultze, R., *Mathematische Berichterstattung in Hitlerdeutschland*, Vandenhoeck & Ruprecht, Göttingen 1993

[Siegmund-Schultze 1997] Siegmund-Schultze, R,. 'The Emancipation of Mathematical Research Publishing in the United States from German Dominance (1878-1945)', *Historia Mathematica* 24 (1997), pp. 135–166

[Siegmund-Schultze 1998] Siegmund-Schultze, R., *Mathematiker auf der Flucht vor Hitler. Quellen und Studien zur Emigration einer Wissenschaft*, Deutsche Mathematiker–Vereinigung/Vieweg, Braunschweig/Wiesbaden 1998

[Simmel 1918] Simmel, G., *Der Konflikt der modernen Kultur: Ein Vortrag*, München 1918; 2nd. edition 1921

[Sirolf 1947] Sirolf, V., *De schuldenaar*, Gottmer, Haarlem 1947; written by Freudenthal under pseudonym

[Skolem 1920] Skolem, Th., 'Logisch-kombinatorische Untersuchungen über die Erfüllbarkeit und Beweisbarkeit mathematischen Sätze nebst einem Theoreme über dichte Mengen', *Skrifter, Videnskabsakademiet i Kristiania* I, No. 4 (1920), pp. 1–36; cited from: [Skolem 1970, pp. 103–136]; English translation in: [Van Heijenoort 1967, pp. 252–263]

[Skolem 1923] Skolem, Th., 'Einige Bemerkungen zur axiomatischen Begründung der Mengenlehre', *5. Kongreß Skandinavischen Mathematiker in Helsingfors, vom 4. bis 7. Juli 1922* (1923), pp. 217–232

[Skolem 1924] Skolem, Th., 'Ein Verfahren zu beliebig angenäherter Bestimmung einer Wurzel einer beliebigen algebraischen Gleichung', *Norsk Matematisk Forening, Skrifter* Series 1, No. 15 (1924)

[Skolem 1926] Skolem, Th., 'Litt om de vigtigste diskussioner i den senere tid angaaende matematikkens grundlag', *Norsk Matematisk Tidsskrift* 8 (1926), pp. 1–13; lecture delivered before the *Norsk Matematisk Forening* on September 30, 1925

[Skolem 1930] Skolem, Th., 'Über die Grundlagendiskussionen in der Mathematik', *Den syvende skandinaviske matematikerkongress i Oslo 19–22 august 1929/Comptes rendus du septième congrès des mathématiciens scandinaves tenu à Oslo 19–22 août 1929*, Brøggers, Oslo 1930, pp. 3–21

[Skolem 1934] Skolem, Th., 'Den matematiske grunnlagforskning', *Norsk Matematisk Tidsskrift* 16 (1934), pp. 75–92; lecture delivered before the *Matematiska Förening* in Uppsala, May 8, 1934

[Skolem 1970] Skolem, Th., *Selected Works in Logic*, Fenstad, J.E. (ed.), Universitetsforlaget, Oslo etc. 1970

[Smoryński 1988] Smoryński, C., 'Hilbert's Programme', *CWI Quarterly* 1 (1988), pp. 3–59

[Solomentsev 1991] Solomentsev, E.D., 'Picard Theorem', in: [Hazewinkel 1988–1993, Vol. 7, pp. 157–158]

[Spengler 1918] Spengler, O., *Der Untergang des Abendlandes. Umrisse einer Morphologie der Weltgeschichte*, Beck, München 1918; *Erster Band: Gestalt und Wirklichkeit* cited from the 1920 11th–14th edition; *Zweiter Band: Welthistorische Perspektiven* cited from the 1922 1st–15th edition

[Spengler 1923] Spengler, O., *Der Untergang des Abendlandes. Umrisse einer Morphologie der Weltgeschichte*, Beck, München 1923, 33rd–47th, revised edition; cited from the 1993 re-edition, DTV, München

[Stammler 1928] Stammler, G., *Begriff Urteil Schluss. Untersuchungen über die Grundlagen und Aufbau der Logik*, Max Niemeyer, Halle/Saale, 1928

[Stark 1922] Stark, J., *Die gegenwärtige Krisis in der deutschen Physik*, Leipzig 1922

[Stehr & Meja 1984] Stehr, N., Meja, V., *Society and knowledge. Contemporary perspectives on the sociology of knowledge*, Transaction, New Brunswick etc. 1984

[Sticker 1922] Sticker, *Die wahre Relativitätstheorie und die Missgriffe Einsteins*, Bielefeld 1922

[Van Stigt 1990] Stigt, W.P. van, *Brouwer's Intuitionism*, North-Holland, Amsterdam etc. 1990

[Van Stigt 1996] Stigt, W.P. van, 'Introduction to *Life, Art, and Mysticism*', *Notre Dame Journal of Formal Logic* 37 (1996), pp. 381–387; introduction to [Brouwer 1996]

[Van Stigt 1998] Stigt, W.P. van, 'Brouwer's Intuitionist Programme', in: [Mancosu 1998, pp. 1–22]

[Struik 1967] Struik, D.J., *A concise history of mathematics*, Dover, New York 1967

[Study 1929] Study, E., 'Die angeblichen Antinomien der Mengenlehre', *Sitzungsberichten der Preussischen Akademie der Wissenschaften, Physikalisch–mathematische Klasse* XIX (1929), pp. 255–267

[Sundholm 1983] Sundholm, G., 'Constructions, proofs and the meaning of logical constants', *Journal of Philosophical Logic* 12 (1983), pp. 151–172

[Szabo 1969] Szabo, M.E., *The Collected Papers of Gerhard Gentzen*, North Holland, Amsterdam & London 1969

[Tagung für exakte Erkenntnislehre 1930] 'Tagung für exakte Erkenntnislehre', *Die Naturwissenschaften* 18 (1930), pp. 1067–1068; announcement of the lectures to be given at the 1930 Königsberg conference

[Theweleit 1977–1978] Theweleit, K., *Männerphantasien. Band 1: Frauen, Fluten, Körper, Geschichte. Band II: Männerkörper. Zur Psychoanalyse des weißen Terrors*, cited from the 1995 re-edition, DTV, München

[Thiel 1972] Thiel, C., *Grundlagenkrise und Grundlagenstreit. Studie über das normative Fundament der Wissenschaften am Beispiel von Mathematik und Sozialwissenschaft*, Anton Hain, Meisenheim am Glan 1972

[Thiel 1984] Thiel, C., 'Konstruktivismus', in: [Mittelstraß 1980–96, Vol. 2, pp. 449–453]

[Thomae 1906A] Thomae, J., 'Gedankenlose Denker', *Jahresbericht der deutschen Mathematiker-Vereinigung* 15 (1906), pp. 434–438

[Thomae 1906B] Thomae, J., 'Erklärung', *Jahresbericht der deutschen Mathematiker-Vereinigung* 15 (1906), pp. 590–592

[Tikhomirov 1993] Tikhomirov, A.N., 'A.N. Kolmogorov', in: [Zdravskovska & Duren 1993, pp. 101–127]

[Toepell 1986] Toepell, M., *Über die Entstehung von David Hilberts 'Grundlagen der Geometrie'*, Göttingen, Vandenhoeck & Ruprecht 1986

[Tonietti 1988] Tonietti, T., 'Four letters from E. Husserl to H. Weyl and their context', in: [Deppert et al. 1988, pp. 343–384]

[Troelstra 1969] Troelstra, A.S., *Principle of Intuitionism*, Springer, Berlin etc. 1969

[Troelstra 1977] Troelstra, A.S., *Choice sequences. A chapter of intuitionistic mathematics*, Clarendon, Oxford 1977

[Troelstra 1978] Troelstra, A.S., 'Commentary to Heyting's papers 'Die formalen Regeln der intuitionistischen Logik' and 'Die formalen Regeln der intuitionistischen Mathematik II, III', in: [Bertin, Bos & Grootendorst 1978, pp. 163–175]

[Troelstra 1981] Troelstra, A.S., 'Arend Heyting and his contribution to intuitionism', *Nieuw Archief voor Wiskunde* (3) 29 (1981), pp. 1–23

[Troelstra 1982] Troelstra, A.S., 'On the origin and development of Brouwer's concept of choice sequence', in: [Troelstra & Van Dalen 1982, pp. 465–486]

[Troelstra 1983] Troelstra, A.S., 'Analysing choice sequences', *Journal of Philosophical Logic* 12 (1983), pp. 197–260

[Troelstra 1986] Troelstra, A.S., 'Introductory note to *1933e*', in: [Gödel 1986–1995, vol. 1, pp. 282–287]

[Troelstra 1990] Troelstra, A.S., *On the early history of intuitionistic logic*, in: [Petkov 1990, pp. 3–17]

[Troelstra & Van Dalen 1982] Troelstra, A.S., Van Dalen, D. (eds.), *The L.E.J. Brouwer Centenary Symposium*, North-Holland, Amsterdam etc. 1982

[Troelstra & Van Dalen 1988] Troelstra, A.S., Van Dalen, D. *Constructivism in Mathematics. An introduction*, two volumes, North Holland, Amsterdam etc. 1988

[Troeltsch 1921] Troeltsch, E., 'Die Revolution in der Wissenschaft', *Schmöller's Jahrbuch* 45 (1921), pp. 1001-1030

[Turing 1936] Turing, A.M., 'On computable numbers, with an application to the Entscheidungsproblem', *Proceedings of the London Mathematical Society*, Series 2, 42 (1936), pp. 230–265

[Turing 1937] Turing, A.M., 'Computability and $\lambda$-definability', *The Journal of Symbolic Logic* 2 (1937), pp. 153–163

[Turner 1996] Turner, J. (ed.), *The Dictionary of Art*, MacMillan, London etc. 1996, 34 volumes

[Ulam et al. 1969] Ulam, S., Kuhn, H., Tucker, A., and Shannon, C., 'John von Neumann, 1903–1957', in: [Flemming & Bailyn 1969, pp. 235–269]

[Van Atten] See under A

[Van Dalen] See under D

[Van Dantzig] See under D

[Van Dooren] See under D

[Van Eeden] See under E

[Van Heijenoort] See under H

[Van der Waerden] See under W

[Van Os] See under O

[Van Stigt] See under S

[Verzeichnis Hilbert-Dissertationen] 'Verzeichnis der bei Hilbert angefertigten Dissertationen', in: [Hilbert 1932–1935, Volume 3, pp. 431–433]

[Vitányi 1988] Vitányi, P.M.B, 'Andrei Nikolaevich Kolmogorov', *CWI Quarterly* 1 (1988) nr. 2, pp. 3–18

[Vollenhoven 1918] Vollenhoven, D.H.Th., *De wijsbegeerte der wiskunde van theïstisch standpunt*, Van Soest, Amsterdam 1918; dissertation

[Vollenhoven 1919A] Vollenhoven, D.H.Th., 'De activiteit der ziel in het rekenonderwijs I', *Paedagogisch tijdschrift voor het christelijk onderwijs* 12 (1919), pp. 97–104

[Vollenhoven 1919B] Vollenhoven, D.H.Th., 'De activiteit der ziel in het rekenonderwijs II', *Paedagogisch tijdschrift voor het christelijk onderwijs* 12 (1919), pp. 105–109

[Vollenhoven 1921] Vollenhoven, D.H.Th., 'Hegel op onze lagere scholen?', *Paedagogisch tijdschrift voor het christelijk onderwijs* 14 (1921), pp. 77–87

[Vollenhoven 1932] Vollenhoven, D.H.Th., *De noodzakelijkheid eener christelijke logica*, Paris, Amsterdam 1932

[Von Freytag] See under F

[Von Mises] See under M

[Von Neumann] See under N

[Voss 1914] Voss, A., *Über die mathematische Erkenntnis*, Teubner, Leipzig etc. 1914

[Vredenduin 1933] Vredenduin, P.G.J., 'De autonomie der wiskunde', *Euclides* X (1933/34), pp. 144–160; lecture delivered in Utrecht on November 27, 1933

[Van der Waerden 1935] Van der Waerden, B.L., 'Nachruf auf Emmy Noether', *Mathematische Annalen* 111 (1935), pp. 469–476

[Waismann 1993] Waismann, F., *Wittgenstein und der Wiener Kreis*, McGuinness, B.F. (ed.), Suhrkamp, Frankfurt am Main 1993, 4. Auflage

[Wang 1967] Wang, H., Introduction to [Kolmogorov 1925], in: [Van Heijenoort 1967, pp. 414–416]

[Wang 1987] Wang, H., *Reflections on Kurt Gödel*, MIT, Cambridge, Massachusetts, etc., 1987

[Wäsche 1926] Wäsche, H., *Grundzüge zu einer Logik der Arithmetik*, Carl Heymanns, Berlin 1926

[Wavre 1924] Wavre, R., 'Y a-t-il une crise des mathématiques? A propos de la notion d'existence et d'une application suspecte du principe du tiers exclu', *Revue de Métaphysique et de Morale* 31 (1924), pp. 435-370

[Wavre 1926A] Wavre, R., 'Logique formelle et logique empirique', *Revue de Métaphysique et de Morale* 33 (1926), pp. 65-75

[Wavre 1926B] Wavre, R., 'Sur le principe du tiers exclu', *Revue de Métaphysique et de Morale* 33 (1926), pp. 425-430

[Weber 1891-92] Weber, H., 'Leopold Kronecker', *Jahresbericht der deutschen Mathematiker-Vereinigung* 2 (1891-92), pp. 5-31

[Weierstrass 1873] Weierstrass, K., *Ansprache bei der Übernahme des Rectorats der Friedrich-Wilhelms-Universität zu Berlin am 15. October 1873*, in: [Weierstrass 1894-1927, vol. III, pp. 331-339]

[Weierstrass 1894-1927] Weierstrass, K., *Mathematische Werke*, Mayer & Müller, Berlin 1894 - Akademische Verlagsgesellschaft, Leipzig 1927, 7 Bände

[Weiss 1929] Weiss, P., 'The nature of systems', *The monist* 39 (1929), pp. 440-472

[Weller 1994] Weller, C., 'Wilhelm Burkamp und der Logische Empirismus', in: [Danneberg et al. 1994, pp. 131-160]

[Weyl 1910] Weyl, H., 'Über die Definitionen der mathematischen Grundbegriffe', *Mathematisch-naturwissenschaftliche Blätter* 7 (1910), pp. 93-95 & 109-113; cited from: [Weyl 1968A, pp. 298-304]

[Weyl 1913] Weyl, H., *Die Idee der Riemannschen Fläche*, Teubner, Leipzig etc. 1913

[Weyl 1918A] Weyl, H., *Das Kontinuum. Kritische Untersuchungen über die Grundlagen der Analysis*, Veit & Comp., Leipzig 1918

[Weyl 1918B] Weyl, H., *Raum · Zeit · Materie. Vorlesungen über allgemeine Relativitätstheorie*, Springer, Berlin 1918

[Weyl 1919] Weyl, H., 'Der circulus vitiosus in der heutigen Begründung der Analysis', *Jahresbericht der deutschen Mathematiker-Vereinigung* 28 (1919), pp. 85-92

[Weyl 1920] Weyl, H., 'Das Verhältnis der kausalen zur statistischen Betrachtungsweise in der Physik', *Schweizerische Medizinische Wochenschrift* 50 (1920), pp. 737-741; cited from: [Weyl 1968A, vol. II, pp. 113-122]

[Weyl 1921] Weyl, H., 'Über die neue Grundlagenkrise der Mathematik', *Mathematische Zeitschrift* 10 (1921), pp. 39-79; cited from: [Weyl 1956, pp. 211-248]; English translation: [Weyl 1998]; lectures delivered at the *Mathematisches Kolloquium* Zürich, 1920

[Weyl 1922A] Weyl, H., 'Die Relativitätstheorie auf der Naturforscherversammlung in Bad Nauheim', *Jahresbericht der deutschen Mathematiker-Vereinigung* 31 (1922), pp. 51-63

[Weyl 1922B] Weyl, H., 'Das Raumproblem', *Jahresbericht der deutschen Mathematiker-Vereinigung* 31 (1922), pp. 205–221

[Weyl 1924] Weyl, H., 'Randbemerkungen zu Hauptproblemen der Mathematik', *Mathematische Zeitschrift* 20 (1924), pp. 131–150

[Weyl 1925] Weyl, H., 'Die heutige Erkenntnislage in der Mathematik', *Symposion* 1 Heft 1 (1925), pp. 1–32; cited from: [Weyl 1968A, Band II, pp. 511–542]; English translation in [Mancosu 1998, pp. 123–142]

[Weyl 1927A] Weyl, H., *Philosophie der Mathematik und Naturwissenschaft*, Oldenbourg, München etc. 1927

[Weyl 1927B] Weyl, H., 'Diskussionsbemerkungen zu dem zweiten Hilbertschen Vortrag über die Grundlagen der Mathematik', *Abhandlungen aus dem Mathematischen Seminar der Hamburgischen Universität* V (1927), pp. 86–88

[Weyl 1929] Weyl, H., 'Consistency in mathematics', *The Rice Institute Pamphlet* 16 (1929), pp. 245–265; cited from: [Weyl 1968A, Band III, pp. 150–170]; lectures delivered at the Rice Institute on May 20, 22, and 23, 1929

[Weyl 1931] Weyl, H., *Die Stufen des Unendlichen*, Gustav Fischer, Jena 1931; lecture delivered at the opening of a conference of the Mathematical Society of the *Universität Jena*

[Weyl 1932A] Weyl, H., *The Open World. Three lectures on the metaphysical implications of science*, Yale University Press, New Haven/Humphrey Milford, Oxford University Press, London 1932; Terry Lectures, delivered presumably in 1931 at Yale University

[Weyl 1932B] Weyl, H., 'Zu David Hilberts siebzigstem Geburtstag', *Die Naturwissenschaften* 20 (1932), pp. 57–58; cited from: [Weyl 1968A, vol. III, pp. 346–347]

[Weyl 1944] Weyl, H., 'David Hilbert and His Mathematical Work', *Bulletin of the American Mathematical Society* 50 (1944), pp. 612–654

[Weyl 1953] Weyl, H., 'Über den Symbolismus in der Mathematik und mathematischen Physik', *Studium generale* 6 (1953), pp. 219–228; cited from: [Reidemeister 1971, pp. 20–38]

[Weyl 1954] Weyl, H., 'Erkenntnis und Besinnung (Ein Lebensrückblick)', *Studia Philosophica*, 1954; cited from: [Weyl 1968A, Band IV, pp. 631–649]

[Weyl 1956] Weyl, H., *Selecta Hermann Weyl*, Birkhäuser, Basel etc. 1956

[Weyl 1968A] Weyl, H., *Gesammelte Abhandlungen*, Chandrasekharan, K. (ed.), Springer, Berlin etc. 1968, I–IV

[Weyl 1968B] Weyl, H., 'Lebenslauf', in: [Weyl 1968A, Band I, p. 87]

[Weyl 1985] Weyl, H., 'Axiomatic Versus Constructive Procedures in Mathematics', Tonietti, T. (ed.), *The Mathematical Intelligencer* 7 (1985), pp. 10–17; 38; previously unpublished, English manuscript, presumably written after 1953

# BIBLIOGRAPHY 429

[Weyl 1998] Weyl, H., 'On the New Foundational Crisis of Mathematics', in: [Mancosu 1998, pp. 86–118]; English translation of [Weyl 1921]

[Whitehead & Russell 1910] Whitehead, A.N., Russell, B., *Principia Mathematica*, vol. 1, Cambridge University Press, Cambridge 1910; cited from the 1978 reprint of the second edition (1927)

[Willink 1998] Willink, B., *De tweede gouden eeuw. Nederland en de Nobelprijzen voor natuurwetenschappen 1870–1940*, Bert Bakker, Amsterdam 1998

[Winternitz 1927] Winternitz, 'Bemerkungen zu Brouwers intuitionistischer Mathematik', lecture delivered in Prague on May 20, 1927; mentioned in *Jahresbericht der deutschen Mathematiker-Vereinigung* 37 (1928), pp. *42–43*

[Wittgenstein 1921] Wittgenstein, L., *Tractatus Logico-Philosophicus. Logisch–Philosophische Abhandlung*, Annalen der Naturphilosophie 14 (1921); cited from: [Wittgenstein 1963, pp. 9–83]

[Wittgenstein 1963] Wittgenstein, L., *Schriften*, Suhrkamp, Frankfurt am Main 1963

[Wolff 1922] Wolff, J., *Over het subjectieve in de wiskunde*, Noordhoff, Groningen 1922; inaugural lecture delivered on October 16, 1922, at the *Rijks-universiteit Utrecht*

[Wright 1980] Wright. C., *Wittgenstein on the Foundations of Mathematics*, Harvard University Press, Cambridge, Massachusetts, 1980

[Zach 1999] Zach, R., 'Completeness before Post: Bernays, Hilbert, and the development of propositional logic', *The Bulletin of Symbolic Logic* 5 (1999), pp. 331–366

[Zanichelli 1928] Zanichelli, N. (ed.), *Atti del Congreso Internazionale dei Matematici Bologna 3–10 settembre 1928*, Tomo I: Rendiconto del congresso, conferenze, Bologna; cited from the 1967 integral reprint, Kraus, Nendeln etc.

[Zariski 1925] Zariski, O., 'Gli sviluppi più recenti della teoria degli insiemi e il principio di Zermelo', *Periodico di matematiche, storia, didattica, filosofia* Ser. IV No. 5 (1925), pp. 57–80

[Zawirski 1932] Zawirski, S., 'Les logiques nouvelles et le champ de leur application', *Revue de Métaphysique et de Morale* 39 (1932), pp. 503–519

[Zdravskovska & Duren 1993] Zdravskovska, S., Duren, P.L. (eds.), *Golden Years of Moscow Mathematics*, American Mathematical Society & London Mathematical Society, USA 1993

[Zermelo 1904] Zermelo, E., 'Beweis, dass jede Menge wohlgeordnet werden kann', *Mathematische Annalen* 59 (1904), pp. 514–516; English translation in: [Van Heijenoort 1967, pp. 139–141]

[Zermelo 1908] Zermelo, E., 'Neuer Beweis für die Möglichkeit einer Wohlordnung', *Mathematische Annalen* 65 (1908), pp. 107-128

[Zeuthen 1896] Zeuthen, H.G., 'Die geometrische Construction als 'Existenzbeweis' in der antiken Geometrie', *Mathematische Annalen* 47 (1896), pp. 222–228

[Zeuthen 1912] Zeuthen, H.G., *Die Mathematik im Altertum und Mittelalter*, Teubner, Berlin 1912

[Ziegenfuss 1949–1950] Ziegenfuss, W. (ed.), *Philosophen-Lexikon. Handbuch der Philosophie nach Personen*, Walter de Gruyter & Co, Berlin 1949–1950, 2 Bände

[Zilsel 1932/33] Zilsel, E., 'Bemerkungen zur Wissenschaftslogik', *Erkenntnis* 3 (1932/33), pp. 143–161

# Index

Abelard, 219
Ackermann, 104, 189n, 246, 276, 277, 282, 365
Adama van Scheltema, 25n, 46, 49n, 53
Ajdukiewicz, 107
Alexander, 50n
Alexandroff, 50n, 237
Ambrose, 111, 112, 345, 369
Archimedes, 120n
Aristotle, 218–220, 281, 290, 320
Artin, 286
Atten, Van, 223n
Auben, 334
Augustinus, 320

Baer, 336
Baire, 2, 8, 9, 11, 12, 16–18
Baldus, 152–154, 220, 229–231, 257, 352, 358, 364
Barrau, 46, 269, 307, 365
Barzin, 260, 290
    and Becker, 262, 295
    and Brouwer, 266, 292, 293, 317n
    and Church, 265, 266, 291
    and De Donder, 292
    and Dresden, 267
    and Errera, 260, 262, 271, 289–291, 294–296, 351
    and Glivenko, 271, 272, 291, 359
    and Heyting, 99, 292–295, 360, 369
    and Lévy, 262, 291
    and Scholz, 295
    and Wavre, 261
    and Wavre and Lévy, 170
    on constructivity, 261
    on intuitionistic logic, 262, 290, 295, 298, 351, 359, 365, 368, 369
    on mathematical existence, 177
    on the excluded middle, 261, 290, 291, 366
Becker, xxi, 145–149, 154, 158, 170–173, 177, 220, 258, 259, 338, 349, 358, 364, 366, 367
    and Barzin and Errera, 262, 295
    and Bieberbach, 149
    and Brouwer, 173
    and Fraenkel, 149
    and Heyting, 277n, 295n, 338n
    and Husserl, 106, 107, 147, 148, 170
    and Mahnke, 164, 178
    and Wavre, 162
    and Wavre and Lévy, 170
    and Weyl, 164, 170n, 172, 178, 258, 270, 329
    on consistency, 171
    on constructivity, 146, 171, 338
    on Hilbert, 144, 259, 352
    on intuitionism, 172, 212, 338
    on intuitionism and formalism, 171
    on intuitionistic logic, 259, 260, 298, 351
    on mathematical existence, 119, 147, 170–172, 212, 359, 365
    on the continuum, 146
    on the excluded middle, 258, 259

Behmann, 104, 105, 153n, 205–207, 225, 364
Behnke, 136n
Belinfante, 366–368
Bell, 367–369
Bellaar-Spruyt, 40
Benacerraf, 107
Benda, 317
Bentley, 60n, 368
Benz, 136n
Bergson, 53, 313, 315, 316n, 317, 317n, 320n, 321, 322, 326, 343, 348, 374
Berkeley, 119, 119n
Bernays, xx, 114, 124, 154, 183, 220, 231, 336, 337, 356, 364
    and Behmann, 205n, 206, 207
    and Beth, 207n
    and Church, 207n
    and Courant, 104
    and Dingler, 168
    and Finsler, 165n
    and Fraenkel, 207n
    and Gentzen, 289
    and Gonseth, 251n
    and Heyting, 101, 177, 274, 277, 278
    and Hilbert, 103, 139n, 141, 142, 225, 306, 336, 356–358
    and Reid, 306n
    and Wavre, 159, 234
    and Weyl, 115, 127
    on Brouwer, 54
    on constructivity, 121n, 124, 142, 143, 154, 165, 205–207, 214, 350
    on formalism, 153n
    on Hilbert's programme, 141, 144n, 364, 367
    on intuitionism, 103, 104, 106, 142
    on mathematical existence, 140
    on the excluded middle, 72, 103, 227, 297
Bernstein, 10, 48, 69, 104, 106, 125, 225, 246, 336, 363
Beth, 207n
Betsch, 176, 269, 365
Bieberbach, xxi, 69, 125, 249, 338, 339, 344, 348, 363–365
    and Becker, 149
    and Bohr, 338
    and Brouwer, 82, 125
    and Heyting, 274
    and Hilbert, 83, 360
    and Hirsch, 189
    and Hjemslev, 250
    and Landau, 339
    and *Deutsche Mathematik*, 339
    on formalism, 125, 250
    on intuitionism, 249, 250, 359
    on intuitionism and formalism, 339
    on mathematical existence, 125, 176, 356
    on the excluded middle, 246, 249
Bishop, 66n, 123
Black, 369
Blaschke, 135, 337
Bloemers, 59
Blumenthal, 48, 84, 85, 137, 337
Böhme, 32
Boethius, 219
Bohr, 151, 337, 338
Bois-Reymond, Du, 37
Bolland, 30
Bolzano, 320
Boole, 219, 220, 290
Boomstra, 364
Borel, E., 2, 8–17, 23, 24, 37, 42, 44, 48, 52, 54, 57, 62, 76, 80, 90, 125, 153, 169, 237
Borel, H., 59
Born, 81n, 323n, 329
Bos, ix
Boutroux, 249
Braithwaite, 113n
Brand, 111, 366

# INDEX 433

Brodén, 101, 112, 175, 176, 268, 364
Brouwer, xx
  -ian counter-examples, 69–72, 77, 80, 357
  and Adama van Scheltema, 25n, 49n, 53
  and Barrau, 46, 307
  and Barzin and Errera, 266, 292
  and Bergson, 53, 317, 321, 322, 343, 348
  and Bernays, 54, 103, 142, 227, 297
  and Bieberbach, 82
  and Carathéodory, 84
  and Carnap, 204
  and De Donder, 292
  and Dingler, 125, 222n
  and Dresden, 156, 233
  and Dubislav, 269n
  and Fraenkel, 1, 100, 127n, 150, 151, 151n, 165, 166, 357, 358
  and Gödel, 206, 208, 210, 282–286, 299, 351, 360
  and Giltay, 34
  and Heyting, 85n, 87n, 273, 274
  and Hilbert, 49, 54, 78, 79n, 81, 83–86, 103, 112, 127n, 145, 153, 180, 226, 228, 229, 232, 243–245, 297, 304, 305, 322, 342, 352, 356, 357
  and Hurewicz, 274
  and Husserl, 86, 106, 349, 360
  and Kagan, 229
  and Kant, 81n
  and Klein, 84n
  and Kohnstamm, 46n, 221
  and Korteweg, 30, 35–37, 39, 46n, 54, 318, 355
  and Mannoury, 28–30, 46n, 52, 59, 60, 112, 232, 265n, 326, 355, 356
  and Menger, 198
  and significs, 59
  and the first act of intuitionism, 39, 355
  and the second act of intuitionism, 60, 356
  and the semi-intuitionists, 13–15, 23, 24, 90
  and the *Mathematische Annalen*, 59, 83–85, 90, 274, 346, 356, 360
  and topology, 48–50, 52, 67, 68, 74, 77, 126, 355, 356
  and Weyl, 70, 86, 105, 107, 119, 126–128, 174, 178, 210, 211, 213, 222–224, 232, 245, 297, 303, 304, 319, 342, 345–347, 353, 356, 357
  and Wittgenstein, 190–198, 213, 350, 360
  on causality, 37, 54, 80
  on choice sequences, 37, 58, 59, 61–63, 81
  on consistency, 41–44, 54, 57, 70, 79, 89
  on constructivity, 35–37, 57, 58, 68, 121, 123
  on Hilbert's programme, 44, 45
  on intuition, 38
  on intuitionism and formalism, 52–58, 77–80, 89, 90, 318, 354, 356, 359
  on Kant, 38, 54
  on language, 27, 32–34, 39, 40, 47, 49, 76, 89
  on logic, 38, 39, 46, 47, 76, 80, 89
  on logicism, 44
  on mathematical existence, 41, 60, 76
  on mysticism, 31, 32
  on natural numbers, 36
  on negation, 48, 72, 89
  on set theory, 43, 58, 60–63, 65, 66, 68, 73, 76, 89, 357
  on solipsism, 27, 32, 47, 66, 89

on the axiom of choice, 44
on the continuum, 36, 66, 69, 70, 74, 76, 80, 81
on the excluded middle, 40, 47, 48, 50–52, 59, 67, 68, 70, 72, 77–79, 89, 90, 355, 358
on the infinite, 44
on the primordial intuition, 38
on time, 69
topology and intuitionism, 50, 51
Brunschvicg, 256
Burali-Forti, 9n, 43n
Burkamp, 180–182, 257, 258, 365, 368

Cantor
's Continuum Hypothesis, 4, 9
's set theory, 1–5, 9, 10, 16, 24, 43n
and Baldus, 153
and Becker, 148
and Brouwer, 38, 40, 43, 44, 54, 60, 68, 70n, 346
and Dedekind, 138n
and E. Borel, 14, 16
and Frege, 138n
and Hilbert, 9, 43n, 56n, 133, 138n, 305
and König, 10
and Kronecker, 5–7
and Mittag-Leffler, 5
and Poincaré, 18
and Weyl, 126
Carathéodory, 48, 67, 84, 85, 230, 249
Cardano, 119
Carnap, xx, 101, 102, 106–108, 108n, 114, 115, 184, 186, 189, 202–204, 206n, 346, 360, 365, 367
and Becker, 149
and Brouwer, 202, 203, 243
and Gödel, 281, 282
and Hahn, 109, 183n
and Hilbert, 368
and Husserl, 106
and Menger, 204

and Reichenbach, 111
and Schlick, 184n, 190n
and the *Gesellschaft für empirische Philosophie*, 110n
and the *Wiener Kreis*, 108n, 109, 202, 204, 376
and Von Neumann, 184, 285n
and Wittgenstein, 192, 192n
on constructivity, 101, 186, 203
on intuitionism, 101, 203
on logicism, 101, 108, 184–187, 214, 347, 360, 367
on mathematical existence, 177
Cassirer, 149, 181n, 366
Castelnuovo, 268
Church, xiv, 112, 207, 260, 265, 298, 368
and Barzin and Errera, 262, 265, 266, 289–291
and Bernays, 207n
and Brouwer, 207
and Errera, 262n, 266, 266n, 274
on constructivity, 200n, 207n
on intuitionism, 112
on intuitionistic logic, 103, 265, 351
on the excluded middle, 102, 265, 360, 366
Chwistek, 364
Clauberg, 53
Cohen, 316n
Colton, xv
Comte, 251
Corner, 283n
Courant, 104, 106, 127n, 336, 337, 364, 365
Cremer, 301

D'Abro, 331–333
Dahms, 312n
Dalen, Van, xiv, 25, 36n, 40, 41, 51, 64n, 68, 77n, 81, 82n, 85, 88, 88n, 119n, 130n, 160n, 237, 324n

# INDEX

Dantzig, Van, 368
Dassen, 369
Dawson, 101, 107, 282, 283
Dedekind, 1, 6, 7, 126, 138n, 173, 305, 340
Denjoy, 88
Derrida, xx
Descartes, 320
Deutschbein, 111, 366
Dijksterhuis, 61
Dilthey, 322
Dingler, 125, 167, 168, 200, 202, 212, 222, 269, 298, 349, 351, 359, 365, 368
Doesburg, Van, 334
Doetsch, 156, 157, 232, 364, 366
Donder, De, 48, 292
Dooyeweerd, 320
Dresden, 113, 154–156, 182, 183, 231, 233, 266, 298, 347, 358, 364, 366, 367
    and Brouwer, 102, 155, 156, 183, 233, 234
    and Church, 265
    and Errera, 267, 290n
    and Heyting, 294
    and Wavre, 162
    on intuitionism, 102, 156, 182
    on intuitionistic logic, 182, 267
    on mathematical existence, 156, 182, 183, 212
    on the excluded middle, 182, 233, 234
Droste, 367
Droysen, xix
Du Bois-Reymond, 62, 160, 284n, 317
Dubislav, 53, 110, 184, 200, 269, 365, 367, 368
Dubucs, 50n
Dummett, 25, 192

Eckehart, 32
Edwards, 5–7

Eeden, Van, 31, 34, 35, 59, 60, 221, 363
Ehrenfest, 229, 308n
Einstein, xvn, xx, 81, 81n, 83–85, 87, 149n, 151, 308n, 323n, 327, 328, 331, 344
El Lissitzky, 334
Engel, 310n
Errera, xiv, xx, 260, 296
    and Barzin, 260, 262, 271, 289–291, 294–296, 351
    and Becker, 262, 295
    and Brouwer, 266, 292, 293, 317n
    and Church, 262n, 265, 266, 274, 291
    and De Donder, 292
    and Dresden, 267, 290n
    and Glivenko, 271, 272, 291, 359
    and Heyting, 99, 292–295, 360, 369
    and Lévy, 262, 291
    and Scholz, 295
    and Wavre, 261
    and Wavre and Lévy, 170
    on intuitionistic logic, 359, 365, 369
    on constructivity, 261
    on intuitionism, 296, 369
    on intuitionistic logic, 262, 290, 295, 298, 351, 368
    on mathematical existence, 177
    on the excluded middle, 261, 290, 291, 366, 368
Euclid, 120, 178, 320
Euler, 235n
Ewald, xiv, 226

Feferman, 281
Feigl, 109, 191, 197, 376
Fermat, 235n
Fichte, 118n, 121, 374
Finsler, 165n, 220, 229–231, 269, 364, 365

Forman, 302, 308n, 313n, 323n, 329–331, 343, 344, 350
Foucault, 340
Fraenkel, xxn, xxi–xxii, 97, 114, 115n, 149–152, 154, 157, 162, 165, 220, 231, 233, 237, 248, 309, 336, 346, 358, 361, 364, 367–369
  *Einleitung in die Mengenlehre*, 125, 149, 150, 154, 166, 167, 179, 213, 231, 248, 253, 357, 360, 363, 364, 366
  and Bernays, 207, 207n
  and Brouwer, 1, 84n, 127n, 150, 151, 165, 166, 357, 358
  and Buchholz, 366
  and Dresden, 156, 182, 183
  and Menger, 200
  and Pasch, 367
  and Scholz, 165
  and Skolem, 140n, 248, 366
  and Von Neumann, 157, 232n
  and Wavre, 162
  and Wavre and Lévy, 170
  and Weyl, 297
  on Becker, 146, 149
  on Burkamp, 181
  on intuitionism, 166, 214, 231, 311n, 320, 322
  on intuitionism and formalism, 100, 101, 151, 154, 167, 249, 352, 360
  on logicism, 108
  on mathematical existence, 150, 151, 157, 158, 166, 167, 176, 199, 213, 214, 349, 350, 359
  on set theory, 150, 151, 166, 248, 358, 364, 365
  on the excluded middle, 151, 231, 233, 248
Franck, 336
Frank, 323n
Frege, 9n, 52, 53n, 122, 138, 138n, 190, 202, 220, 305, 375

Freudenthal, 50, 85, 86, 112, 286, 286n, 368
Freytag, Von, 119
Frizell, 102, 363
Fueter, 128
Furtwängler, 281, 283

Galavotti, 112n
Galilei, 320
Gan, 334
Garcia, 113, 369
Gauss, 120, 173, 182n
Gehrcke, 328
Geiger, 118n, 172n
Gentzen, 289
Gillies, 198
Giltay, 34
Glivenko, 209, 210, 241, 262, 271–273, 277, 278, 288–293, 298, 299, 331, 351, 359, 366, 367
Gödel, 112, 188, 207–210, 254, 281–283, 285, 286, 342, 369
  's completeness theorem, 138n, 208, 283, 367
  's first incompleteness theorem, 183, 284, 360
  's second incompleteness theorem, 41, 284, 347
  's translation, 209, 212, 239, 271, 286–289, 299, 361
  and Bernays, 103
  and Brouwer, 81, 90, 282, 283, 299, 351, 360
  and Carnap, 281, 282
  and Carnap and Von Neumann, 184
  and Gentzen and Bernays, 289
  and Hahn, 188n
  and Heyting, 279, 285n
  and Kolmogorov, 237, 288, 289
  and Kreisel, 62n
  and Menger, 204, 208, 214, 283n, 287
  and the *Wiener Kreis*, 109, 281, 376

## INDEX

and Wittgenstein, 283
on Behmann, 206
on constructivity, 124, 200n, 207n, 209, 210
on Hilbert's programme, 104, 285
on intuitionism, 285, 288
on intuitionistic logic, 253n, 259, 260n, 263n, 286, 351, 368, 369
on mathematical existence, 208, 209
on the excluded middle, 283

Goenner, 327n
Goethe, 258
Gonseth, xxi, 85, 128, 129, 162, 179, 251–253, 359, 365, 366
Gordan, 122
Grattan-Guinness, 126
Gray, 18
Grelling, 104–106, 110, 225, 231, 364, 366
Groot, De, 88n
Gropius, 334

Haan, De, 59, 125, 221, 356, 363, 364
Hacker, 190
Hadamard, 10–12, 48, 112, 151, 168, 251, 365
Härlen, 257, 366
Hahn, 48, 58, 108, 108n, 109, 183n, 184, 184n, 188, 212, 283, 286, 310, 367, 376
Hamel, 104–106, 366
Hardy, 48, 366
Hartmann, 270, 366
Hartwich, 310
Haupt, 367
Hausdorff, 10, 69, 341
Hecke, 85, 135, 136n
Hedrick, 104–106, 182, 368
Hegel, 374
Heidegger, 87n, 106, 120, 170, 312, 359, 366

Heisenberg, 329
Heiss, 112, 270, 366
Hellinger, 104, 106, 149, 364
Hempel, 110
Hensel, 151n
Herbrand, 112, 207n, 288, 367, 368
Herglotz, 145
Hermann, 334
Hersch, xviiin
Herzberg, A., 110, 366
Herzberg, L., 110, 110n, 111, 359, 366
Heyting, xxi, 101, 102, 107, 115, 123, 124, 200, 236n, 271, 273–275, 280n, 331, 342, 365–369
's dissertation, 177, 268
's formalisation of intuitionistic logic, 99, 116, 213, 271, 273, 274, 276–280, 287, 293, 296, 298, 299, 311, 351, 359, 360, 367
and Barzin and Errera, 99, 262, 266, 289, 292, 294, 295, 351, 360, 368, 369
and Becker, 260, 277n, 295n, 338n
and Bernays, 274n, 278
and Brouwer, 85n, 87n, 274
and Dijksterhuis, 61
and Dresden, 156
and Freudenthal, 286n
and Gentzen, 289n
and Glivenko, 241, 271, 271, 272n, 273, 273n
and Gödel, 209, 210, 279, 285n, 286
and Gödel's incompleteness theorem, 286
and Hirsch, 189
and Kolmogorov, 237, 238, 277, 279–281, 289, 293
and Lévy, 263, 292
and Reidemeister, 184n
and Scholz, 295
and the semi-intuitionists, 8n
and Van der Waerden, 62n

and Wavre, 162
　　and Wavre and Lévy, 170
　　and Weyl, 131
　　on Brouwer, 50, 62, 86, 89
　　on choice sequences, 63
　　on consistency, 276
　　on constructivity, 123, 185, 186
　　on formalisation, 275
　　on formalism, 101, 177
　　on intuitionism, 27n, 101, 184, 326n, 360
　　on language, 275
　　on logic, 201, 275
　　on mathematical existence, 119, 131, 177, 187, 201, 202, 340n
　　on negation, 277
　　on the excluded middle, 277
Hilbert, xxi, 52, 133, 355, 364–366
　　-school, 104–106
　　and Ackermann, 246, 282
　　and Becker, 145
　　and Bergson, 322, 343, 348
　　and Bernays, 103, 106, 115, 139n, 141, 142, 227, 356, 357
　　and Brouwer, 48, 49, 54, 74, 78, 79n, 81, 83–86, 88–90, 100, 101, 112, 127, 127n, 145, 229, 232, 244, 245, 305, 356, 357
　　and Cantor, 9, 43n
　　and Einstein, 84, 87
　　and formalism, xiv, 153, 162, 163, 213, 242, 244, 246, 297, 298, 352, 374
　　and Frege, 138, 138n
　　and Gödel, 103, 208, 210, 284, 285, 347
　　and Gordan, 122
　　and Korteweg, 86n
　　and Kronecker, 134, 144n
　　and the *Mathematische Annalen*, 83–85, 360
　　and Weyl, 101, 126, 132, 134, 141, 155, 173, 174, 210, 212–214, 229, 232, 297, 305, 348, 355, 357
　　and *Lebensphilosophie*, 322, 325, 343
　　on consistency, 138–140, 145, 163, 212, 214, 226, 357
　　on constructivity, 141, 145, 165, 173, 174, 214, 350
　　on logic, 359
　　on mathematical existence, 122, 137–140, 144, 163, 164, 173, 175, 212
　　on proof theory, 141, 143
　　on the excluded middle, 103, 226–228, 241, 243–246, 297
　　on the infinite, 162, 241, 358
　　on the intuitionistic revolution, 304, 306, 342, 348
Hippocrates, 120n
Hirsch, 112, 189, 367–369
Hitler, 92, 312, 335, 337, 344, 358
Hjelmslev, 250
Hölder, 57n, 112, 145, 146, 176, 231n, 269, 298, 309n, 364, 365
Hofmann, 368
Holl, De, 30
Hopf, 50n, 336
Horák, 365
Huntington, 139
Hurewicz, 274
Hurwitz, 9
Husserl, 106, 147n, 148, 170, 322, 359, 375
　　and Becker, 106, 145–148, 170, 259
　　and Brouwer, 62n, 86, 106, 173, 349, 360
　　and Carnap, 106
　　and Geiger, 172n
　　and Heidegger, 87n, 106, 170n
　　and Hilbert, 138n
　　and Kaufmann, 106
　　and Lipps, 106
　　and Mahnke, 106, 173n
　　and Menger, 321

and Schmidt, 106
and Weyl, 106, 126, 132n, 149n

Iongh, De, 63, 86n

Jahnke, 88
James, 313
Jaspers, 120
Jongejan, 60
Jørgensen, 156, 162, 368
Joseph, 149
Jourdain, 10

Kagan, 229
Kant, 38, 53, 54, 57, 81, 118n, 120, 121, 318, 320, 374
Kapp, 357
Kaufmann, 106, 109, 203n, 205, 205n, 206, 367, 368
Kienzler, 193, 194
Kleene, 200n, 207, 207n
Klein, xxi, 49, 53, 54, 59, 84n, 127n, 133, 249, 317, 318, 340
Kline, xv, 120
Klomp, 329
Koebe, 48, 69, 88
König, 9, 10
Koetsier, 50, 51
Kohnstamm, 46n, 88, 221, 269, 270, 365
Kolmogorov, 112, 123, 237, 241, 271, 279, 280, 289, 298, 331
　's translation, 239, 240, 273, 288, 298
　and Becker, 260
　and Glivenko, 271, 272
　and Gödel, 288, 289
　and Heyting, 277, 281, 289, 293
　and Lévy, 168
　and Wavre, 254
　on intuitionism, 238, 279
　on intuitionistic logic, 112, 236, 238, 239, 241, 271, 280, 298, 299, 351, 361, 368

　on the excluded middle, 237, 238, 240, 241, 245, 358, 365
Korselt, 112, 363
Korteweg, 25n, 28, 30, 35, 36, 36n, 37, 39, 45, 46n, 54, 86n, 229, 273, 318, 355
Koschmieder, 364
Kreisel, 27n, 62n, 124, 180
Kronecker, xxii, 1, 2, 4–7, 11, 23, 51, 121n
　and Baldus, 153
　and Barzin and Errera, 261
　and Bernays, 142
　and Brodén, 175
　and Brouwer, 1, 76, 90, 348
　and Cantor, 5
　and Fraenkel, 157
　and Hilbert, 23, 134, 144n, 226, 304, 305
　and Husserl, 106
　and Pierpont, 267
　and Weierstraß, 5, 6
　on irrational numbers, 6, 7, 23
　on mathematical existence, 119
　on natural numbers, 6, 348
　on set theory, 6, 7, 23
Kroner, 317
Kummer, 4, 106
Kynast, 367

Landau, 69, 136, 141, 151, 336, 337, 339
Landry, 271, 366
Larguier des Bancels, 176, 270, 365
Lebesgue, 2, 8–13, 15–18, 76, 88, 121, 159
Leibniz, 119, 219, 320
Lenard, 328
Lenoir, 256
Leśniewski, 367
Leupold, 155
Levi, 113, 364
Levi-Civita, 48, 151

Lévy, 255, 256, 260, 290, 291
   and Barzin and Errera, 262, 291
   and Dresden, 267
   and Errera, 296
   and Heyting, 131, 201, 292, 294
   and Reymond, 256
   and Wavre, 161n, 168–170, 254, 255, 358
   and Zermelo, 243
   on intuitionistic logic, 263–265, 298, 351, 365, 366
   on mathematical existence, 169
   on the excluded middle, 168, 254, 255, 365
Lewis, 286
Lietzmann, 104–106, 114, 117, 176, 217, 268, 270, 359, 365, 366
Lindelöf, 125
Lipps, 106, 268, 365, 366
Litt, 315n, 331n, 332
Löwy, 269, 365, 369
Lombroso, 87
London, 110, 364
Loor, De, 74, 120, 197, 247
Lorenzen, 123n
Lützen, xiv
Lukács, 312
Łukasiewicz, 107, 274, 367
Lusin, 17
Luzin, 237

Mach, 332
Mahnke, 106, 164, 173n, 178, 366
Majer, 113n, 224
Mancosu, xiv, 52n, 225n
Mannoury, 28, 29, 29n, 74, 92, 112, 274, 356, 363, 364, 368
   and Brouwer, 28–30, 46, 46n, 52, 59, 60, 89, 153, 265n, 346, 355, 356
   and Heyting, 273, 274
   and Korteweg, 28
   and Menger, 200
   on intuitionism and formalism, 52, 112
   on intuitionistic logic, 326
   on mathematical existence, 29
   on the excluded middle, 232
Marcus, 112, 366
Markov, 123
Mauve, 34, 46
McGuinness, 193n
Mehrtens, xv, xvi, xviin, 24n, 56n, 96n, 107n, 233, 250, 301, 302, 307, 330n, 340–343, 352, 353
Melle, Von, 135
Menger, 79, 112, 114, 198, 320, 365–368
   and Bergson, 321
   and Brouwer, 88, 198n
   and Carnap, 204
   and Dingler, 200
   and Gödel, 204, 208, 209, 214, 281, 283n, 287, 361
   and Hahn, 310
   and Husserl, 321
   and the *Wiener Kreis*, 109
   and Wittgenstein, 190n, 191
   on constructivity, 166, 199, 200, 207, 214, 215, 350, 359, 367, 368
   on intuitionism, 112
   on mathematical existence, 168, 199, 200, 212, 349
Messer, 316n, 322
Mill, Van, 50, 51
Minkowski, 348n
Misch, 322, 322n
Mises, Von, 110, 189, 308n, 323n
Mittag-Leffler, 5
Mittelstraß, 27n
Mondriaan, 334
Monk, 190n
Mussolini, 82

Nagel, 367
Natorp, 316n

Nelson, 52n, 141, 306, 322
Neugebauer, 279
Neumann, Von, 100–102, 107, 112, 115, 157, 158, 184, 189, 364
  and Brouwer, xv, 345
  and Carnap, 184, 285n
  and Fraenkel, 232n
  and Gödel, 184, 281n, 285
  and Hilbert, 179, 352, 366
  and Skolem, 248
  and the *Gesellschaft für empirische Philosophie*, 110
  and Weyl, xv, 232, 345
  on constructivity, 154, 157, 158, 186, 214
  on formalism, 185, 360, 367
  on intuitionism, 101, 112
  on mathematical existence, 180, 186
  on the excluded middle, 231
Neurath, 108, 108n, 109, 184
Newton, xv, 119, 320
Nietzsche, 313
Noether, 69, 336, 337
Nohl, 322

Oikonomou, 113, 268, 365
Oldenbarnevelt, Van, 27n
Orlov, 366
Ornstein, 59, 151
Os, Van, 320n, 369
Ostrowski, 135–137, 364
Otto, 365

Pacotte, 316n
Painlevé, 82
Palmer, xv
Pascal, 320
Pauli, 333n
Peano, 22, 40, 59, 122
Peijpers, 30
Peirce, 144n, 219
Pesch, Van, 30

Petzoldt, 110, 112, 149, 256, 308n, 332, 359, 365, 366
Pierpont, 102, 112, 182, 266, 267, 366
Pitcher, 191n
Planck, 86, 117, 331
Plato, 118n, 218, 320, 376
Pöggeler, 171n
Poincaré, 1, 2, 18–24, 38, 48, 52–54, 56, 57, 76, 80, 90, 119, 125, 134, 167, 226, 320, 321
Pólya, 69, 128, 129, 324, 357
Posy, 81n
Putnam, 107

Radon, 135
Ramsey, xxi, 101, 108, 112, 192, 193, 248, 250, 251, 257n, 307, 347, 365
Reichenbach, 107, 110, 110n, 111, 143n, 184, 337, 359, 367
Reid, 136n, 306n
Reidemeister, xxi, 109, 135, 136, 136n, 184, 184n, 188, 226n, 336, 337, 364, 366, 376
Reymond, 179n, 253n, 256, 366, 368
Richard, 125
Richardson, 191
Richter, 334
Rickert, 52n, 312–316, 319, 325, 326
Ridder, 118n
Riemann, 340
Ringer, 109, 305, 306n, 307, 308n, 309, 316n
Rivier, xiv, 176, 268, 270, 270n, 365, 367
Rodchenko, 334
Rowe, 133, 310
Russell, 52, 108, 184, 220, 314n, 347, 375
  's paradox, 9
  and Becker, 148
  and Behmann, 205
  and Brouwer, 38, 40, 44
  and Carnap, 109

and Doetsch, 157
and Frege, 220
and Hahn, 109
and Heyting, 277
and Hilbert, 225
and Kolmogorov, 237
and Lewis, 286
and Poincaré, 18, 19
and Ramsey, 250
and Schlick, 281
and Skolem, 247
and Von Neumann, 100
and Whitehead, 72, 107n
and Wittgenstein, 190, 192, 195, 197
on intuitionism, 108n
on logic, 44
on mathematical existence, 122

Scheler, 52n
Schlick, 108, 183n, 184, 184n, 190n, 192, 192n, 193, 281, 285, 376
Schmeidler, 337n, 367
Schmidt, 50n, 104–106, 189, 367
Schmitz, 221n
Schoenflies, 5, 5n, 10, 58, 59, 69, 112, 364
Scholz, xxi, 114, 165, 188, 219, 248, 271, 279, 295, 365, 368, 369
Schopenhauer, 34, 53, 78, 81, 313
Schouten, 84n, 88
Schroder, 219, 220
Schrödinger, 329
Schuh, 274
Schur, 69, 189, 336, 337
Schwarz, 6
Sheldrake, 29n
Sieg, 79n, 164
Siegmund-Schultze, 306, 337
Simmel, 314
Skolem, xxi, 113n, 122, 140, 175, 243, 246–248, 307, 365, 367
Socrates, 217, 320
Souslin, 17

Spengler, 311, 323–325, 329, 343, 350, 357, 374
Stammler, 366
Stark, 328
Stigt, Van, 27n, 53, 73
Study, xiv, 61, 81n, 112, 248, 309, 320n, 360, 367
Sundholm, 131

Tarski, 109, 274
Theweleit, xxn
Thomae, 52, 53n
Troelstra, A., 25, 124, 160n, 274n, 287n
Troelstra, P.J., 306
Trommler, 334
Turing, 123, 200n, 205, 207, 207n

Urysohn, 237

Veblen, 139
Vollenhoven, 125, 221n, 320, 321, 363, 368
Voss, 112, 363
Vredenduin, 369
Vries, Hk. de, 25, 81, 273
Vries, J. de, 81

Waals, Van der, 30, 221
Waerden, Van der, 62, 337
Wäsche, 365
Waismann, 109, 191, 192n, 193, 194
Wang, 239n
Wavre, 113, 118n, 158, 162, 231, 234, 242, 246, 292, 298, 309n, 347, 352, 358, 364, 365
and Barzin and Errera, 261
and Becker, 260
and Bergson, 320n
and Du Bois-Reymond, 160n
and Gonseth, 251
and Hahn, 188
and Lévy, 168–170, 201, 254, 255, 358
on constructivity, 161

on intuitionistic logic, 254
on mathematical existence, 158–161, 212–214, 349
on the excluded middle, 234–236, 254, 255, 257n, 365
Weierstraß, 4–6, 106, 162, 173, 241
Weiss, 367
Weitzenböck, 69, 145
Weyl, xxi, 125, 164, 364–368
  *Über die neue Grundlagenkrise*, 94, 115, 125, 128, 132, 212, 222, 296, 308n, 345, 346, 357
  and Becker, 170n, 172, 178, 329
  and Bernays, 127, 142
  and Brouwer, 58, 70, 86, 99, 105, 116, 126–128, 222, 224, 245, 304, 345–347, 356, 357
  and Heyting, 201, 213
  and Hilbert, 105, 126, 132, 134, 141, 153, 164, 173, 174, 180, 212, 214, 226, 228, 229, 232, 297, 304, 305, 348, 352, 355, 357
  and Husserl, 106, 126, 132n, 149n
  and Klein, 127n
  and Pauli, 333n
  and phenomenology, 149
  and Pólya, 128, 129, 357
  and Spengler, 323n, 324, 343, 350
  and Weitzenböck, 145
  and Wittgenstein, 195
  and *Lebensphilosophie*, 319, 331, 344
  conversion to intuitionism, 127, 128, 303, 345, 346
  on constructivity, 125, 131, 155, 178, 210–212
  on intuitionism and formalism, 155, 174, 210, 211, 213, 232, 297, 352, 353, 358–360
  on mathematical existence, 125, 130, 131, 155, 178, 212, 297, 345

on the excluded middle, 222–224, 232, 270, 296, 345
on the intuitionistic revolution, 302–304, 342, 346
Weyland, 328
Whitehead, 72, 107, 109, 122, 205, 237, 247, 250, 277
Wien, 323n
Wiener, 137
Windelband, 312
Winternitz, 366
Wittgenstein, 183, 190, 191, 193, 197, 198, 342, 347
  and Ambrose, 111, 112
  and Brouwer, 81, 90, 108, 190–193, 197, 213, 350, 360
  and Carnap, 203n, 204
  and Gödel, 281, 283
  and Menger, 190n
  and Reidemeister, 109
  and Russell, 192
  and Schlick, 190n
  and the *Wiener Kreis*, 109, 376
  and Waismann, 193
  and Weyl, 195
  on constructivity, 196
  on mathematical existence, 194–197, 213, 350
  on the excluded middle, 198
Wolff, 220, 229–231, 274, 306, 364
Wright, 190

Zach, 139n, 154n, 225n, 227
Zariski, 177, 178, 268, 365
Zawirski, 107, 368
Zeno, 219
Zermelo, 1, 10–12, 12n, 14, 19, 20, 24, 44, 48, 52, 54, 91, 115n, 149, 243, 341
Zeuthen, 120
Zilsel, 184, 209, 368

# Dankwoord/ Acknowledgements

Gedurende de afgelopen vier jaar hebben velen op verschillende manieren aan dit proefschrift bijgedragen. De volgende personen wil ik op deze plaats met name bedanken.

Voor het bekommentariëren van eerdere delen van het manuskript: Mark van Atten, Harm Dorren, Sylvester van Koten, Dineke Oldenwening, Teun Koetsier en Göran Sundholm, en in het bijzonder Elwin Lammers, voor de historische check.

Voor de praktische ondersteuning: Ria Bekkering, Elly Broekhoven en Toos Raats voor het aanvragen van alle oude artikelen en boeken, zowel degene die hierin terecht zijn gekomen als degene waar niks in bleek te staan, en voor het doorgeven van de stapels faxen; André de Meijer, Patrick Schoo en Freek Wiedijk voor hulp op computergebied; en Wilberd van der Kallen, Roderik Lindenbergh en Martijn van Manen voor het hardere LaTeX-werk.

Ik wil prof. em. De Iongh bedanken voor het interview dat ik hem mocht afnemen. Ook ben ik dank verschuldigd aan wijlen prof. Van der Waerden voor het interview.

I was pleasantly surprised by the open attitude of researchers in the academic fields of history and foundations of mathematics to newcomers in the field, and I am grateful to the following persons, from both inside and outside these circles, for answering my questions, providing information, and discussing the subject: Eisso Atzema, David Auben, Umberto Bottazzini, Ada Colau, John Dawson, Yvonne Dold, Peter van Emde Boas, Moritz Epple, Solomon Feferman, Jens Erik Fenstad, David Fowler, Richard Gill, Donald Gillies, Catherine Goldstein, Ivor Grattan-Guinness, Jan Hogendijk, Oksana Maslovskaya, Herbert Mehrtens, Eckart Menzler-Trott, Paul van Praag, Michael Rabin, Gianni Rigamonti, Karl Schuhmann, Christov Scriba, Wilfried Sieg, Reinhard Siegmund-Schultze, Skúli Sigurdsson, Mark Steiner, Christian Thiel, Anne Troelstra, Tito Tonietti, and Ferdinand Verhulst.

# DANKWOORD/ACKNOWLEDGEMENTS

I am indebted to the following persons who provided help with the more difficult parts of the various languages used for the dissertation: for German: Ulrike Schmülling; English: John Edwards; French: San Vũ-Ngọc; Norwegian/Danish: Herman Beun; Greek: Martijn van Manen; and Rumanian: Marius Crainic. Furthermore, I would like to thank Miss McNab for thoroughly checking the English in part of the dissertation.

I would like to thank the following persons and institutions for providing access to archive materials and for their kind permission to quote from the archival sources (if applicable):

**Académie des Sciences archives:** Mme. Claudine Pouret, Institut de France, Académie des Sciences, Service des Archives;

**Bernays and Weyl Nachlaß:** Dr Beat Glaus and Herr Rudolf Mumenthaler, Wissenschaftshistorische Sammlungen, ETH-Bibliothek;

**Brouwer archive:** Prof. Dr Dirk van Dalen, Universiteit Utrecht;

**Carnap Nachlaß:** Dr Brigitte Uhlemann, Universität Konstanz, Zentrum Philosophie und Wissenschaftstheorie, and the University of Pittsburgh Libraries, Special Collections Department;

**Einstein archive:** Mr. Ze'ev Rosenkranz, The Albert Einstein Archives;

**Feigl Nachlaß:** Mr. C. Kenneth Waters, Minnesota Center for Philosophy of Science;

**Private A. Fraenkel archive:** Prof. Dr Benjamin Fraenkel, The Hebrew University;

**Official A. Fraenkel archive:** Mr. Rafael Weiser, The Jewish National and University Library, Department of Manuscripts and Archives;

**Gonseth collection:** Mme. Danielle Mincio, Bibliothèque cantonale et universitaire, and Prof. Dr Doris Jakubec, Fondation Gonseth;

**Heyting Nachlaß:** Prof. Dr Anne Troelstra, Universiteit van Amsterdam;

**Hilbert and Klein Nachlaß:** Dr Helmut Rohlfing, Niedersächsische Staats- und Universitätsbibliothek Göttingen, Abteilung Handschriften und seltene Drucke;

**Hilbert and Reidemeister Personalakte:** Dr Ulrich Hunger, Georg-August-Universität Göttingen, Universitätsarchiv;

**Schmidt Nachlaß:** Dr W. Schultze, Humboldt-Universität zu Berlin, Universitätsbibliothek, Archiv;

**Scholz *Nachlaß*:** Prof. Dr Wolfram Pohlers and Herr Enno Folkerts, Westfälische Wilhelms-Universität, Institut für mathematische Logik und Grundlagenforschung.

Zeer dankbaar ben ik mijn begeleiders Henk Bos en Dirk van Dalen voor de goede en plezierige samenwerking. De unieke kombinatie van Henks altijd kritische en gedetailleerde blik en Dirks geweldige kennis van Brouwer en het intuïtionisme heeft een essentiële bijdrage geleverd aan dit proefschrift.

Tenslotte wil ik mijn goede vriend Bas Dorren bedanken voor zijn niet aflatende ondersteuning.

### Addition at the commercial edition

I am indebted to the following persons who helped me making this book an improved version of my dissertation, by their critical remarks, suggestions and help: Arthur Bakker, Daniel Beckers, N.G. de Bruijn, Bas Dorren, Yvonne Hartwich, Teun Koetsier, Paolo Mancosu, Colin McLarty, Herbert Mehrtens, Volker Remmert, Norbert Schappacher, Dirk Schlimm, Liesbeth de Wreede, and Erik van Zwet. I am especially grateful to Mark van Atten, Ivor Grattan-Guinness, Fred Muller and Jan Stegeman for thoroughly commenting on (parts of) the dissertation. Also, I would like to thank Dirk van Dalen, David Rowe and Erhard Scholz for their remarks on the text, their encouragement, and the contacts with Birkhäuser. From the side of Birkhäuser, Edgar Klementz ensured that all the arrangements for publishing the book were in place.

I would like to thank once more the above-mentioned persons and institutions for their renewed permission to quote from the respective archival sources. The following persons and institutions should be added to the list:

**Church collection:** Mr. A. Church, jr.;

**Errera archive:** M. Didier Devriese, Archives de l'Université Libre de Bruxelles;

**Study correspondence:** Mrs. M. Klos, Bibliothek des mathematischen Instituts der Justus-Liebig-Universität Giessen;

**Study manuscripts:** Prof. W. Scharlau, Westfälische Wilhelms-Universität Münster, Mathematisches Institut.

Finally, I am most grateful to Lousewies van der Laan for her dedication in commenting, as a non-mathematician, on almost the whole manuscript, her patience with my work in the evenings and week-ends, and, above all, her unflagging support.

Het leven is een toovertuin. Met wonder zacht blinkende bloemen, maar tusschen de bloemen loopen de kaboutermannetjes, daar ben ik zoo bang voor, die staan op hun kop en het ergste is, dat ze mij toeroepen, dat ik ook op mijn kop moet gaan staan, een enkele maal probeer ik het, en schaam me dood; maar soms roepen dan de kabouters, dat ik het heel erg goed doe, en toch ook een echter kabouter ben. Maar dat laat ik me in geen geval ooit wijsmaken.[5]

L.E.J. Brouwer[6]

---

[5]'Life is a magic garden. With wondrously softly shining flowers, but between the flowers are the gnomes, of whom I am so afraid, they stand on their head and the worst thing is that they call to me that I should also stand on my head, I try it now and then, and die of shame; but sometimes the gnomes then call that I am doing very well, and that I am indeed a real gnome, too. But by no means will I ever swallow that.'

[6]Letter from Brouwer to Adama van Scheltema, 07/09/1906; cited from [Van Dalen 1984A, p. 68]